Advances in
Heterocyclic Chemistry

Volume 49

Editorial Advisory Board

R. A. Abramovitch, *Clemson, South Carolina*
A. Albert, *Canberra, Australia*
A. T. Balaban, *Bucharest, Romania*
A. J. Boulton, *Norwich, England*
H. Dorn, *Berlin, G.D.R.*
J. Elguero, *Madrid, Spain*
S. Gronowitz, *Lund, Sweden*
T. Kametani, *Tokyo, Japan*
O. Meth-Cohn, *South Africa*
C. W. Rees, FRS, *London, England*
E. C. Taylor, *Princeton, New Jersey*
M. Tišler, *Ljubljana, Yugoslavia*
J. A. Zoltewicz, *Gainesvillle, Florida*

Advances in
HETEROCYCLIC CHEMISTRY

Edited by

ALAN R. KATRITZKY, FRS

Kenan Professor of Chemistry
Department of Chemistry
University of Florida
Gainesville, Florida

Volume 49

ACADEMIC PRESS, INC.
Harcourt Brace Jovanovich, Publishers
San Diego New York Boston
London Sydney Tokyo Toronto

This book is printed on acid-free paper. ∞

COPYRIGHT © 1990 BY ACADEMIC PRESS, INC.
All Rights Reserved.
No part of this publication may be reproduced or transmitted in any form or by any means, electronic or mechanical, including photocopy, recording, or any information storage and retrieval system, without permission in writing from the publisher.

ACADEMIC PRESS, INC.
San Diego, California 92101

United Kingdom Edition published by
ACADEMIC PRESS LIMITED
24-28 Oval Road, London NW1 7DX

LIBRARY OF CONGRESS CATALOG CARD NUMBER: 62-13037

ISBN 0-12-020649-8 (alk. paper)

PRINTED IN THE UNITED STATES OF AMERICA
90 91 92 93   9 8 7 6 5 4 3 2 1

Contents

PREFACE ix

Condensed 4-Thiazolidinones
HRUSHI K. PUJARI

I. Introduction 3
II. Synthesis 3
III. Molecular Spectra 88
IV. Molecular Dimensions 98
V. Reactions 99
VI. Useful Applications of Condensed 4-Thiazolidinones 105
References 106

Advances in Amination of Nitrogen Heterocycles
HELMUT VORBRÜGGEN

I. Introduction 118
II. Theory and General Mechanisms: Addition–Elimination Mechanism 118
III. Reactivity Factors in Aminations 119
IV. Aminations via Leaving Groups Derived from Hydroxy-N-Heterocycles .. 125
V. Miscellaneous Aminations 173
VI. Comparison of Different Amination Methods 183
References 185

Saturated Bicyclic 6/5 Ring-Fused Systems with Bridgehead Nitrogen and a Single Additional Heteroatom
TREVOR A. CRABB, DAVID JACKSON, AND ASMITA V. PATEL

I. Introduction 194
II. Perhydropyrazolo[1,2-*a*]pyridazines 196
III. Perhydropyrazolo[1,5-*a*]pyridines 198
IV. Perhydro-isoxazolo[2,3-*a*]pyridines 200
V. Perhydropyrrolo[1,2-*b*]pyridazines 207

VI. Perhydroimidazo[1,5-*a*]pyridines 208
VII. Perhydro-oxazolo[3,4-*a*]pyridines 214
VIII. Perhydrothiazolo[3,4-*a*]pyridines 220
IX. Perhydropyrrolo[1,2-*c*]pyrimidines 223
X. Perhydropyrrolo[1,2-*c*][1,3]oxazines 224
XI. Perhydro-imidazo[1,2-*a*]pyridines 226
XII. Perhydro-oxazolo[3,2-*a*]pyridines 229
XIII. Perhydrothiazolo[3,2-*a*]pyridines 238
XIV. Perhydropyrrolo[1,2-*a*]pyrazines 242
XV. Perhydropyrrolo[2,1-*c*][1,4]oxazines 249
XVI. Perhydropyrrolo[2,1-*c*][1,4]thiazines 254
XVII. Perhydropyrrolo[1,2-*a*]pyrimidines 255
XVIII. Perhydropyrrolo[2,1-*b*][1,3]oxazines 258
XIX. Perhydropyrrolo[2,1-*b*][1,3]thiazines 259
XX. Conformational Analysis of Saturated 6/5 Ring Systems with Bridgehead Nitrogen and a Single Additional Heteroatom 259
References 267

Synthesis of Condensed 1,2,4-Triazolo[3,4-*z*] Heterocycles
MOHAMMED A. E. SHABAN AND ADEL Z. NASR

I. Introduction 279
II. Condensed 1,2,4-Triazolo-azoles 280
III. Condensed 1,2,4-Triazolo-diazoles 283
IV. Condensed 1,2,4-Triazolo-oxazoles 290
V. Condensed 1,2,4-Triazolo-thiazoles 292
VI. Condensed 1,2,4-Triazolo-selenazoles 299
VII. Condensed 1,2,4-Triazolo-triazoles 300
VIII. Condensed 1,2,4-Triazolo-oxadiazoles 305
IX. Condensed 1,2,4-Triazolo-thiadiazoles 307
X. Condensed 1,2,4-Triazolo-tetrazoles 309
XI. Condensed 1,2,4-Triazolo-azines 310
XII. Condensed 1,2,4-Triazolo-diazines 318
XIII. Condensed 1,2,4-Triazolo-oxazines 341
XIV. Condensed 1,2,4-Triazolo-thiazines 344
XV. Condensed 1,2,4-Triazolo-triazines 347
XVI. Condensed 1,2,4-Triazolo-thiadiazines 358
XVII. Condensed 1,2,4-Triazolo-tetrazines 361
XVIII. Condensed 1,2,4-Triazolo-azepines 362
XIX. Condensed 1,2,4-Triazolo-diazepines 364
XX. Condensed 1,2,4-Triazolo-triazepines 367
References 369

Advances in Pyridazine Chemistry
MIHA TIŠLER AND BRANKO STANOVNIK

I. Introduction 385
II. Synthetic Methods 386

III.	Transformations of Pyridazines	404
IV.	Theoretical Aspects and Physical Properties	424
V.	Polymers	432
VI.	Biological Activity	433
VII.	Other Applications	439
VIII.	Analysis	439
	References	440

Preface

Volume 49 of *Advances in Heterocyclic Chemistry* contains five chapters. Professors M. Tišler and B. Stanovnik of the University of Ljubljana, Yugoslavia cover the progress in pyridazine chemistry since 1979, the date of their previous publication in Volume 24 of *Advances*.

Professor Vorbrüggen of Berlin has contributed a survey of the amination of nitrogen heterocycles, which deals particularly with aminations in which an amino group replaces some other functionality on the heterocyclic ring. Thus, this review complements our recent review in Volume 44 dealing with the Tschitschibabin reaction, in which a hydrogen atom on the heterocyclic ring is replaced by an amino group.

The other three chapters in the present volume all deal with bicyclic heterocycles. Dr. H. K. Pujari of Kurukshetra University in India describes condensed 4-thiazolidinones, and Professor M. A. E. Shaban and A. Z. Nasr of Alexandria University in Egypt survey the synthesis of condensed 1,2,4-triazolo heterocycles. Finally, Professor T. A. Crabb of Portsmouth, England has covered saturated bicyclic 6/5 ring-fused systems with a bridgehead nitrogen and a single additional heteroatom. None of these groups of bicyclic heterocycles has been comprehensively reviewed before.

A. R. KATRITZKY

Condensed 4-Thiazolidinones

HRUSHI K. PUJARI

*Department of Chemistry, Kurukshetra University
Kurukshetra 132 119, Haryana, India*

I. Introduction	3
II. Synthesis	3
A. Cyclization Procedures Involving the 2- and 3-Positions of 4-Thiazolidinone	4
1. Pyrrolo[2,1-b]thiazol-3(2H)-ones	4
2. Imidazothiazol-3(2H)-ones	5
3. Pyrazolo[3,2-b]thiazol-3(2H)-ones	13
4. Thiazolo[2,3-b]thiazol-3(2H)-one	14
5. Thiazolo[2,3-b]oxazol-5(6H)-one	15
6. Thiazolo-s-triazol-5(6H)-ones	15
7. Thiazolo[3,2-a]pyridin-3(2H)-ones	17
8. Thiazolo[3,2-a]pyrimidin-3(2H)-ones	18
9. Thiazolo[2,3-c][1,4]thiazin-3(2H)-ones	23
10. Thiazolo[2,3-c][1,2,4]benzothiadiazin-3(2H)-ones	24
11. Thiazolo-triazin-3(2H)-ones	25
12. Thiazolo[2,3-c]-[1,2,4]benzotriazin-3(2H)-ones	28
13. Thiazolo[3,2-b]-[1,2,4,5]tetrazin-3(2H)-ones	29
14. Oxospiro[cycloalkane-1,3'(4'H)-(2H)thiazolo[3,2-b]-[1,2,4,5]tetrazines]	30
15. Oxospiro[indan-1,3'(4'H)thiazolo[3,2-b]-[1,2,4,5]tetrazines]	31
16. Thiazolo[3,2-a]-[1,3]diazepin-3(2H)-ones	31
17. Thiazolobenzodiazepinones	32
18. Thiazolo[2,3-d]-[1,5]benzoxazepin-1(2H)-ones	35
19. Thiazolo[3,2-a]-[1,3]diazocin-3(2H)-ones	36
20. Thiazolo[2,3-b]-[1,3]benzodiazocin-3(2H)-one	36
21. Thiazolo[3,2-a]-[1,3]diazonin-3(2H)-ones	37
22. Thiazolo[3,2-a]-[1,3]diazecin-3(2H)-ones	37
23. Thiazolo[3,2-a]-[1,3]diazacycloundecan-3(2H)-ones	38
24. Thiazolo[3,2-a]-[1,3]diazacyclododecane-3(2H)-ones	38
25. Thiazolo[3,2-a]-[1,3]diazacyclotridecane-3(2H)-ones	39
26. Bis-(thiazolo)-[3,4-b:3',2'-d]-s-triazol-3(2H)-ones	39
27. Pyrrolo[2,3:4',5']pyrrolo[2',1'-b]thiazol-3(2H)-ones	40
28. Thiazolo[3,2-a]thiopyrano[4,3-d]pyrimidin-3(2H)-ones	40
29. Thiazolo[3,2-a]indol-3(2H)-ones	41
30. Thiazolo[3,2-a]benzimidazol-3(2H)-ones	42
31. Thiazolo[2,3-b]imidazo[5,4,-b]pyridin-3(2H)-ones	54
32. Thiazolopurin-3(2H)-ones	55
33. Thiazolo[3,2-a]thiazolo[5,4-d]pyrimidin-3(2H)-ones	57

34. Thiazolo[3,2-a]thieno[2,3-d]pyrimidin-3(2H)-ones 57
35. Thiazolo[3,2-a]-[1]benzothieno[2,3-d]pyrimidin-3(2H)-ones 58
36. Thiazolo[3,2-a]-s-triazolo[3,4-b]-[1,3,4]thiadiazin-6(7H)-one 59
37. Thiazoloquinazolinones 60
38. Thiazolo[2,3-a]isoquinolin-3(2H)-ones............................. 70
39. Naphthimidazo[2,1-b]thiazol-3(2H)-ones 71
40. Thiazolo[3',2':1,2]imidazo[4,5-f]quinolin-10(9H)-one................. 73
41. Thiazolo[3,2-a]perimidin-3(2H)-ones............................... 74
42. Thiazolo[2,3-a]benzo(f)isoquinolin-17(16H)-ones 75
43. Thiazolo[3,2-a]aceperimidin-3(2H)-one............................ 75
44. Thiazolo[2',3'-b]benzofuro[2,3-f]benzimidazol-3(2H)-one............. 76
B. Cyclization Procedures Involving the 2- and 5-Positions of
   4-Thiazolidinone 76
   1. Hexahydro-1,4-epithiopyrimidin-3-ones 77
   2. Tetrahydro-1,4-epithioazepin-3,7-diones............................ 78
   3. Hexahydro-1,4-epithiopyridin-3-ones 79
   4. Tetrahydro-4,7-epithiofuro[3,4-c]pyridin-5-ones 82
   5. Tetrahydro-4,7-epithiopyrrolo[3,4-c]pyridin-5-ones 83
C. Cyclization Procedures Involving the 2,3- and 2,5-Positions of
   4-Thiazolidinone 84
   1. 6,8a-Epithioimidazo[1,2-a]pyrimidin-5,7-diones 84
   2. 5,7a-Epithiopyrido[1,2-c]quinazolin-4-ones 84
   3. 4,6a-Epithiopyrido[2,1-a]phthalazin-3-ones 86
   4. 4,11a-Epithiopyrrolo[3,4-c]pyrido[2,1-b]benzothiazol-1,3,5-triones...... 86
   5. 3,10a-Epithiopyrido[2,1-b]benzothiazol-4-one 87
III. Molecular Spectra of Condensed 4-Thiazolidinones........................ 88
   A. Ultraviolet Spectra 88
   B. Infrared Spectra 90
   C. ^{1}H-NMR Spectra... 91
   D. ^{13}C-NMR Spectra 93
   E. Mass spectra 96
IV. Molecular Dimensions 98
   X-ray Diffraction 98
V. Reactions of Condensed 4-Thiazolidinones 99
   A. Reactions with Electrophiles 99
      1. Alkylation 99
      2. Diazo Coupling... 100
      3. Reaction with Nitrosoarenes 101
      4. Reaction with Acids 101
   B. Reactions with Nucleophiles.. 102
      1. Amination 102
      2. Reaction with Alkali 102
      3. Reaction with Alcohols... 102
      4. Reaction with Grignard Reagents 103
      5. Reaction with Diazomethane 103
   C. Halogenation 104
   D. Oxidation 104
   E. Reduction 104
   F. Thionation 105

VI. Useful Aplications of Condensed 4-Thiazolidinones	105
A. Biological Properties	105
B. Dyestuffs	105
References	106

I. Introduction

The first comprehensive review on 4-thiazolidinones (**1**) appeared in 1961 and was written by Brown (61CRV463). Subsequent reviews dealt with the uses of **1** as stabilizers for polymeric materials (64MI1) and as preparations and uses of rhodanines (2-thiono-4-thiazolidinones) (78RRC820, 78RRC1152). The synthesis, reactions (78AHC83), and biological activity of **1** (81CRV175) have been highlighted.

The systematic review by Mosby of bridgehead nitrogen heterocycles refers to the early work on five different condensed 4-thiazolidinones (with the general structure of **2**) and covered the literature up to 1960 (61HC). Since then, a considerable amount of work has been published but only a brief review by this author has appeared (80MI1). A 1984 publication simply describes five such systems of **2**, namely, imidazo[2,1-*b*]thiazol-3(2*H*)-one, thiazolo[2,3-*b*]oxazol-5(6*H*)-one, thiazolo[2,3-*b*]thiazol-3(2*H*)-one, thiazolo[3,2-*a*]perimidin-3(2*H*)-one, and thiazolo[3,2-*a*]benzimidazol-3(2*H*)-one (84MI1). The last heterocyclic system has also been briefly mentioned by Tennant (80HC), and the first has been described by Preston (86HC). Condensed 4-thiazolidinones (**2–4**) are conspicuously absent in Metzger's three-volume work on thiazole chemistry (79HC). Consequently, an attempt has been made in this review to present in a systematic fashion the work encompassing systems (**2–4**). The literature is covered up to 1987.

II. Synthesis

The synthesis of condensed 4-thiazolidinones can be achieved through a cyclization procedure involving any two positions of **1**. The involvement of the 1- and 4-positions is, however, obviously ruled out. Cyclizations involving the 2- and 3-positions of **1** have been the mainstay for the synthesis of condensed 4-thiazolidinones (**2**). A small quantity of elegant work on the condensed 4-thiazolidinones involving the 2- and 5-positions (**3**) as well as 2,3- and 2,5-positions (**4**) of **1** has been reported. The synthesis of **2–4**, in general, follows well-documented procedures with some modifications for the preparations of **1** (61CRV463). In this review the synthesis of **2–4** is discussed under different subheadings depending on the positions of **1** that are involved in the cyclization procedure and also on the nucleus to which **1** is fused.

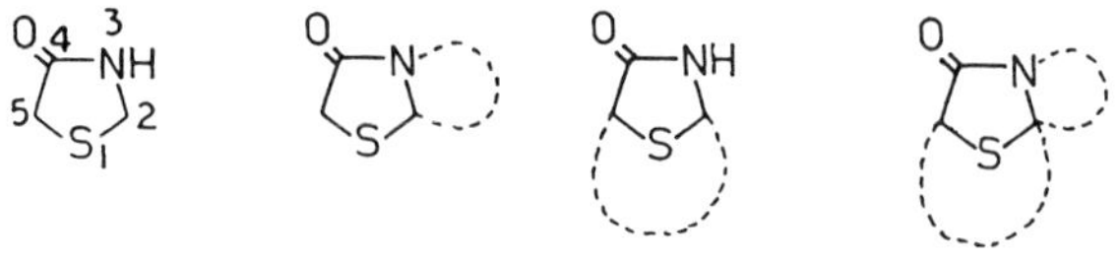

A. Cyclization Procedures Involving the 2- and 3-Positions of 4-Thiazolidinone

1. *Pyrrolo[2,1-b]thiazol-3(2H)-ones*

Synthesis of this system, as depicted in Scheme 1, has been reported by Russian scientists (86KGS1690). Reaction of 5-*n*-propylpyrrolinyl-2-thione (**5**) with chloroacetic acid in benzene gives **7** in 50% yield. The reaction proceeds via the intermediate acid **6**, which undergoes cyclization to furnish the bicyclic compound **7**. The facile acid-catalyzed ring closure of thiazole is demonstrated by the fact that when **5** is heated with chloroacetic acid in the presence of sodium methoxide, the acid **10** is the only isolable product, which on treatment with hydrogen chloride affords **7**. The bicyclic compound **7** reacts with *p*-dimethylaminobenzaldehyde in the presence of piperidine to yield 5,6-dihydro-2-*p*-dimethylaminobenzylidene-5-*n*-propylpyrrolo[2,1-*b*]thiazol-3(2*H*)-one (**9**), presumably via **8**.

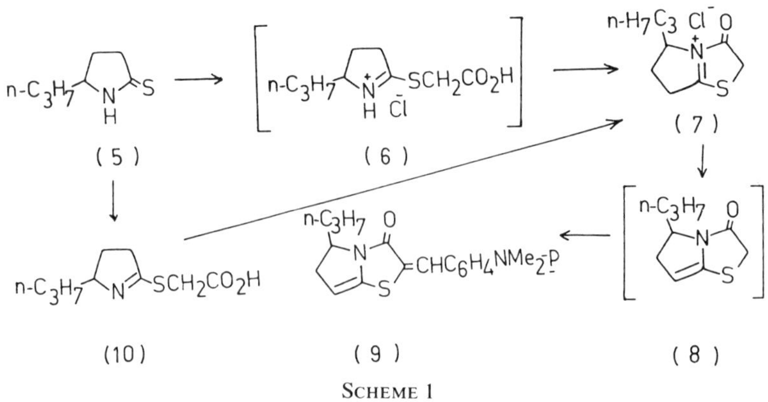

Scheme 1

2. Imidazothiazol-3(2H)-ones

Many examples of imidazo[2,1-*b*]thiazol-3(2*H*)-one (**11**), imidazo[5,1-*b*]thiazol-3(2*H*)-one (**14**), and their partially (**12, 15**) and completely saturated (**13**) derivatives are known. Interesting biological activities have been found.

(11)   (12)   (13)

(14)   (15)

a. *Imidazo[2,1-b]thiazol-3(2H)-ones.* The synthesis of imidazo[2,1-*b*]thiazol-3(2*H*)-one (**11**) can be accomplished by two routes. The first and most common is one in which a thiazole ring is built onto an imidazole nucleus by the reaction of imidazolyl-2-thione with an α-halogenoacid (or its ester). The second route in which an imidazole ring is built onto a thiazole ring is achieved by the reaction of 2,4-diaminothiazole with an α-halogenoketone followed by oxidative cyclization of the intermediate.

i. *From an imidazole.* The simplest synthesis of imidazo[2,1-*b*]thiazol-3(2*H*)-one (**18**) is by the reaction of imidazolyl-2-thiones (**16**) with α-halogenoacids (or esters) and subsequent cyclization of the intermediate acids (esters) **17** with acetic anhydride [61ZOB3267; 72JPR785; 73IJC747; 74JPR147; 78IJC(B)329] (Scheme 2).
In the case of unsymmetrical imidazolylthiones (**16**, R^1 = Ph, R^2 = H), the structure of cyclized product **18a** as the 5-phenyl-isomer has been settled by a comparison. The cyclization product obtained from intermediate **17a** (R^5 = H) was not found to be identical to the 6-phenyl-isomer (**27**) obtained through an unequivocal synthesis [62LA(657)113] (see *From a thiazole* in the following section). In a variation, the reaction of **16** (R^1 = Ph, R^2 = H) with ethyl chloroacetate and aromatic aldehydes in the presence of pyridine and piperidine in dry ethanol furnishes arylidene derivatives of bicyclic compound **19** (81JIC1117). The assignment of the phenyl group to position C-5 is, however, based on analogy to earlier work (61ZOB3267).

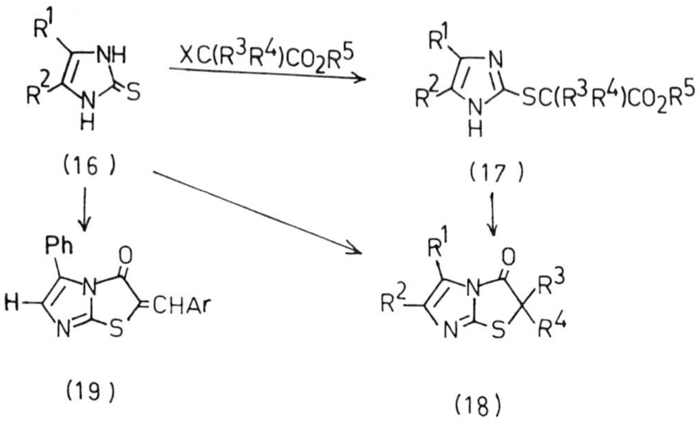

SCHEME 2

	R^1	R^2	R^3	R^4
a	Ph	H	H	H
b	p-$O_2NC_6H_4$	H	H	H
c	Ph	Ph	H	H
d	p-MeOC_6H_4	p-MeOC_6H_4	H	H
e	Ph	Ph	CH_3	H
f	Ph	Ph	CF_3	F

The easy ring-opening quality of hexafluoro-1,2-epoxypropane in the presence of a nucleophile has been used to accomplish the synthesis of imidazo[2,1-b]thiazol-3(2H)-one. Thus, the sodium salt of 4,5-diphenylimidazolyl-2-thione (16, $R^1 = R^2 = $ Ph) reacts with hexafluoro-1,2-epoxypropane in acetonitrile to furnish 5,6-diphenyl-2-fluoro-2-trifluoromethylimidazo[2,1-b]thiazol-3(2H)-one (18f) in a one-step synthesis (78BCJ3091).

The reaction of imidazolyl-2-thione (20) with dimethyl acetylenedicarboxylate (DMAD) in methanol yielded a product for which structure 21 was originally proposed [71JCS(C)3602] but was later modified to imidazo-thiazinone (22) on the basis of its ^{13}C-NMR spectrum, which was found to be similar to that of 24 rather than 23 [81JCS(P1)415] (Scheme 3). The absence of coupling between the lactam carbon and H_A in the ^{13}C-NMR spectrum of the product obtained from 20 and DMAD supported structure 22. A coupling constant, $J(^{13}C, H_A)$, of ~6.4 Hz is expected from the alternate thiazolidinone structure 21 (78HCA607). The reaction, however, when carried out in acetonitrile gives a mixture of 21 and 22 in the ratio of 1:2. These could not be resolved by column chromatography on alumina, but their structures were confirmed by NMR spectral data

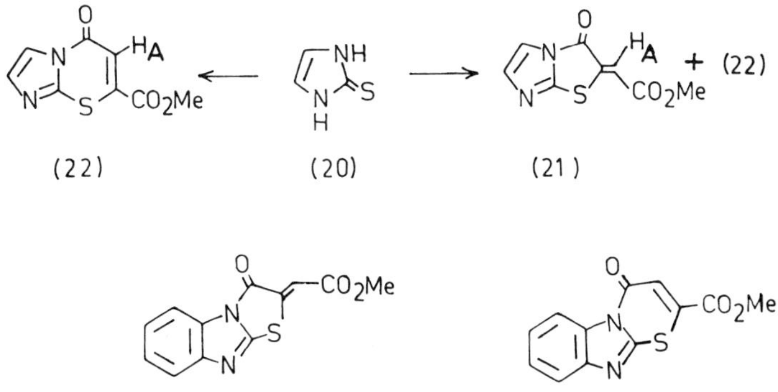

SCHEME 3

[81JCS(P1)415]. Surprisingly, 4,5-dihydroimidazolyl-2-thione, under identical conditions, yields solely the 5,6-dihydro derivative of **21** [71JCS(C)3602] whose ^{13}C-NMR spectrum strongly resembles that of **23** [81JCS(P1)415].

ii. *From a thiazole.* There is only one example depicting the synthesis of imidazo[2,1-*b*]thiazol-3(2*H*)-one through the second route. Thus, 2,4-diaminothiazole (**25**) reacts with phenacylbromide to give the intermediate ketone **26** that, on oxidative cyclization with nitric acid, furnishes the bicyclic compound **27** [62LA(657)113] (Scheme 4).

b. *Dihydroimidazo[2,1-b]thiazol-3(2H)-ones.* The synthesis of **12** has also been achieved through two routes: either by annulating a thiazole ring onto an imidazoline or vice-versa. In the former case, dihydroimidazolyl-2-thione reacts with an α-halogenoacid or its equivalent to furnish **12** directly, but in some cases, the intermediates are isolated, characterized, and then cyclized to give **12**. In the latter path, 2-imino-4-

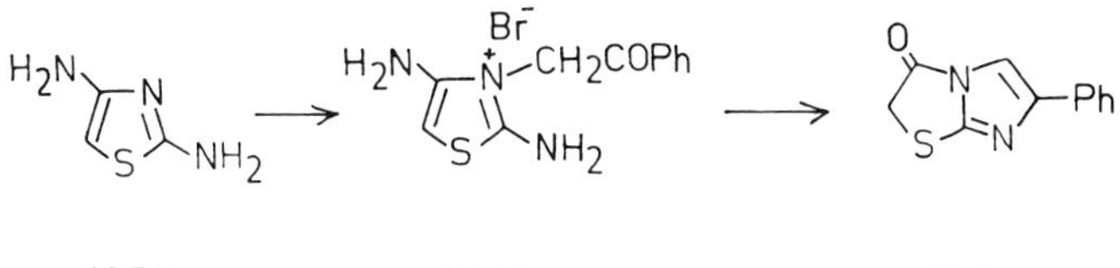

SCHEME 4

thiazolidinone is condensed with appropriate alkylating agents followed by cyclization of the intermediate to yield the desired bicyclic product **12**.

i. *By the first route.* Arylidene derivative **29** of bicyclic compound **12** was first prepared by the reaction of 4,5-dihydroimidazolyl-2-thione (**28**) with ethyl chloroacetate and aromatic aldehydes in the presence of pyridine and piperidine (56JOC24) (Scheme 5). Compound **29** has also been synthesized by cyclization with acetic anhydride in the presence of aromatic aldehydes of acid **30** obtained from the condensation of **28** with the sodium salt of chloroacetic acid. The reaction of **28** with chloroacetic acid in hydrochloric acid, however, gives 3-(β-aminoethyl)thiazolidin-2,4-dione hydrochloride (**32**), characterized through IR spectral data (1667 and 1724 cm^{-1} for both C=O groups) and chemical transformations. The formation of **32** has been reasonably presumed to proceed through the bicyclic intermediate **31** (56JOC193). Imidazoline-2-thiolacetic acid hydrochloride (**33**) was earlier wrongly assigned to structure **32** (42JA2706). Several arylidene derivatives (**29**) have been prepared by the reaction of **28** with chloroacetic acid and aldehydes in the presence of anhydrous sodium acetate in a one-step synthesis (59MI1; 61ZOB1635; 68ZOR179]; the method followed is essentially similar to that reported earlier (56JOC24).

Stephen and Wilson (26JCS2531; 29MI1) and Limar (59MI1) claimed to have synthesized 5,6-dihydroimidazo[2,1-*b*]thiazol-3(2*H*)-one (**35**) by reacting **28** with ethyl chloroacetate in the presence of pyridine or anhydrous sodium acetate. The compound reported by Limar was not characterized and the work of Stephen and Wilson could not be repeated by Campaigne and Wani who could, however, synthesize **35** by reacting **28** with ethyl

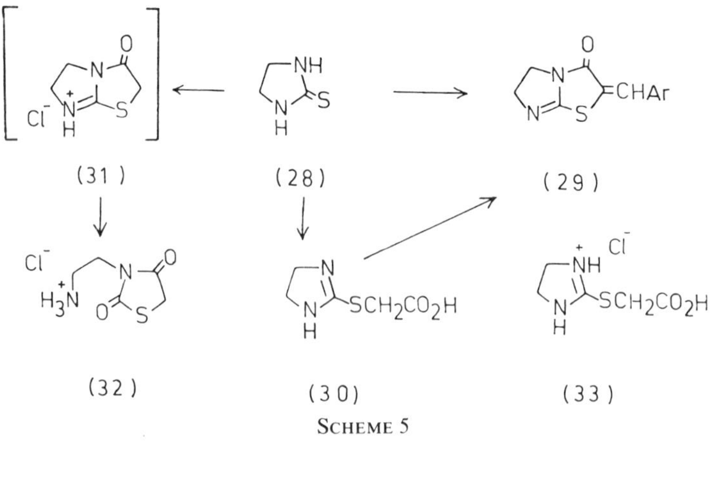

SCHEME 5

chloroacetate and subsequently treating the intermediate **34** with ammonia in anhydrous ethanol (64JOC1715). The hydrochloride (**31**) has been obtained two ways, each involving a two-step synthesis. Thus, **28** reacts with chloroacetic acid in the presence of sodium acetate to give the acid **36**, which, on treatment with concentrated HCl, yields **31**. Alternately, **28** and ethyl chloroacetate in presence of anhydrous sodium acetate in ethanol give **37** which is converted to **31** with concentrated HCl (64JOC1715). A one-step synthesis of **31** has been achieved by heating **28** with ethyl chloroacetate at 150° for 4 hours (68JIC710) or with chloroacetic acid in glacial acetic acid (71KGS93) (Scheme 6).

The use of 4-methyl analog **38** in place of **28** in the previous reaction results in the formation of thiazolidin-2,4-dione derivatives (**40, 41**) instead of the desired bicyclic compound **39**. Thus, the reaction of **38** with ethyl chloroacetate and with ethyl α-bromopropionate gives **40** and **41**, respectively, and is characterized by twin peaks at 1670 and 1750 cm^{-1} in **40** and 1665 and 1732 cm^{-1} in **41** for both C=O groups, as well as chemical transformation of **40** to **43** which is also obtained from **42** as shown in Scheme 7. The formation of **40** and **41** is presumed to have proceeded through **39**. Compound **38**, however, upon refluxing with ethyl chloroacetate in the presence of pyridine in anhydrous ethanol and subsequent heating with aldehyde and piperidine, furnishes **42** (70IJC885). The reaction of **38** with chloroacetic acid in aqueous medium give **40** (70IJC885) and not **44** as reported earlier (42JA2706).

Under identical conditions, **31** is found to be stable, while **39** (R = H, X = Cl) is labile. This difference is probably due to the presence of the methyl on the imidazole ring. To test the effect of a methyl group present on the thiazole ring on the stability of bicyclic compound **45**, the reaction of **28** with ethyl α-bromopropionate was carried out. Compound **46** was

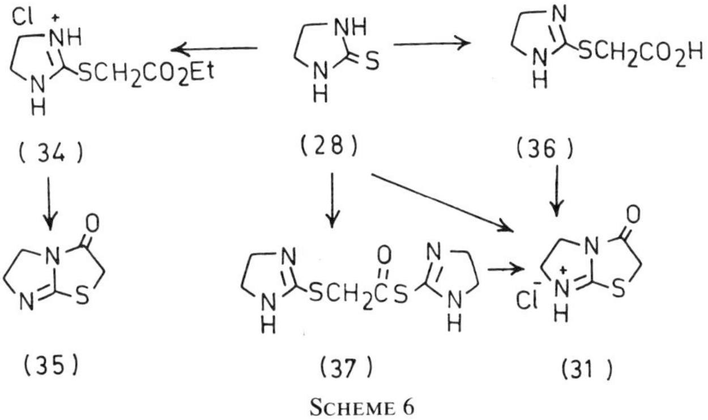

SCHEME 6

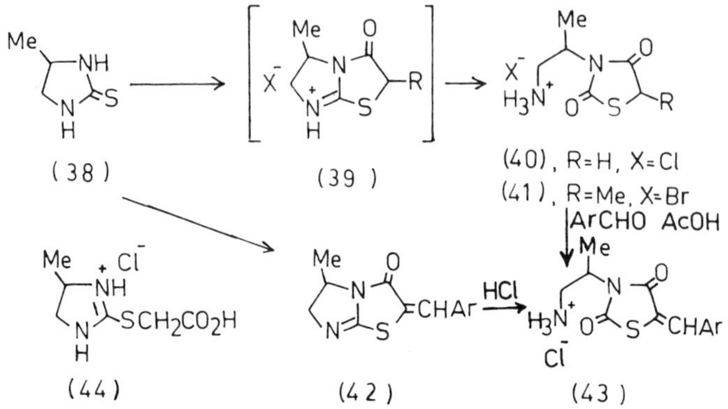

SCHEME 7

obtained, which suggests that when a methyl group is present either on the imidazole or on the thiazole ring, the bicyclic compound undergoes ring cleavage to give thiazolidin-2,4-diones (72IJC766) (Scheme 8).

Bicyclic system **12**, having a group bulkier than methyl, is quite stable. Thus, **28**, on reaction with α-bromoacids (esters) or 2-bromolactone and subsequent basification, give **47a–e** [71JCS(C)3602, 80EUP2978]. With aryl/alkyl(cyano)methylbenzene sulfonate [RCH(CN)OSO$_2$Ph] followed by acid-catalyzed hydrolysis, **28** affords **47a,f–j** (60USP2933497; 61JOC2715). When the sulfonic ester is not sufficiently reactive, the use of the iodide ion in the reaction mixture is indispensable. The reaction of **28** with 2,2-dicyano-1-phenyloxirane also furnishes **47a** (81S981) (Scheme 9).

The reaction of **28** with acetylene dicarboxylic acid or DMAD furnishes **48a** [71JCS(C)3602] and **48b** [71JCS(C)3602; 77MI1], respectively (Scheme 10). Compound **48b**, synthesized by this reaction, was earlier formulated as imidazo-thiazinone (**49**) (67CJC939, 67CJC953) on the basis of mass spectral studies of the monocyclic system (thiazinone) obtained from thiourea and DMAD, whose structure is still in dispute. Evidence has been furnished that the reaction of DMAD with thiourea gives methyl (2-imino-

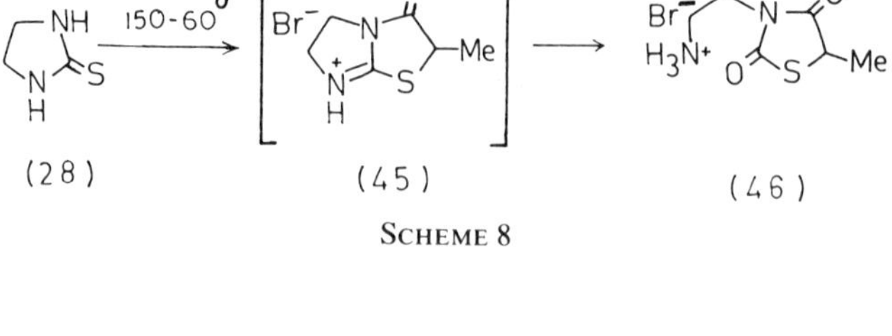

SCHEME 8

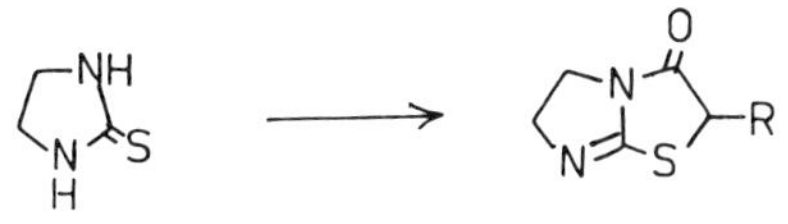

(28) → (47)

	R		R
a.	Ph	f.	o-ClC$_6$H$_4$
b.	(CH$_2$)$_2$CO$_2$Et	g.	2,4-Cl$_2$C$_6$H$_4$
c.	(CH$_2$)$_3$CO$_2$Et	h.	3,4-Cl$_2$C$_6$H$_4$
d.	(CH$_2$)$_3$CO$_2$H	i.	(CH$_2$)$_2$Ph
e.	(CH$_2$)$_2$OH	j.	(CH$_2$)$_2$C$_6$H$_{11}$

SCHEME 9

4-thiazolidinone-5-ylidene)acetate [71CI(L)705]. The correct structural assignment as **48** for the reaction product is now based on an unequivocal synthesis involving the reaction of **35** with glyoxalic acid to give **48a**, which on esterification with diazomethane affords **48b** (Scheme 10). Structure **48b** is further confirmed by its unambiguous synthesis from methyl 4-thiazolidinone-5-ylideneacetate [71JCS(C)3602] (see synthesis by second route). The reaction of **38** with DMAD similarly furnishes the methyl analog of **48b**, except the position of the methyl group on the imidazole ring is uncertain (77MI1).

The bicyclic products obtained from unsymmetrical substrates such as thiohydantoins (**50, 53,** and **56**) can be represented by either structures **51** or **52, 54** or **55,** and **57** or **58**, respectively. The data are inadequate to confirm either structure [71JCS(C)3602; 76ZN(B)111; 81S981] (Scheme 11).

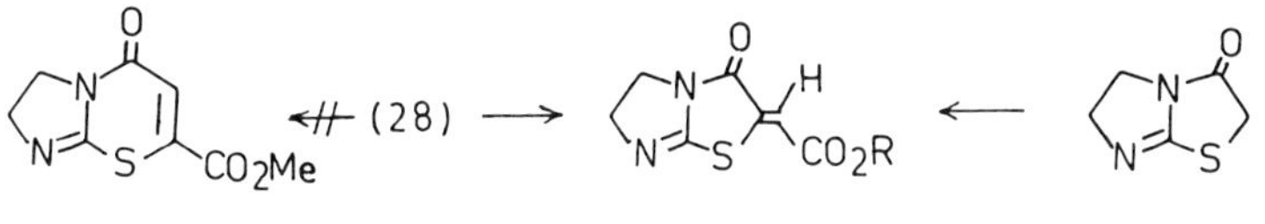

(49)   ↛ (28) →   (48)   ←   (35)

a, R = H
b, R = Me

SCHEME 10

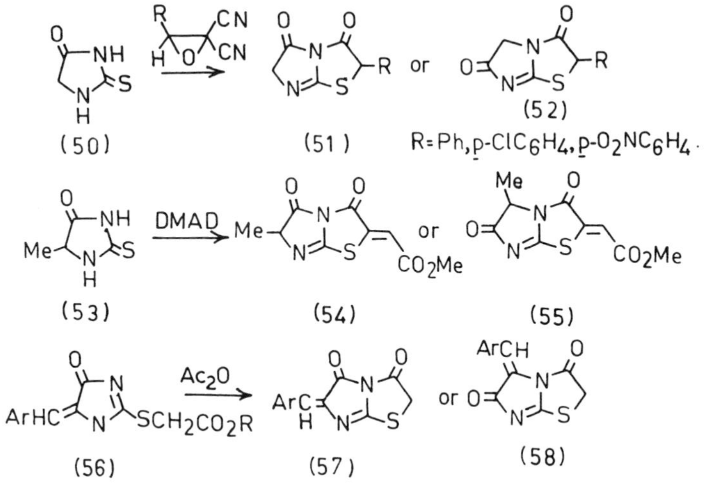

SCHEME 11

ii. *By the second route.* In the second route, the synthesis of **12** can be accomplished starting from 2-imino-4-thiazolidinone. Thus **59** reacts with 1,2-dibromoethane to give **48b** [71JCS(C)3602]. The condensation of **60** with chloroacetic acid yields the acid **61**, which on cyclization with hot acetic acid or concentrated sulfuric acid, affords **62**. That the alkylation of **60** with chloroacetic acid takes place at the exocyclic nitrogen atom is confirmed by the hydrolysis of **61**, which results in the formation of 5-arylidenethiazolidin-2,4-dione (75IJC238). An allyl functionalized group present at a suitable position of the thiazolidinone ring can undergo bromination–dehydrobromination to give the bicyclic compound. Thus **63**, on treatment with bromine and potassium acetate, furnishes **64** (73KGS424) (Scheme 12).

c. *Tetrahydroimidazo[2,1-b]thiazol-3(2H)-one.* Methyl (Z)-5,6,7,7a-tetrahydro-3-oxoimidazo[2,1-b]thiazol-2-ylidene acetate (**67**), the only example of this series, has been synthesized by the reaction of 4,5-dihydro-1-methylimidazolyl-2-thione (**65**) with DMAD in methanol [81JCS(P1)415] (Scheme 13). Compound **67** is of some interest for ^{13}C-NMR spectroscopy since it has an sp^3-carbon atom (C-7a) attached to one sulfur, one oxygen, and two nitrogen atoms (see Section III). The formation of **67** is presumed to have proceeded through the intermediate **66**.

d. *Imidazo[5,1-b]thiazol-3(2H)-ones.* 5-Methyl-7-nitroimidazo[5,1-b]thiazol-3(2H)-one (**69**, R = Me) and its 5-ethyl analog have been ob-

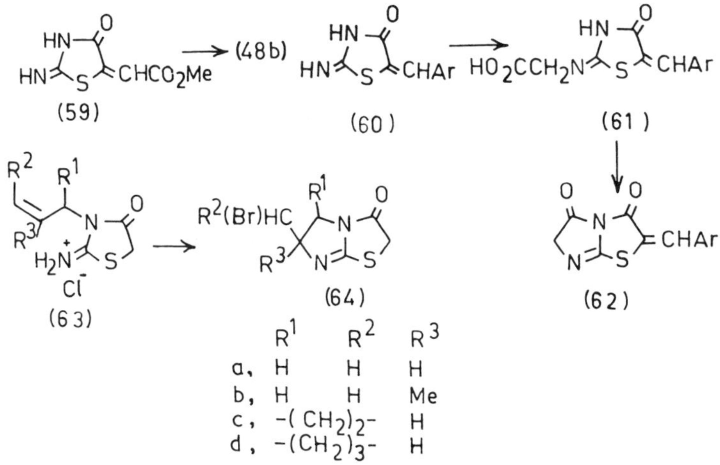

Sec. II.A] CONDENSED 4-THIAZOLIDINONES 13

SCHEME 12

	R^1	R^2	R^3
a,	H	H	H
b,	H	H	Me
c,	$-(CH_2)_2-$		H
d,	$-(CH_2)_3-$		H

tained in excellent yields by heating the appropriate acids (**68**) wit acetic anhydride for 7hr at 100°C (67KGS93) (Scheme 14).

e. *Dihydroimidazo[5,1-b]thiazol-3(2H)-ones*. The acid **70** on cyclization with acetic anhydride, affords 7-arylazo-5,6-dihydro-5-oxoimidazo[5,1-b]thiazol-3(2H)-one (**71**) [81ZN(B)501] (Scheme 15).

3. *Pyrazolo[3,2-b]thiazol-3(2H)-ones*

The only example of this system has been reported in a patent (83EGP156815). 5-Amino-1-benzoyl/substituted benzoyl-4-ethoxycarbonylpyrazolyl-3-thiones (**72**) react with either ethyl 3-anilino or substituted anilinocrotonates in boiling ethanol in the presence of metallic sodium to give 2-(α-anilino/substituted anilino)ethyl-5-benzoyl/substituted ben-

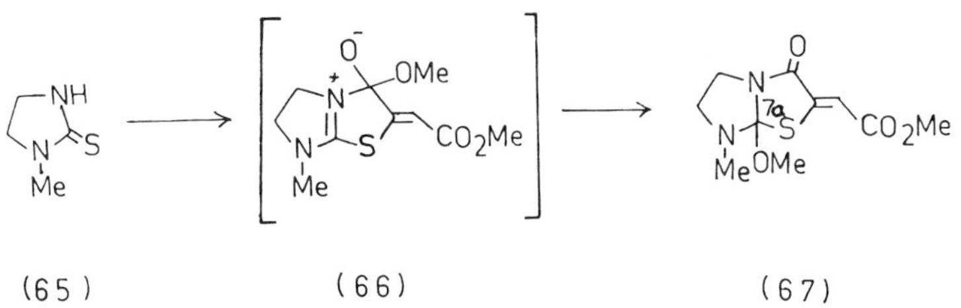

SCHEME 13

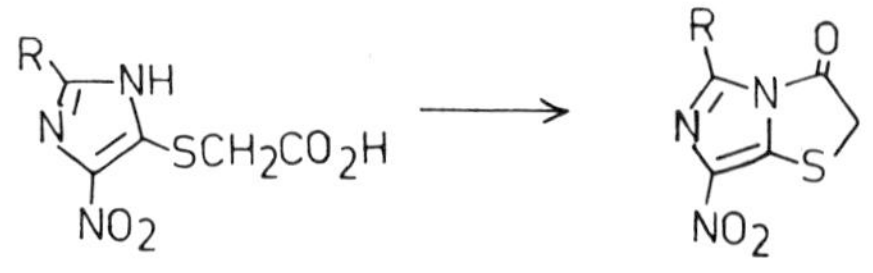

(68) → (69)

R = Me, Et

SCHEME 14

zoyl-7-ethoxycarbonyl-6-iminopyrazolo[3,2-*b*]-thiazol-3(2*H*)-ones (73) (Scheme 16).

4. *Thiazolo[2,3-b]thiazol-3(2H)-one*

Reaction of 2-phenyl-2-thiazoline (**74**) and ketene in liquid sulfur dioxide gives 5,6-dihydro-7a-phenyl-7a*H*-thiazolo-[2,3-*b*]thiazol-3(2*H*)-one-1,1-dioxide (**75**) (Scheme 17). The cycloaddition was earlier presumed to have proceeded through a 1,4-dipolar species (**76**) (67CC935), but later a 1,3-dipolar species (**77**) was suggested. The adduct **77** reacts with **74** in a [3 + 2] cycloaddition to give bicyclic compound **75** (69JHC729). The possibility of an equilibrium between an open 1,3-dipolar ion (**77**) and cyclic structure **78** cannot be ruled out. A 1,3-dipolar system without a double bond is rare. Huisgen postulates such an intermediate in the thermolysis of ethylene carbamate with *N*-benzylideneaniline to yield 2,3-diphenyloxazolidine [63AG(E)565]. More chemical and physical data are needed before the nature of the adduct is firmly established.

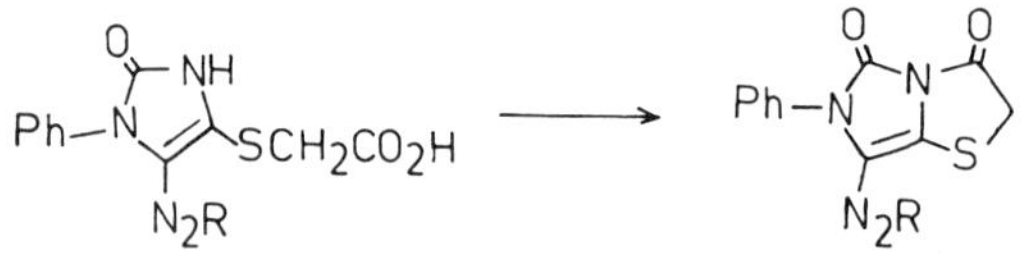

(70) → (71)

R = Ph, 4-MeC$_6$H$_4$, 4-ClC$_6$H$_4$, 4-O$_2$NC$_6$H$_4$, 2-MeC$_6$H$_4$, 3-MeC$_6$H$_4$

SCHEME 15

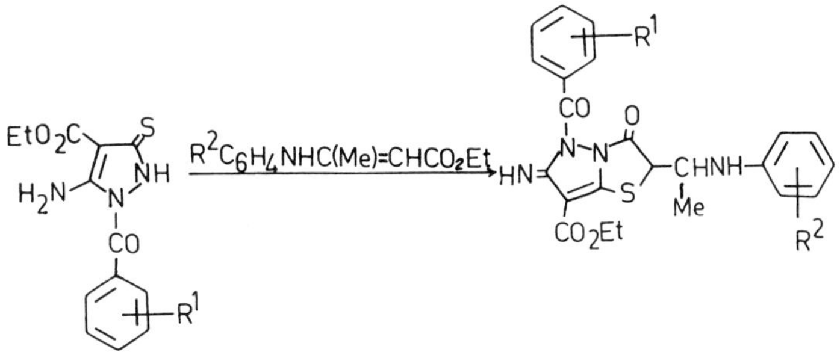

(72) → (73)

SCHEME 16

5. *Thiazolo[2,3-b]oxazol-5(6H)-one*

2-(*p*-Nitrophenyl)-2-oxazoline (**79**) similarly reacts with ketene in liquid sulfur dioxide to furnish 2,3-dihydro-7a-*p*-nitrophenyl-7a*H*-thiazolo[2,3-*b*]oxazol-5(6*H*)-one-7,7-dioxide (**80**) (69JHC729) (Scheme 18).

6. *Thiazolo-s-triazol-5(6H)-ones*

Fusion of *s*-triazole and 4-thiazolidinone rings can be effected in two possible ways, as exemplified by thiazolo[3,2-*b*]-*s*-triazole-5(6*H*)-one (**83**) and thiazolo[2,3-*c*]-*s*-triazole-5(6*H*)-one (**85**). Only scanty work on these condensed thiazolidinone systems is available, although a considerable amount of work on their thiazole counterpart has been reported.

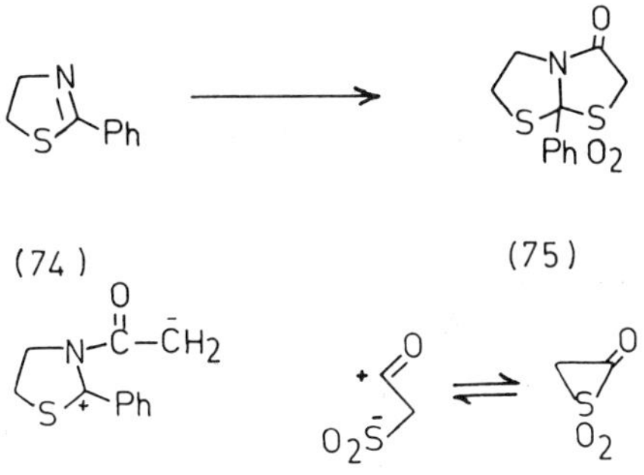

SCHEME 17

(79) → (80)

SCHEME 18

a. *Thiazolo[3,2-b]-s-triazol-5(6H)-ones*. Reaction of 3-aryl-s-triazolyl-5-thione (**81**) with chloroacetic acid in acetic acid in the presence of sodium acetate gives the acid **82** which, on treatment with acetic anhydride or phosphoryl chloride, yields a cyclized product that has been formulated as 2-arylthiazolo[3,2-b]-s-triazol-5(6H)-ones (**83**) (76JPR12). The arylidene derivatives **84** have been obtained by condensing **83** with aromatic aldehydes in acetic anhydride. This is done by refluxing the acid **82** with aromatic aldehydes in acetic anhydride or by refluxing the thione **81** with chloroacetic acid and aromatic aldehydes in the presence of anhydrous sodium acetate in acetic acid and acetic anhydride. The last method gives better yields [76JPR12; 86IJC(B)776] (Scheme 19). The structural assignments for the cyclization products obtained from the acid **82** and the arylidene product obtained from the thione **81** are **83** (76JPR12) and **84** [86IJC(B)776], respectively, rather than the other possible isomeric structure **85** and its arylidene derivative. These assignments were based on analogy to earlier work on their thiazole counterparts (71JOC10) and triazolo-pyrimidines (65JA1980). The final structural assignment for the cyclized product, therefore, must await unequivocal synthesis or strong spectral evidence.

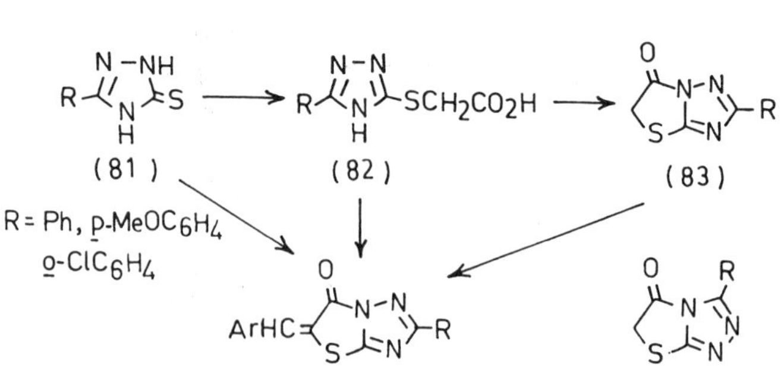

R = Ph, p-MeOC$_6$H$_4$, o-ClC$_6$H$_4$

SCHEME 19

b. *Thiazolo[2,3-c]-s-triazol-5(6H)-ones.* 2-Phenylhydrazino-4-thiazolidinone (**86**, R = R^1 = H) and its 5-substituted derivatives, undergo cyclization on reaction with formaldehyde or aromatic aldehydes to give 2,3-dihydro-2-phenylthiazolo[2,3-c]-s-triazol-5(6H)-one (**87**, R = R^1=R^2 = H) and its 3,6-disubstituted derivatives, respectively [76ZN(B)380] (Scheme 20).

7. Thiazolo[3,2-a]pyridin-3(2H)-ones

Japanese workers were the first to use the reactive properties of nitrile compounds to synthesize thiazolo[3,2-a]pyridin-3(2H)-ones (77S839). Ethyl α-cyanocinnamate (**88**, R^3 = Ph, R^4 = CO$_2$Et) reacts with ethyl mercaptoacetate in the presence of triethylamine to give 5-amino-2-benzylidene-6,8-bis(ethoxycarbonyl)-7-phenyl-7H-thiazolo[3,2-a]pyridin-3(2H)-one (**89**, R^1 = R^3 = Ph, R^2 = R^4 = CO$_2$Et), which is also obtained from the reaction of 5-benzylidene-2-ethoxycarbonylmethyl-2-thiazolin-4-one (**90a**) with ehtyl α-cyanocinnamate (77S839). Reaction of **90a,b** with benzylidenemalononitrile (**88**, R^3 = Ph, R^4 = CN) and of **90c** with ethyl α-cyanocinnamate furnish the respective bicyclic compounds **89** [83ZN(B)781] (Scheme 21). Similar results were also obtained from **88** (R^3 = furan-2-yl- or thiophen-2-yl) (86M105).

Reaction of **91** and **92** with **88** furnishes the arylidene products **89** [81S635; 84ZN(B)824; 87IJC(B)216], whereas the condensation of **92a** with cyanoacetic acid hydrazide yields the parent bicyclic product **93** (84JHC1885). Arylidene product **89** is presumed to have been obtained via the intermediate 5-benzylidene derivative of **91** and **92** formed by the addition of the active methylene in **91** and **92** to the activated double bond of benzylidene malononitrile with concomitant elimination of malononitrile, which subsequently adds to the activated double bond of the interme-

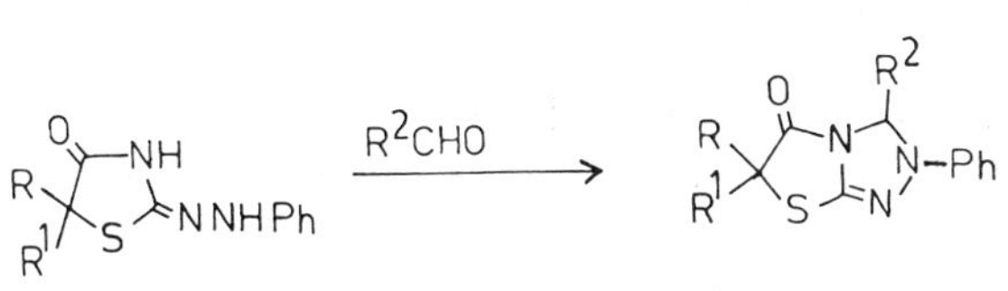

(86)   →R^{2}CHO→   (87)

R=H, R^1=H, Ph$_2$CH
RR^1= PhCH, p-MeOC$_6$H$_4$CH, o-ClC$_6$H$_4$CH
R^2= H, Ph, p-MeOC$_6$H$_4$

SCHEME 20

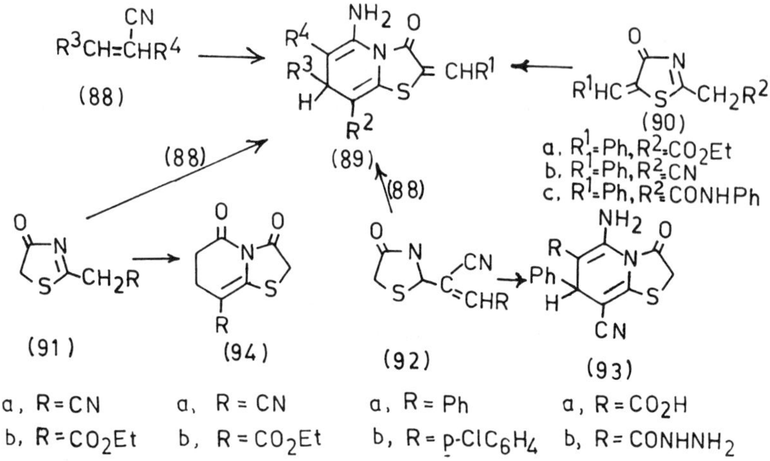

SCHEME 21

diate to give **89**. In the reaction of **92b** with cyanoacetic acid hydrazide, the latter adds to the activated double bond of **92b** and is followed by cyclization to give **93a** via hydrolysis of the intermediate **93b**. Similarly, **91a,b** react with ethyl acrylate to afford **94a,b,** respectively (83H1021).

Synthesis of thiazolo[3,2-*a*]pyridin-3(2*H*)-one can also be accomplished by an alternate route in which a thiazole ring is built onto a pyridine ring. Thus, copper-catalyzed reaction of piperidine-2-thione (**95**) with methyl diazoacetate affords 5,6-dihydro-7*H*-thiazolo[3,2-*a*]pyridin-3(2*H*)-one **97**) (80LA168) (Scheme 22). During the reaction uncyclized intermediate (**96**) could not be isolated, it evidently cyclized immediately to the bicyclic product (**97**), whereas with the pyrrolidine-2-thione the corresponding uncyclized product was isolated (80LA168) (Scheme 22).

8. *Thiazolo[3,2-*a*]pyrimidin-3(2H)-ones*

The synthesis of thiazolo[3,2-*a*]pyrimidin-3(2*H*)-ones has also been accomplished by two routes. The first, commonly followed, is one in which a

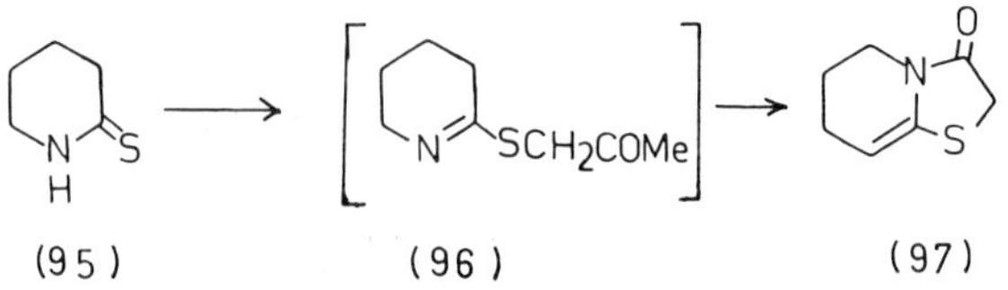

SCHEME 22

thiazole ring is built onto a pyrimidine nucleus by the reaction of pyrimidine-2-thiones with α-halogenoacids (ester or acid halides). The second route, in which a pyrimidine ring is built onto a thiazole nucleus, involves the reaction of 2-imino-4-thiazolidinones with malonic acid chloride or ethyl acrylate followed by cyclization of the intermediates isolated in some cases.

a. *From a Pyrimidine.* Stephen and Wilson in 1928 reported 3,4,5,6-tetrahydropyrimidine-2-thione (**98**) did not react with ethyl chloroacetate in ethanol in the presence of ethoxide ion, but in pyridine medium, the uncyclized compound **99** was isolated (28JCS1415). Van Allan in 1956 first synthesized and characterized the bicyclic compound **100a**. Thus the thione **98**, on reaction with ethyl chloroacetate at 145°C, yielded 4,5,6,7-tetrahydrothiazolo[3,2-*a*]pyrimidin-3(2*H*)-one hydrochloride (**100a**) (56JOC24). The reaction of **98** with ethyl α-bromopropionate, 1-cyano-3(phenyl)propylbenzenesulfonate (PhCH$_2$CH$_2$CH(CN)OSO$_2$Ph), acetylenedicarboxylic esters, and diethyl dichloromalonate gave **100b** (72IJC766), **101** (R = CH$_2$CH$_2$Ph) (60USP2933497), **102** (77MI1), and spiro compound (**103**) (71IJC1216), respectively. The free base **101** (R = H) has been synthesized by reacting **98** with chloroacetic acid (83JIC970) or chloroacetyl chloride (81MIP1) in the presence of a base such as sodium acetate or by the basification of the hydrochloride **100a** with potassium carbonate (83JIC970). 2-Alkyl-substituted bicyclic compounds (**101**, R = Me, Et, Pr, Pr-i) are obtained from the reaction of **98** with appropriate halogeno acids or halogeno acid chlorides (81MIP1) (Scheme 23).

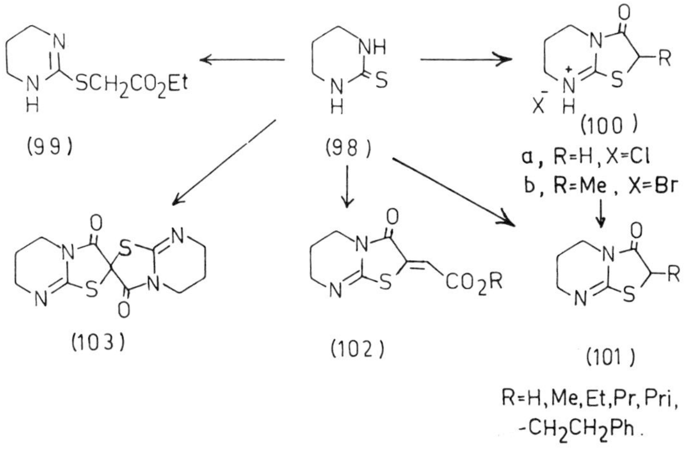

SCHEME 23

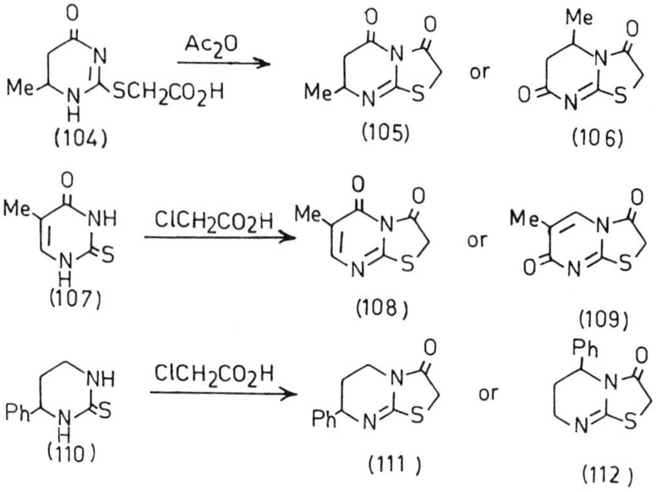

SCHEME 24

The bicyclic compounds obtained from unsymmetrical pyrimidinethiones (**104, 107,** and **110**) can be represented by either structures **105** or **106** (50BRP634951), **108** or **109** (83MI3), and **111** or **112** (75FRP223371, 75GEP2317109; 76JAP74126828), respectively (Scheme 24). No characterization of the cyclized products was attempted although cyclized product obtained from **104** was used in the preparation of merocyanine dyes (50BRP634952).

4,6,6-Trimethyl-1,6-dihydropyrimidine-2-thione (**113**), on condensation with aromatic aldehydes in ethanolic KOH, gives arylidene derivatives **114** which react with chloroacetic acid in the presence of anhydrous sodium acetate in acetic anhydride to afford a cyclized product formulated as **115** (81MI1) that can also be represented by the alternate structure **116** (Scheme 25). Similarly, the thione **113** reacts with chloroacetic acid and aromatic aldehydes in the presence of anhydrous sodium acetate in acetic acid and acetic anhydride to furnish a cyclized product for which structure **117** has been assigned without any evidence (81MI1). The cyclized product can also be represented by the other isomeric structure **118**. The bicyclic compound (**115** or **116**) condenses with aromatic aldehydes to give arylidene derivatives that are also obtained in a single-step synthesis involving the reaction of **114** with chloroacetic acid and aromatic aldehydes in the presence of anhydrous sodium acetate in acetic acid and acetic anhydride.

The thione **119**, on reaction with chloroacetic acid, gives a mixture of the cyclized product **120** as well as the uncyclized acid **121**, the former being

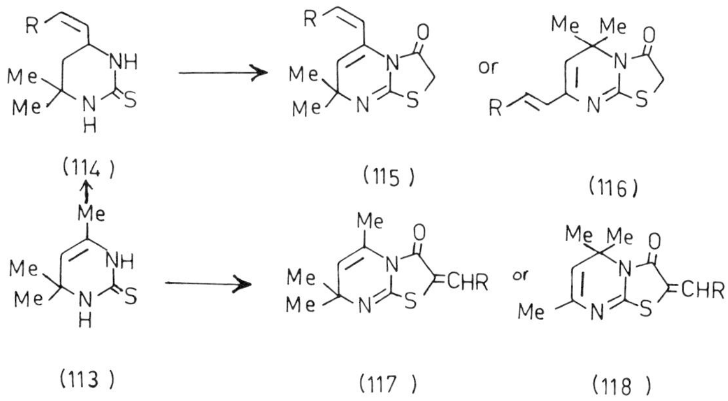

SCHEME 25

formed only in 10 percent yield. The mixture was resolved by column chromatography. When the reaction was carried out in the presence of boron trifluoride etherate in methanol, the cyclized compound **120** was the only isolable product [87IJC(B)556]. The assignment of structure **120**, and not its isomeric structure **122**, to the cyclized compound was based on the ^{1}H-NMR data that revealed a downfield shift of proton H_A (∂ 0.73) in the cyclized product **120** (R = Ph) when compared to that in **119** (R = Ph). Such a downfield shift of H_A is not expected from alternative structure **122** (Scheme 26).

Thiones **123** react with chloroacetic acid in the presence of anhydrous sodium acetate in acetic acid and acetic anhydride to give a cyclized product for which structure **124** has been assigned (82MI1). The cyclized product could be isomeric structure **125**. The same workers similarly obtained a product formulated as **126** from the reaction of the thione **123** (n = 6) with chloroacetic acid (81MI2). Compound **126** reacts with aromatic aldehyde to furnish the arylidene product **127** which is also obtained from the reaction of thione **123** (n = 6), chloroacetic acid, and aldehydes as well as from the condensation reaction of thiazolidinone **124** (n = 6)

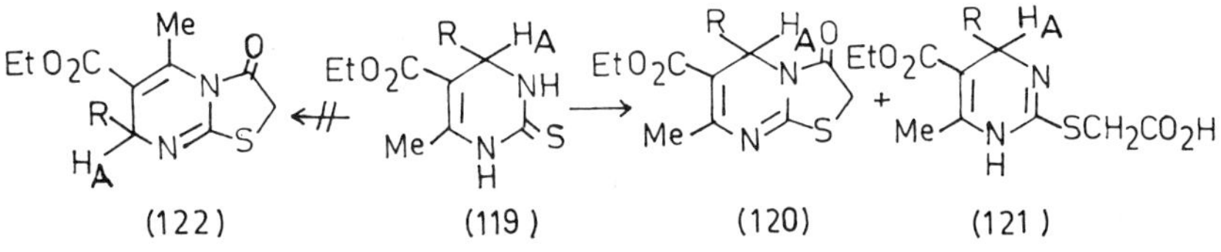

SCHEME 26

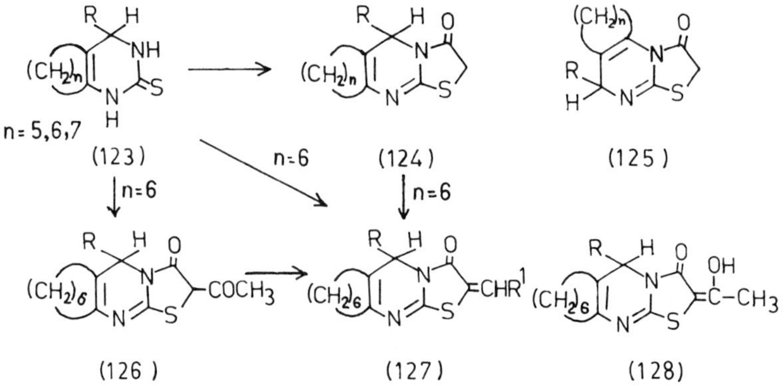

SCHEME 27

with aldehydes. No evidence was cited in support of structures **124, 126,** and **127.** Deacetylation of **126** occurred during its condensation with aldehydes (81MI2). Although **126** has one acetyl carbonyl group and one lactam carbonyl group, it shows only one carbonyl peak at 1680 cm^{-1} in its IR spectrum, which makes structure **126** untenable. The product may exist as enolic structure **128,** which is consistent with the IR data (Scheme 27).

Egyptian workers following the same reaction conditions reported the synthesis of cycloalkanethiazolopyrimidine systems (78MI1; 79MI1; 81MI3). Thiones **129** react with chloracetic acid to give thiazolidinones **130** that condense with aldehydes to yield arylidene thiazolidinones **131.** Compound **131** is also obtained in a single step from the reaction of **129** with chloroacetic acid and aldehydes in the presence of anhydrous sodium acetate in acetic acid and acetic anhydride (Scheme 28). Here also the regiochemistry of the cyclized products is uncertain.

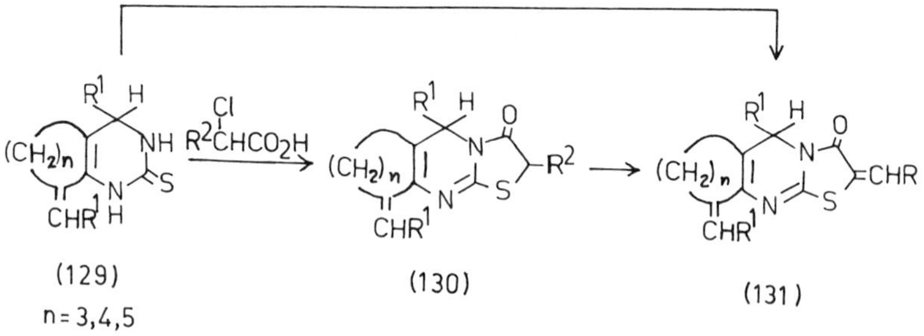

SCHEME 28

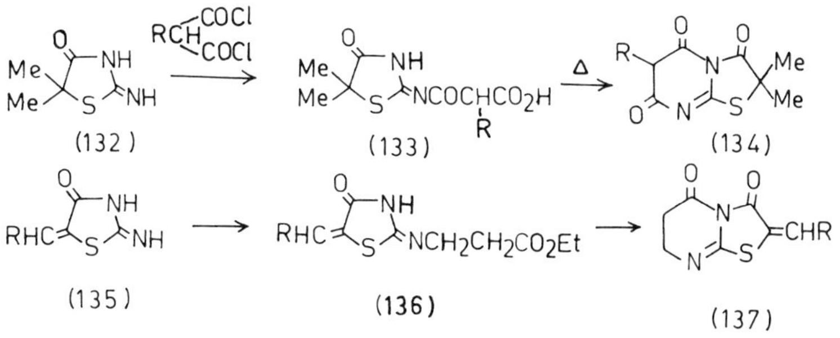

SCHEME 29

b. *From a Thiazole.* There are only two instances that describe the synthesis of thiazolo[3,2-*a*]pyrimidin-3(2*H*)-ones through this route. 5,5-Dimethyl-2-imino-4-thiazolidinone (**132**) reacts with alkylmalonyl chloride to give the acids **133** which, on pyrrolytic dehydration, furnish 2,2-dimethyl-6-alkyl-4,5,6,7-tetrahydro-2*H*-thiazolo[3,2-*a*]pyrimidin-3,5,7-triones (**134**) (50JCS1127). A better result is obtained by heating the acid in a vacuum. Similarly, the bicyclic compound **137** is prepared by reacting 5-arylidene-2-imino-4-thiazolidinones (**135**) with ethyl acrylate and then cyclizing the resulting intermediate esters **136** with acetic anhydride (75IJC238) (Scheme 29).

9. *Thiazolo[2,3-c][1,4]thiazin-3(2H)-ones*

A method involving the facile addition of mercaptoacetic acid across the double bond of an azomethine for the one-step preparation of 4-thiazolidinones has been used in the synthesis of thiazolo[2,3-*c*][1,4]thiazine-3(2*H*)-one. Thus, 2*H*-3,4-dihydro-1,4-thiazines (**138**) react with alkylmercaptoacetic acid at 100°C under nitrogen to give thiazolo[2,3-*c*][1,4]thiazin-3(2*H*)-ones (**139**) Similarly, 7*H*-hexahydro-1,4-benzothiazine [**138**, $RR^1 = (CH_2)_4$] condenses with mercaptoacetic acid to yield **139** [$RR^1 = (CH_2)_4$, $R^2 = R^3 = H$] (64M1335) (Scheme 30).

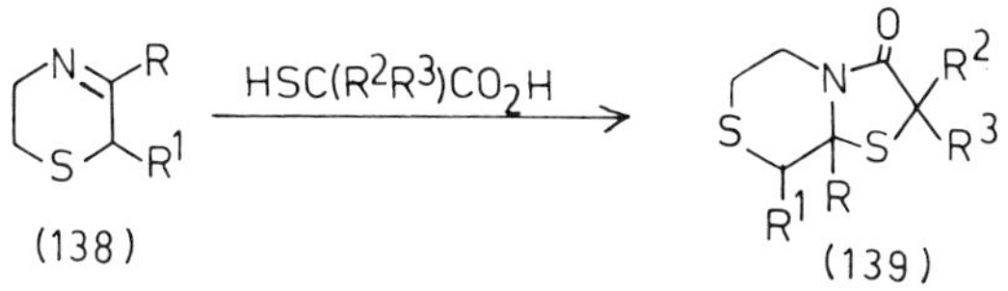

SCHEME 30

10. Thiazolo[2,3-c][1,2,4]benzothiadiazin-3(2H)-ones

The synthesis of a thiazolo[2,3-c][1,2,4]thiadiazin-3(2H)-one has so far not been reported. However, the synthesis of its benz-analog (**144**) has been achieved as illustrated in Scheme 31. The thione **140a** reacts with ethyl α-chloroacetoacetate to give β-ketoester **141a** which, on treatment with acetic anhydride and pyridine, yields 6-chloro-2-ethoxycarbonyl-3-methylthiazolo[2,3-c][1,2,4]benzothiadiazin-9,9-dioxide (**142a**). However, when the cyclization was carried out in acetic anhydride without pyridine, a different compound identified by spectral data as 6-chloro-2-(α-hydroxyethylidene) thiazolo[2,3-c][1,2,4]benzothiadiazin-3(2H)-one-9,9-dioxide acetate (**143a**) was obtained. The ^{1}H-NMR spectrum of the product does not show an ethyl ester group, but it does show two methyl singlets at δ 2.32 and 2.72, while IR spectrum exhibits two carbonyl absorptions at 1727 cm^{-1} (five-membered lactam formed with cyclization taking place at the ester carbon atom) and 1770 cm^{-1} (acetoxy carbonyl group). The cyclization onto the nitrogen adjacent to the benzene ring rather than onto the sulfonamide nitrogen in **141a** was supported by the large downfield signal of H_A (δ 8.95 in **143a**) thus ruling out alternate structure **145a** from which such a down-field signal is not expected (75JHC1207). Refluxing **143a** in ethanol causes hydrolysis of the labile acetate group to furnish the enol **144a**. Compounds **143b,c** have also been synthesized by the same procedure from the respective starting materials **140b,c** (70USP3475425) (Scheme 31).

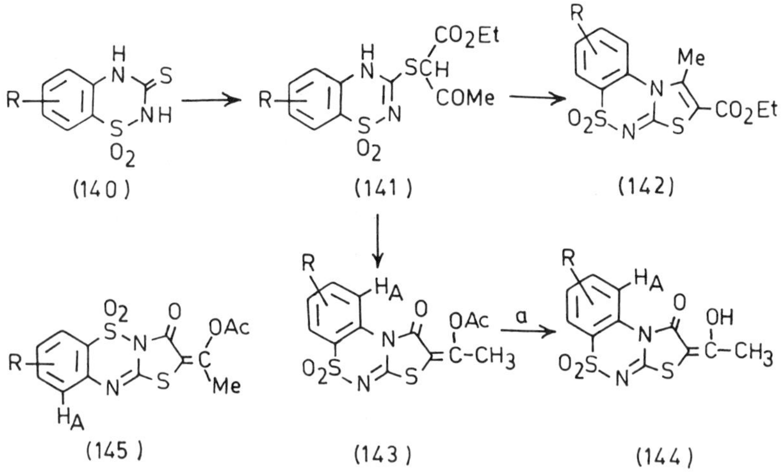

a, R = 6-Cl; b, R = 7-Me; c, R = 8-OMe

SCHEME 31

11. Thiazolo-triazin-3(2H)-ones

a. *Thiazolo[1,2,4]triazin-3(2H)-ones.* Fusion of 4-thiazolidinone and 1,2,4-triazine nuclei can be affected in two ways represented by thiazolo[2,3-c]-[1,2,4]-triazin-3(2H)-one (**148**) and thiazolo[3,2-b]-[1,2,4]-triazin-3(2H)-one (**157**). Synthesis of both the systems have been accomplished by building the 1,2,4-triazine nucleus onto the thiazolidinone nucleus or vice-versa.

i. *Thiazolo[2,3-c]-[1,2,4]-triazin-3(2H)-ones.* 2-Hydrazino-4-thiazolidinone (**146**, R = H) reacts with chloroacetic anhydride to afford **147** which, on pyrrolysis, readily cyclizes to give 2H,7H-5,6-dihydrothiazolo[2,3-c]-[1,2,4]-triazin-3,6-dione (**148**) (72ZOR1722). Reaction of phenyl derivative **146** (R = Ph) with ethyl bromoacetate gives cyclized product 2H,7H-7-phenyl-5,6-dihydrothiazolo[2,3-c]-[1,2,4]-triazin-3,5-dione (**149**) (84MI2) (Scheme 32).

Reaction of 1-methyltetrahydro-1,2,4-triazine-3-thione (**150**, R = Me) with ethyl bromoacetate gives 7H,7-methyl-5,6-dihydrothiazolo[2,3-c]-[1,2,4]-triazin-3(2H)-one (**151**, R = Me, R^1 = H) (70JHC1231). Similarly, **150** (R = Me) reacts with α-cyanobenzyl *p*-toluene sulfonate [PhCH(CN)OTS] or α-cyanobenzyl bromide [PhCH(CN)Br] to afford **151** (R = Me, R^1 = Ph) (71JHC621). Structure **151** was proved by its reduction with diborane to **152**, which was also obtained by an independent synthesis involving the reaction of **153** with thionyl chloride followed by base-catalyzed cyclization (70JHC1231; 71JHC621). To ascertain whether the methyl group at the 1-position in **150** (R = Me) influences the direction of cyclization, the thione **150** (R = H) was allowed to condense with α-cyanobenzyl-*p*-toluene sulfonate and ethyl α-bromophenylacetate. In both cases **151** (R = H, R^1 = Ph) was obtained. Compound **150**, however, on reaction with 1,2-dibromoethane, furnished another bicyclic system (**154**). The difference in the behavior of **150** toward dibromoethane and other reagents such as ethyl bromoacetate and α-cyanobenzyl *p*-toluene

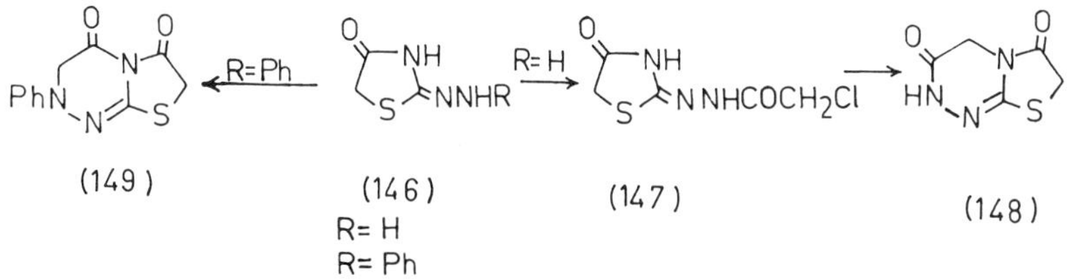

(149)   (146)   (147)   (148)
        R= H
        R= Ph

SCHEME 32

sulfonate is due to the presence of an sp^3-functionalized carbon in the former and an sp^2-functionalized carbon in the latter (71JHC621) (Scheme 33).

ii. *Thiazolo[3,2-b]-[1,2,4]-triazin-3(2H)-ones*. Reaction of 6-methyl-5-oxo-1,2,4-triazine-3-thione (**155**, R = Me) with bromoacetic acid in the presence of metallic sodium in ethanol gave a product, 2*H*,7*H*-6-methylthiazolo[3,2-*b*]-[1,2,4]-triazin-3,7-dione, for which structure **157** was assigned (68ACH191). Later, when the reaction was repeated the product was found to be the uncyclized acid **156**, which then was cyclized with acetic anhydride to bicyclic **157** (71JHC1011). Structure **157** was established by its unequivocal synthesis; the cyclization of pyruvic acid hydrazone (**158**) in acetic acid/acetic anhydride (68ACH191). Thiones (**155**) react with chloroacetic acid and aromatic aldehydes in the presence of anhydrous sodium acetate in acetic acid and acetic anhydride to give 2-arylidene derivatives **159** in a single step (74JPR163) (Scheme 34).

The reaction of thione **155** with DMAD gives a cyclized product that can be represented by any of the four possible structures **160–163** [84JCS(P1)2707] (Scheme 35). The proton-decoupled ^{13}C-NMR spectrum did not show close similarities to either five-membered thiazolidinone (**21**) or six-membered thiazinone (**22**) (Scheme 3) [81JCS(P1)415]. The three-bond ^{13}C,H-coupling constants (78HCA607) of 4.9–6.2 Hz exhibited by the cyclized product could be explained as coupling between an amide carbonyl carbon and H_A, only present in structures **160** and **161** in cis configurations (75OMR617; 76JOC3863; 78OMR197; 80OM200; 81OMR316) The lack of vicinal coupling $^3J(^{13}C,H)$ between the ester carbonyl carbon and H_A ruled out the six-membered thiazinone structures **162**

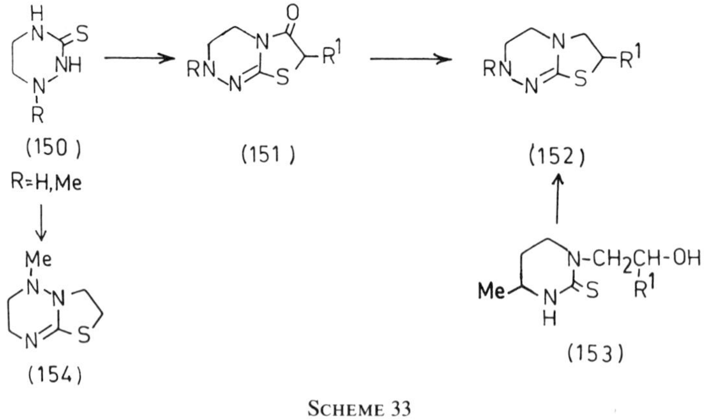

(150)   (151)   (152)
R=H,Me

(154)   (153)

SCHEME 33

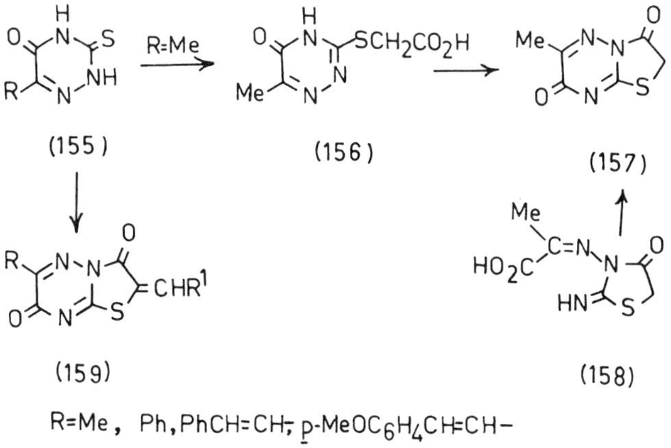

SCHEME 34

R=Me, Ph, PhCH=CH, p-MeOC$_6$H$_4$CH=CH–

and **163**. Structure **160** and not **161** was finally assigned to the cyclized product on the basis of a comparison of the chemical shift of the carbonyl carbon of the triazine ring with that of compounds of known structure [84JCS(P1)2707]. This is the first example in which ^{13}C-NMR chemical shifts have been used to settle the orientation of cyclization.

Reaction of 4,5-dihydro-5,6-diphenyl-1,2,4-triazine-3-thione (**164**) with chloroacetic acid gave a product for which structure **167** was assigned by Ali *et al.* (75IJC109) by analogy with the work on **168** of Bogachev and Fomenko (66URP175968) and also of Trepanier and Kreiger on **151** (70JHC1231). Later, **167** was modified to 6,7-diphenyl-7*H*-thiazolo[3,2-*b*]-[1,2,4]-triazin-3(2*H*)-one (**165**) [77IJC(B)46]. Unfortunately, Ali and co-

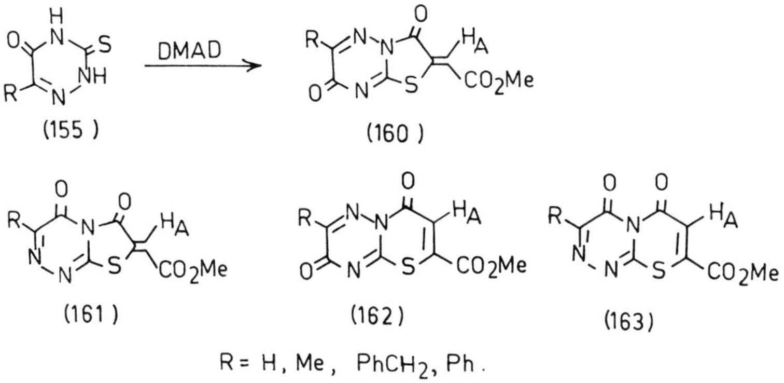

R = H, Me, PhCH$_2$, Ph.

SCHEME 35

workers (75IJC109) are not justified in citing the structural analogy of the Russian workers (66URP175968) whose work itself lacks supporting evidence. Comparison with the work of Trepanier and Krieger (70JHC1231) seems to be invalid because **164** and **150** (R = Me) cannot be compared because of the presence of a C-5 phenyl group in **164** and an N^1-methyl group in **150** (R = Me). In fact, the work of American workers, if studied in depth, would suggest that the reaction of **164** with chloroacetic acid should result in the formation of **165** rather than **167**. Cyclization in a pyrimidine series where steric repulsion due to a substituent plays a prominent role has also been reported (67JHC577). The structural assignment of **165** for the cyclized product obtained from **164** was confirmed by ^{1}H-NMR data. The benzyl proton in **166**, obtained from the reaction of **164** with 1,2-dibromoethane, resonates at δ 5.95, and the signal at δ 5.84 exhibited by the cyclized product confirms structure **165** and rules out structure **167** since the benzyl proton in **167** would resonate at a lower field than 5.95 because of the deshielding effect of the carbonyl group [77IJC(B)46]. The stability of the intermediate formed (not isolated) in the reaction of **164** with chloroacetic acid also supports structure **165** [77IJC(B)46] (Scheme 36).

b. *Thiazolo[3,2-a]-[1,3,5]-triazin-3(2H)-ones.* 7H-5,7-Disubstituted-5,6-dihydrothiazolo[3,2-a]-[1,3,5]-triazin-3(2H)-ones (**170**) are obtained by cyclization of the acids (**169**) in a straightforward manner [76ZN(B)1397] (Scheme 37).

12. *Thiazolo[2,3-c]-[1,2,4]benzotriazin-3(2H)-ones*

2,3-Dihydro-1,2,4-benzotriazine-3-thiones (**171a,b**) react with halogenoacid in the presence of fused sodium acetate in anhydrous ethanol to

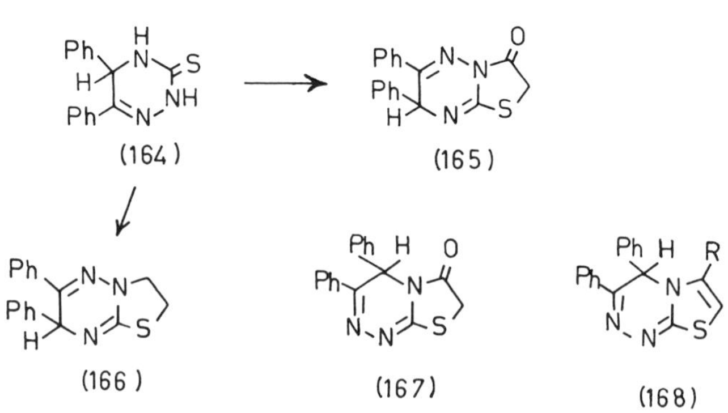

SCHEME 36

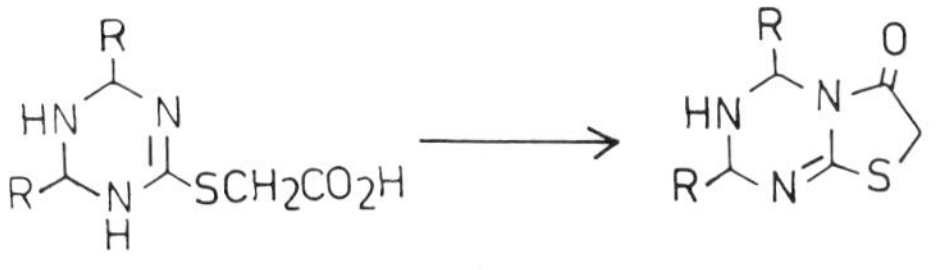

Sec. II.A] CONDENSED 4-THIAZOLIDINONES 29

SCHEME 37

R=OH, R=NHR[1], R[1]=Ph, m-MeC$_6$H$_4$, CH$_2$Ph, NHPh.

give the acids **172a,b** which, on treatment with acetic anhydride and pyridine, furnish cyclized products. Structure 9H-thiazolo[2,3-c]-[1,2,4]-benzotriazin-3(2H)-one (**173**) and not the alternate structure, 5H-thiazolo[3,2-b]-[1,2,4]-benzotriazin-3(2H)-one (**174**), was assigned to the cyclized products because of the chemical shifts of H$_A$ in **175** obtained from the reaction of **171** and 1,2-dibromoethane. H$_A$ in **175b** resonates at δ 7.02 and the signal at δ 7.52 (due to the deshielding effect by the carbonyl group) exhibited by the cyclized product obtained from acid **172b** supports structure **173b**. Such a downfield shift would not be expected from **174b** (80H149) (Scheme 38).

13. *Thiazolo[3,2-b]-[1,2,4,5]tetrazin-3(2H)-ones*

Reaction of the dithione **176** with ethyl chloroacetate gives 7H-6-thioxo-5,6-dihydrothiazolo[3,2-b]-[1,2,4,5]-triazin-3(2H)-one (**177**) and a

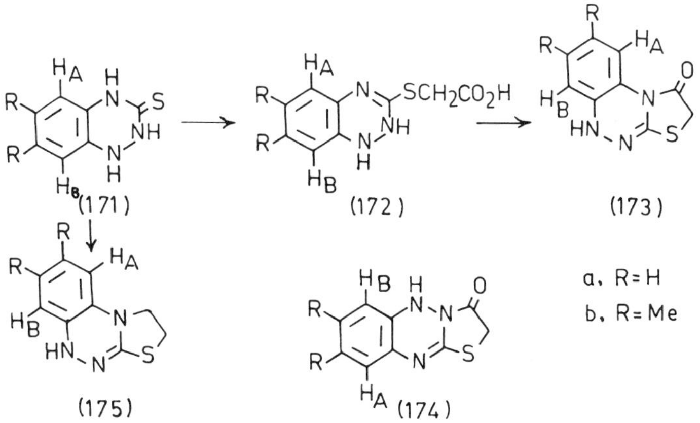

a, R=H
b, R=Me

SCHEME 38

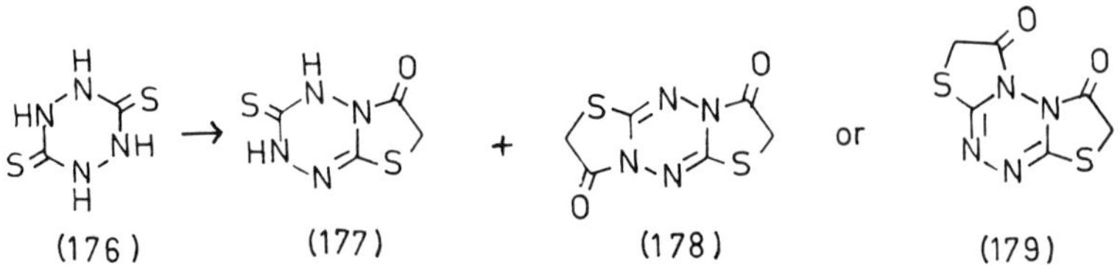

dilactam which can be represented by either **178** or **179**. No attempt was made to characterize the dilactam (61ACS1575) (Scheme 39).

14. *Oxospiro[cycloalkane-1,3'(4'H)-(2H)thiazolo[3,2-b]-[1,2,4,5]tetrazines]*

Reaction of tetraazaspirocycloalkanethiones (**180a–g**) with chloroacetic acid and sodium acetate in ethanol results in the facile synthesis of 6'(7'H)-oxospiro[cycloalkane-1,3'(4'H)-[2H]thiazolo[3,2-b]-[1,2,4,5]-tetrazines] (**181a–g**) confirmed by IR, NMR, and mass spectra. The thiones **180**, on condensation with ethyl chloroacetate and aromatic aldehydes in the presence of pyridine and piperidine, affords the arylidenethiazolidinones **182** which are also obtained from **181** and aldehydes [81IJC(B)296; 82IJC(B)315; 85IJC(B)1227; 86IJC(B)354, 86IJC(B)812; 87IJC(B)437, 87IJC(B)739] (Scheme 40).

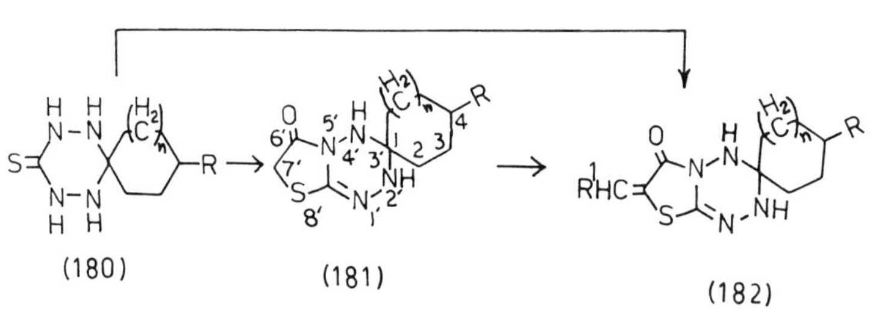

R	n		R	n
a, H	0		e, H	3
b, H	1		f, H	5
c, Me	1		g, H	7
d, H	2			

SCHEME 40

15. Oxospiro[indan-1,3'(4'H)-(2H)thiazolo[3,2-b]-[1,2,4,5]-tetrazines]

3-Phenyl- (**183a**) and 3-methylspiro[indan-1,3'[1,2,4,5]tetrazine]-6'-thione (**183b**), obtained from the reaction of thiocarbohydrazine with 3-phenylindan-1-one and 3-methylindan-1-one, react with chloroacetic acid to give 3-phenyl- (**184a**) and 3-methyl-6'(7'H)-oxospiro[indan-1,3'(4'H)-[2H]thiazolo[3,2-b]-[1,2,4,5]tetrazine] (**184b**), respectively, in good yields (87UP1,87UP2). The methyl analog **184c** is also synthesized from the reaction of **183b** with α-chloropropionic acid (87UP2). Structures (**184a–c**) are confirmed by IR, NMR, and mass spectral data (Scheme 41).

16. Thiazolo[3,2-a]-[1,3]diazepin-3(2H)-ones

Reaction of 4,5,6,7-tetrahydro-1,3-diazepine-2-thione (**185**) with ethyl chloroacetate and ethyl α-bromopropionate at 140-45°C for 30 min gives 5,6,7,8-tetrahydrothiazolo[3,2-a]-[1,3]diazepin-3(2H)-one hydrochloride (**186a**) and its 2-methyl analog (**186b**), respectively (69AJC2697) (Scheme 42). The reaction of thione (**185**) with chloroacetic acid when carried out in an aqueous medium gave 3-(δ-aminobutyl)thiazolidine-2,4-dione hydrochloride (**187**). The formation of **187** is presumed to be via **186a**, which undergoes hydrolysis by hydrochloric acid generated in the reaction to afford **187**. The identity of **187** was proved by its conversion to the benzylidene derivative which was also obtained by the acid hydrolysis of benzylidene derivative of **186a** (69AJC2697). Compound **185** reacts with DMAD to furnish a cyclized product for which the thiazolidinone structure **188** and not the alternate thiazinone structure **189** was assigned on the basis of ^{13}C-NMR spectral data [77JCS(P2)1070; 81JCS(P1)415].

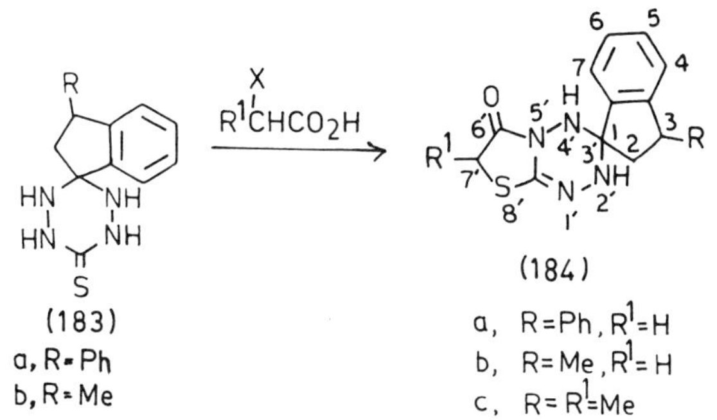

SCHEME 41

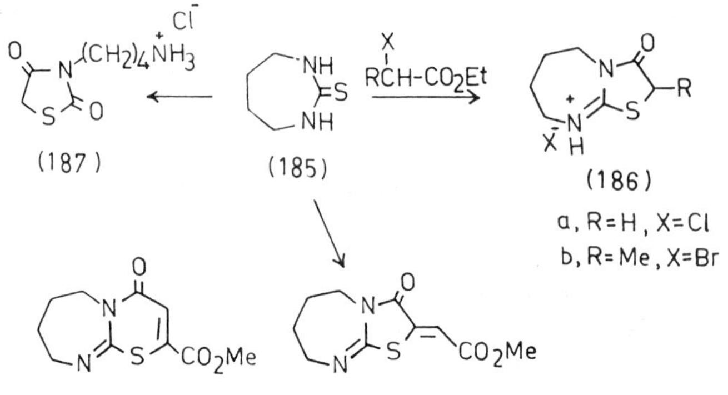

SCHEME 42

17. Thiazolobenzodiazepinones

All four possible structural isomers of thiazolobenzodiazepinones, namely, thiazolo[1,3]benzodiazepinone, thiazolo[1,4]benzodiazepinone, thiazolo[1,5]benzodiazepinone, and thiazolo[2,4]benzodiazepinone have been synthesized.

a. *Thiazolo[2,3-b]-[1,3]benzodiazepin-3(2H)-ones.* 4,5-Dihydro-5-methyl-1,3-benzodiazepine-2-thione (**190**) reacts with ethyl chloroacetate and ethyl α-bromo-*n*-hexanoate to give 5,6-dihydro-6-methylthiazolo[2,3-*b*]-[1,3]benzodiazepin-3(2*H*)-one (**191a**) and its 2-*n*-butyl analog (**191b**); the latter is isolated as its hydrochloride (69JHC491). The structural assignment for cyclized product **191** was based on ^{1}H-NMR spectral data; the product did not show a large downfield shift of an aromatic proton (H$_A$) which rules out the other alternate structures **192**. A downfield shift of H$_A$ would have been observed because of the deshielding effect of the carbonyl group on peri-proton (H$_A$) had the ring closure occurred on the other nitrogen atom (Scheme 43).

b. *Thiazolo[1,4]benzodiazepinones.* The method involving the facile addition of mercapto acetic acid across the double bond of the azomethine in a one-step synthesis of 4-thiazolidinone has been used in the syntheses of (i) thiazolo[3,2-*a*]-[1,4]benzodiazepin-1(2*H*)-one (**194**), (ii) thiazolo[3,2-*d*]-[1,4]benzodiazepin-3(2*H*)-one (**198**), and (iii) *s*-triazolo[4,3-*a*]-thiazolo[3,2-*d*]-[1,4]benzodiazepin-3(2*H*)-ones (**200, 202**).

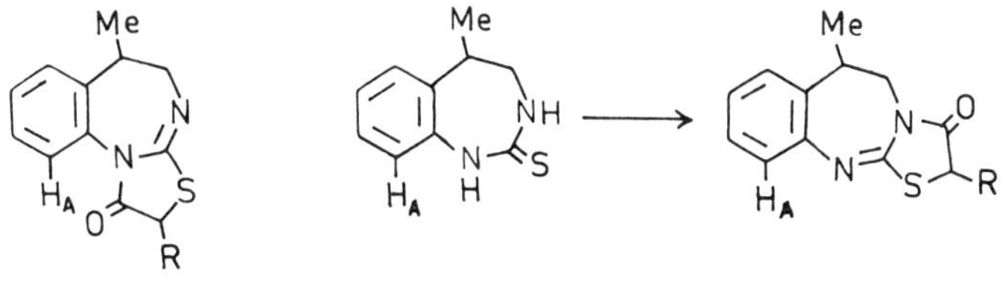

SCHEME 43

(192)　(190) → (191)

a, R = H
b, R = CH$_3$(CH$_2$)$_3$

Refluxing 7-chloro-5-phenyl-3H-1,4-benzodiazepine (**193**) with mercaptoacetic acid in benzene for 23 hr affords a mixture of 3a,4-dihydro-8-chloro-6-pyenylthiazolo[3,2-*a*]-[1,4]benzodiazepin-1(2H)-one (**194**) and the dilactam (**195**) which are separated by column chromatography on silica gel (74JOC167) (Scheme 44). Coffen and co-workers have also reported that benzodiazepine containing a sulfide bridge (**196**), obtained by the reaction of **193** with hydrogen sulfide, reacts with mercaptoacetic acid in tetrahydrofuran to furnish only the monolactam **194** (75USP3850948; 76USP3906001). However, with 1,4-benzodiazepinone as a substrate, a much longer reaction time was needed. 7-Chloro-1-methyl-5-phenyl-1,3-dihydro-2H-1,4-benzodiazepin-2-one (**197**), on refluxing with mercaptoacetic acid in benzene for five days, furnished the desired condensed thiazolidinone (**198**) (74JOC167).

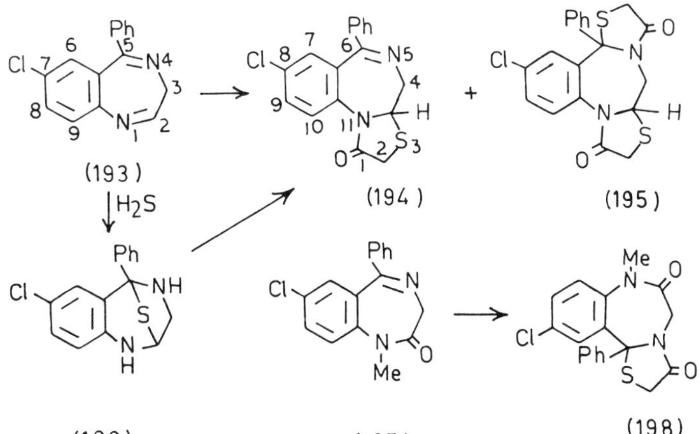

SCHEME 44

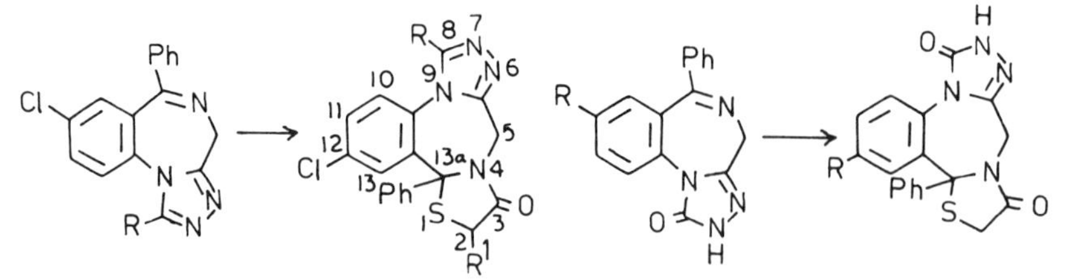

(199)   (200)   (201)   (202)

R = H, Me
R¹ = H, Me

R = Cl, Me, Et,
Me$_2$NCH$_2$CH$_2^-$

SCHEME 45

s-Triazolo[4,3-a]-[1,4]benzodiazepines (**199, 201**) condense with mercaptoacetic acid in benzene or xylene to furnish the corresponding tetracyclic thiazolidinone compounds **200** (74JOC167; 76JAP74109398) and **202** (75JAP7412699) (Scheme 45). The easy lactam ring formation leading to the synthesis of **194, 195, 198, 200,** and **202** is not surprising in view of the work describing the formation of the amide from mercaptoacetic acid and ethylamine (71JPR849).

Mesoionic thiazolones obtained by the reaction of cyclic thioamide and α-halogenoacid possess an acidic hydrogen atom. This hydrogen migrates to the carbon α with respect to the carbonyl functionality. Thus 3H-7-chloro-5-phenyl-1,4-benzodiazepine-2-thione (**203**) reacts with bromoacetic acid, and subsequent cyclization of intermediate **204** with acetic anhydride and triethylamine affords 8-chloro-6-pyenylthiazolo[3,2-a]-[1,4]benzodiazepin-1(2H)-one (**205**) (76USP3897446) (Scheme 46).

c. *Thiazolo[3,2-a]-[1,5]benzodiazepin-1(2H)-one.* In a similar fashion, 3H-4-amino-3-ethoxycarbonyl-1,5-benzodiazepine-2-thione (**206**) reacts with methyl α-bromopropionate to furnish 5-amino-4-ethoxy-carbonyl-

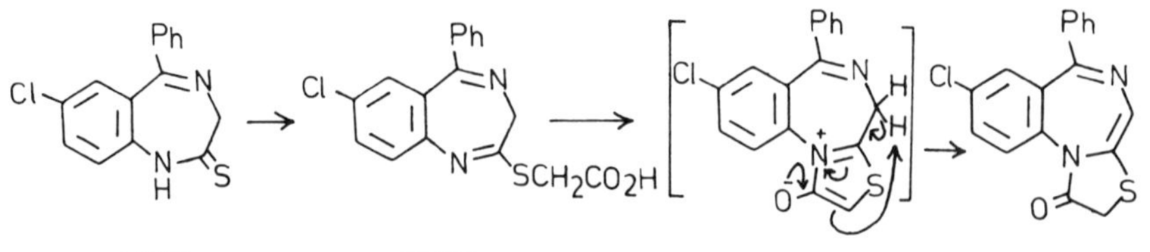

(203)   (204)   (205)

SCHEME 46

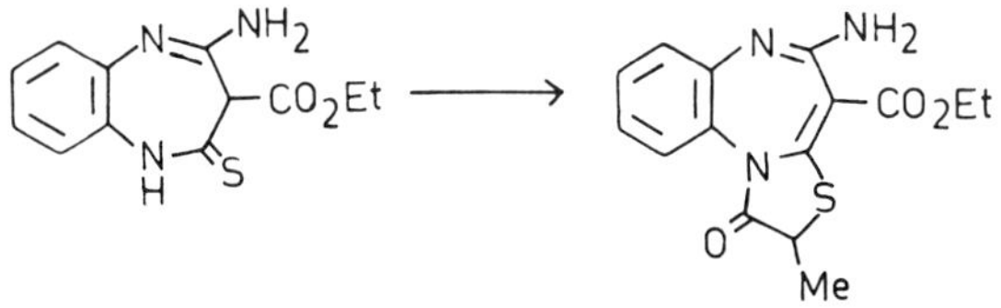

[Structures 206 and 207]

(206) → (207)

SCHEME 47

2-methylthiazolo[3,2-*a*]-[1,5]benzodiazepin-1(2*H*)-one (**207**) (74E-GP105235) (Scheme 47).

d. *Thiazolo[3,2-b]-[2,4]benzodiazepin-3(2H)-ones*. Reaction of 1*H*-4,-5-dihydro-2,4-benzodiazepine-3-thione (**208**) with ethyl chloroacetate and ethyl α-bromo-*n*-hexanoate gives 5,10-dihydrothiazolo[3,2-*b*]-[2,4]benzodiazepin-3(2*H*)-one (**209a**) isolated as a hydrochloride and its 2-*n*-butyl analog (**209b**), respectively as expected (68JHC609) (Scheme 48).

18. *Thiazolo[2,3-d]-[1,5]benzoxazepin-1(2H)-ones*

Synthesis of thiazolo[2,3-*d*]-[1,5]benzoxazepin-1(2*H*)-one has been achieved by the condensation of [1,5]benzoxazepinethione with α,β-bifunctional compounds. 2-Amino-3-ethoxy-carbonyl-1,5-benzoxazepine-4-thione (**210**) reacts with chloroacetic acid (or its ethyl ester) to give 4-ethoxycarbonyl-5-iminothiazolo[2,3-*d*]-[1,5]benzoxazepin-1(2*H*)-one (**211a**) (74EGP105235; 84MI3). Similarly, **210** reacts with ethyl α-halopropionate, ethyl 2-chloro-3-anilinocrotonate, and oxalyl chloride to afford **211b** and **212a,b**, respectively (84MI3) (Scheme 49).

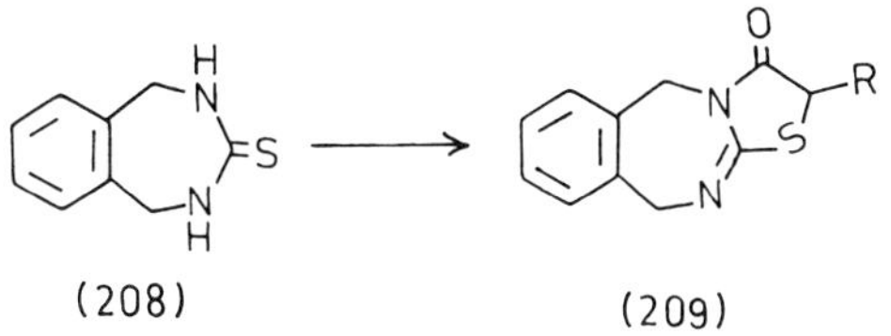

[Structures 208 and 209]

(208) → (209)

a, R = H
b, R = CH$_3$(CH$_2$)$_3$

SCHEME 48

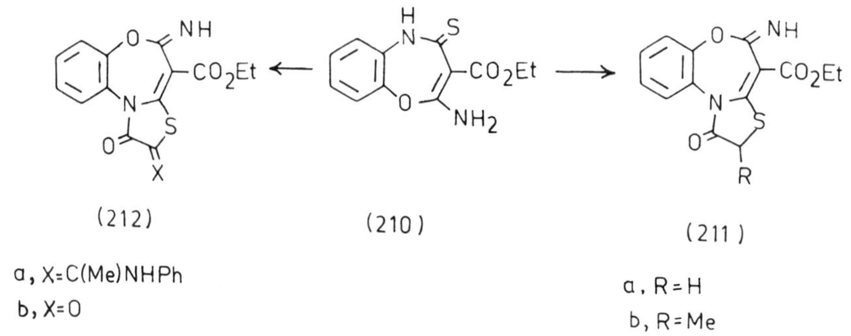

SCHEME 49

(212) a, X=C(Me)NHPh; b, X=O

(210)

(211) a, R=H; b, R=Me

19. Thiazolo[3,2-a]-[1,3]diazocin-3(2H)-ones

Dehuri and Nayak (83JIC970) reported that 3,4,5,6,7,8-hexahydro-1,3-diazocine-2-thione (213), obtained from 1,5-dibromopentane and thiourea in boiling ethanol, reacts with chloroacetic acid or its ethyl ester followed by basification to give 5H-6,7,8,9-tetrahydrothiazolo[3,2-a]-[1,3]diazocin-3(2H)-one (214). The reaction of 213 with ethyl chloroacetate was not smooth and the yield was low. The thione 213, on reaction with ethyl chloroacetate and aromatic aldehydes in the presence of pyridine and piperidine, furnishes the arylidene product 215 which is also obtained from 214 and aromatic aldehydes. No spectral data are cited to confirm the thiazolidinone structures 214 and 215 (Scheme 50).

20. Thiazolo[2,3-b]-[1,3]benzodiazocin-3(2H)-one

There is only one publication reporting the synthesis of this system. Reaction of 3,4,5,6-tetrahydro-1,3-benzodiazocine-2-thione (216) with

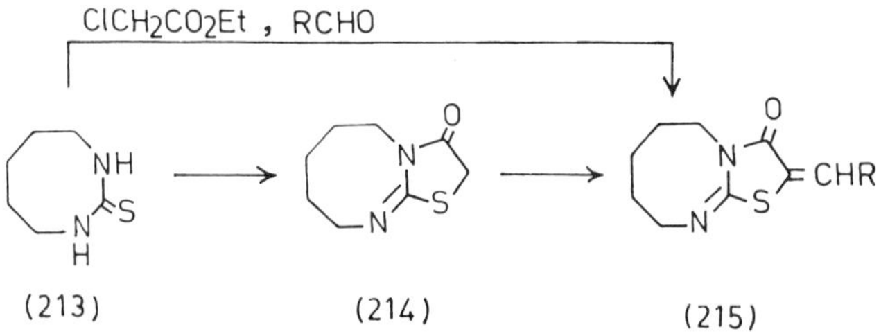

SCHEME 50

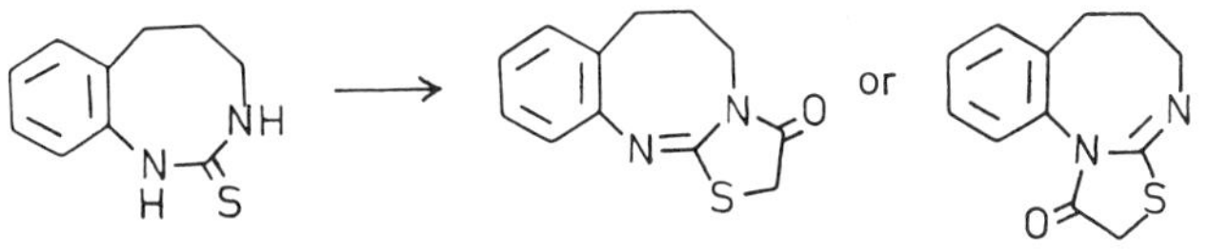

SCHEME 51

chloroacetic acid gave a cyclized product for which the structure 5H-6,7-dihydrothiazolo[2,3-b]-[1,3]benzodiazocin-3(2H)-one (217) was assigned without any evidence (77MI3) (Scheme 51). The cyclized product could also be represented by the alternate structure 218.

21. *Thiazolo[3,2-a]-[1,3]diazonin-3(2H)-ones*

1H-Hexahydro-1,3-diazonine-2-thione (219), obtained from the condensation of 1,6-dibromohexane and thiourea, reacts with chloroacetic acid to give 5,6,7,8,9,10-hexahydrothiazolo[3,2-a]-[1,3]diazonin-3(2H)-one (220). The thiazolidinone (220) is also obtained, albeit, in low yield by heating thione 219 and ethyl chloroacetate, followed by basification. Compound 219, on reaction with ethyl chloroacetate and aromatic aldehydes in the presence of pyridine and piperidine, furnishes directly the arylidene thiazolidinone 221 which is also obtained by condensation of 220 with aldehydes (83JIC970) (Scheme 52). No spectral data are reported to confirm the thiazolidinone structures.

22. *Thiazolo[3,2-a]-[1,3]diazecin-3(2H)-ones*

Reaction of octahydro-1,3-diazecine-2-thione (222) with chloroacetic acid or ethyl chloroacetate gives 6,7,8,9,10,11-hexahydro-5H-thiazolo-[3,2,-a]-[1,3]diazecin-3(2H)-one (223). As in earlier cases, the reaction of

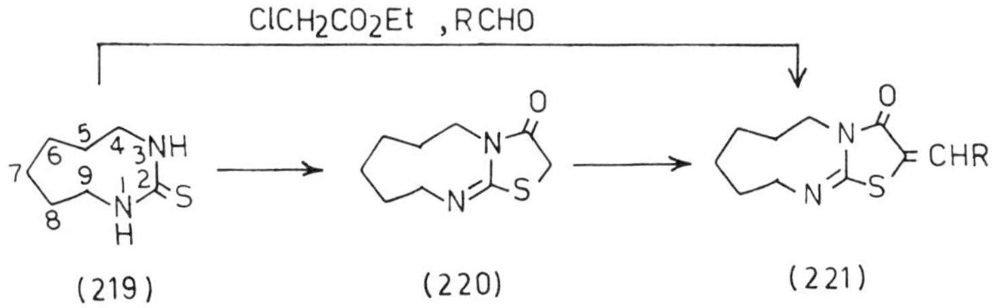

SCHEME 52

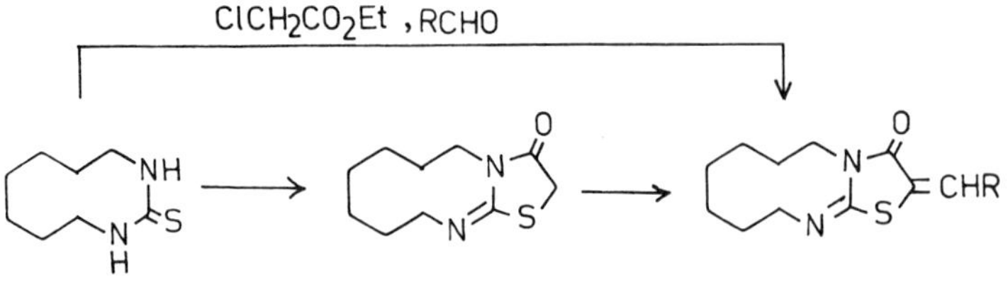

SCHEME 53

222 with ethyl chloroacetate was not smooth and the yield was low. Compound **222** reacts with ethyl chloroacetate and aldehydes in the presence of pyridine and piperidine to yield arylidene thiazolidinones (**224**), which are also obtained from **223** and aldehydes (83JIC970) (Scheme 53). Thiazolidinone structures **223** and **224** have not been characterized by spectral data.

23. *Thiazolo[3,2-a]-[1,3]diazacycloundecan-3(2H)-ones*

1,3-Diazacycloundecane-2-thione (**225**) reacts with chloroacetic acid or its ethyl ester to give thiazolo[3,2-*a*]-[1,3]-diazacycloundecan-3(2*H*)-one (**226**). The thione (**225**), on reaction with ethyl chloroacetate and aldehydes, gives **227** which is also obtained by the condensation of **226** with aldehydes (83JIC970) (Scheme 54). No spectral data are cited.

24. *Thiazolo[3,2-a]-[1,3]diazacyclododecane-3(2H)-ones*

Condensation of 1,3-diazacyclododecane-2-thione (**228**) with chloroacetic acid or its ethyl ester yields thiazolo[3,2-*a*]-[1,3]diazacyclododecan-3(2*H*)-one (**229**). Thione **228**, on reaction with ethyl chloroacetate and aldehydes, furnishes **230** which is also obtained from **229** and aldehydes (83JIC970) (Scheme 55). No spectral data are cited.

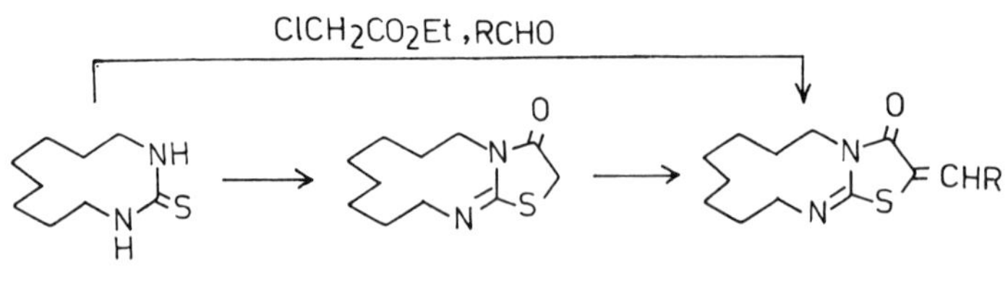

SCHEME 54

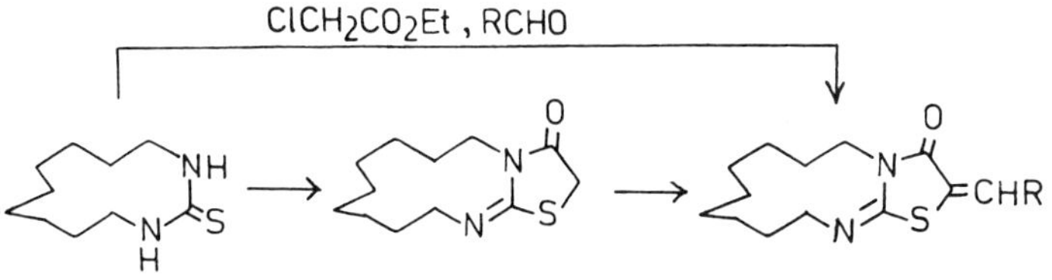

(228) (229) (230)

SCHEME 55

25. Thiazolo[3,2-a]-[1,3]diazacyclotridecane-3(2H)-ones

1,3-Diazacyclotridecane-2-thione (**231**) reacts with chloroacetic acid or ethyl chloroacetate to furnish thiazolo [3,2-*a*]-[1,3]diazocyclotridecane-3(2*H*)-one (**232**). Compound **231**, on condensation with ethyl chloroacetate and aldehydes in the presence of pyridine and piperidine, yields arylidene-thiazolidinone (**233**) which is also obtained from **232** and aldehydes (83JIC970) (Scheme 56). No spectral data are recorded.

Although the synthesis of condensed thiazolidinones (**214, 220, 223, 226, 229** and **232**) are accomplished in a straightforward manner, structures must await spectral studies. Moreover, structures of thiones **213, 219, 222, 225, 228** and **231** used as the starting materials are not established beyond doubt.

26. Bis-(thiazolo)-[3,4-b:3',2'-d]-s-triazol-3(2H)-ones

Bis-(4-thiazolidinones) (**234**), on treatment with concentrated sulfuric acid, undergo cyclodehydration to furnish 7*H*-bis-(thiazolo)-[3,4-*b*:3', 2'-*d*]-*s*-triazol-3(2*H*)-ones (**235**) (85JIC147) (Scheme 57). Compounds **235** in general were found to be more potent herbicides than their precursors, **234**, suggesting that **235** must be compact and possess planar structure.

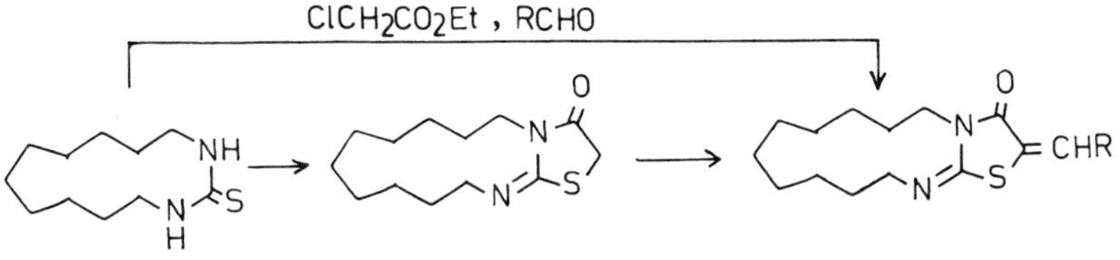

(231) (232) (233)

SCHEME 56

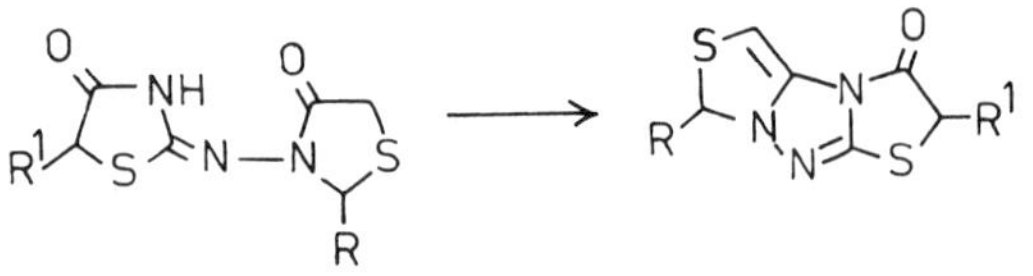

(234) → (235)

R^1 = H, Me
R = alkyl, aryl

SCHEME 57

This conforms with earlier observations that compact size and planarity of a molecule often enhance its herbicidal activities [56N(L)1042; 63MI1; 76T615].

27. Pyrrolo[2,3:4',5']pyrrolo[2',1'-b]thiazol-3(2H)-ones

Reaction of 4-thiazolones (**236**) with *N*-arylmaleimides gives 5,8-disubstituted pyrrolo[2,3:4',5']pyrrolo[2',1'-*b*]thiazol-3,6-(2*H*,7*H*)-diones (**237**) (83H1021) (Scheme 58).

28. Thiazolo[3,2-a]thiopyrano[4,3-d]pyrimidin-3(2H)-ones

2,6-Diphenyl-3,5-bis-(arylidene)-thiopyrane-4-one (**238**), on condensation with thiourea in basic medium, gives thione **239** which reacts with α-halogenoacids to furnish a cyclized product for which the structure 2-*H*/aryl-5-aryl-9-arylidene-6,8-diphenylthiazolo[3,2-*a*]thiopyrano[4,3-*d*]-pyrimidin-3(2*H*)-ones (**240**) was assigned (85MI1) (Scheme 59). The other possible isomeric structure **241** could also represent the cyclized product.

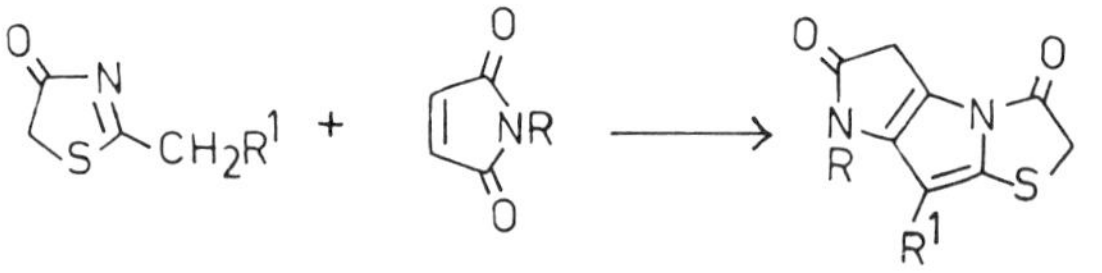

(236) (237)

R = Ph, p-MeC$_6$H$_4$
R^1 = CN, CO$_2$Et

SCHEME 58

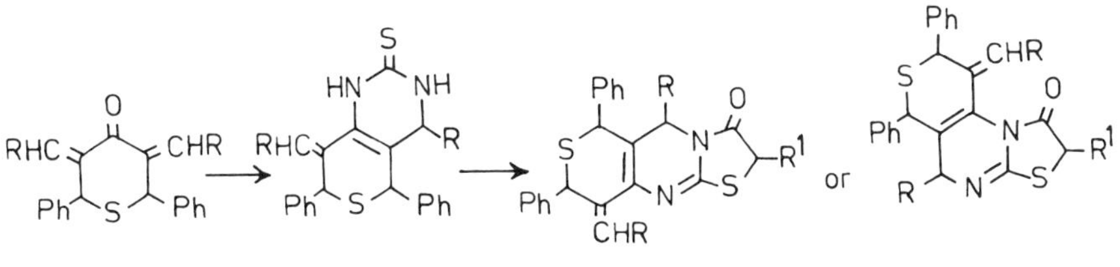

SCHEME 59

29. Thiazolo[3,2-a]indol-3(2H)-ones

Synthesis reported by American workers of the reaction of γ-oxo-indol-1-butanoic acids with thionyl chloride serves as a convenient method for this relatively inaccessible tricyclic system (79JOC3994). The reaction of γ-oxo-3-[(methoxycarbonyl)methyl]indol-1-butanoic acid (**242a**) with thionyl chloride followed by treatment with methanol gives (Z)-methyl 2,3-dihydro-2-(2-methoxycarbonylethylidene)-3-oxo-thiazolo[3,2-a]indol-9-acetate (**243a**). Similarly, reaction of **242b** with thionyl chloride and subsequent treatment with methanol and neopentyl alcohol gives **243b** and **243c**, respectively (Scheme 60). The structural assignment of **243**, as opposed to thiazinone structure **244**, was based on ^{1}H- and ^{13}C-NMR spectra along with a comparison to 2-ethoxycarbonylethylidenethiazolo[3,2-a]-benzimidazol-3(2H)-one (**23**). In the ^{1}H-NMR spectrum of **243b**, the appearance of H_A as a multiplet of δ 8.0 (which is close to the corresponding value δ 7.9 for **23**) supports structure **243b** as opposed to the alternate isomeric structure **244b** whose δ H_A is predicted to be at lower field (75KGS47; 79TL53). In ^{13}C-NMR, the chemical shift and long-range C—H coupling constants in the unsaturated 1,4-dicarbonyl moiety of **243b** favorably tally with that of **23**. Again the value of the coupling constant between amide carbon and olefinic proton found in **243b** is similar to that obtained in **23**, suggesting the Z-geometry in **243** is similar to that in **23** (79TL53) (See Section III).

A straightforward synthesis of this system involves the addition of mercaptoacetic acid or its ester across the double bond of the azomethine linkage. 3,3-Dimethyl-2-ethoxycarbonyl-1-indolene (**245**) reacts with appropriate mercaptocaids (or esters) to give the corresponding thiazolo[3,2-a]indol-3(2H)-ones (**246**) (74LA206) (Scheme 60).

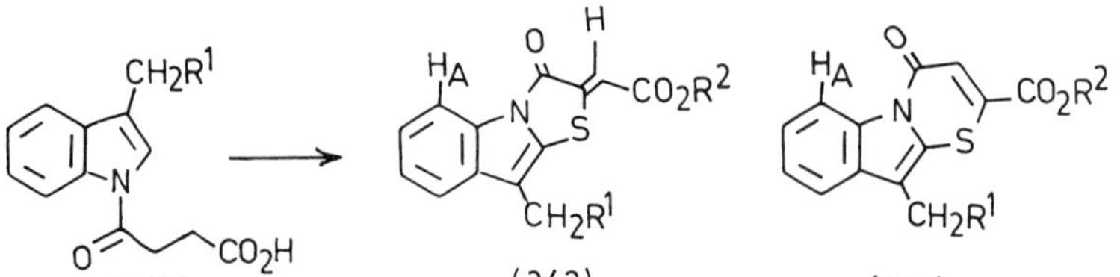

(242) (243) (244)

a, $R^1 = CO_2Me$
b, $R^1 = H$

a, $R^1 = CO_2Me$, $R^2 = Me$
b, $R^1 = H$, $R^2 = Me$
c, $R^1 = H$, $R^2 = CH_2CMe_3$

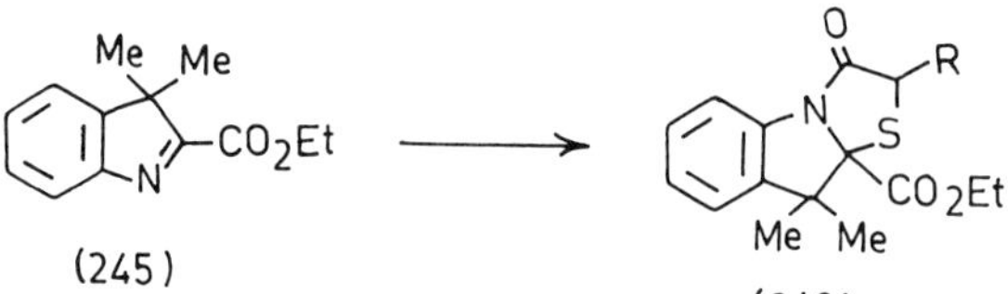

(245) (246)

SCHEME 60

30. *Thiazolo[3,2-*a*]benzimidazol-3(2*H*)-ones*

The synthesis of thiazolo[3,2-*a*]benzimidazol-3(2*H*)-one (**249a**) was first reported by Stephen and Wilson (26JCS2531; 29MI1) in 25% yield by cyclization with metallic sodium in refluxing benzene of ethyl/methyl benzimidazo-2-thiolacetate (**248,** R = H, R^1 = Et/Me) obtained from benzimidazolyl-2-thione (**247**) with ethyl/methyl chloroacetate. Later, Duffin and Kendall (50BRP634951; 51JCS734; 56JCS361) obtained **249a** in 64% yield by the cyclodehydration of the acid **248** (R = R^1 = H) with acetic anhydride in pyridine. The method of Duffin and Kendall has been followed by other workers for the synthesis of **249a** (63ZOB945; 65ZOB1276) and **249** (R = alkyl) (68CPB2167; 69IJC769). Although a few minutes of heating acid **248** in a mixture of acetic anhydride and pyridine for cyclodehydration is enough, unnecessary prolonged heating of acid **248** (R = R^1 = H) for 5 hr is reported for the synthesis of **249a** (75JIC1193). Compound **249a** is also obtained in 98% yield by heating acid **248** (R = R^1 = H) with dicyclohexycarbodiimide (DCC) in pyridine, (68KGS443), refluxing the ester **248** (R = H, R^1 = Et) in *o*-dichlorobenzene, albeit in poor yield (30%) (56JOC24), or by oxidizing **250** with chromium trioxide in pyridine (67CJC2903). The synthesis of **249a** from **250,** obtained from the thione **247** and chloroacetaldehyde, fixes the position of the hydroxyl group at the 3-position in **250** which, on dehydration with phosphoryl chloride in pyridine, gives thiazolo[3,2-*a*]benzimidazole (**251**). This establishes unambiguously the involvement of sulfur rather than nitrogen in the condensation reaction between the thione **247** and α-halogenocarbonyl compounds.

Reaction of **247** with chloroacetic acid in the presence of sodium acetate in ethanol gave a product with a melting point of 212°C formulated as **249a** (60MI1; 61UKZ503) which, on treatment with hydrochloric acid, yielded a compound with a melting point of 182°C for which structure 2,4-thiazolidinedione (**252**) was assigned (61UKZ503) without any evidence. However, on the basis of the melting points, the compounds with melting points of 212°C and 182°C are probably the acid **248** [R = R^1 = H; melting point = 211–212°C (65ZOB1276) or 215°C (69IJC769)] and the cyclized product **249a** [melting point = 181°C (56JCS361)], respectively, instead of **249a** and **252**, respectively, as reported by Russian workers (60MI1; 61UKZ503) (Scheme 61).

The thione **247** reacts with chloroacetic acid and aldehydes in a mixture of acetic acid and acetic anhydride to give arylidene thiazolidinone (**253**) [86IJC(B)776] which can also be obtained by the reaction of aldehydes with acid **248** (R = R^1 = H) (56JOC24; 72MI1) or with thiazolidinone **249a** (56JCS361).

Likewise, thiones **254a–d** react with chloroacetic acid; subsequent heating of the intermediate acids **255a–d** with acetic anhydride in pyridine

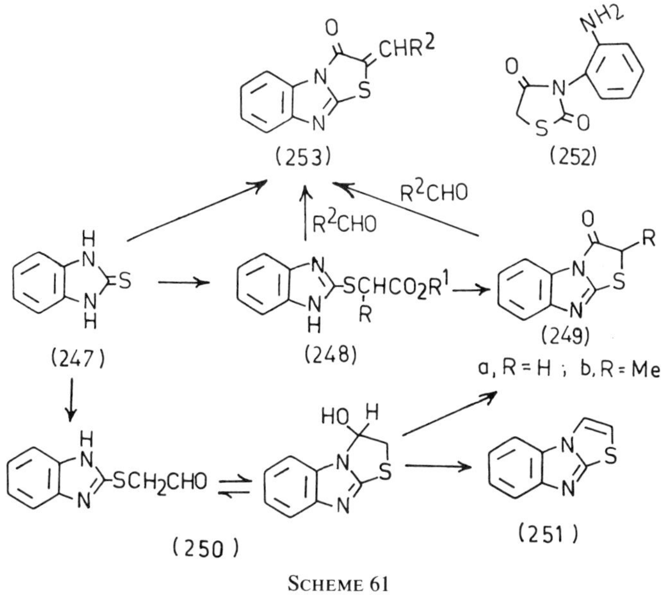

SCHEME 61

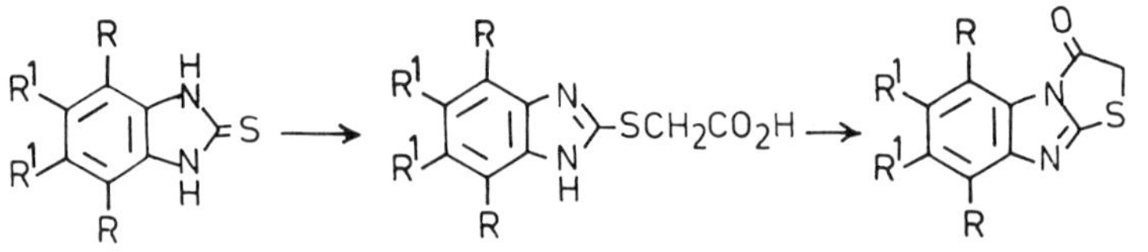

Scheme 62

a, R=Me, R¹= H; b, R=H, R¹= Me; c, R=OMe, R¹= H; d, R = H, R¹= OMe.

SCHEME 62

furnishes the corresponding thiazolo[3,2-*a*]benzimidazol-3(2*H*)-ones (**256a–d**) [70IJC10; 71KGS822; 86IJC(B)267] (Scheme 62).

Reaction of benzimidazolyl-2-thione (**247**) with (phenylimino)oxalic acid dichloride [69MI2; 71KGS471) and (arylhydrazono)oxalic acid dichloride (69MI2; 70URP256774; 71KGS930) gives 2-phenyliminothiazolo[3,2-*a*]-benzimidazol-3(2*H*)-one (**257**, R = Ph) and 2-arylhydrazinothiazolo-[3,2-*a*]benzimidazol-3(2*H*)-ones (**257**, R = R¹C₆H₄NH, R¹ = *o*-NO, R¹ = *o*-MeO, R¹ = *p*-Me) respectively (Scheme 63).

The easy ring-opening behavior of hexafluoro-1,2-epoxypropane in the presence of a nucleophile has also been useful as a synthetic tool for the preparation of thiazolo[3,2-*a*]benzimidazol-3(2*H*)-ones. Sodium salt of benzimidazolyl-2-thione (**247**) reacts with hexafluoro-1,2-epoxypropane in an aprotic solvent, such as dioxane or CH₃CN, to give 2-fluoro-2-(trifluoromethyl)-thiazolo[3,2-*a*]benzimidazol-3(2*H*)-one (**260**) (81S981). The sodium salt of thione **247**, acting as a nucleophile, readily attacks the

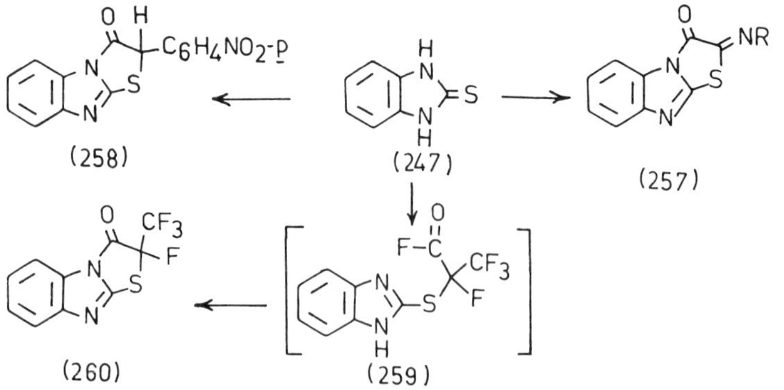

SCHEME 63

central carbon atom of hexafluoro-1,2-epoxypropane to give acyl-fluoride intermediate **259** which undergoes intramolecular cyclization to afford **260** (Scheme 63).

If mesoionic thiazolone, initially formed by the reaction of cyclic thioamide with dicyanooxirane, possesses an acidic hydrogen atom at the α-position to the carbon atom located between the sulfur and nitrogen, this hydrogen atom migrates onto the carbon bearing the aryl group to give thiazolo[3,2-*a*]benzimidazol-3(2*H*)-one. Thione **247** reacts with 2,2-dicyano-3-*p*-nitrophenyl-oxirane to afford 2-*p*-nitrophenylthiazolo[3,2-*a*]-benzimidazol-3(2*H*)-one (**258**) in a single step (81S981) (Scheme 63).

2-Acetonylmercapto-1-alkoxycarbonylbenzimidazole, in the presence of sodium hydride or organic bases such as 1,5-diazabicyclo[4,3,0]-5-nonene (DBN) and triethylenediamine (TED), yields the N → C ester transfer products, while its aromatic ester counterpart gives thiazolidinone where the N → C ester transfer product is an intermediate, although not isolated. 2-Acetonylmercapto-1-phenoxycarbonylbenzimidazole (**262**), obtained by the successive reactions of **247** with chloroacetone and the intermediate ketone (**261**) with phenyl chloroformate, in the presence of sodium hydride in tetrahydrofuran (THF), gives 2-acetylthiazolo[3,2-*a*]benzimidazol-3(2*H*)-one (**264**) (77TL275). Structure **264** was confirmed by its conversion to **266** with excess ethyl chloroformate via enol **265**, which undergoes esterification, and also by its independent synthesis involving the reaction of **267** with phenyl acetoacetate in the presence of sodium hydride in THF. This fact suggests that **263**, the N → C ester transfer product, is involved as an intermediate in the formation of **264** from **262** (Scheme 64).

Sulfide **269** (R = H), obtained from benzimidazolyl-2-thione (**268a**) and ethyl α-chloroacetoacetate, undergoes cyclization with acetic anhydride and pyridine to give thiazole (**270**) (64JOC865). This reaction was reinvestigated to determine whether sulfide **269** (R = H), in acetic anhydride in the absence of pyridine, would provide an alternative thiazolidinone as obtained in the benzothiadiazine series (**143**). Heating sulfide **269** (R = H) in acetic anhydride in the absence of pyridine yielded two products identified as **270** and **271** (R = H, R^1 = Ac). Refluxing the latter in ethanol, as anticipated, caused hydrolysis of the enol acetate to give **265** (70USP3475424; 75JHC1207) (Scheme 65). Similarly, thiazolidinone (**271**, R = Cl, R^1 = COEt) was obtained by heating **269** (R = Cl) in propionic anhydride (70USP3475424), although the regiochemistry of the chlorine atom in the benzene ring is uncertain.

Thione adds to the unsaturated carbon—carbon linkage of maleic anhydride probably by a Michael-type reaction, subsequent cyclization results in the synthesis of 4-thiazolidinone derivatives. The reaction of thi-

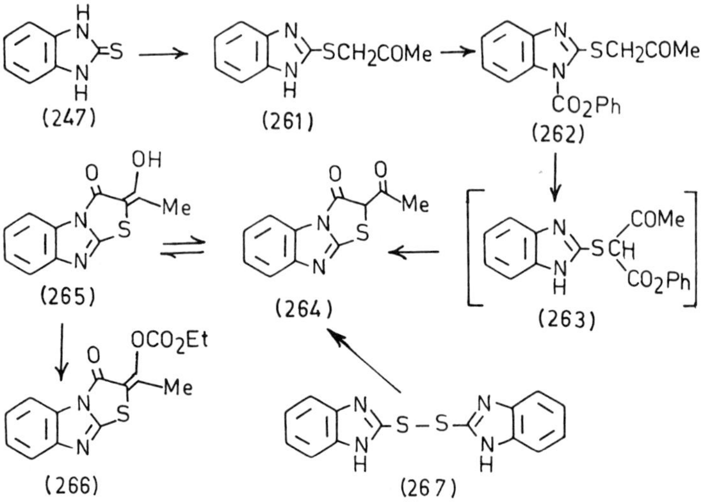

SCHEME 64

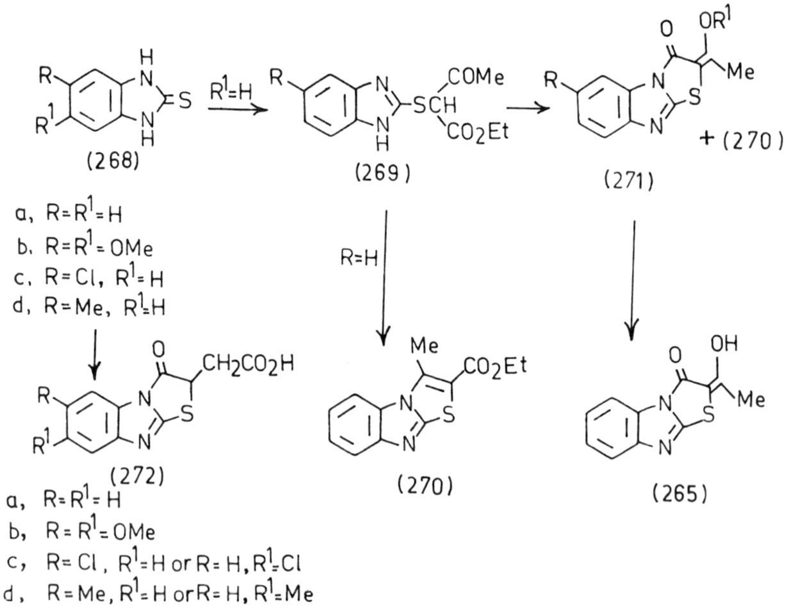

SCHEME 65

ones **268a–d** with maleic anhydride in boiling dioxane gives condensed thiazolidinones (**272a–d**) (75JAP7495997) (Scheme 65). However, in the case of **272c,d,** the regiochemistry of the substituent in the benzene ring is not secure.

Reaction of thioureas with acetylene dicarboxylic esters represents one of the knotty problems encountered in 4-thiazolidinone chemistry. The reaction of N,N'-disubstituted thioureas with DMAD has been claimed to give thiazinones (67CB3671, 67CJC939) or thiazolidinone (64JA107; 73CPB270). The confusion has, however, been removed by settling in favor of thiazolidinone on the basis of chemical evidence [71CI(L)705], X-ray analysis (70CC890), and ^{13}C-NMR spectroscopy by application of a C—H spin coupling constant (78HCA607). But a benzimidazolyl-2-thione, like imidazolyl-2-thione, reacts with acetylene dicarboxylic ester to give condensed thiazolidinones and/or condensed thiazinones, depending on the solvent used and the reaction conditions as illustrated in Scheme 66. The reaction of **247** with DMAD in acetic acid was first studied by Grinblat and Postovskii (61ZOB394) and then by Liu and co-workers (77MI1); only 2-(methoxycarbonyl)methylidenethiazolo [3,2-*a*]benzimidazol-3(2*H*)-one (**23**) was isolated. The reaction was repeated by English workers (78TL2621) who obtained **23** in addition to a second cyclized product. The

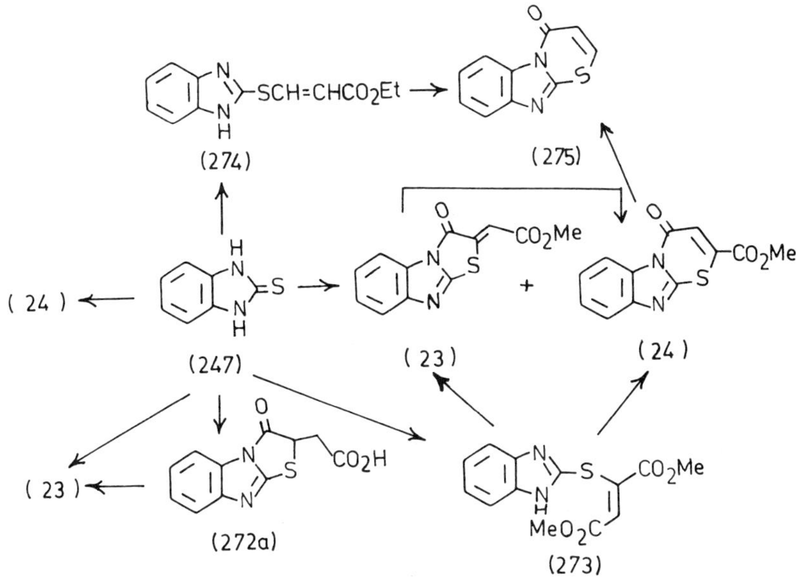

SCHEME 66

structure of **23** was confirmed by X-ray crystallography as well as by independent synthesis from **247** and maleic anhydride in glyme followed by sequential treatment of **272a** with diazomethane in THF and bromine in acetic acid. Six-membered thiazinone structure **24** was tentatively suggested on the basis of ^{1}H-NMR spectroscopy for the second cyclized product, which they were unable to isolate cleanly.

Much insight came from the work of Wade (79JOC1816), who studied the reaction of **247** with DMAD under different conditions and characterized the products. Reaction of **247** with DMAD in methanol at refluxing temperature for 19 hr, or kept at room temperature for several days, gave exclusively structure **24** in ~90% yield. Hydrolysis and thermal decomposition of **24** yielded a compound that is identical with **275**, independently obtained by the addition of ethyl propiolate to **247** followed by thermal ring closure of the intermediate ester **274**. Prolonged reaction of **247** with DMAD in methanol yielded exclusively structure **24**; reaction for a shorter time gave a mixture of three compounds, **23, 24,** and **273**, the exact ratio of which depended on the reaction time, temperature, and the solvent used as revealed by ^{1}H-NMR spectroscopy. When the reaction was carried out for a very short time (8 min) at low temperature ($-1°C$), the initial Michael addition product **273** was the only isolable product. The reaction in an NMR tube confirmed the simultaneous formation of both cyclized products **23** and **24** from the uncyclized product **273**. A good yield of structure **23** (70%) was isolated by careful preparative liquid chromatography by using ethyl acetate–benzene from the mixture of **23** and **24** obtained from the reaction of **247** and DMAD in THF carried out at room temperature overnight. The rearrangement of **23** to **24** was demonstrated by refluxing **23** in methanol overnight; this affected complete conversion of **23** to **24**. Conducting the rearrangement in an NMR tube in deuterated methanol furnished deuterated ester **24**, suggesting that rearrangement involves reversion of **23** to **273** by methanolysis of the amide bond in **23** and subsequent recyclization with the other ester group to give **24**. That this rearrangement is sensitive to base was shown by adding methoxide ion. The conversion of **23** to **24** was complete within a few minutes compared to several hours in the absence of added base. This is reflected again in the exclusive formation of **24** in 5 min from the reaction of **247** and DMAD in methanol in the presence of methoxide ion at room temperature. The formation of **23** and **24** from the reaction of **247** with DMAD in acetic acid or methanol has also been reported by Indian workers (79TL53).

Acheson and Wallis [81JCS(P1)415] reported that **247** reacted with DMAD in dry acetonitrile to give **23** only, while in dry methanol the only product isolated was **24**. They further observed that **23** was converted to **24** by refluxing in dry methanol, and this rearrangement was catalyzed by basic impurities since it did not take place if the methanol contained a few drops of acetic acid. The reaction of thione **247** and DMAD in a mixture of acetonitrile and water gave uncyclized product **273,** which could be cyclized to **23** in dry acetonitrile or to **24** in dry methanol. Structures **23** and **24** have been established by ^{13}C-NMR spectroscopy which offers a method of distinguishing thiazolidinone from thiazinone structures (see section III). Surprisingly, Acheson and Wallis in their publication [81JCS(P1)415] did not refer to the important work of Wade (79JOC1816).

The reaction of **247** with DMAD in methanol results in the exclusive formation of thiazinone **24**. The reaction can be catalyzed by sodium methoxide. In the absence of added base, thiazolidinone **23** is formed in addition to compound **24**. This is in contrast to the findings in the reactions of thioureas with DMAD in which thiazolidinones are the only isolable products. The difference may be due to the enhanced stability of the 6,5,6-ring system because of pseudoaromaticity and to the inherent strain present in the 6,5,5-ring system, or it may be due to the difference in basicity between a benzimidazolyl-2-thione and thioureas.

Reaction of thione **276** (R = H) with diethyl acetylenedicarboxylate in acetic acid or ethanol similarly gives both thiazolidinone (**277**) and thiazinone (**278**) as expected (79TL53) (Scheme 67). Condensation of **276** (R = Me, Cl, NO_2, CO_2H) and **280** with acetylenedicarboxylic esters gives a cyclized product in each case for which respective thiazolidinone structures **279** and **281** were assigned (77MI1). This work warrants reinves-

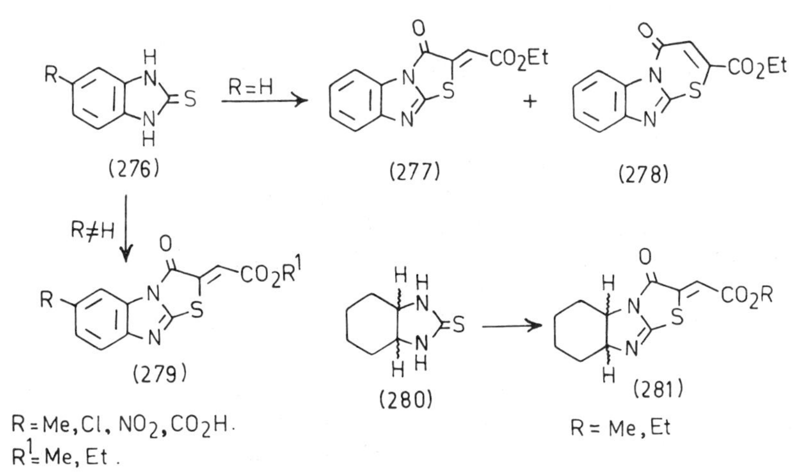

R = Me, Cl, NO_2, CO_2H.
R^1 = Me, Et.

R = Me, Et

SCHEME 67

tigation in view of the findings of Wade (79JOC1816) and of Acheson and Wallis [81JCS(P1)415]. Moreover, the regiochemistry of the substituents in **279** and the stereochemistry of the A/B ring junction in **281** are uncertain.

Cyclization of 2-(phenacylthio benzimidazoles, obtained from the reaction of unsymmetrical benzimidazolyl-2-thiones (**282**) with phenacyl halides can give rise to two isomeric cyclized products, depending on the nitrogen atom involved in the ring closure. No systematic studies on the orientational preference shown by substituents in the benzene ring during the cyclization leading to the synthesis of 3-arylthiazolo[3,2-*a*]benzimidazoles have been made, probably because of the difficulties of establishing the position of substituents in the benzimidazole nucleus. However, much success in securing the orientation on cyclization in the case of their thiazolidinone counterparts has been achieved by ^{1}H-NMR spectroscopy.

In an extensive work on the orientation of cyclization leading to the synthesis of thiazolo[3,2-*a*]benzimidazol-3(2*H*)-ones, studies were directed to ascertain whether the cyclization was governed by an electron donating–withdrawing nature of the substituents or by the steric factor only. Benzimidazolyl-2-thiones (**282a–k**) react with chloroacetic acid to give acids **283a–k** which, on treatment with acetic anhydride in pyridine, furnish in all cases a single products for which structure **284a–k,** and not the other possible isomeric structure **285a–k,** was assigned on the basis of ^{1}H-NMR spectral data [70IJC10; 79IJC(17B)572; 80IJC(B)1035; 81IJC(B)294; 84JIC1053; 85IJC(B)1224; 86IJC(B)807; 87IJC(B)532, 87UP3, 87UP4; 88IJC(B)121] (Scheme 68).

On the other hand, acids **289a–g,** obtained from thiones **288a–g** and chloroacetic acid, on cyclization with acetic anhydride in pyridine or Dowtherm-A give, both isomers **290a–g** and **291a–g** [81CPB1876; 86IJC(B)807; 87UP5]; although it was earlier reported that acid **289a–d,** on cyclization with acetic anhydride in pyridine, yielded only one isomer for which structures, **290a, 291b, 290c** and **290d** were assigned on the basis of ^{1}H-NMR spectroscopy [70IJC10; 72IJC274; 73IJC1119; 78IJC(B)478]. Acid **289e** gave both isomers **290e** and **291e** which could not be separated, but their identities were confirmed by ^{1}H-NMR spectral data (74TH1). It seems quite possible that one of the isomers was lost during the work-up (Scheme 69).

The structural assignment of cyclized products **290a–f** and **291a–f** by Japanese workers (81CPB1876; 87UP1) was based on the downfield shift (~0.40 δ) of the C_5-H protons (H_B in **290** and H_A in **291**) compared to that of the corresponding protons of acids **289a–f.** This method, however, leaves some ambiguity, especially in the case of **290e** in which the downfield shift

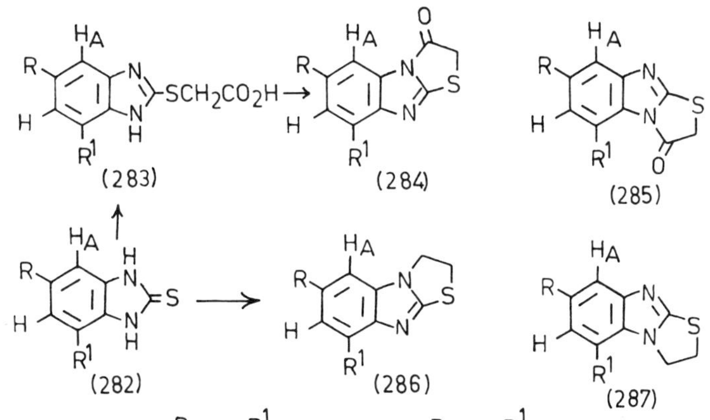

Sec. II.A]   CONDENSED 4-THIAZOLIDINONES   51

SCHEME 68

	R	R¹		R	R¹
a,	H	Me	g,	Br	Cl
b,	H	NO₂	h,	Cl	Br
c,	H	Cl	i,	Br	Me
d,	H	F	j,	Me	Br
e,	Br	Br	k,	Me	Me
f,	Cl	Cl			

of C_5-H was small. For further confirmation of the cyclized products, the Japanese workers advanced excellent chemical evidence that involved the Raney–nickel desulfurization of cyclized products **290a,c–f** and **291a,c–f**. This resulted in the formation of 1-acetyl-5(or 6)-substituted benzimidazoles whose identities were proved by direct comparison with authentic samples (82CPB2714).

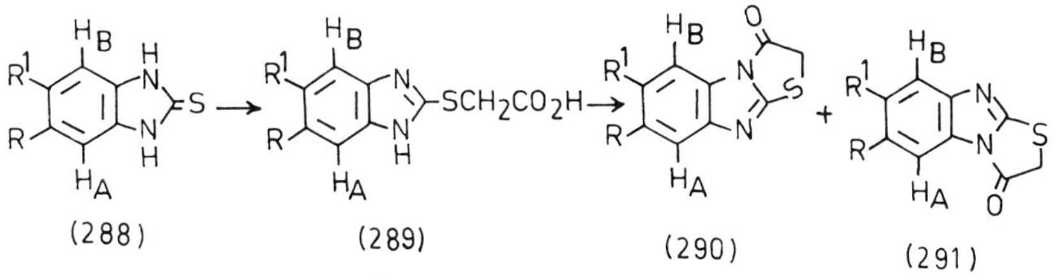

	R	R¹		R	R¹
a,	H	Cl	e,	H	NO₂
b,	H	Br	f,	H	CO₂Me
c,	H	Me	g,	Br	OMe
d,	H	OMe			

SCHEME 69

The structural assignment for cyclized products **284, 285, 290,** and **291** in the author's laboratory was based mainly on two approaches. The first deals with the comparison of the chemical shifts of the peri-proton (H_A) in **284e–g,i,j** with the corresponding proton of **286e–g,i,j** or **287e–g,i,j** obtained from the reaction of thiones (**282e–g,i,j**) with 1,2-dibromoethane [80IJC(B)1035; 81IJC(B)294; 84JIC1053; 87IJC(B)532; 88IJC(B)121]. Proton H_A in **284e–g,i,j** resonates downfield when compared to that in **286e–g,i,j** (or **287e–g,i,j**), while H_A in **285e–g,i,j** resonates at approximately the same position as that in **286e–g,i,j** (or **287e–g,i,j**). This deshielding has its origin in the magnetic anisotropy of the carbonyl group with little contribution from the rest of the ring.

The second approach concerns the comparative studies of the observed and calculated chemical shifts of the aromatic protons of cyclized products **284** or **285** and **290** or **291**. The calculated chemical shifts of the aromatic protons are derived by taking into consideration the shielding or deshielding effect (69MI1) of substituents on the chemical shifts compared to the corresponding protons of the parent compound **249a**. This approach was used to establish structures **284a** [88IJC(B)121], **284b** [86IJC(B)807], **284c** [79IJC(17B)572] **284d** (87UP3), **284e** [88IJC(B)121], **284f** [87IJC(B)532], **284g** (84JIC1053), **284h** [85IJC(B)1224], **284j** [88IJC(B)121], **284k** (87UP4), **290a** [88IJC(B)121], **290g** [86IJC(B)807], and **291g** [86IJC(B)807]. The isomer whose calculated chemical shifts of the aromatic protons tally with the observed chemical shifts is taken to be the correct one. This method gives good results in all condensed thiazolidinones synthesized in the author's laboratory and provides an easy and reliable method of distinguishing both isomers (see Section III).

The structural assignment for **290** and **291** was also arrived at by comparing the chemical shifts of aromatic protons H_A and H_B of **290g** and **291g** with those of **289g** In structure **290g** H_B is shifted downfield, while H_A in **291g** is shifted downfield compared to the corresponding protons of the acid **289g**. Structure **290a** also was confirmed directly from the most downfield signal due to H_B which showed meta coupling. This lends further support to the method of calculation.

Structure **290d** was earlier assigned on the basis of the comparison between the observed and calculated chemical shifts of the aromatic protons of cyclized products **290d**. The calculated chemical shifts were derived from **249a** which was used as a reference compound (72IJC274). The same conclusion was drawn when **256d** was used as a reference compound [86IJC(B)267], confirming the validity of the method of calculation used by these Indian workers.

Structure **284b,** later proved to be correct [86IJC(B)807], was assigned to the cyclized product obtained from acid **283b** on a wrong presumption

(84JIC89). The most downfield signal was wrongly considered to be due to H-5 in **284b,** as it is deshielded by the carbonyl group and such a downfield signal was not possible from the alternate structure **285b.** In fact, the most downfield signal in **284b** would be due to H-7 because of the deshielding effect of the nitro group on the ortho proton (H-7). A nitro group deshields ortho protons to a greater extent than meta or para protons (69MI1), and this value is much greater than that obtained from the deshielding effect of the carbonyl group. This important factor was overlooked by these workers (84JIC89). Similarly, acid **289e,** on cyclization with acetic anhydride in pyridine, was reported to give only one isomer for which structure **290e** (77JHC1093) or **291e** (84JIC89) was assigned without any supporting evidence.

Thione **292a** reacts with chloroacetic acid and aromatic aldehydes in a mixture of acetic anhydride and acetic acid to give a cyclized product for which structure **293a** was assigned [86IJC(B)776], whereas thione **292b** under similar conditions is reported to have yielded the other isomer **294b** (67MI1) (Scheme 70). In both cases, no evidence, spectral or chemical, was cited to support the structures. In view of the findings of earlier work on unsymmetrical thiones, formation of both isomers **293a,b** and **294a,b** is possible and warrants reinvestigation.

After having acquired data for the cyclization of acids **283a–k** and **289a–g,** the former giving only one isomer **(284a–k),** and the latter giving both isomers **(290a–g** and **291a–g),** the rationale for the preferred orientation was suggested [88IJC(B)121]. Acids **283a** (having an electron-donating methyl group), **283b** (having an electron-withdrawing nitro group), and **283c,d** (having dual character halogen atoms), on cyclization, yield thiazolidinones **284a–d,** respectively suggesting that the cyclization took place in the same direction regardless of the nature of the substituent. This proves the electron-donating or electron-withdrawing nature of the substituent does not affect the mode of cyclization, but rather cyclization is governed

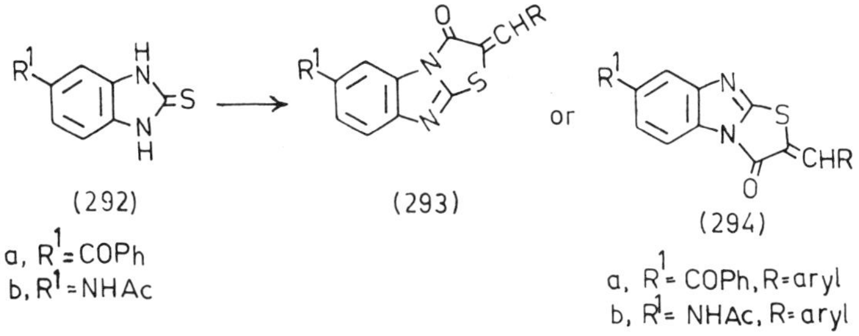

(292)
a, R^1 = COPh
b, R^1 = NHAc

(293)

or

(294)
a, R^1 = COPh, R = aryl
b, R^1 = NHAc, R = aryl

SCHEME 70

by steric factors only. Thus, cyclization takes place at a *meta*-nitrogen wherever there is an ortho-substituent regardless of the nature of the substituent. This is also found to be true in the case of disubstituted acids (**283e–k**) having one ortho-substituent, where the cyclization takes place at a *meta*-nitrogen, yielding **284e–k,** respectively. Acids **289a–g** in which both *ortho*-positions are free, thus devoid of any such steric hindrances, are expected on cyclization to give both isomers **290a–g** and **291a–g**. Indeed this was found to be true, lending further support to the rationale that the cyclization was governed by the steric factors alone and not by the nature of the substituents.

31. *Thiazolo[2,3-*b*]imidazo[5,4-*b*]pyridin-3(2*H*)-ones*

There are only two publications that describe this system. Imidazo[5,4-*b*]pyridine-2-thione (**295a**) reacts with chloroacetic acid to give a cyclized product for which structure thiazolo[2,3-*b*]imidazo[5,4-*b*]pyridine-3(2*H*)-one (**296**) was assigned. Thione **295a** adds to the C═C double bond of maleic anhydride by a Michael-type reaction to yield cyclized product **297** (83AP985). Reaction of **295a,b** with chloroacetic acid and aromatic aldehydes in a mixture of acetic anhydride and acetic acid gives arylidene derivatives **298a,b** [86IJC(B)776] (Scheme 71). The compounds for which

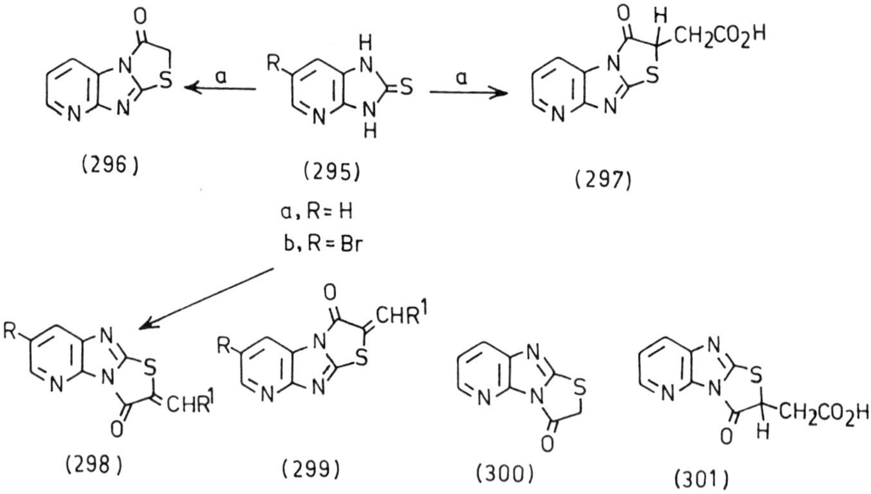

(296) (295) (297)

a, R = H
b, R = Br

(298) (299) (300) (301)

a, R = H, R[1] = aryl
b, R = Br, R[1] = aryl

SCHEME 71

structures **296–298** were assigned could be represented by other alternate structures **300**, **301**, and **299**, respectively. The data are inadequate to decide which structures are better.

32. Thiazolopurin-3(2H)-ones

Of all the possible ways of fusing a purine ring onto a thiazole ring, only the syntheses of two, namely, thiazolo[2,3-f]purin-3(2H)-one (**304**) or thiazolo[2,3-e]purin-3(2H)-one (**305**) and thiazolo[2,3-b]purin-3(2H)-one (**307**, **311**) have been reported.

a. *Thiazolo[2,3-f]purin-3(2H)-one.* 2-Methyl-6-oxopurine-8-thione (**302**) reacts with chloroacetic acid to give acid **303** which, on refluxing in acetic anhydride for 30 min, affords a cyclized product for which structure 7-mehtylthiazolo[2,3-f]purin-3,5(2H,6H)-dione (**304**) was assigned (59JOC1410) (Scheme 72). The structural assignment was based on analogy with 5-ethyl-2-(β-hydroxyethyl)-3-methylthiazolo[2,3-f]purine (36JCS1559), whose structure in turn was wrongly based on 3-methylthiazolo[3,2-a]benzimidazole. Purine-8-thione and benzimidazolyl-2-thione cannot be compared because of the simple fact that the former is an unsymmetrical thione, whereas the latter is symmetrical. Hence, the cyclized product could be represented by another alternate thiazolo[2,3-e]purine structure (**305**).

b. *Thiazolo[2,3-b]purin-3(2H)-ones.* Condensation of purine-2-thiones (**306a–d** and **310**) with ethyl α-chlorophenylacetate in refluxing pyridine for 6 hr furnishes 9α-ethoxy-9-methyl- (**307a**), 9α-ethoxy-9-phenyl- (**307b**), 6-benzyl-9α-ethoxy-9-methyl- (**307c**), 6-benzyl-7,9-dimethyl-9α-ethoxy- (**307d**), and 5,9-dimethyl-9α-ethoxy-6-methylthio-2-phenylthiazolo[2,3-b]purine-3,8(2H,9H)-dione (**311**), respectively [84IJC(B)316] (Scheme 73). The appearance of a strong carbonyl band around 1740 cm^{-1} in the IR spectra of the products supports the cyclized structures **307a–d** and **311**.

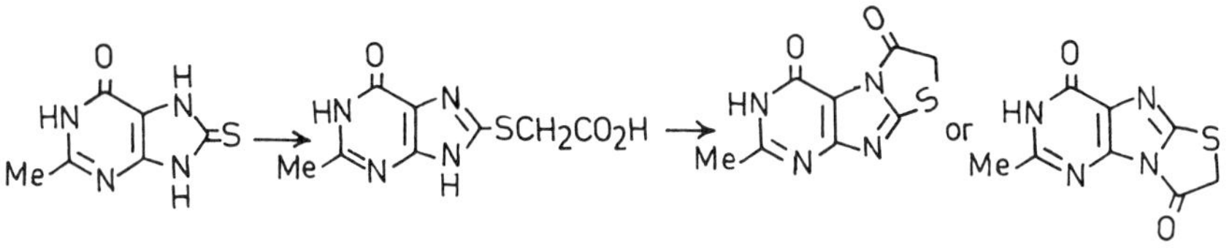

(302)   (303)   (304)   (305)

SCHEME 72

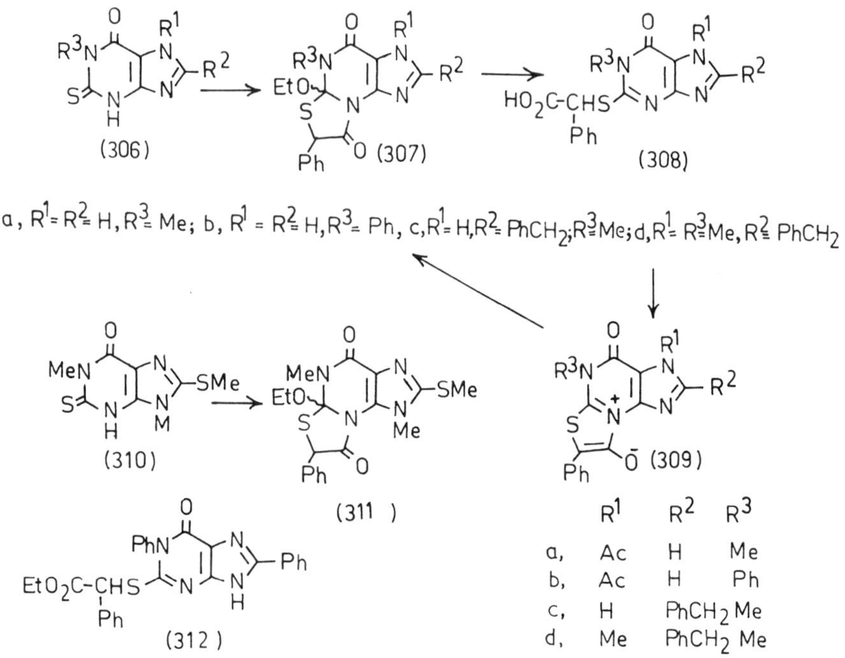

SCHEME 73

a, $R^1 = R^2 = H, R^3 = Me$; b, $R^1 = R^2 = H, R^3 = Ph$; c, $R^1 = H, R^2 = PhCH_2, R^3 = Me$; d, $R^1 = R^3 = Me, R^2 = PhCH_2$

	R^1	R^2	R^3
a,	Ac	H	Me
b,	Ac	H	Ph
c,	H	$PhCH_2$	Me
d,	Me	$PhCH_2$	Me

Convincing evidence came from the mass spectral data. The molecular ion peak at m/z 406 (M$^{·+}$, 89%) and fragmented ions at m/z 361 (M−OEt$^+$, 100%) and at m/z 360 (M−EtOH$^{·+}$, 75%) are tenable with the cyclic structures **307b** rather than open structure **312**. Mild hydrolysis of **307a–d** in sodium hydroxide at room temperature affords acids **308a–d** which, on heating in a mixture of acetic anhydride and pyridine for a few minutes, results in the synthesis of mesoionic compounds **309a–d**. Elemental analyses as well as the absence of an N−H band in the IR spectra of **309a,b** suggests that acetylation of the mesoionic compounds during or after cyclodehydration took place, resulting in the synthesis of **309a,b**. Mesoionic compounds **309a–d** in refluxing ethanol through facile ethanol addition (as well as deacetylation in case of **309a,b**) furnish back the same thiazolidinones (**307a–d**). Facile transformation of **309a,b** to **307a,b** in refluxing ethanol with concomitant loss of an acetyl group unequivocally confirms that formation of **309** does not involve any deep seated structural change during cyclodehydration.

33. Thiazolo[3,2-a]thiazolo[5,4-d]pyrimidin-3(2H)-ones

Condensation of thiazolo[5,4-d]pyrimidine-6-thiones (**313a–d**) with ethyl α-chlorophenylacetate or methyl α-chlorophenylacetate in the presence of pyridine or potassium carbonate gives products formulated as thiazolo[3,2-a]thiazolo[5,4-d]pyrimidin-3(2H)-ones (**314a–f**) on the basis of γ-lactam carbonyl absorption at ~1740 cm^{-1} in the IR spectra of **314a–d** [83IJC(B)243]. The observed γ-lactam carbonyl absorption at low wave number (~1720 cm^{-1}) in **314e–f** is probably due to the long range conjugation effect of the amidine residue transmitted via a ring-sulfur atom on the lactam C=O bond [71ACR1; 75JCS(P2)1294]. The isolation of identical ethanol adducts (**314b,c**) from mesoionic compound **318** in refluxing ethanol or methanol supports the assigned cyclic structures **314** in preference to open chain esters. Acid **315b**, obtained by mild alkaline hydrolysis of **314b**, on treatment with warm acetic anhydride–pyridine, furnishes a yellow mesoionic product **318** confirmed through physical properties such as a blue shift in the λ_{max} with increasing polarity of the medium and disappearance of its yellow color (MeCN) on protonation. Acid **315a**, similarly obtained from **314d**, upon warming with acetic anhydride–perchloric acid gives perchlorate (**316**), which undergoes acylation with acetic anhydride and pyridine or p-nitrobenzoylchloride in the presence of triethylamine to give **317a,b**, respectively; these exhibit a blue shift with increasing polarity of the medium, thus confirming the mesoionic structure for **317a,b**. The acyl derivatives **317a,b**, on refluxing in methanol, affords a compound identical (thin layer chromatography and IR) to **314d**, also obtained from direct condensation of **313a** with methyl α-chlorophenylacetate. The transformation evidently occurs through loss of an acyl residue in **317a,b** with concomitant addition of methanol followed by prototropic rearrangement to furnish **314d** (Scheme 74).

34. Thiazolo[3,2-a]thieno[2,3-d]pyrimidin-3(2H)-ones

5,6-Dimethyl-3-phenyl-4-oxothieno[2,3-d]pyrimidine-2-thione (**319**) reacts with ethyl α-chlorophenylacetate in the presence of anhydrous potassium carbonate in boiling dimethylformamide (DMF)–acetone for 12 hr to give 6,7-dimethyl-2,9-diphenyl-9α-ethoxy-9-oxothiazolo[3,2-a]thieno[2,3-d]pyrimidin-3(2H)-one (**321**) in 90% yield [81IJC(B)538] (Scheme 75). The absorption at 1745 cm^{-1} (five-membered lactam C=O) in addition to that at 1690 cm^{-1} (six-membered C=O) in the product supports the cyclic structure **321**. The formation of **321** must have been through the initially formed unstable mesoionic compound **320**, which adds 1 mol of ethanol, liberated during reaction, to afford **321**.

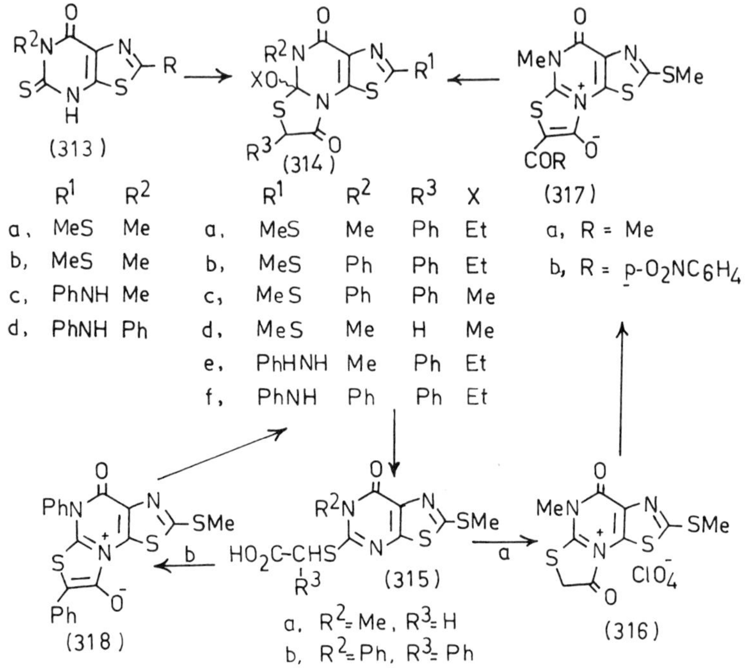

SCHEME 74

35. Thiazolo[3,2-a]-[1]benzothieno[2,3-d]pyrimidin-3(2H)-ones

The synthesis of thiazolo[3,2-*a*]-[1]benzothieno[2,3-*d*]pyrimidin-3(2*H*)-ones (**323**) is accomplished in two ways. The first involves direct condensation of thione with α-chloroester, while the second, the cyclization of the acid through a mesoionic compound, is affected as illustrated in Scheme 76 [81IJC(B)538]. 3-Methyl(phenyl)-4-oxo-5,6,7,8-tetrahydro[1]benzothieno[2,3-*d*]pyrimidine-2-thiones (**322a,b**) react with ethyl α-chlorophe-

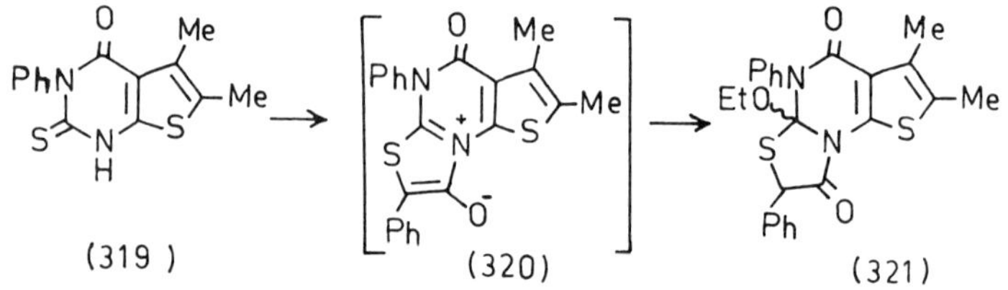

SCHEME 75

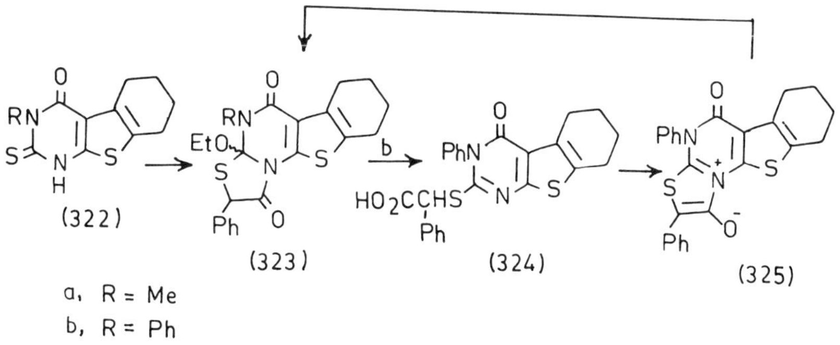

SCHEME 76

a, R = Me
b, R = Ph

nylacetate in the presence of anhydrous potassium carbonate in boiling DMF–acetone to afford 11α-ethoxy-11-methyl(phenyl)-2-phenyl-10-oxo-6,7,8,9-tetrahydrothiazolo[3,2-a]-[1]benzothiano[2,3-d]pyrimidin-3(2H)-ones (**323a,b**) in almost quantitative yield. The absorption at 1745 cm^{-1} due to the five-membered lactam carbonyl group of structure **323a,b** lends support for the cyclic structure. The isolation of the identical ethanol adduct **323b** from the mesoionic compound **325** in refluxing ethanol confirms the cyclic structure of the products obtained by the direct condensation of **322** with ethyl α-chlorophenylacetate. The mesoionic compound **325** is synthesized by acetic anhydride–pyridine mediated cyclization of acid **324** which is obtained by the mild alkaline hydrolysis of cyclized product **323b**.

36. *Thiazolo[3,2-*a*]-*s*-triazolo[3,4-*b*]-[1,3,4]thiadiazin-6(7H)-one*

The practical utility of the facile addition of mercaptoacetic acid across the double bond of an azomethine has been used in the synthesis of thiazolo-triazolothiadiazin-6(7H)-one (**328**). Thus, 7H-3-methyl-s-triazolo[3,4-b]thiadiazine hydrochloride (**327**), obtained by reacting 1-amino-2-methyl-s-triazolyl-2-thione (**326**) with chloroacetaldehyde diethylacetal, reacts with mercaptoacetic acid in the presence of p-toluenesulfonic acid in boiling anhydrous benzene for 40 hr using a Dean–Stark water separator followed by basification with sodium bicarbonate to furnish 8αH,9H-3-methylthiazolo[3,2-a]-[1,3,4]triazolo[3,4-b][1,3,4]-thiadiazin-6(7H)-one (**328**) (74IJC287) (Scheme 77). The absorption at 1725 cm^{-1} (lactam carbonyl) in product **328** supports the cyclic structure in preference to the open-chain acid structure.

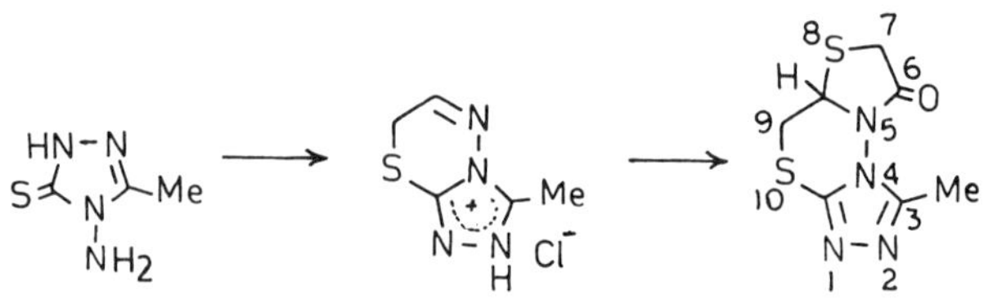

(326) (327) (328)

SCHEME 77

37. Thiazoloquinazolinones

Of three possible thiazoloquinazolinone systems possessing the 4-thiazolidinone unit, namely, 5H-thiazolo[2,3-b]quinazolin-3(2H)-one (**329**), 5H-thiazolo[3,2-a]quinazolin-1(2H)-one (**330**), and 10bH-thiazolo[3,2-c]quinazolin-3(2H)-one (**331**), only the synthesis of the first two systems has been achieved.

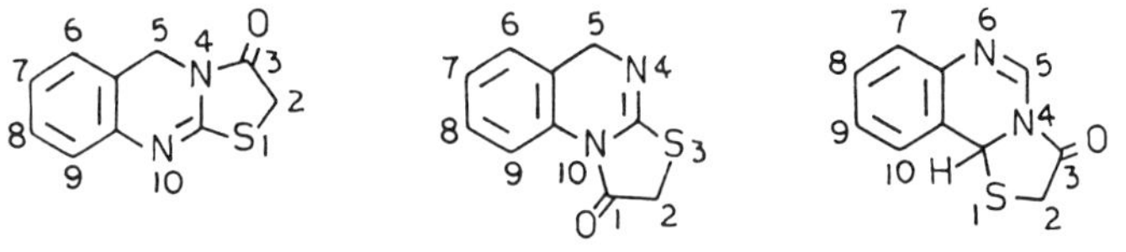

(329)   (330)   (331)

a. *5H-Thiazolo[2,3-b]quinazolin-3(2H)-ones.* Synthesis of 5H-thiazolo[2,3-b]quinazolin-3(2H)-one has been accomplished by two alternate routes. The first route, in which a thiazolidinone ring is built onto a quinazoline ring, involves the reaction of 3,4-dihydroquinazoline-2-thiones with chloroacetic acid and subsequent cyclization of the intermediate acid. The second route, in which a quinazoline ring is built onto a thiazolidinone ring, deals with the reaction of 2-alkylmercapto-4-thiazolones with anthranilic acid followed by cyclization of the intermediate in situ. The latter route provides an unequivocal synthesis of **329**, whereas in the former route, the possibility of the formation of **329** or **330** or both exits.

i. *From a quinazoline.* The first example of this system was synthesized by Kendall and Duffin who cyclized acid **333** with acetic anhydride in pyridine and obtained a product for which the structure 2H-thiazolo[2,3-b]quinazolin-3,5-dione (**334**) was assigned without any evidence

(50BRP634951). No precise data were cited for this cyclized product which could be represented by another alternate structure, 2H-thiazolo[3,2-a]quinazolin-1,5-dione (**336**). Structure **334** was condensed (at the site of the methylene group at the 2-position) with pseudoindolium and benzothiazolium salts to yield sensitizing dyes (50BRP634952). Reaction of **332** with ethyl chloroacetate yields a cyclized product that could be represented by either **334** or **336**. Here also, no adequate data are recorded in support of either structure (85AP502). Dhatt and Narang (54JIC787) repeated the work of Kendall and Duffin (50BRP634951) and found that acid **333**, on heating with acetic anhydride–pyridine mixture, underwent decomposition. However, the treatment of acid **333** with acetic anhydride in the absence of pyridine for 6 hr at 80°C gives a cyclized product confirmed by a molecular ion peak [M]$^{\cdot+}$ at m/z 218 [77ZN(B)94]. Structure **334** was assigned to the cyclized product on the basis of the comparison with an authentic sample of the other isomer **336**, obtained through an unequivocal synthesis (See Scheme 89) (69IJC881). Isomer **336** was not found to be identical with cyclized product **334** obtained from acid **333**. Thione **332** reacts with chloroacetic acid and aromatic aldehydes in the presence of anhydrous sodium acetate in acetic acid–acetic anhydride at refluxing temperature for 4 hr to furnish directly in a single step the arylidene product **335**, which is also obtained from **334** and aromatic aldehyde (Scheme 78).

6-Bromo-4-oxoquinazolinyl-2-thione (**337a**) reacts with chloroacetic acid to give acid **338a** which, on treatment with acetic anhydride–pyridine,

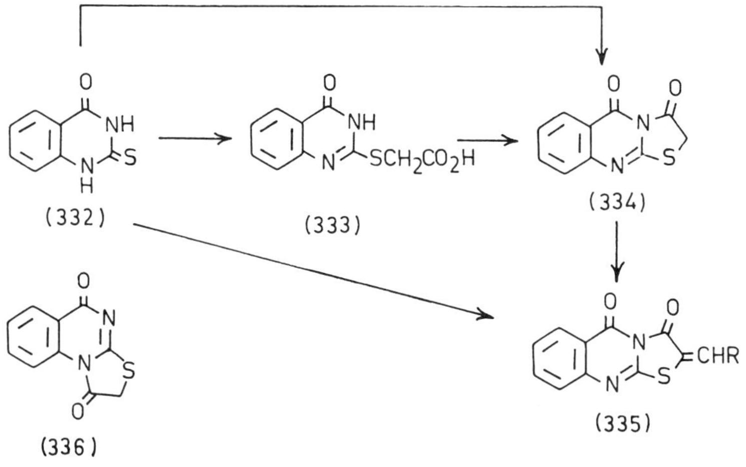

SCHEME 78

affords a cyclized product for which the structure 7-bromo-2H-thiazolo[2,3-b]quinazolin-3,5-dione (**339a**) was assigned on the basis of IR and ^{1}H-NMR spectral data in preference to the other isomer, 7-bromo-2H-thiazolo[3,2-a]quinazolin-1,5-dione (**341a**) [78IJC(B)537] (Scheme 79). The absorption of both carbonyl groups at higher frequencies (1685 and 1790 cm^{-1})in the cyclized product shows the presence of the —CO—N—CO— group (68JHC179), lending support to structure **339a**. In view of the equivalence of the protons ortho to bromine and nonequivalence of the protons adjacent to nitrogen in **339a** and **341a**, the two structures could be distinguished by the NMR signal exhibited by the aromatic proton adjacent to nitrogen. This proton in the cyclized product has a normal value of ∂ 7.40 which supports structure **339a**. The corresponding proton in **341a** would resonate at a lower field than ∂ 7.40 because of the deshielding effect of the carbonyl group at a peri-proton.

However, unlike **339a**, the structure of the cyclized product obtained from acid **338b** could not be established on the basis of IR spectral data. The absorption of carbonyl groups at 1680 and 1775 cm^{-1} and the molecular ion peak [M]$^{\cdot+}$ at m/z 232 in the cyclized product suggests cyclization had indeed taken place, yet the spectral data are not helpful in deciding in favor of either **339b** or **341b**. Structure **339b** in preference to **341b** has, however, been assigned to the cyclized product on the basis of ^{1}H-NMR studies (84JIC1050). Reaction of **337b** with 1,2-dibromoethane gives a single product (confirmed by the molecular ion peak [M]$^{\cdot+}$ at m/z 218) which could be represented by either **340** or **342**. In either structure, the signal at δ 2.18 is assignable to methyl protons. The appearance of the

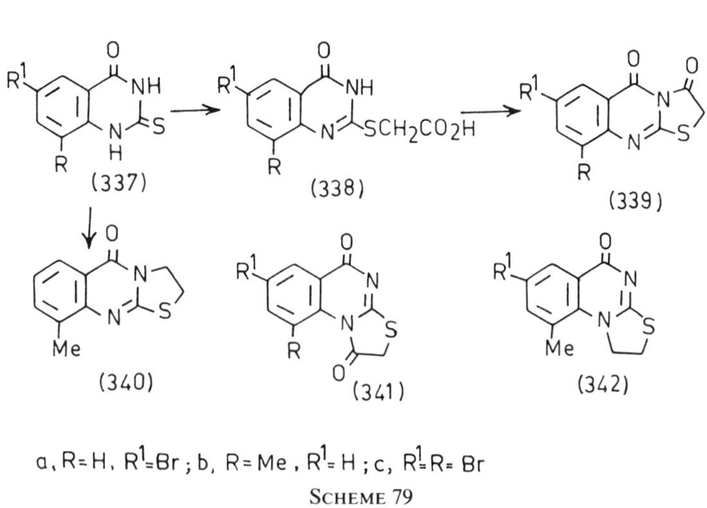

a, R=H, R^1=Br ; b, R=Me , R^1= H ; c, R^1=R= Br

SCHEME 79

Sec. II.A]   CONDENSED 4-THIAZOLIDINONES   63

methyl signal of δ 2.20 (almost of the same value as δ 2.18) in the ^{1}H-NMR spectrum of the cyclized product obtained from acid **338b** led to the assignment of structure **339b**. Had **341b** been correct, then the methyl protons would have shown a downfield shift due to the deshielding effect of the carbonyl group of the thiazolidinone ring. The magnetic anisotropic effect of a similar carbonyl group deshields the peri-methyl protons (70IJC10). Structure **340** in preference to **342** was assigned to the product obtained from the reaction of **337b** and 1,2-dibromoethane on analogy to the work of Howard and Klein who obtained **340** (H in place of Me) from **332** and 1,2-dibromoethane, while reaction of ethyl N-thiocarbamoylanthranilate with 1,2-dibromoethane furnished **342** (H in place of Me) (62JOC3701). Similarly, **339c** is obtained from **337c** via **338c** using the same series of reactions [78IJC(B)537] (Scheme 79).

Synthesis of 2H-thiazolo[2,3-b]quinazolin-3,5-diones (**344a–e**) has been reported in different ways starting from **343a–e, 345a–e, 346a–e,** and **347a–e,** while **344f** is obtained from **343f** as illustrated in Scheme 80 (83H1549). 2-Carboxyphenylthioureas (**343a–e**), on refluxing with chloroacetic acid in absolute ethanol in the presence of anhydrous sodium acetate for 25 hr give cyclized products (**344a–e**). Carrying out the reaction for a shorter period (6 hr) gives 2-(o-carboxyphenyl)imino-4-thiazolidinones (**345a–e**) which, on further heating for 19 hr, could be cyclized to **344a–e**. 2-(N-Cyano)anthranilic acids (**346a–e**), obtained by

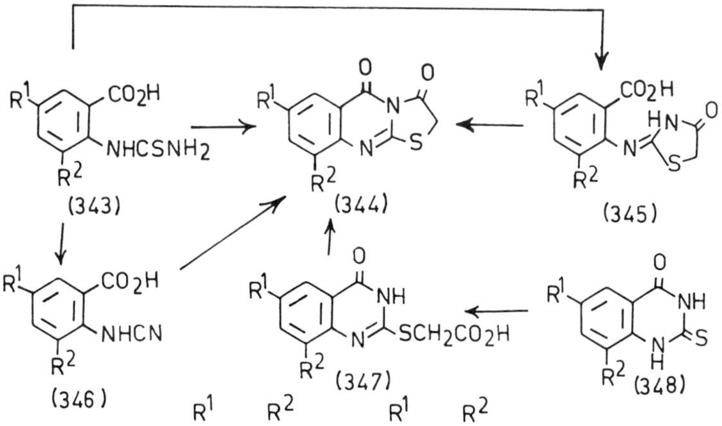

	R^1	R^2		R^1	R^2
a,	H	H	d,	I	H
b,	Cl	H	e,	I	I
c,	Cl	Cl	f,	NO$_2$	H

SCHEME 80

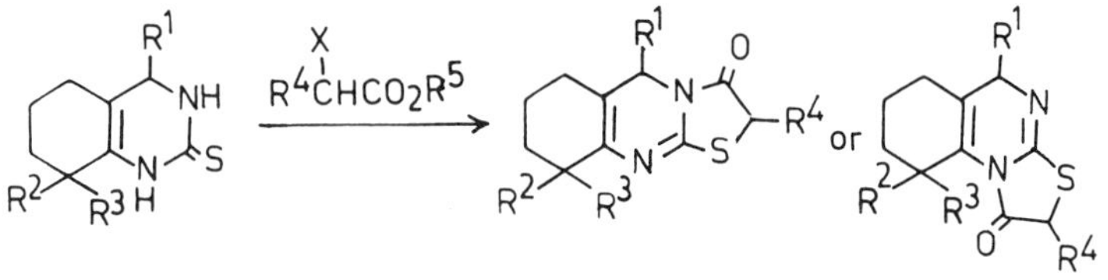

(349)   (350)   (351)

R^1 = alkyl, aryl; R^2R^3 = 2H, CHaryl; R^4 = H, Me.

SCHEME 81

the reaction of **343a–e** with lead diacetate, condense with mercaptoacetic acid to afford **344a–e**. Acids **347**, obtained by the condensation of thiones **348a–e** with chloroacetic acid, furnish **344a–e** upon treatment with acetic anhydride–pyridine. The structural assignment of **344a–e** for the cyclized products was based on their synthesis from **345a–e** and also from **346a–e**. Structure **344a**, obtained by three different methods, melts at 294–295°C, while in a fourth method it is *reported* to have a melting point of 193°C.[1] The melting point reported by Egyptian workers [77ZN(B)94] for **344a** is 209°C; their structure was proved by the molecular ion peak and its nonidentical comparison with the other isomer **336** (Scheme 78). Although Indian workers (83H1549) refer to the publication by Egyptian workers, they did not comment on the difference in melting points of **344a**. In view of this and in the absence of NMR and mass spectral studies, the work depicted in Scheme 80 warrants reinvestigation.

Thiones **349** react with chloroacetic acid, α-bromopropionic acid, or their esters to give cyclized products for which structure **350** was assigned (78MI1; 81ACH197; 83MI2) (Scheme 81). The cyclized product could also be represented by alternate structure **351**. Data are inadequate to decide in favor of either.

Reaction of 3,4-dihydroxyquinazoline-2-thione (**352a**) with ethyl chloroacetate and *N*-phenylmaleimide gives in each case a cyclized product formulated as 5*H*-thiazolo[2,3-*b*]quinazolin-3(2*H*)-one (**329**) and 3(2*H*)-oxo-5*H*-thiazolo[2,3-*b*]quinazolin-2-acetanilide (**353**), respectively (83AP569). Structures **352a,b** similarly react with maleic anhydride and acetylene dicarboxylic acid or esters to give cyclized products **354a,b** and **355a,b,** respectively (83AP569; 86JHC1359, 86USP4588812). Since the starting thiones (**352a,b**) are unsymmetrical, their reactions with ethyl chloroacetate, maleic anhydride, and *N*-phenylmeleimide could give one

---

[1] It seems possible that the reported value of 193°C is a misprint.

or both isomers depending on the direction of cyclization. Hence, the cyclized products **329, 353–355** could be represented by alternate structures **330, 356–358**, respectively. No attempt was made to settle this question (Scheme 82).

Reaction of 4-oxoquinazolinyl-2-thione (**332**) with DMAD gives four major products (**359–362**) and one minor product (**363**) whose structures were unequivocally established by ^{13}C-NMR spectroscopy and X-ray crystallographic analysis [86JCS(P1)2095] (Scheme 83). Surprisingly, no condensed thiazolidinone (**364** or **365**) was isolated from the reaction mixture. This is in contrast to the behavior of other thiones (**20, 28, 53, 65, 155, 185, 247, 276** ($R = H$), **276, 280,** and **352**) that react with DMAD to give the corresponding condensed thiazolidinones (**21, 48, 54, 67, 160, 188, 23, 277, 279, 281,** and **357**).

ii. *From a thiazole.* Reaction of 5-arylhydrazino-2-methylmercapto-4-thiazolones (**366a**) and 5-arylidene-2-ethylmercapto-4-thiazolones (**366b**)

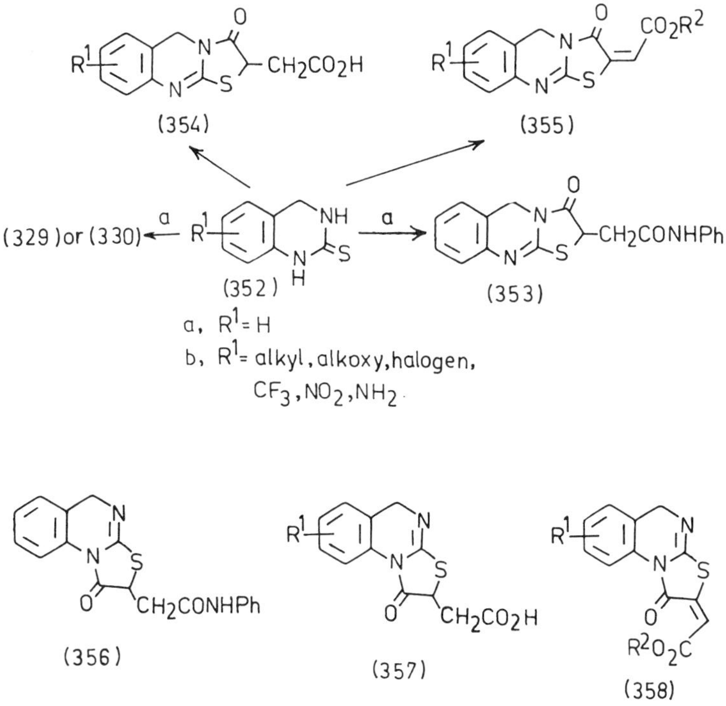

SCHEME 82

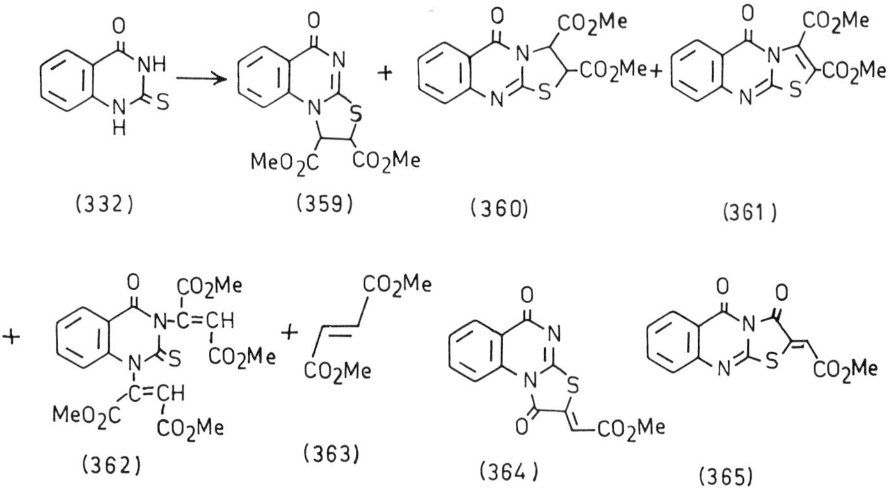

SCHEME 83

with anthranilic acid afford 2-arylhydrazino-2-*H*-thiazolo[2,3-*b*]quinazolin-3,5-diones (**368a**) and 2-arylidene-2*H*-thiazolo[2,3-*b*]quinazolin-3,5-diones (**368b**), respectively (83AP394, 83MI1) (Scheme 84). No intermediate uncyclized product (**367**) was isolated.

b. *5H-Thiazolo[3,2-a]quinazolin-1(2H)-ones*. The synthesis of 5*H*-thiazolo[3,2-*a*]quinazolin-1(2*H*)-ones has also been accomplished by two methods. In the first, a thiazolidinone ring is built onto a quinazoline ring, while in the second, a quinazoline ring is built onto the thiazolidinone ring formed in situ.

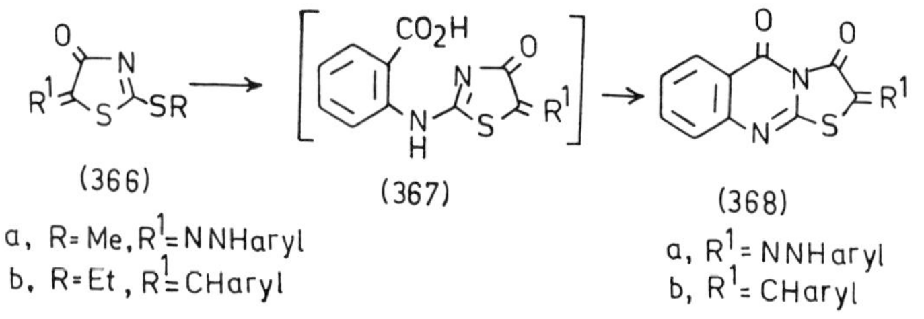

SCHEME 84

i. *From a quinazoline.* 3-Phenyl-4-quinazolone-2-thiolacetic acid (**369a**) and α-(3-phenyl-4-quinazolone-2-thiol)propionic acid (**369b**), on treatment with acetic anhydride–perchloric acid, are smoothly converted to the respective perchlorate salts **370a,b** which are fairly stable in dry condition and slowly decompose under storage [79IJC(18B)39; 82IJC(B)517]. The perchlorate salts (**370a**), on boiling with acyl chloride in the presence of triethylamine, undergo acylation to form mesoionic ketones (**372b**) which, on boiling with reagent-grade chloroform (usually containing ethanol) or with ethanol-free chloroform and ethanol or methanol, undergoes deacylation to furnish 3a-ethoxy(or methoxy)-4-phenylthiazolo[3,2-*a*]quinazolin-3,5-dione (**371a**) [82IJC(B)517]. Acid **369c**, on treatment with acetic anhydride–pyridine at room temperature, affords the mesoionic compound **372a** which, in refluxing ethanol, yields **371c**. The participation of ethanol in this reaction is confirmed from the fact that the mesoionic compound **372a**, when boiled in ethanol-free chloroform, did not undergo any reaction, and **372a** was recovered unchanged. The incorporation of ethanol, which results in the formation of **371a–c**, was confirmed by ^{1}H-NMR spectroscopy. Condensed 4-thiazolidinones (**371a,c**) are also obtained by refluxing 3-phenyl-4-oxoquinazolin-2-thione (**373**) with an appropriate haloester in the presence of anhydrous potassium carbonate in acetone [79IJC(18B)39; 82IJC(B)517]. Structure **370b**, also in boiling methanol or ethanol, affords **371b**. Structure **371c**, on mild alkaline hydrolysis, undergoes cleavage of the thiazolidinone ring to furnish acid **369c** [79IJC(18B)39] (Scheme 85).

Reaction of 4-oxoquinazolinyl-2-thiones (**374a–f**) with chloroacetic acid or ethyl bromoacetate gives the acids or esters **375a–f** which, on reductive cyclization with sodium borohydride in sodium hydroxide, afford thiazolo[3,2-*a*]quinazolin-1,5-diones (**376a–f**) (70JIC579, 70JIC1059; 78JIC928) (Scheme 86). The work of Singh and Chaudhary (70JIC579) was repeated by Talukdar and co-workers [82IJC(B)517] who found that 3-phenyl-4-oxoquinazoline-2-thiolacetic acid or its ethyl ester (**375a**) did not yield the cyclized product **376a**. If the formation of **376** is correct, then the reported synthesis of the thiazolidinone ring in alkaline medium would be a rare case. In view of this and the findings of Talukdar *et al.* [82IJC(B)517] coupled with a lack of spectral evidence for products **376a–f**, the work reported in Scheme 86 warrants reinvestigation.

6-Methyl-4-oxoquinazolinyl-2-thiolacetic acid (**378**), obtained by the reaction of 6-methyl-4-oxoquinazoline-2-thione (**377**) and chloroacetic acid, upon treatment with acetic anhydride–pyridine, yields a cyclized product confirmed by the molecular ion peak $[M]^{\cdot+}$ at m/z 232 and for which structure 7-methyl-2*H*-thiazolo[3,2-*a*]quinazolin-1,5-dione (**379**) was assigned on the basis of IR and ^{1}H-NMR data in preference to possible

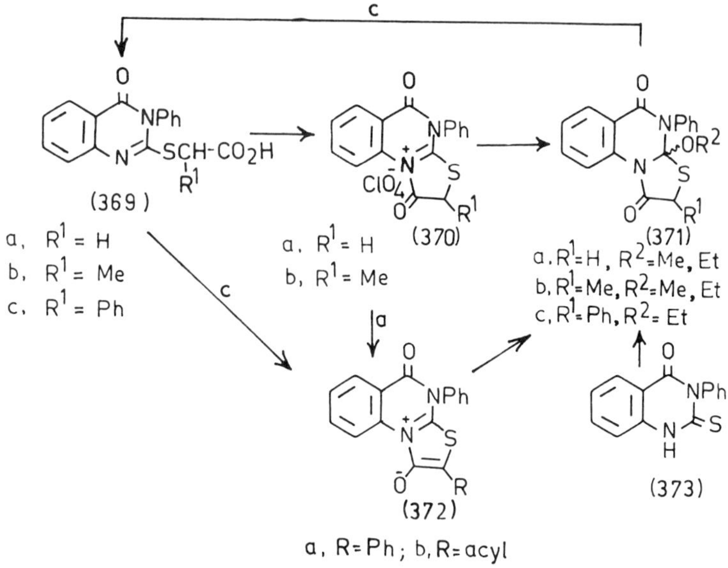

SCHEME 85

a, R=Ph; b, R=acyl

alternate isomer **380** (82TH1) (Scheme 87). The absorption of two carbonyl groups at low frequencies (1665 and 1760 cm^{-1} indicates the absence of the —CO—N—CO group (68JHC179) in the cyclized product, supporting structure **379**. Because of the equivalence of the protons (H$_B$ and H$_C$) ortho to the methyl group and the nonequivalence of proton H$_A$ in **379** and

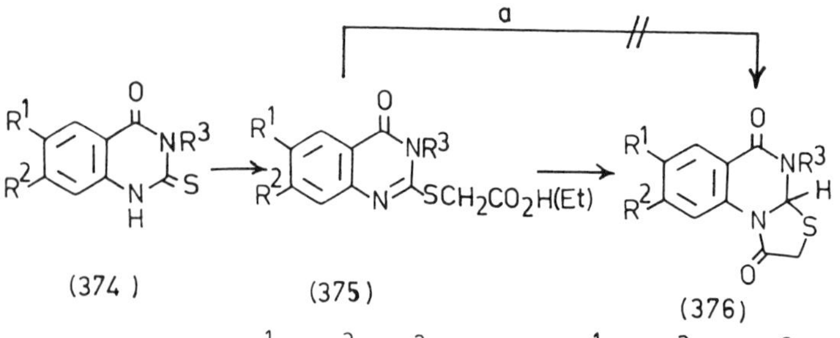

	R^1	R^2	R^3		R^1	R^2	R^3
a,	H	H	Ph	d,	OMe	o-alkyl	Ph
b,	OH	OMe	Ph	e,	F	H	aryl
c,	OMe	OMe	Ph	f,	H	F	aryl

SCHEME 86

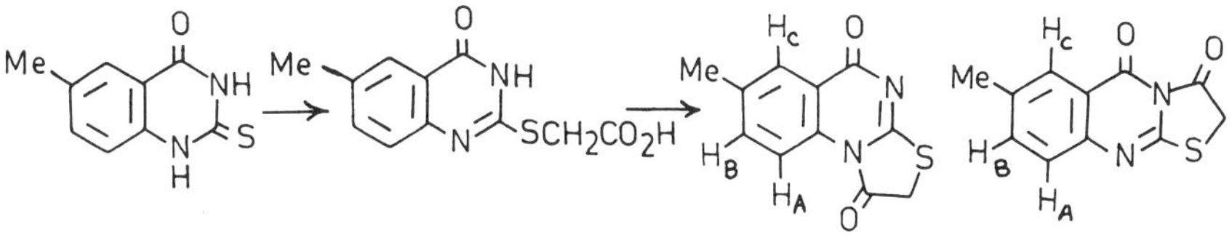

Sec. II.A] CONDENSED 4-THIAZOLIDINONES 69

(377)   (378)   (379)   (380)

SCHEME 87

380, the two structures could be distinguished by the NMR signal exhibited by H_A. The most downfield signal at δ 8.96 (exhibiting ortho coupling, J_{AB} = 9 Hz) is assigned to H_A in **379** because of the anisotropic effect of the carbonyl group of the thiazolidinone ring. Had **380** been the correct structure for the cyclized product, the most downfield signal would have been due to H_C exhibiting meta splitting.

ii. *From a thiazolidinone.* Reaction of ethyl 5-chloro-N-(2-chloroacetyl) anthranilate (**381**) with potassium thiocyanate in ethanol gives ethyl(6-chloro-1,4-dihydro-4-oxoquinazolinyl)-2-thiolacetate (**385**) in good yield. The formation of **385** was explained via the first intermediate **382** which had undergone one cyclization to give **383** followed by a second cyclization to furnish 7-chloro-2H-thiazolo[3,2-a]quinazolin-1,5-dione (**384**). The thiazolidinone ring in **384** was, however, unstable in the alkaline reaction mixture (potassium thiocyanate of excess molar ratio was used) and reacted with ethanol to give **385**. Ester **385**, on alkaline hydrolysis, gives acid **347b**, which undergoes cyclization upon heating in acetic anhydride to afford a cyclized product for which structure **384** was assigned (68JHC185) (Scheme 88). The structural assignment of the cyclized product as **384** in preference to **344b** was made because the conversion of **381** to **385** via intermediate **384** (not isolated at this stage) settles the orientation of ring closure in the formation of **384** from acid **347b**. However, the reasoning advanced by the investigators (68JHC185) is not tenable simply because the synthesis of **384** from **381** is unequivocal, while that of **384** from **347b** is not, in which case, the possibility of getting isomer **344b** also exists. Moreover, acid **347b**, on cyclization with acetic anhydride in the absence of pyridine, gives **384** (68JHC185), while with acetic anhydride–pyridine, it gives **344b** (83H1549). But the melting points of the acid **347b** reported by both groups of investigators are quite different [225–227°C (68JHC185) and 160°C (83H1549)]. No NMR data of the cyclized products (**344b** and **384**) are cited in either publication, hence, the structural confirmation must await stronger evidence.

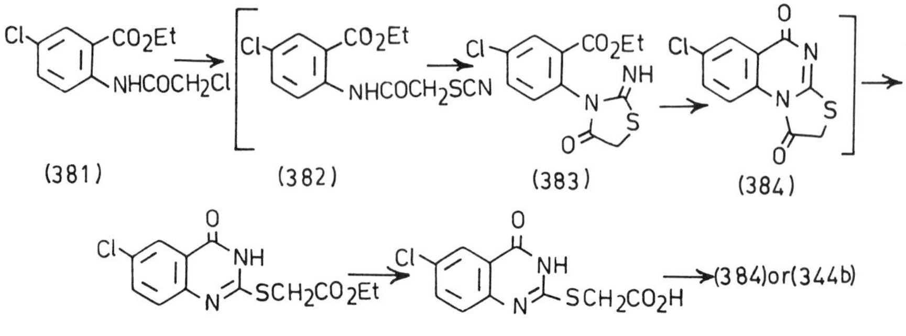

(381) (382) (383) (384) (385) (347b)

SCHEME 88

Reaction of *N*-(2-chloroacetyl)anthranilic acid (**386**) with thiourea in ethanol at reflux temperature gives 2*H*-thiazolo[3,2-*a*]quinazolin-1,5-dione (**336**) via intermediates **387** and **388** (not isolated) as formulated in Scheme 88 (69IJC881). Unlike the case of **384**, the thiazolidinone ring in **336** did not undergo cleavage because the reaction medium was not alkaline (Scheme 89).

38. *Thiazolo[2,3-*a*]isoquinolin-3(2*H*)-ones*

Of two possible thiazoloisoquinolinones, only thiazolo[2,3-*a*]isoquinolin-3(2*H*)-one (**389**) has been synthesized. The synthesis of thiazolo[3,2-*b*]isoquinolin-1(2*H*)-ones (**390**) is yet to be achieved.

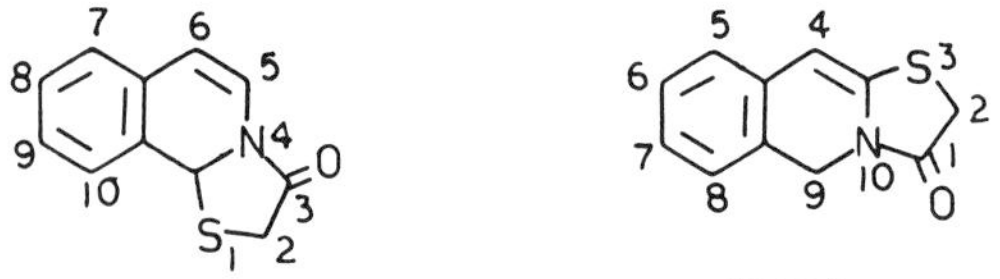

(389) (390)

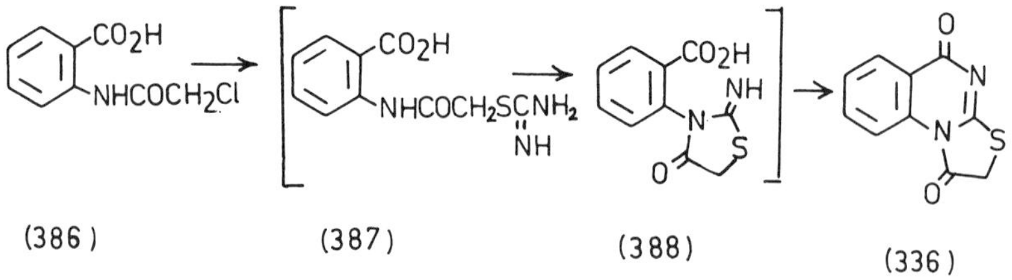

(386) (387) (388) (336)

SCHEME 89

Sec. II.A] CONDENSED 4-THIAZOLIDINONES 71

The facile addition of mercaptoacetic acid across the double bond of an azomethine for a one-step synthesis of 4-thiazolidinone has also been exploited in the synthesis of thiazolo[2,3-*a*]isoquinolin-3(2*H*)-ones (**392**). Thus, 3,4-dihydroisoquinolines (**391a–d**), on heating with mercaptoacetic acid in benzene or toluene for several hours (11–50 hr) using a Dean–Stark water separator, give **392a–c** (67IJC221) and **392d** (68TH1), respectively. Dihydroisoquinoline (**391a**) with mercaptoacetic acid in refluxing ethanol gives a second product **393** in addition to **392a** (67IJC221). The use of the methyl ester instead of the acid considerably shortens the reaction period. Structures **391e,f** react with methyl mercaptoacetate in refluxing ligroin for 4 hr to furnish the desired thiazolo[2,3-*a*]isoquinolin-3(2*H*)-ones (**392e,f**) (66AP846) (Scheme 90).

39. *Naphthimidazo[2,1-b]thiazol-3(2H)-ones*

The synthesis of naphthimidazo[2,1-*b*]thiazol-3(2*H*)-ones, in which a 4-thiazolidinone nucleus has fused to both naphth[2,3-*d*]imidazole and naphth[1,2-*d*]imidazole ring systems, has been reported.

a. *Naphth[2',3':4,5]imidazo[2,1-b]thiazol-3(2H)-ones.* Naphth[2',3': 4,5]-imidazo[2,1-*b*]thiazol-3(2*H*)-one (**396**) and its 6,7,8,9-tetrahydro derivative have been synthesized in a straightforward manner. The reaction of 1*H*-naphth[2,3-*d*]imidazolyl-2-thione (**394**) with chloroacetic acid and subsequent cyclization of the intermediate acid **395** with acetic anhydride–pyridine gives **396** (58JCS1974). Similarly, the 6,7,8,9-tetrahydro derivative of **396** is obtained from 5,6,7,8-tetrahydro-1*H*-naphtho-[2,3-*d*]imidazole through the same series of reactions (59JCS3332) (Scheme 91).

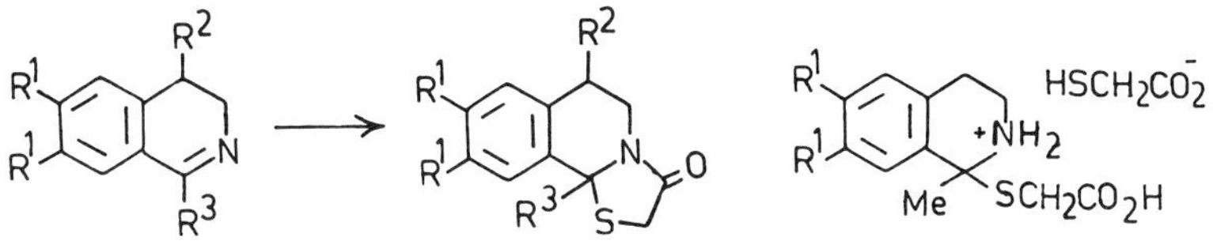

(391)   (392)   (393)

	R^1	R^2	R^3			R^1	R^2	R^3
a,	H	H	Me		d,	H	Me	Me
b,	OMe	H	Me		e,	OMe	H	H
c,	OMe	H	Ph		f,	H	H	H

SCHEME 90

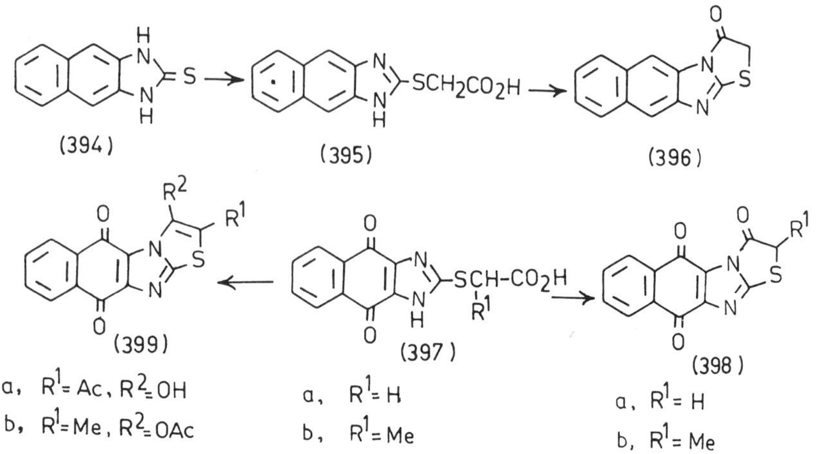

SCHEME 91

1*H*-4,9-Dioxonaphth[2,3-*d*]imidazolyl-2-thiolacetic acid (**397a**) and α-(1-*H*-4,9-dioxonaphth[2,3-*d*]imidazolyl-2-thiol)propionic acid (**397b**) with acetic anhydride in the absence of pyridine undergo cyclodehydration to give condensed thiazolidinones (**398a,b**), respectively. With acetic anhydride–pyridine, these acids (**397a,b**) yield the corresponding condensed thiazoles (**399a,b**) which have an acetyl group (82NKK456) (Scheme 91).

b. *Naphth[1',2':4,5]imidazo[2,1-b]thiazol-3(2H)-one*. Acid **401** or its methyl ester, obtained from the reaction of 1*H*-naphth[1,2-*d*]imidazolyl-2-thione (**400**) and chloroacetic acid or ethyl chloroacetate, on cyclization with acetic anhydride, gave a product for which the structure naphth[1',2':4,5]imidazo[2,1-*b*]thiazol-3(2*H*)-one (**402**) was assigned (72KGS399) (Scheme 92). The cyclized product could also be accommodated by isomeric structure **403**. No adequate data were cited to decide in favor of either structure.

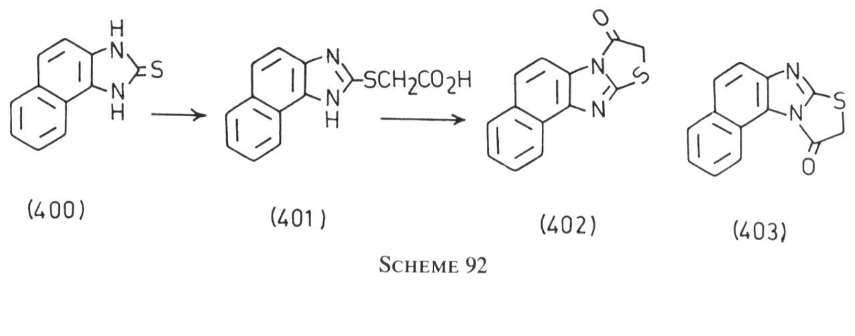

SCHEME 92

40. Thiazolo[3',2':1,2]imidazo[4,5-f]quinolin-10(9H)-one

Reaction of 3H-imidazo[4,5-f]quinoline-2-thione (**404**) with chloroacetic acid gives acid **405** which, on treatment with a mixture of acetic anhydride and pyridine, undergoes cyclization to furnish a product confirmed by the appearance of a band at 1735 cm^{-1} (N—C=O) and molecular ion peak [M]$^{+}$ at m/z 241. Thiazolo[2',3':2,3]imidazo[4,5-f]quinolin-6(7H)-one (**406**), in preference to the other possible isomer thiazolo[3',2':1,2]-imidazo[4,5-f]quinolin-8(7H)-one (**409**), was assigned to the cyclized product on the basis of the comparative studies of proton signals of the cyclized product (**406** or **409**) with those of **407** or **408** (obtained by the reaction of **404** and 1,2-dibromoethane), as well as with the proton signals of acid **405** [86IJC(B)264] (Scheme 93).

Reaction of **404** with 1,2-dibromoethane gives a single compound that may be represented by either structure **407** or **408**. In either structure the most downfield signal at δ 9.61 is due to H$_A$, and the signal at δ 10.62 (a value sufficiently higher than 9.61) is due to the deshielding effect of the carbonyl group on the peri-proton (H$_A$). The cyclized product obtained from acid **405** is in conformity with structure **406**. Such a downfield signal for a peri-proton cannot be expected from **409**. Had structure **409** been correct, the proton H$_E$ would have been deshielded by the carbonyl group by ~δ 0.45, a value obtained from the thiazolo[3,2-a]benzimidazol-3(2H)-one system [87IJC(B)532]. In the cyclized product obtained from acid **405**, H$_E$ resonates at δ 8.56, almost the same region as in **407** (or **408**). Exactly

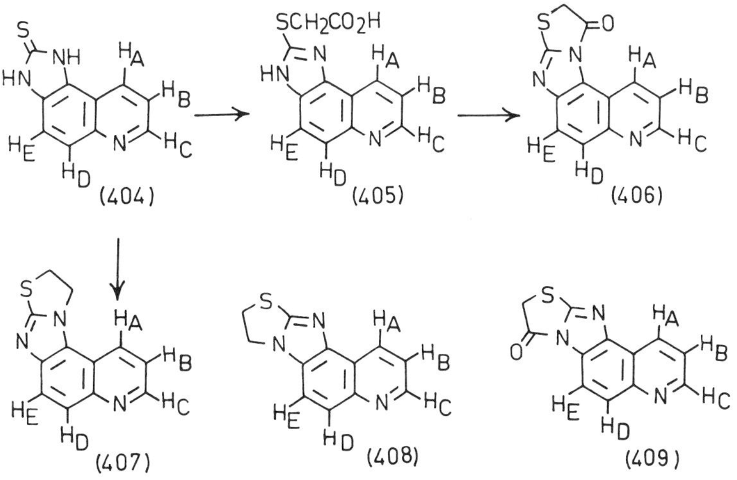

SCHEME 93

the same conclusion was also drawn by the comparison of the chemical shifts of the respective aromatic protons in acid **405** and its cyclized product **406** or **409**. In acid **405**, the signals at δ 9.82, 8.45, 9.37, 8.70, and 8.59 were assigned to H_A, H_B, H_C, H_D, and H_E protons, respectively. The signals at δ 10.62 (in place of δ 9.82 obtained in the acid) in the cyclized product must be due to H_A since this proton is deshielded by a carbonyl group, supporting structure **406**.

41. *Thiazolo[3,2-a]perimidin-3(2H)-ones*

Reaction of perimidine-2-thione (**410**) with chloroacetic acid gives perimidine-2-thiolacetic acid (**411**) which, on treatment with acetic anhydride–pyridine undergoes cyclodehydration to give thiazolo[3,2-*a*]-perimidin-3(2*H*)-one (**412**) (69IJC767), which was confirmed by IR absorption at 1710 cm^{-1} (N − C = O). Structure **412** is also obtained in one step by the reaction of **410** with ethyl chloroacetate (79AP776). Thione **410** reacts with DMAD and oxalyl chloride to afford 2-methoxycarbonylmethylidenethiazolo[3,2-*a*]perimidin-3(2*H*)-one (**413**) (76AP928) and 2*H*-thiazolo[3,2-*a*]perimidin-3,5-dione (**414**) (79AP776), respectively (Scheme 94). In view of the conflicting findings reported from the reaction of cyclic thiones with DMAD as discussed earlier, cyclized product **413** obtained from **410** and DMAD should be subjected to ^{13}C-NMR spectral analysis for structural confirmation.

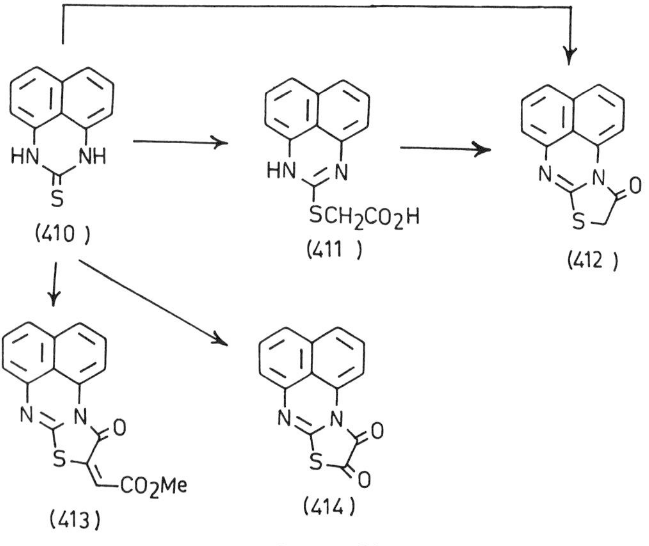

SCHEME 94

42. Thiazolo[2,3-a]benzo(f)isoquinolin-17(16H)-ones

The synthesis of steroidal hormone analogs containing a thiazolidinone ring (418) has been achieved in a simple and elegant manner starting from amine as depicted in Scheme 95 [69IJC735; 71JCS(C)266]. The reaction of amines 415a,b with ethyl formate gives amides 416a,b which, on Bischler–Naperalski cyclization, afford dihydrobenzo(f)isoquinolines (417a,b). Condensation of bases 417a,b with mercaptoacetic acid gives the desired (±)14H-11,12-dihydrothiazolo[2,3-c]benzo(f)isoquinolin-17(16H)-one (418a) and its 3-methoxy analog (418b), respectively (steroid numbering followed). This sequence of reactions represents the first total synthesis of a heterosteroid incorporating a 4-thiazolidinone nucleus. Demethylation of 418b with molten pyridine hydrochloride furnishes 419. The stereochemistry of heterosteroid 418b has been secured through ^{1}H-NMR spectral studies (see Section III, C and III, D).

43. Thiazolo[3,2-a]aceperimidin-3(2H)-one

Aceperimidine-2-thione (420) reacts with chloroacetic acid in ethanolic potash to furnish directly thiazolo[3,2-a]aceperimidin-3(2H)-one (422) (confirmed by the appearance of lactam carbonyl absorption at 1720 cm^{-1} and by the molecular ion peak [M]$^{·+}$ at m/z 266) instead of the uncyclized acid (421) which is obviously the intermediate in the reaction [87IJC(B)693] (Scheme 96).

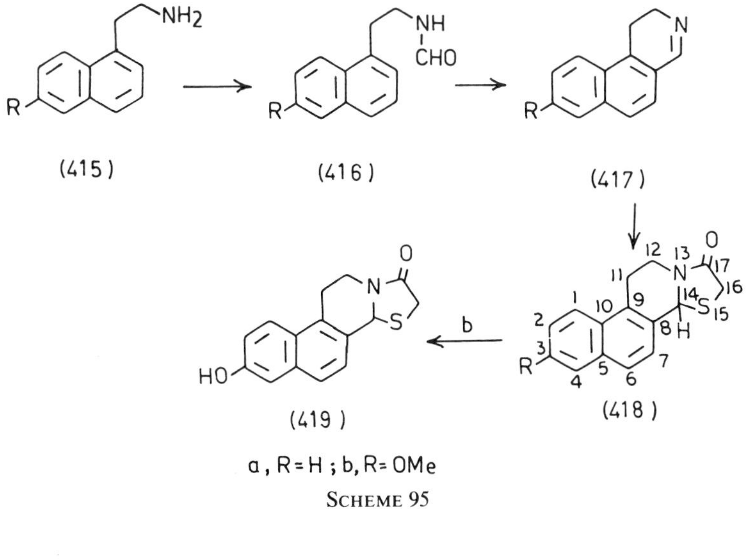

a, R = H ; b, R = OMe

SCHEME 95

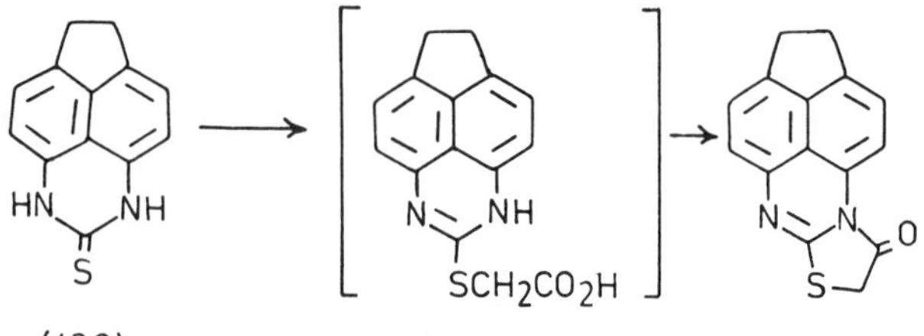

(420) (421) (422)

SCHEME 96

44. *Thiazolo[2',3'-*b*]benzofuro[2,3-*f*]benzimidazol-3(2*H*)-one*

Reaction of benzofuro[2,3-*f*]benzimidazole-2-thione (**423**) with chloroacetic acid gives acid **424** which, on cyclization with acetic anhydride, yields a cyclized product for which structure thiazolo[2',3'-*b*]benzofuro [2,3-*f*]benzimidazol-3(2*H*)-one (**425**) was assigned (77MI2); however, alternate structure **426** cannot be ruled out in the absence of adequate data (Scheme 97).

B. Cyclization Procedures Involving the 2- and 5-Positions of 4-Thiazolidinone

Simple and elegant syntheses of five different heterocyclic systems, namely, hexahydro-1,4-epithiopyrimidin-3-ones, tetrahydro-1,4-epithioazepin-3,7-diones, hexahydro-1,4-epithiopyridin-3-ones, tetrahydro-4,7-

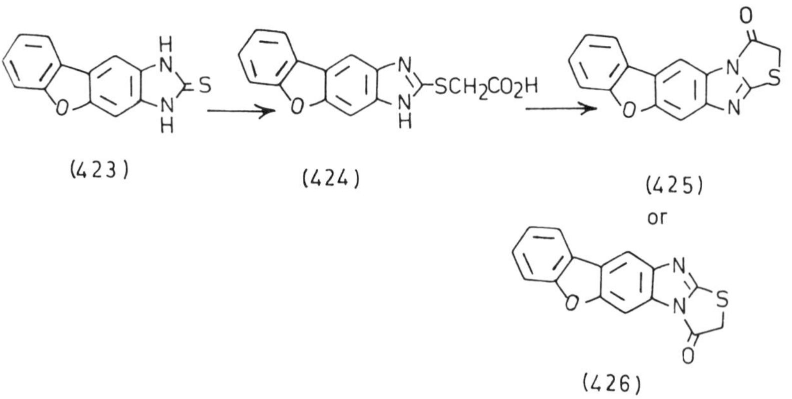

(423) (424) (425) or (426)

SCHEME 97

epithiofuro[3,4-*c*]pyridin-5-ones and tetrahydro-4,7-epithiopyrrolo[3,4-*c*]pyridin-5-ones, have been reported from Potts' laboratory to occur through a cyclization procedure involving the 2- and 5-positions of 4-thiazolidinone (**1**) (74JOC3631; 76JOC813, 76JOC818).

1. *Hexahydro-1,4-epithiopyrimidin-3-ones*

Reaction of *anhydro*-2-aryl-4-hydroxy-3-phenylthiazolium hydroxide (**427**) with isothiocyanates or isocyanates in hot benzene gives a crystalline cycloadduct for which the structure hexahydro-1,4-epithiopyrimidin-3-oxo-5-thiones/ones (**428**) was assigned on the basis of spectral data (76JOC813) in preference to hexahydro-1,4-pithiopyrazin-3-oxo-6-thiones/ones (**429**) and ylide structure (**430**) (Scheme 98). The IR spectrum of the cycloadduct is devoid of OH, SH, or NH absorption but shows the carbonyl and thiocarbonyl absorption. The ^{1}H-NMR spectrum of the cycloadduct, obtained from **427a** and phenyl isothiocyanate, is very simple. It consists of 15 aromatic protons at ∂ 7.63 and a singlet proton at ∂ 12.06; these are exceptionally low field. In the cycloadduct from **427a** and phenyl

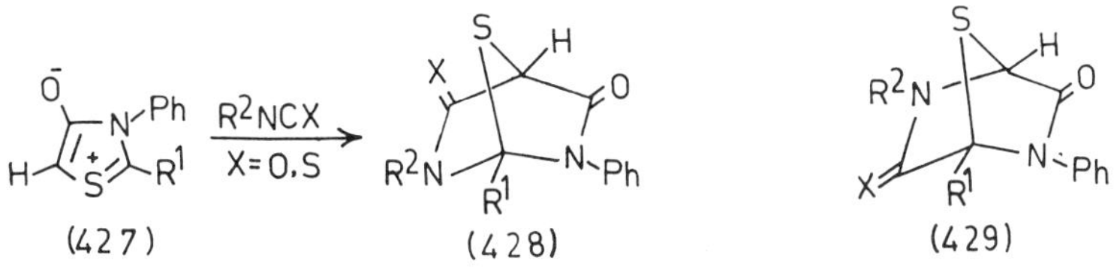

a, R1 = Ph
b, R1 = p-ClC6H4

	R^1	R^2	X		R^1	R^2	X
a,	Ph	Ph	S	f,	p-ClC$_6$H$_4$	Ac	S
b,	Ph	Ph	O	g,	p-ClC$_6$H$_4$	EtSO$_3$	O
c,	p-ClC$_6$H$_4$	PhCO	O	h,	p-ClC$_6$H$_4$	p-ClC$_6$H$_4$CO	O
d,	p-ClC$_6$H$_4$	p-MeOC$_6$H$_4$SO$_2$	O	i,	p-ClC$_6$H$_4$	H	O
e,	p-ClC$_6$H$_4$	Cl$_3$CCO	O				

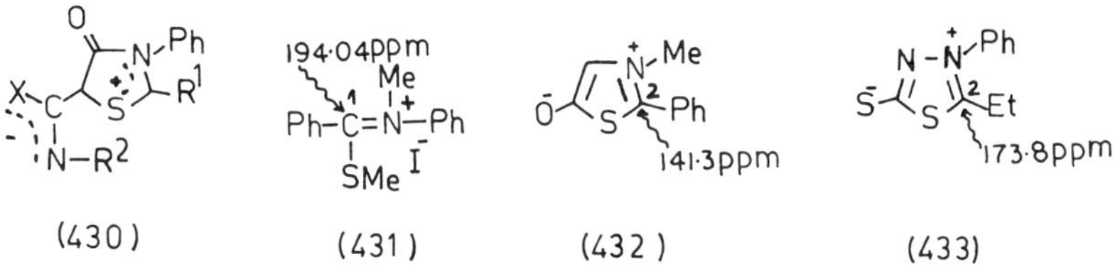

SCHEME 98

isocyanate, the singlet proton moved upward to ∂ 10.27, albeit still a low value. Structure **428** was proposed in which the H-4 is strongly deshielded by both C=S and C=O groups at the 3- and 5-positions in **428a** and by both C=O groups at the 3- and 5-positions in **428b**. Thus, **428** resonates at very low field. Since the C=S group is known to exert a stronger deshielding effect than the C=O group [68JCS(C)1777; 71JOC1846], such a downward shift is tenable with structure **428a**. The alternate structure **429**, representing a different mode of addition of isocyanate (or thiocyanate) to **427**, is discarded by the exceptionally low chemical shift of H-4. Again the upfield shift of H-4 from **428a** and **428b** is not expected since the C=O or C=S group at the 6-position in **429** has no deshielding influence on the H-4 proton.

Ylide structure **430**, which cannot be ruled out by IR and ^{1}H-NMR data was, however, eliminated on the basis of chemical reaction and ^{13}C-NMR data. The adducts obtained from **427** and isocyanates (isothiocyanates) are remarkably inert to protonation with perchloric acid, and alkylation with suitable reagents. This observation is in contrast to the general behavior of an ylide structure [71JCS(B)1648], hence assigning structure **430** for the cycloadduct is ruled out. Comparison of the corresponding carbon chemical shifts in compounds containing C=N$^+$, an essential feature of an ylide structure such as *S*-methylthioamide derivative **431** (C-1 resonates at 194.04 ppm), *anhydro*-5-hydroxy-3-methyl-2-phenylthiazolium hydroxide (**432**, C-2 resonates at 141.3 ppm), and *anhydro*-2-ethyl-5-mercapto-3-phenyl-1,3,4-thiadiazolium hydroxide (**433**, C-2 resonates at 173.8 ppm) with the adduct **428f** (C-1 resonates at 125.78 ppm), obtained from **427b** and acetyl isothiocyanate, rules out ylide structure **430**. Similarly, hexahydro-1,4-epithiopyrimidin-3,5-diones (**428b–e, g–i**) are synthesized from **427** and appropriate isocyanates.

2. *Tetrahydro-1,4-epithioazepin-3,7-diones*

The [3 + 3] cycloaddition of the three-carbon species of cyclopropene to the mesoionic rings offers a method for synthesizing six-membered or larger ring systems through possible thermal ring expansion. Thus *anhydro*-2,3-diphenyl-4-hydroxythiazolium hydroxide (**434a**), a masked thiocarbonyl ylide 1,3-dipole, reacts with diphenylcyclopropenone in refluxing benzene to give a stable 1:1 adduct established by the appearance of a molecular ion peak [M]$^{\cdot+}$ in the mass spectrum of the adduct. The adduct can be represented by any one of the structures **435–438** formed by different modes of addition of diphenylcyclopropenone to **434** (76JOC818) (Scheme 99). A carbonyl absorption at 1650 and 1725 cm^{-1} (both C=O groups) in the adduct, obtained from **434a** and diphenylcyclopropenone,

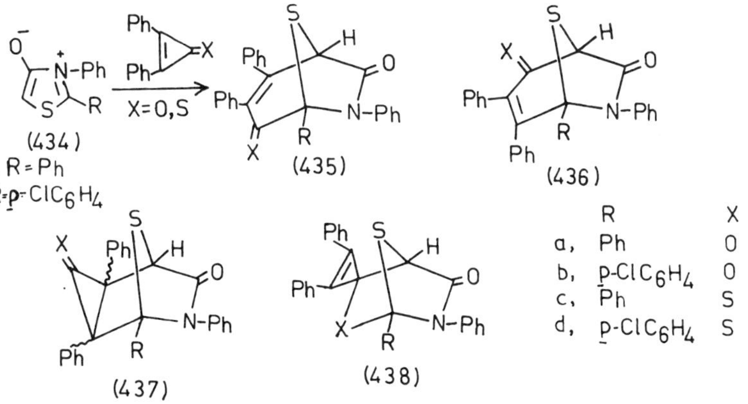

SCHEME 99

obviously rules out structures **437a** and **438a** but supports two other structures, **435a** and **436a**. The structural assignment of the former in preference to the latter for the adduct was based on ^{1}H-NMR data. A singlet proton at ∂ 6.1 in the adduct, obtained by the reaction of **434a** with diphenylcyclopropenone, is compatible with the structure of tetrahydro-1,2,5,6-tetraphenyl-1,4-epithioazepin-3,7-dione (**435a**) rather than **436a** because this bridgehead proton (H-4), located in the deshielding zones of the two flanking carbonyl groups in the latter, would have resonated much further downfield. Although there is a change in the geometry of the 1,4-epithioazepine system (**435a**) compared to the 1,4-epithiopyrimidine system (**428a**) (76JOC813), it is insufficient to cause a chemical shift change from δ 10.27 to δ 6.10. Similarly **435b–d** are obtained from the cycloaddition of diphenylcyclopropenone and its thione analog with **434**.

3. *Hexahydro-1,4-epithiopyridin-3-ones*

Facile reaction of electron-deficient olefins (dipolarophiles) such as *trans*-dibenzoylethylene, dimethyl fumarate, furanonitrile, methyl vinyl ketone, ethyl crotonate, ethyl acrylate, ethyl methacrylate, and dimethyl maleate with a mesoionic compound containing a masked thiocarbonyl ylide skeleton gives stable 1 : 1 cycloadducts. The structure of the cycloadduct was established by its carbonyl absorption in IR spectra and the molecular ion peak [M]$^{·+}$. The stereochemistry of the cycloadducts was, however, secured by NMR spectra (74JOC3631) (Scheme 100).

Reaction of *anhydro*-2-*p*-chlorophenyl-4-hydroxy-3-phenylthiazolium hydroxide (**439a**) and *anhydro*-4-hydroxy-2,3,5-triphenylthiazolium

(439)			(440)				(441)	
R^1	R^2		R^1	R^2	R^3	R^4	R^5	R^6
a, p-ClC$_6$H$_4$	H	a,	p-ClC$_6$H$_4$	H	PhCO	H	H	PhCO
b, Ph	Ph	b,	Ph	Ph	PhCO	H	H	PhCO
c, Ph	H	c,	p-ClC$_6$H$_4$	H	H	H	H	Ac
		d,	p-ClC$_6$H$_4$	H	H	Ac	H	H
		e,	Ph	Ph	H	H	H	Ac
		f,	p-ClC$_6$H$_4$	H	H	H	H	CO$_2$Et
		g,	p-ClC$_6$H$_4$	H	CO$_2$Et	H	H	H
		h,	Ph	Ph	Me	H	H	CO$_2$Et
		i,	Ph	Ph	H	H	Me	CO$_2$Et
		j,	p-ClC$_6$H$_4$	H	CO$_2$Me	H	H	CO$_2$Me
		k,	Ph	Ph	CO$_2$Me	H	H	CO$_2$Me
		l,	Ph	Ph	CO$_2$Me	H	H	CO$_2$Me
		m,	p-ClC$_6$H$_4$	H	H	CO$_2$Me	H	CO$_2$Me
		n,	Ph	Ph	H	CO$_2$Me	H	CO$_2$Me
		o,	Ph	H	H	CO$_2$Me	H	CO$_2$Me
		p,	Ph	Ph	CN	H	H	CN

SCHEME 100

hydroxide (**439b**) with *trans*-dibenzoylethylene in hot benzene gives stable 1:1 cycloadducts to which structure 1-*p*-chlorophenyl-5α,6β-dibenzoyl-2-phenyl-1,2,3,4,5,6-hexahydro-1α,4α-epithiopyridin-3-one (**440a**) and 5α,6β-dibenzoyl-1,2,4-triphenyl-1,2,3,4,5,6-hexahydro-1α,4α-epithiopyridin-3-one (**440b**) were assigned rather than structures 1-*p*-chlorophenyl-5β,6α-dibenzoyl-2-phenyl-1,2,3,4,5,6-hexahydro-1α,4α-epithiopyridin-3-one (**441a**) and 5β,6α-dibenzoyl-1,2,4-triphenyl-1,2,3,4,5,6-hexahydro-1α,4α-epithio-pyridin-3-one (**441b**), respectively.

Reaction of **439a** with methyl vinyl ketone yields a single product for which structure **440c** was assigned from a group of four possible structures, namely, **440c**, **441c**, **440d**, and **441d**, that can be obtained depending on the mode of addition. However, reaction of **439b** with methyl vinyl ketone forms two isomers **440e** and **441e**. Structure **440f** was assigned to the cycloadduct obtained from **439a** and ethyl acrylate in preference to other alternate structures **441f, 440g,** and **441g**. Reaction of **439b** with ethyl crotonate and with ethyl methacrylate gives **440h,i** rather than **441h,i**. Condensation of **439a–c** with dimethyl fumerate gives 5-exo-6-endo products **440j–l,** whereas with dimethylmaleate, endo-products **440m,o** are obtained.

The addition of furanonitrile to **439b** in refluxing benzene for 89 hr did not yield 1:1 cycloadduct **440p,** but gave the sulfur-free compound, 4,5-dicyano-1,3,6-triphenylpyridin-2-one (**443f**) which is presumably formed by eliminating H_2S from **440p**. However, when the reaction was carried out for a shorter time (26 hr), cycloadduct **440f** in 70% yield and the sulfur-free compound **443f** were both isolated.

The stereochemistry of the cycloadducts was secured firmly by the chemical shifts, multiplicity, and coupling constant values shown by the protons at the 4-, 5-, and 6-positions of the adducts. The ^{1}H-NMR spectra of all the adducts except **440i** showed normal patterns of splitting. Cycloadducts **440h** and **440i,** obtained from the reaction of **439b** with ethyl crotonate and ethyl methacrylate, respectively, differ from one another only in the position of the methyl substituent. This small structural difference, however, results in a significant and interesting change in their NMR spectra. In **440h,** a normal quartet–triplet pattern for the ethoxy group was observed, while in **440i** the ethoxy carbonyl methyl protons resonated as a triplet; The methylene protons, however, did not appear as a normal quartet, rather, they appeared as a complex 14-line pattern. In this cycloadduct (**440i**), the chiral center at C-6, the bulkiness of methyl group at C-6, and the inherent rigidity possessed by the bicyclic system prevent free rotation about O—CH_2 bond, thus behaveing like an ABX_3 pattern. Overlapping resonance reduces to an observable 14-line multiplet, although theoretically a 16-line multiplet is expected.

Reaction of mesoionic compound **439a–d** with acetylenic compounds did not yield the desired stable 1:1 cycloadducts but gave pyridine and thiophene derivatives (Scheme 101). Thus, reaction of **439a,c,d** with DMAD gives corresponding pyridines **443a,c,d,** presumably obtained via cycloadducts **442a,c,d** after elimination of elemental sulfur (74JOC3631). The reaction of the trisubstituted mesoionic compound **439b** with DMAD (74JOC3631), dibenzoylacetylene (74JOC3631), and acetylene dinitriles (74JOC3627), however, yields in each case two products: pyridones **443b,e,f** and thiophenes **444a–c** via unstable cycloadducts **442b,e,f**. The former is obtained after splitting off elemental sulfur, while the latter is obtained eliminating phenylisocyanate from the cycloadducts (Scheme 101).

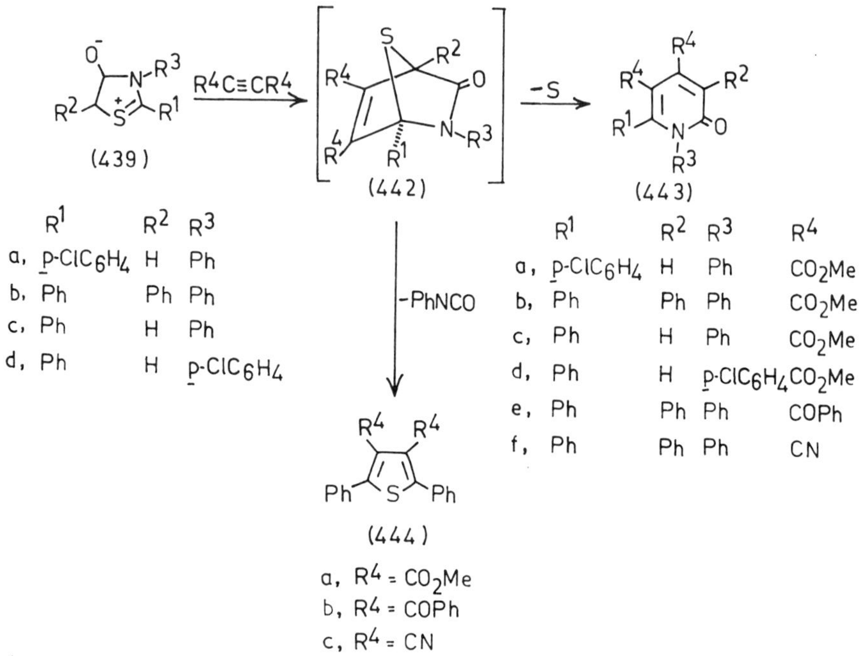

SCHEME 101

4. Tetrahydro-4,7-epithiofuro[3,4-c]pyridin-5-ones

Addition of **439a,b** to maleic anhydride at room temperature in dry benzene affords cycloadducts 3aα-H,7aα-H-7-p-chlorophenyl-6-phenyl-tetrahydro-4α,7α-epithiofuro[3,4-c]pyridin-1,3,5-trione (**445a**) and 3aα-H,7aα-H-tetrahydro-4,6,7-triphenyl-4α,7α-epithiofuro[3,4,-c]pyridin-1,3,5-trione (**445b**), respectively (74JOC3631) (Scheme 102). The endo structure (**445**), in preference to the exo structure (**446**), was assigned to the cycloadducts on the basis of a small trans coupling ($J = 1.5$ Hz) shown by H-4 in **445a**. The adduct **445a** undergoes methanolysis with diazomethane to give the cis-diester **440m**. On the other hand, **445b** with diazomethane yields both cis-isomer **440n** and trans-isomer **440k**, the trans-isomer being the minor product. The formation of trans-diester **440k** may be ascribed to the reaction of cis-diester **440n** with an excess of diazomethane which gives rise to methoxide ion and then epimerises C-3a of the cycloadduct **445b**. The formation of the cis-diesters (**440m** and **440n**) corroborates a consistency in stereochemical configuration that exists within the cycloadducts obtained from the mesoionic system.

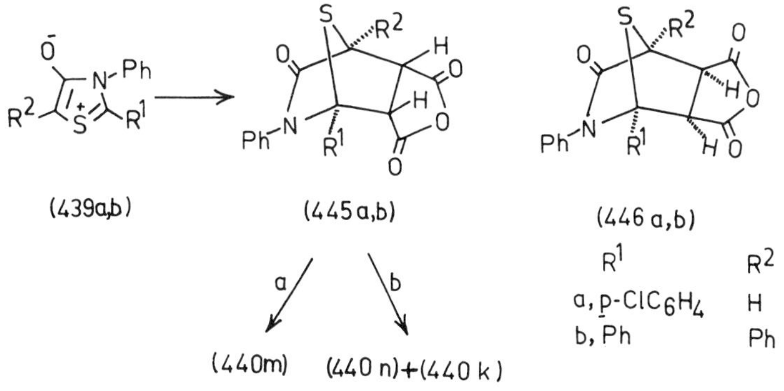

5. Tetrahydro-4,7-epithiopyrrolo[3,4-c]pyridin-5-ones

Reaction of **439a–c** with N-phenylmaleimide in refluxing benzene furnishes stable 1:1 cycloadducts which may be represented by either structures **447a–c** or **448a–c** (74JOC3631) (Scheme 103). The endo-structure, 4aα-H,7aα-H-2,6-diphenyl-7-R^1-4-R^2-tetrahydro-4α,7α-epithiopyrrolo[3,4-c]pyridin-1,3,5-trione (**447a–c**) was preferred to the alternate exo-structure **448a–c** on analogy to the similar cycloadducts obtained from the reaction of 1,3-dihydro-benzo[c]thiophene-2-oxide (**449**)

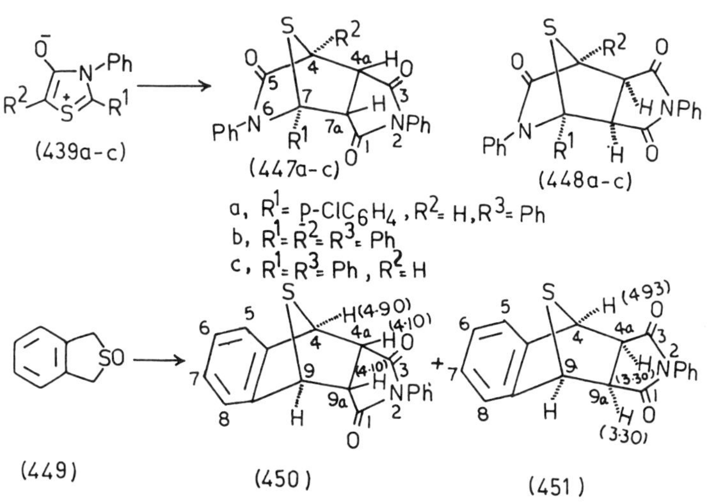

Scheme 103

with *N*-phenylmaleimide in which a mixture of endo- (**450**) and exo- (**451**) isomers was obtained (66JA4112). The ^{1}H-NMR spectrum of the cycloadduct obtained from **439a** and *N*-phenylmaleimide shows two doublets at δ 4.40 (7a-H) and δ 4.70 (4-H), and one doublet of doublet at δ 4.05 (4a-H). These doublets are closer to the chemical shifts of the corresponding protons of the endo-isomer (**450**) than that of the exo-isomer (**451**), which supports structure **447**.

C. Cyclization Procedures Involving the 2,3- and 2,5-Positions of 4-Thiazolidinone

Syntheses of different heterocyclic systems in which the 2,3- and 2,5-positions of 4-thiazolidinones (**1**) are attached to different rings have also come from Potts' laboratory.

1. 6,8a-Epithioimidazo[1,2-a]pyrimidin-5,7-diones

Reaction of *anhydro*-3-hydroxy-7-methyl-2-phenylimidazo[2,1-*b*]thiazolium hydroxide (**452**) with isocyanates in boiling benzene readily provides the stable 1:1 cycloadducts, hexahydro-1-methyl-6-phenyl-8-R-6,8a-epithioimidazo[1,2-*a*]pyrimidin-5,7-diones (**453a–c**). Structures were characterized by IR, ^{1}H-NMR, ^{13}C-NMR, and mass spectral data; the latter indicated that all components of the reactants were retained in the adduct (79JOC3803) (Scheme 104). With maleimides, maleic anhydride, furanonitrile, DMAD, or dibenzoylacetylene, compound **452** did not yield the desired cycloadducts (**454, 456, 458**), but gave sulfur-free compounds **455 and 457,** presumably via the cycloadducts, with the elimination of H_2S or elemental sulfur as illustrated in Scheme 104.

2. 5,7a-Epithiopyrido[1,2-c]quinazolin-4-ones

Reaction of *anhydro*-3-hydroxy-2-phenylthiazolo[3,2-*c*]quinazolin-4-ium hydroxide (**459**) with ethyl acrylate in boiling xylene overnight gives pyrroloquinazoline (**461a**) and carbonyl sulfide. The reaction, when carried out for 5 hr in the absence of solvent (xylene), furnished a 1:1 cycloadduct, 4*H*-5,6,7,7a-tetrahydro-7-ethoxycarbonyl-5-phenyl-5,7a-epithiopyrido[1,2-*c*]quinazolin-4-one (**460a**), although the yield was only 15% (84CC213). Cycloadduct **460a** could not be converted to the rearranged product **461a** even upon prolonged heating in xylene. The reaction of **459** with dimethyl fumarate, however, gives both **460b** and **461b,** which

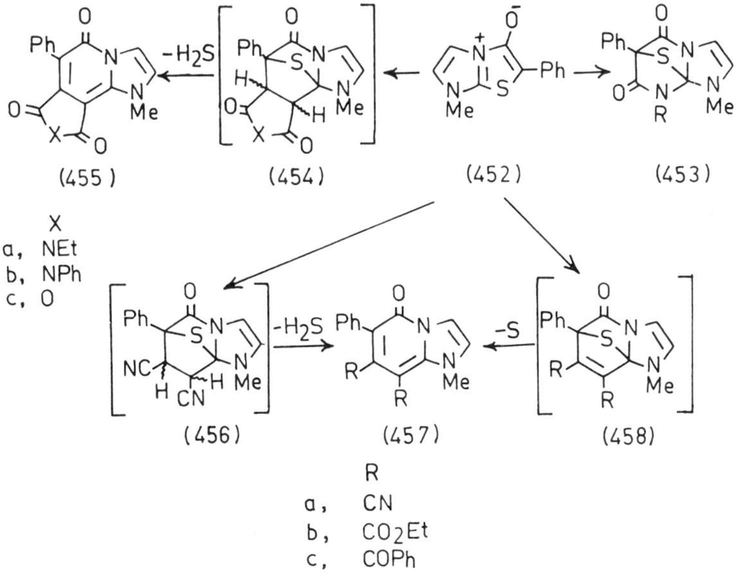

Sec. II.C] CONDENSED 4-THIAZOLIDINONES 85

SCHEME 104

	X
a,	NEt
b,	NPh
c,	O

	R
a,	CN
b,	CO₂Et
c,	COPh

were separated by high pressure liquid chromatography. Structures **460** and **461** were characterized by IR, ¹H-NMR, and mass spectra. Mesoionic ring compound **459**, on the other hand, reacts with DMAD in boiling toluene to give pyridone **462** obtained by expulsion of sulfur from the labile, which was presumably the intermediate (Scheme 105).

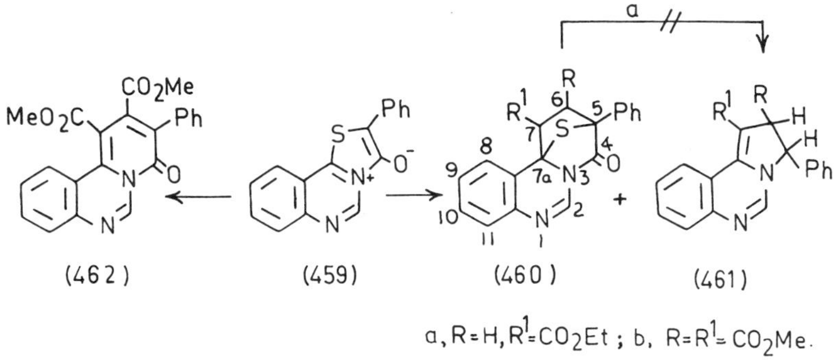

a, R=H, R¹=CO₂Et ; b, R=R¹=CO₂Me.

SCHEME 105

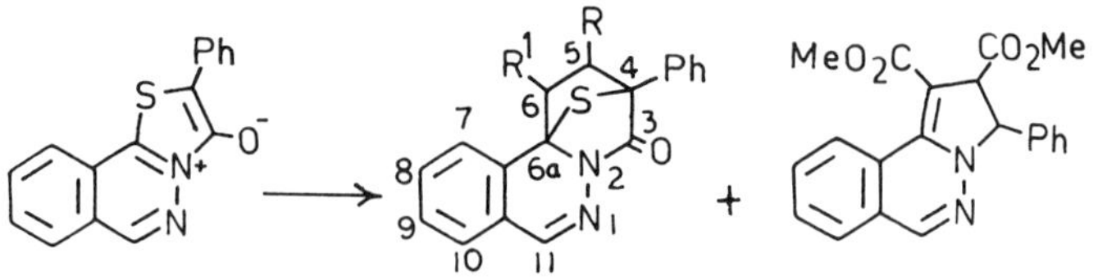

(463) (464) (465)

a, R = H, R¹ = CO₂Et
b, R = R¹ = CO₂Me

SCHEME 106

3. *4,6a-Epithiopyrido[2,1-*a*]phthalazin-3-ones*

anhydro-3-Hydroxy-2-phenylthiazolo[2,3-*a*]phthalazin-2-ium hydroxide (**463**) reacts with ethyl acrylate to give 3*H*-4,5,6,6a-tetrahydro-6-ethoxy-carbonyl-4-phenyl-4,6a-epithiopyrido[2,1-*a*]phthalizin-3-one (**464a**), while with dimethyl fumarate in boiling xylene, both the cycloadduct **464b** and the rearranged product **465** are isolated. The rearranged product (**465**) is formed by the extrusion of elemental sulfur from **464b** (84CC213) (Scheme 106).

4. *4,11a-Epithiopyrrolo[3,4-*c*]pyrido[2,1-*b*]benzothiazol-1,3,5-triones*

anhydro-3-Hydroxy-2-phenylthiazolo[2,3-*b*]benzothiazolium hydroxide (**466**), possessing a masked thiocarbonyl ylide dipole, readily reacts with maleimides in boiling toluene to give stable 1 : 1 cycloadducts **467a,b** in good yield (78JOC2697) (Scheme 107). The spectral data (IR, ^{1}H-NMR,

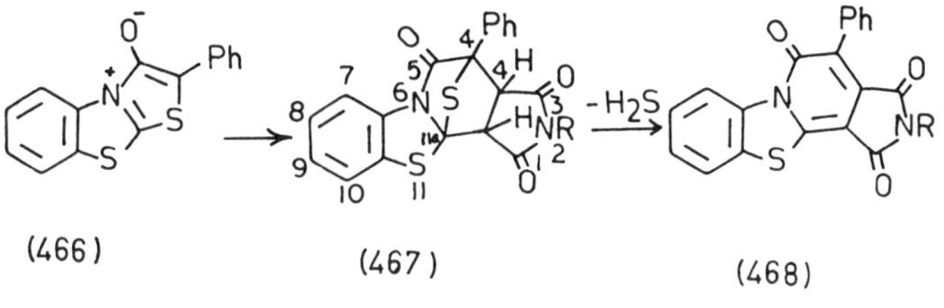

(466) (467) (468)

a, R = Et; b, R = Ph

SCHEME 107

and mass) are consistent with the assigned structures **467a,b**. The endo-configuration was assigned to the cycloadduct on analogy with cycloadduct **447**, obtained from **439** with maleimide. Complete conversion of **467** to **468** could be affected by refluxing the former in xylene.

5. *3,10a-Epithiopyrido[2,1-b]benzothiazol-4-one*

Cycloaddition of furanonitrile with **466** readily furnishes primary adduct **469** in 90% yield. Although IR, ^{1}H-NMR, and mass spectra of the adduct supported the assigned structure (**469**), the exo–endo relationship of 1,2-protons could not be established beyond doubt (78JOC2697) (Scheme 108). Unlike **467**, cycloadduct **469** is thermally stable and did not lose H_2S on refluxing in xylene for 40 hr. Even treatment of **469** with sodium methoxide, a process known to cause elimination of H_2S in the cycloadducts obtained from a monocyclic system (74JOC3619), did not afford pyridone **470**. DMAD with **466** yields pyridone **472**, presumably via cycloadduct **471** with extrusion of elemental sulfur.

The easy formations of cycloadducts **453, 460, 464, 467,** and **469** from the cycloadditon of dipolarophiles with mesoionic ring compounds are the first observed examples where addition occurs at a bridgehead carbon atom. The bicyclic adducts with ring nitrogen atoms are strained molecules, and if appropriate, these adducts attain aromatic character through the extrusion of a small molecule such as H_2S or elemental sulfur.

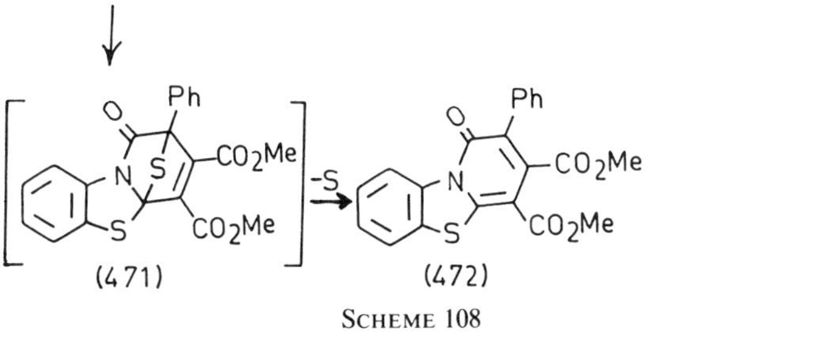

SCHEME 108

III. Molecular Spectra of Condensed 4-Thiazolidinones

A. ULTRAVIOLET SPECTRA

The UV absorption spectra of condensed 4-thiazolidinones have been reported by few workers, and critical studies have been made by fewer still [56JOC24; 64JOC1715; 74JOC3631; 76JOC813, 76JOC818; 77ZN(B)94; 78JOC2697]. The UV spectra are less useful than IR and NMR spectra for characterizing condensed 4-thiazolidinones. The product obtained from **334** and benzenediazonium chloride has been formulated as hydrazone **473** rather than azo **474** on the basis of UV spectral data [77ZN(B)94]. The UV spectrum of the product shows a maximum at 410 nm which provides evidence of a hydrazone structure since phenylazo compounds absorb strongly at 270–280 nm, while phenylhydrazones absorb above 320 nm (60JA2909; 62JA3514; 64JOC2959). On the other hand, the presence of strong bands at 382 and 400 nm in the UV spectrum of the product obtained from **249a** an aryldiazonium chloride has been cited as evidence for the azo structure **475** as opposed to hydrazone structure **476** (68KGS1008).

The critical studies of the UV absorption spectra of **478–480** are made using **477** as a reference compound in which the principal site of UV absorption resides in the α,β-unsaturated carbonyl group. The UV absorption curve of 2-arylidene-5,6-dihydroimidazo[2,1-*b*]thiazol-3(2*H*)-one (**478**) is similar to 5-arylidene-3-ethyl-2-ethylimino-4-thiazolidinone (**477**). Thus, **477a,b** in methanol have absorption maxima λ = 326 nm (log ϵ = 4.40) and λ = 348 nm (log ϵ = 4.13) while **478a,b** in methanol have max λ = 320 nm (log ϵ = 4.46) and λ = 342 nm (log ϵ = 4.12). In **478**, both the nitrogen atoms are joined by a two-carbon ethylene chain that induces strain in the molecule, resulting in a small (6 nm) hypochromic shift of the maximum. However, a less strained molecule (**479**) in which both the nitrogen atoms are joined by a three-carbon propylene chain has absorption maxima λ = 350 nm (log ϵ = 4.44), comparable with that of **477b**. In **480**, both the nitrogen atoms are fused through a phenylene group which splits the absorption spectrum into two bands, λ = 300 nm (log ϵ = 4.42) and λ = 357 nm (log ϵ = 4.17), indicating considerable contribution of the phenylene group (56JOC24) (Scheme 109).

The intensity of the imidazole band of low wavelength is increased, while the $\pi \rightarrow \pi^*$ band in 2,3-dihydrothiazolo[3,2-*a*]benzimidazole is unaffected by the introduction of a carbonyl group. This is evidenced by the UV absorption spectra of thiazolo[3,2-*a*]benzimidazol-3(2*H*)-ones (56JCS361). The effect of conjugation in 2-arylidenethiazolo[3,2-*a*]benzimidazol-3(2*H*)-ones is shown by UV absorption which consists of three intense bands at 257–273 nm, 281–296 nm and 319–390 nm, which are

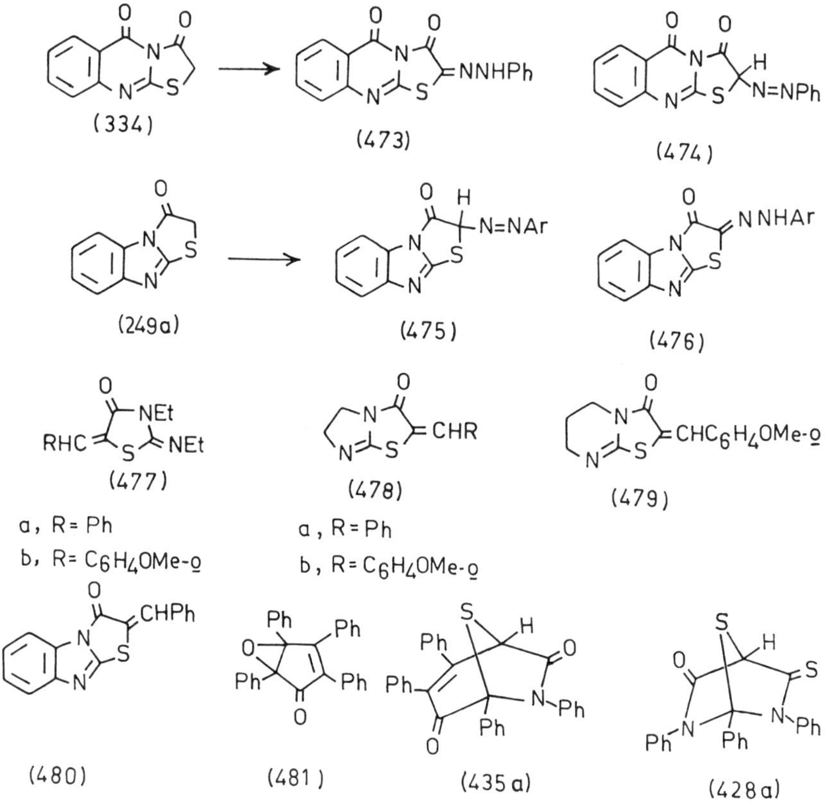

SCHEME 109

probably due to amido, benzimidazole, and K-band transitions, respectively (61ZOB1635). On the other hand, methoxy/ethoxycarbonylmethylidenethiazolo[3,2-a]benzimidazol-3(2H)-ones exhibit UV absorption with comparatively shorter wavelengths of 210–240 nm and 260–320 nm (72MI1).

4,5-Epoxy-2,3,4,5-tetraphenylcyclopenten-1-one (**481**) has UV absorption maxima $\lambda = 233$ nm (log $\epsilon = 4.23$) and $\lambda = 338$ nm (log $\epsilon = 3.85$), whereas cycloadduct **435a** exhibits its absorption maxima at 205 nm (log $\epsilon = 4.64$), 246 nm (log $\epsilon = 4.45$), 300 nm (log $\epsilon = 4.26$), and 353 nm (log $\epsilon = 4.17$). The shift to longer wavelengths is attributed to interaction of the bridge sulfur atom with the amide carbonyl chromophore and α,β-unsaturated system (76JOC818). Similar interactions between the bridge sulfur and amide carbonyl as well as between the thiocarbonyl group and bridge sulfur are observed in **428a** (76JOC813).

B. INFRARED SPECTRA

Infrared spectra of condensed 4-thiazolidinones have been reported by numerous workers and display structurally and diagnostically useful features. The five-membered amide carbonyl band, usually found in the region between 1730–1740 cm^{-1}, is strong and typically characteristic of the thiazolidinone structure. The carbonyl absorption reported is as high as 1790 cm^{-1} [78IJC(B)537] and as low as 1685 cm^{-1} [71JCS(C)266]. Unsaturation of condensed 4-thiazolidinones conjugated with the carbonyl group, as in arylidene derivatives, produces a shift to lower energies in the carbonyl absorption [69AJC2697, 69IJC767, 69IJC769; 70IJC10; 78IJC(B)329; 82IJC(B)315; 85IJC(B)1227; 86IJC(B)354; 86IJC(B)812; 87IJC(B)693, 87IJC(B)739, 87UP1]. The amide carbonyl absorption is also shifted by conjugation with imine unsaturation at an α-position with respect to the carbonyl group (68KGS1008; 69MI2; 71KGS471, 71KGS822, 71KGS930). This lends additional support to the condensed 4-thiazolidinone structures.

The exceptionally low IR carbonyl absorption in **482a** (R = H, alkyl, aryl, alkoxy, OH, NH$_2$) is explained by the resonance interaction between the amidine nitrogen at the 4-position and the carbonyl group at the 2-position, possessing true vinylogous amide character which is depicted in **482a ↔ 482b** (Scheme 110) (64JOC865; 66USP3225059; 67CJC2903; 69URP230823; 71KGS393). On the other hand, this nitrogen at the 4-position does not seem to have any effect on the carbonyl group at the 3-position in **249;** the amide carbonyl absorbs at high frequencies (1730–1740 cm^{-1}). The lowering of the carbonyl absorption in thiazolo[3,2-*a*]benzimidazol-3(2*H*)-ones by an arylhydrazino substituent at the α-position to the carbonyl group implies the existence of hydrazone tautomer **476** (69MI2; 71KGS471, 71KGS822, 71KGS930). The presence of the azo form (**475**) had been claimed solely on the absence of NH absorption and the presence of IR azo bands at 1550–1580 cm (68KGS1008).

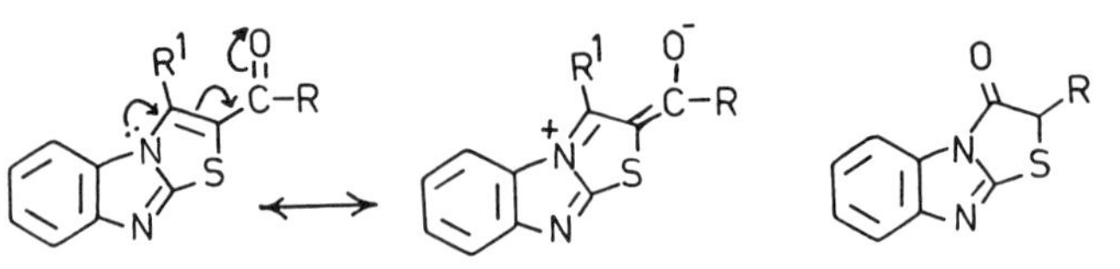

(482a)   (482b)   (249)

SCHEME 110

C. ^{1}H-NMR Spectra

The ^{1}H-NMR spectra of many condensed 4-thiazolidinones have been reported. The characteristic methylene protons (SCH$_2$) of condensed 4-thiazolidinones give a two-proton singlet in the region δ 4.0–4.6. The fact that the carbonyl group deshields the peri-proton has been used to arrive at the correct structural assignments of the cyclized products [69JHC491; 72IJC274; 73IJC1119; 77IJC(B)46; 78IJC(B)478; 79IJC(17B)572; 80H149; 80IJC(B)1035; 81CPB1876, 81IJC(B)294; 84JIC1053; 85IJC(B)1224; 86IJC(B)267, 86IJC(B)807; 87IJC(B)532, 87IJC(B)556, 87UP3, 87UP5; 88IJC(B)121]. The deshielding has its origin in the magnetic anisotropy of the carbonyl group with little contribution from the rest of the thiazolidinone ring.

The structural ambiguity of cyclized products is settled two different ways. One method, first reported by Japanese workers (81CPB1876) and then by others [86IJC(B)264, 86IJC(B)807; 87UP3] is based on comparative studies of the chemical shifts of the peri-proton of the cyclized product (**284, 290, 291,** and **406**) with the corresponding protons of the acid (**283, 289,** and **405**) (Table I). The downfield shift value of H-5 in thiazolo[3,2-a]benzimidazol-3(2H)-ones (**284, 290** and **291**) is ~0.40 ppm. The Japanese workers (82CPB2714), however, mentioned that this method left some ambiguity in **284a** and **290b,e,f**, especially in the case of **290e** where the downfield shift of H-5 was too small (δ 0.17). For further confirmation, they advanced excellent chemical evidence involving the Raney–Nickel desulfurization of the cyclization products, which resulted in the substituted benzimidazoles whose identities were proved by direct comparison with authentic samples.

The second approach, developed at Pujari's laboratory, involves comparative studies of the observed chemical shifts with the calculated chemical shifts of the aromatic protons of the cyclized products. This approach leaves no ambiguity. The calculated chemical shifts of the aromatic protons of the cyclized products are derived by taking into consideration the shielding (or deshielding) effect of the substituents (69MI1) on the values of the corresponding protons of the parent compound (**249a**) which is used as a reference. In the case of **284a** and **290d,** compounds **256a** and **256d** were used, respectively, as reference compounds in addition to the parent compound **249a**; exactly the same structural assignments were found. This method is used successfully to establish the orientation of cyclization of nuclear substituted thiazolo[3,2-a]benzimidazol-3(2H)-ones formed during a ring-closure reaction of unsymmetrical benzimidazolyl-2-thiolacetic acids. Of the two possible structural isomers, the one showing a closer relationship between the observed and calculated chemical shifts of

TABLE I
DOWNFIELD SHIFT VALUE OF peri-PROTON IN
CYCLIZED PRODUCTS **284, 290, 291**, AND **406**
DOWNFIELD FROM THE CORRESPONDING
PROTONS OF ACIDS **283, 289**, AND **405**

Compounds	Chemical shifts (ppm) (DMSO-d_6)	Difference ($\triangle$ppm)
284a	7.65	0.53[b]
283a	7.12	
284d	8.06[a]	0.45[c]
283d	7.61[a]	
284k	7.86[a]	0.49[d]
283k	7.37[a]	
290a	7.79	0.33[b]
289a	7.46	
290b	7.87	0.29[b]
289b	7.58	
290c	7.64	0.43[b]
289c	7.21	
290d	7.34	0.40[b]
289d	6.94	
290e	8.48	0.17[b]
289e	8.31	
290f	8.28	0.26[b]
289f	8.02	
290g	7.79[a]	0.46[e]
289g	7.33[a]	
291a	7.82	0.41[b]
289a	7.41	
291b	7.77	0.40[b]
289b	7.37	
291c	7.69	0.40[b]
289c	7.29	
291d	7.72	0.42[b]
289d	7.30	
291e	7.98	0.39[b]
289e	7.59	
291f	7.90	0.40[b]
289f	7.50	
291g	8.39[a]	0.39[e]
289g	8.00[a]	
406	10.62[a]	0.60[f]
405	9.82[a]	

[a] Solvent: TFA.
[b] 81CPB1876.
[c] 87UP3.
[d] 87UP4.
[e] 86IJC(B)807.
[f] 86IJC(B)264.

the aromatic protons was chosen as the correct isomer (Table II). Table II, depicting studies in 18 such cases, clearly reveals the general applicability of this method.

The deshielding influence of the carbonyl group at the 3-position on methylene protons at the 2-position in the condensed 4-thiazolidinones is also observed. The signal of the methylene protons appeared at low field compared to that of the dihydro series of condensed thiazoles (Table III). This offers proof that the acid has undergone cyclization resulting in the synthesis of condensed 4-thiazolidinones, although it does not help in deciding the orientation of cyclization.

The stereochemistry of heterosteroid **418b** has been secured by ^{1}H-NMR spectral data [71JCS(C)266]. The large difference (~1.5 ppm) observed in the geminal proton (H-12) absorption suggests a semichair for ring C of this heterosteroid [71JCS(C)262]. The SCH$_2$ protons (H-16) of **418b**, unlike other condensed 4-thiazolidinones, exhibit an AB quartet around δ 3.7. One component of the AB quartet is submerged in the intense methoxy signal, but the other three show fine splitting. That the AB quartet results from the interaction of H-14 across the sulfur atom is confirmed, since irradiation of H-14 eliminates this AB quartet. This type of coupling is also dependent on steric factors. The difference in the magnitude of further splitting found in the A and B protons of the spectrum suggests that H-16 α- and β-protons are differently disposed towards the H-14. An exact value ($J_{AB} = 15$ Hz) for the geminal coupling of the H-16 protons was calculated from the double-resonance simplified AB spectrum [71JCS(C)266]. The coupling constant for methylene protons flanked by sulfur and an amide group is deduced for the first time; this value may be useful for correlative work on geminal coupling constants (69T4681).

D. ^{13}C-NMR Spectra

^{13}C-NMR spectral data are available for a limited number of condensed 4-thiazolidinones [76JOC813; 78HCA607; 79IJC(18B)479, 79JOC3803, 79JOC3994, 79TL53; 81JCS(P1)415; 84JCS(P1)2707].

The constitutional isomers between condensed 4-thiazolidinones (**483**) and condensed 4-thiazinones (**484**), obtained from the reaction of cyclic thiones with acetylenic dicarboxylic esters, are distinguished by ^{13}C-NMR spectroscopy. These isomers have been differentiated on the basis of C,H-spin coupling constants first demonstrated by Swiss workers (78HCA607) then, by others [79IJC(18B)479, 79JOC3994, 79TL53; 84JCS(P1)2707]. The magnitude of coupling of the lactam carbon with H$_A$ ($^3J_{CO,H_A}$ = ~6 Hz) found in **483** serves to differentiate it from thazinone

TABLE II
Observed and Calculated Chemical Shifts (ppm) for H-5 and H-8 Protons in Cyclized Products[a]

Compounds	Solvent	H-5[b] Calc.	H-5[b] Obs.	H-5[b] Diff.	H-8 Calc.	H-8 Obs.	H-8 Diff.	Reference
284a	CDCl$_3$	7.71	7.66	0.05	—	—	—	88IJC(B)121
285a	CDCl$_3$	—	—	—	7.51	7.66	0.15	
284a[c]	CDCl$_3$	7.10[d]	7.18	0.08	7.26[e]	7.18	0.08	88IJC(B)121
285a[c]	CDCl$_3$	7.02[d]	7.18	0.16	7.34[e]	7.18	0.16	
284b	DMSO-d$_6$[f]	8.23	8.15	0.08	—	—	—	86IJC(B)807
285b	DMSO-d$_6$	—	—	—	7.93	8.15	0.22	
284c	DMSO-d$_6$	7.90	7.84	0.06	—	—	—	79IJC(17B)572
285c	DMSO-d$_6$	—	—	—	7.60	7.84	0.24	
284d	TFA[g]	8.00	8.06	0.06	—	—	—	87UP3
285d	TFA	—	—	—	7.55	8.06	0.51	
284e	DMSO-d$_6$ + CDCl$_3$	8.07	8.02	0.05	—	—	—	88IJC(B)121
285e	DMSO-d$_6$ + DCDl$_3$	—	—	—	7.82	8.02	0.20	
284f	TFA	8.20	8.17	0.03	—	—	—	87IJC(B)532
285f	TFA	—	—	—	7.75	8.17	0.42	
284g	CDCl$_3$ + TFA	8.34	8.24	0.10	—	—	—	88IJC(B)121
285g	CDCl$_3$ + TFA	—	—	—	7.93	8.24	0.31	
284h	DMSO-d$_6$ + CDCl$_3$	7.85	7.83	0.02	—	—	—	85IJC(B)1224
285h	DMSO-d$_6$ + CDCl$_3$	—	—	—	7.60	7.83	0.23	
284i	TFA	8.26	8.20	0.06	—	—	—	88IJC(B)121
285i	TFA	—	—	—	7.81	8.20	0.39	
284j	TFA	8.02	7.97	0.05	—	—	—	88IJC(B)121
285j	TFA	—	—	—	7.57	7.97	0.40	
284k	TFA	7.87	7.86	0.01	—	—	—	87UP4
285k	TFA	—	—	—	7.42	7.86	0.44	
290a	CDCl$_3$	7.91	7.92	0.01	7.63	7.51	0.12	88IJC(B)121
291a	CDCl$_3$	7.83	7.92	0.09	7.71	7.51	0.20	
290b	DMSO-d$_6$	8.12	7.79	0.33	7.47	7.61	0.14	78IJC(B)478
291b	DMSO-d$_6$	7.77	7.61	0.16	7.82	7.79	0.03	
290d	DMSO-d$_6$	7.46	7.47	0.01	7.60	7.57	0.03	72IJC274
291d	DMSO-d$_6$	7.80	7.57	0.23	7.26	7.47	0.21	
290d[h]	DMSO-d$_6$	7.42	7.47	0.05	7.58	7.57	0.01	86IJC(B)267
291d[h]	DMSO-d$_6$	7.76	7.57	0.19	7.24	7.47	0.23	
290g[i]	TFA	7.66	7.79	0.13	7.90	8.01	0.11	86IJC(B)807
291g[i]	TFA	8.35	8.39	0.04	7.21	7.36	0.15	

[a] Cyclized products were **284, 285, 290,** and **291**; Compound **249a** was used as the reference compound.
[b] Calc. calculated ; Obs., observed; Diff., difference.
[c] Compound **256a** was used as the reference compound.
[d] Value refers to the H-6 proton.
[e] Value refers to the H-7 proton.
[f] DMSO-d$_6$, dimethylsulfoxide-d$_6$.
[g] TFA, trifluoroacetic acid.
[h] Compound **256d** is used as the reference compound.
[i] Calculated values tally with observed values in both isomers, **290g** and **291g**, obtained from **289g**.

TABLE III
CHEMICAL SHIFTS OF SCH_2 PROTONS IN CONDENSED 4-THIAZOLIDINONES[a] AND IN THE CORRESPONDING DIHYDRO SERIES

Compounds	SCH_2 (ppm) (TFA)	References
173b	4.33	80H149
175b	3.92	
181	3.84	81IJC(B)296
181 (CH_2 in place of CO)	3.18	
284c	4.44	78TH1
286c	3.93	79IJC(17B)572
284d	4.86	87UP3
286d	4.41	
284e[b]	4.64	88IJC(B)121
286e[b]	4.00	
284f	4.85	87IJC(B)532
286f	4.37	
284h[c]	4.75	84JIC1053
286h[c]	4.34	
284i	4.82	80IJC(B)1035
286i	4.32	
284j	4.77	81IJC(B)294
286j	4.40	
339b	4.06	84JIC1050
339b (CH_2 in place of CO)	3.49	
379	4.65	82TH1
379 (CH_2 in place of CO)	3.90	
406	5.00	86IJC(B)264
407	4.58	

[a] Compounds **173b, 181, 284, 339, 379,** and **406**.
[b] Solvent: DMSO-d^6 + $CDCl_3$.
[c] Solvent: TFA + $CDCl_3$.

structure **484** in which the coupling constant $^2J(CO,H_A) \leq 1$ Hz and is often not resolved. The ^{13}C-NMR spectra of **483** and **484** show several marked differences. The nonbenzenoid C—H resonance in **483** is at higher field than in **484**; the ester carbonyl carbon and the carbon attached to the sulfur are at lower field in **483** than in **484**. these data also offer a method of distinguishing one from the other [81JCS(P1)415]. The configuration of trisubstituted exocyclic double bonds has been established on the basis of two-bond and three-bond C—H spin coupling constants (78HCA607; 79JOC3994, 79TL53). Coupling constant ($^3J CO, H_A$) of the amide carbonyl carbon and the olefinic proton (H_A) in **483** having a value of ~6 Hz suggests the fumarate geometry (Z-isomer) of the olefinic linkage.

(483) (484) (485) (67)

Adduct **67** is of interest since it possesses an sp^3 carbon atom attached to two nitrogens, one oxygen, and one sulfur atom. This leads to the very low field ^{13}C resonance when compared to the sp^3 carbon atom (C-2) in **485**, thus demonstrating that a carbon attached to four heteroatoms is deshielded [81JCS(P1)415].

E. MASS SPECTRA

Mass fragmentation of a few condensed 4-thiazolidinones, namely, 2-arylidenethiazolo[3,2-*a*]pyridin-3(2*H*)-ones (**89**) (78MI2; 79MI2),6′(7*H*)-oxospiro[cycloalkane-1,3′(4′*H*)-[2*H*]thiazolo[3,2-*b*][1,2,4,5]tetrazines] (**181a,b,d–g**) [87IJC(B)739], thiazolo[3,2-*a*]benzimidazol-3(2*H*)-ones (**256, 284, 290,** and **291**) (85TH1–85TH3;87UP4), and thiazolo[3,2-*a*]-aceperimidine (**422**) [87IJC(B)693] have been studied, although structures of several other condensed 4-thiazolidinones have been confirmed by their molecular ion peaks [M]$^{\cdot +}$.

The mass spectral fragmentation of condensed 4-thiazolidinones follows several pathways. In addition to different modes of cleavage, depending on the types of nuclei attached to the thiazolidinone ring, the fragmentation of the thiazolidinone ring follows more or less the same paths (*a–e*) as depicted in Scheme 111. The molecular ion [M]$^{\cdot +}$ (**486**) loses elements of CO, CH$_2$CO, and SCH$_2$CO via paths *a, b,* and *c* resulting in the formation of [M—CO]$^{\cdot +}$, [M—CH$_2$CO]$^{\cdot +}$ and [M—SCH$_2$CO]$^{\cdot +}$ ions, respectively. The ion [M—CO]$^{\cdot +}$ subsequently decomposes to [M—CO—CS]$^{\cdot +}$ by elimination of CS. The ion [SCH$_2$CO]$^{\cdot +}$ appears at *m/z* 74. The 1,2- thiazetidonyl cation at *m/z* 88, formed via path *d* from [M]$^{\cdot +}$ (**486**), loses CO to give ion H$_2$CNS$^+$ at *m/z* 60 which in turn, on successive splitting of two hydrogen radicals, gives HN$^{\cdot +}$=C=S at *m/z* 59 and N$^+$=C=S at *m/z* 58, respectively. The ion at *m/z* 88 undergoes decomposition to give ion CH$_2$=S$^{\cdot +}$ at *m/z* 46, which loses one hydrogen radical to yield thietenyl cation CH$_2$=S$^+$ at *m/z* 45. The prominent cleavage via path *e* found in the mass spectra of 3-alkyl/arylthiazolo[3,2-*a*]benzimidazoles (69JHC797) is observed in some condensed 4-thiazolidinones (Scheme 111).

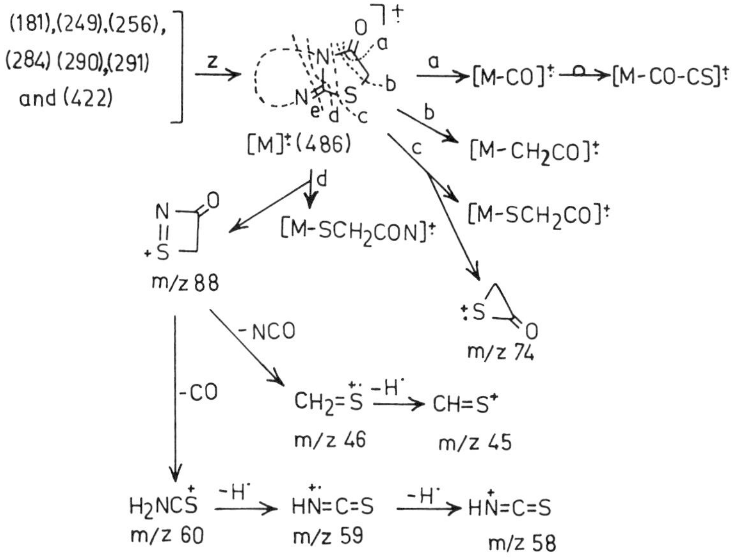

SCHEME 111

[Fragmentation scheme showing [M]⁺ (486) with labeled bond cleavages a–e leading to:
- (181), (249), (256), (284), (290), (291) and (422) via pathway z
- [M−CO]⁺ → [M−CO−CS]⁺ via path a
- [M−CH₂CO]⁺ via path b
- [M−SCH₂CO]⁺ via path c, giving m/z 74 (:S=C=O related structure)
- [M−SCH₂CON]⁺ via path d
- m/z 88 structure (N=C−O, ⁺S ring) then −NCO → CH₂=⁺S˙ (m/z 46) −H˙→ CH=S⁺ (m/z 45)
- −CO from m/z 88 → H₂N⁺CS (m/z 60) −H˙→ H⁺N=C=S (m/z 59) −H˙→ HN⁺=C=S (m/z 58)]

In the mass spectra of spiro[alkane-thiazolo-tetrazine] (**181a,b,d–g**), the molecular ion peaks in **181a,b** are the base peaks. On the other hand in the mass spectra of **181d–g**, the molecular ion peaks are not the base peaks, suggesting that **181d–g** are not stable under electron impact. This is quite understandable since cyclopentane and cyclohexane rings in **181a** and **181b**, respectively, are more stable in comparison to higher-membered cycloalkane rings **181d–g**. In another competing path, the molecular ions (from **181**), through the cleavage of one bond of the cycloaklane ring, yield ions that undergo McLafferty rearrangement as well as loss of methyl and hydrogen radicals through a four-centered mechanism to furnish the corresponding fragmented ions [87IJC(B)739].

In the mass spectra of thiazolo[3,2-*a*]benzimidazol-3(2*H*)-ones (**249a, 256, 284, 290,** and **291**), the molecular ion peaks are the base peaks in all cases except in **284f** where the molecular ion peak is the second most intense peak. In **256c,d** and **284d,k** the molecular ions undergo cleavage via paths *b–e*, while **284e–i, 290b,** and **291b** do not undergo cleavage directly via these paths, but these paths are in operation after molecular ions lose one halogen atom. With **256c,d, 290g** and **291g**, the molecular ions undergo successive demethylation and decarboxylation resulting in the contraction of a six-membered ring to a five-membered ring to give the [M—CH₃—CO]⁺ ion. Similar loss of a methyl radical followed by CO has been

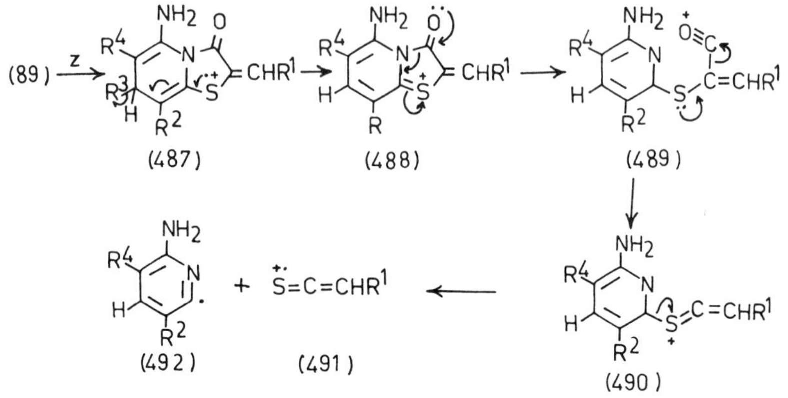

SCHEME 112

reported in the mass spectra of aromatic methyl ethers (63T2233). The mass fragmentation patterns were substantiated by shifts in the peaks (corresponding ions) of substituted groups or by following the isotopic cluster of peaks (ions containing bromine or chlorine atoms). Most of the fragmentation modes postulated in the mass spectra of condensed 4-thiazolidinones have been supported by metastable transitions.

With arylidenethiazolo[3,2-*a*]pyridin-3(2*H*)-ones (**89**), the molecular ion peaks are found to be highly intense. The [M]$^{\cdot+}$ ion (**487**) loses an aryl group (R^3) at the 7-position to give fragment ion **488** as the base peak. Ion **490**, obtained after the split of CO from **488** via **489**, decomposes to fragmented ion **491**. This is the only ring cleavage observed in all the compounds of this series (**89**). Molecular ion **487** also loses the alkoxycarbonyl group (R^4) at the 6-position. The temperature dependence of mass spectra was tested, and the increase in the fragment ion intensity was observed only in the thiazolidinone ring-opening fragmentations (79MI2). The fragmentation patterns were deduced from shifts in the peaks by substituted groups and by deuterated derivatives. Some of the patterns were also confirmed by the metastable peaks (78MI2; 79MI2) (Scheme 112).

IV. Molecular Dimensions

X-Ray Diffraction

A single crystal X-ray study (78TL2621) of one of the products obtained from the reaction of benzimidazolyl-2-thione (**247**) with DMAD confirms thiazolidinone structure **23**, which was earlier tentatively assigned on the basis of chemical evidence (61ZOB394; 77MI1). This is an isolated instance of X-ray analysis of a condensed 4-thiazolidinones.

V. Reactions of Condensed 4-Thiazolidinones

A. REACTIONS WITH ELECTROPHILES

The methylene carbon atom in a condensed 4-thiazolidinone flanked by a sulfur atom and a carbonyl group possesses enhanced nucleophilic activity and attacks an electrophilic center with ease. If the structure permits, the reaction product loses a molecule of water, and an unsaturated derivative is formed. The reaction is carried out in the presence of a base which abstracts a methylene proton. It is the anion thus formed that attacks the electrophilic center. Generally, the anion condenses with aromatic aldehydes, nitroso compounds, aryldiazonium salts, and ethyl orthoformate, as well as undergoing Vilsmeier–Haack and Mannich reactions.

1. *Alkylation*

Condensed 4-thiazolidinones (**493**) react with aromatic aldehydes in the presence of a base such as piperidine (58JOC24; 63ZOB945; 64JOC1715; 65ZOB1276; 67KGS894, 67MI1; 69AJC2697; 70IJC10; 70LA132; 71KGS93; 71KGS822), anhydrous sodium acetate, [58JOC897; 68JIC710; 69IJC767, 69IJC769; 70IJC10, 70IJC885; 72JPR785; 76ZN(B)111; 77JHC1093, 77ZN(B)94; 78IJC(B)329; 80MI2; 82IJC(B)315; 84JIC89; 85IJC(B)1227; 86IJC(B)354, 86IJC(B)812; 87IJC(B)693, 87IJC(B)739], triethylamine (50BRP634951; 51JCS734; 56JCS361), or DCC-pyridine mixture (65ZOB1276) to afford arylidene derivatives (**494**) in good to excellent yields (A = imidazole, benzimidazole, triazole, pyrimidine, triazine, benzodiazepine, pyridoimidazole, purine, quinazoline, perimidine, diazepine, aceperimidine, spirocycloalkanotetrazine or spiroindanotetrazine nucleus). The activity of the methylene hydrogens is further illustrated by their reaction with ethyl orthoformate in the presence of acetic anhydride to give ethoxymethylidene derivative **495** in moderate yield (A = benzimidazole or naphthoimidazole) (50BRP63495; 51JCS734; 56JCS361; 79UKZ1096; 80MI2). The easy formation of aminoalkenylated derivative **496** from **493** (A = benzimidazole) and DMF in the presence of phosphoryl chloride (Vilsmeier–Haack reaction) suggests the nucleophilic character of the methylene carbon (65MI1). The nucleophilic reactivity of the methylene group in **493** is also demonstrated by its ability to undergo alkylation with formaldehyde and a variety of primary and secondary amines (Mannich reaction) giving alkyl/dialkylaminomethyl derivatives **497** in moderate yield (A = imidazole or benzimidazole nucleus) [69JMC962; 71KGS822; 81ZN(B)501]. There is only one instance each of a functionalized aldehyde (glyoxalic acid) (78TL2621) and a ketone (acetone) (63ZOB945) reacting with **493** (A = benzimidazole). The two reactions resulted in the formation of **498** and **499**, respectively (Scheme 113).

2. *Diazo Coupling*

The methylene carbon in **493** couples readily with aryldiazonium salts in weakly acidic or basic medium to give arylhydrazones (**500**) in high yields (A = imidazole, benzimidazole, triazine, imidazopyridine, or quinazoline nucleus) [63ZOB945; 65ZOB1276; 67KGS894, 67M11; 68KGS1008; 70-LA1321; 71KGS822; 72JPR785; 74JPR163; 76ZN(B)1397; 77ZN(B)94; 81ZN(B)501; 83AP985; 86PHA324] (Scheme 113). The UV and IR spectroscopic data of the reaction products suggest the existence of hydrazone form **500** along with its tautomeric azoform.

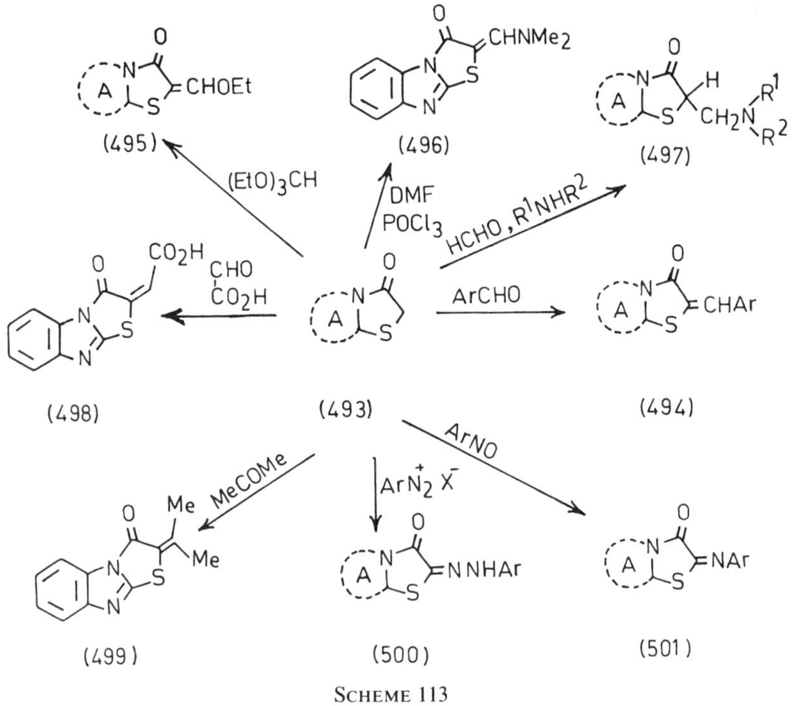

SCHEME 113

3. Reaction with Nitrosoarenes

The methylene carbon in **493** is capable of reacting with nitrosoarenes in the presence of a base such as sodium carbonate or piperidine to give arylimino derivatives **501** (A = imidazole or benzimidazole nucleus) (63ZOB945; 65ZOB1276; 67KGS894, 67MI1; 71KGS93, 71KGS822) (Scheme 113).

A comparative study on the reactivity of the methylene hydrogen atoms in sulfur–nitrogen heterocycles towards an electrophilic reagent as exemplified by their reactions with aldehydes was undertaken. It was found that the reactivity decreased in the order of **249a** > **35** > **502** > **503**. Compound **504** is as unreactive as **503**. This clearly demonstrates that the

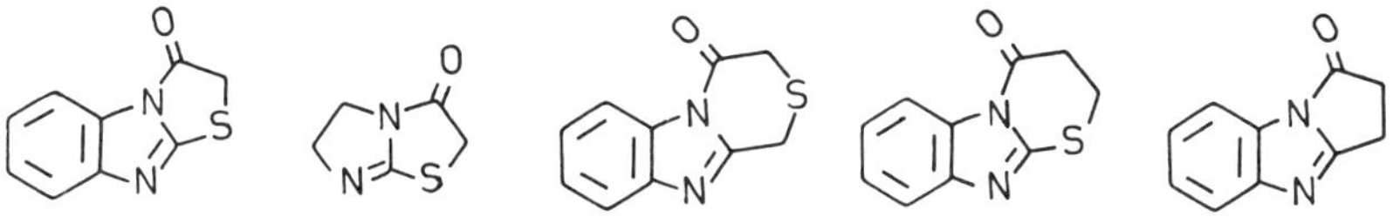

(249a)   (35)   (502)   (503)   (504)

adjacent sulfur is thus a prerequisite condition for the activity of the methylene group situated at the α-position with respect to lactam carbonyl group (70MI1).

4. Reaction with Acids

Condensed 4-thiazolidinones (**505**) having an azo-methane linkage, and on heating with acid, undergo ring fission at the C=N double bond to give thiazolidin-2,4-dione derivatives (**507**) (A = imidazole or benzimidazole nucleus) (61UKZ503, 61ZOB1635; 80EUP2978). Structure **505** is first protonated with an acid to form the quaternary salt **506,** which behaves as an ambident electrophile and becomes the target of nucleophilic attack. The ring fission takes place by nucleophilic attack at the carbon flanked by sulfur and nitrogen and gives rise to thiazolidin-2,4-dione derivative **507** (Scheme 114).

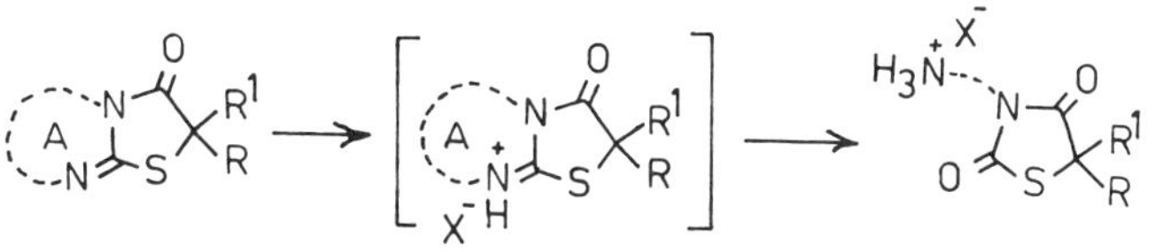

(505)   (506)   (507)

SCHEME 114

B. REACTIONS WITH NUCLEOPHILES

1. Amination

Condensed 4-thiazolidinone (**508**) behaves like an amide as illustrated by its reaction with carbonyl reagents such as hydrazines (58JOC897; 70-LA132; 72JPR785; 74JPR147) and amines (72JPR785; 74JPR147; 76JPR168) resulting in the fission of the thiazolidinone ring to give **509** (A = imidazole or benzimidazole nucleus) and **510** (A = imidazole or quinazoline nucleus), respectively (Scheme 115).

2. Reaction with Alkali

The condensed thiazole system (devoid of a carbonyl function) is stable to alkaline hydrolysis, whereas in the condensed 4-thiazolidinones (**508**), the carbonyl function imparts instability and the amide linkage undergoes fission in alkaline medium to give thioacids **511** [50BRP634951; 51JCS734; 56JCS361, 56JOC24; 72M11; 75JAP7552065; 79IJC(18B)39; 81IJC(B)538; 81JCS(P1)415; 83IJC(B)243; 84IJC(B)316] (Scheme 115).

3. Reaction with Alcohols

2-Fluoro-2-trifluoromethylthiazolo[3,2-*a*]benzimidazol-3(2*H*)-one (**508**; A = benzimidazole nucleus, R^1 = F, R^2 = CF_3), on heating in alcohol even for a short period, undergoes alcoholysis to give ester **512**. The electron-withdrawing nature of the fluorine and trifluoromethyl groups at

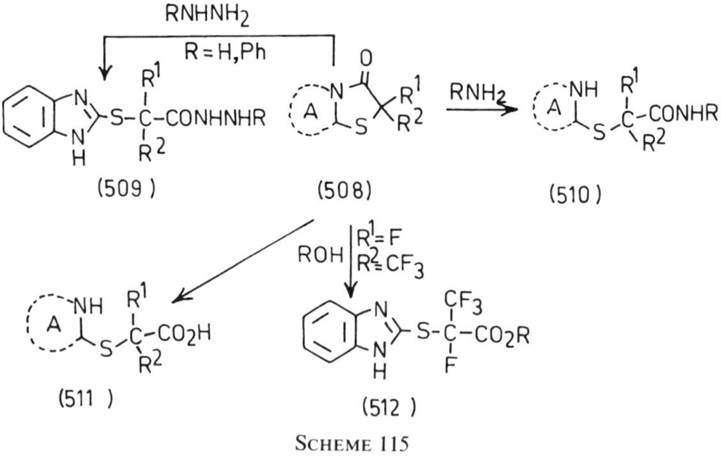

SCHEME 115

the 2-position makes the thiazolidinone ring labile towards mild nucleophilic attack (78BCJ3091) (Scheme 115).

4. Reaction with Grignard Reagents

The mode of reaction of condensed 4-thiazolidinones with Grignard reagents depends on the nature of substituents at the methylene carbon atom. Thus, arylidene derivative **513** reacts with a Grignard reagent to yield Michael-type adduct **514** (A = imidazole or benzimidazole nucleus) by addition of the Grignard reagent to the C=C double bond rather than to the carbonyl group (70LA132; 72JPR785; 76JPR168). However, in **186** and **518**, where Michael addition is not possible, these compounds undergo normal Grignard reaction to furnish the corresponding hydroxy products (**517** and **519** in good yield (70LA132; 74FRP2164490) (Scheme 116).

5. Reaction with Diazomethane

Reaction of **513** (A = benzimidazole nucleus) with diazomethane yields methylated product **515**. The formation of **515** may be rationalized by the initial formation of a Michael adduct which, on subsequent loss of nitrogen and prototropic rearrangement, gives rise to **515** (70LA132) (Scheme 116).

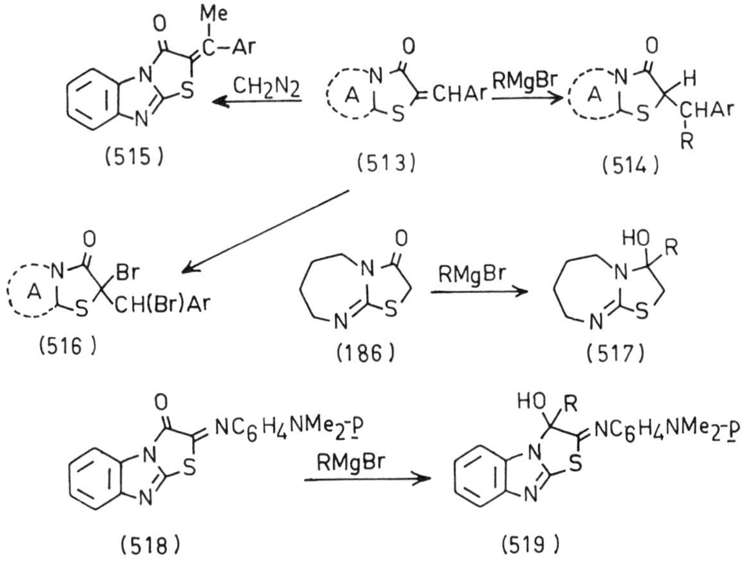

SCHEME 116

C. Halogenation

Facile addition of bromine to the C=C double bond in **513** results in the synthesis of dibromo adducts (**516;** A = imidazole, benzimidazole, perimidine, or diazepine nucleus) (68JIC710; 69AJC2697, 69IJC767, 69IJC769; 70IJC10, 70IJC885; 81JIC1117) (Scheme 116).

D. Oxidation

The inertness of the 4-thiazolidinone ring towards oxidizing agents is demonstrated by the isolation of thiazolo[3,2-*a*]benzimidazol-3(2*H*)-one (**249a**) as the oxidation product during oxidation of 3-hydroxy-2,3-dihydrothiazolo[3,2-*a*]benzimidazole (**250**) by chromium trioxide in pyridine (67CJC2903; 68CPB2167). Structure **447c** and **440k,** on peracid oxidation, afford the corresponding sulfoxides (74JOC3631), whereas attempted peracid oxidation of **249a** resulted in ring fission (70LA132).

E. Reduction

The amide carbonyl group in **151** is reduced by diborane to give deoxocompound **152** (70JHC1231). Reduction of **186** with phenyl lithium gives **517** (R = Ph) (73GEP2160655). The reductive cyclization of 2-(*o*-nitrobenzylidene)thiazolo[3,2-*a*]benzimidazol-3(2*H*)-one (**520**) with zinc dust in acetic acid to give quinolino[3,2:5,4']thiazolo[3',2':1,2]benzimidazole (**521**) provides the means for constructing the thiazoloquinoline ring system (79MI3). Similarly, reduction with sodium dithionite of the hydrazino group in **522** to the corresponding 2-amino derivative **523** serves as a method for synthesizing 2-amino derivatives (70JMC1018) (Scheme 117).

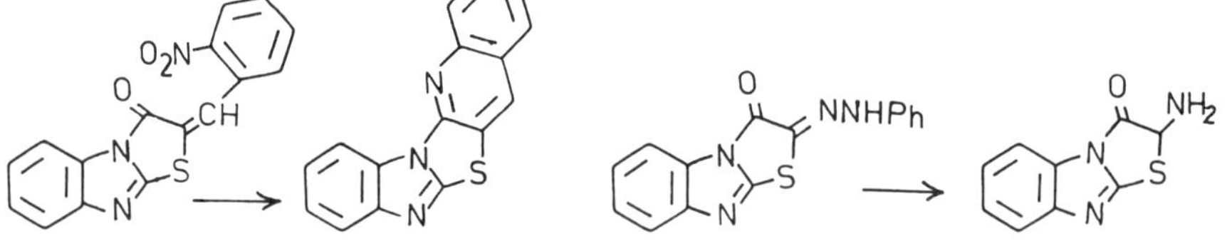

(520)   (521)   (522)   (523)

Scheme 117

F. THIONATION

Conversion of a carbonyl group to a thione group in condensed 4-thiazolidinone can be affected by phosphorus pentasulfide, as exemplified by the synthesis of thiazolo[2,3-c]-[1,4]thiazin-3(2H)-thiones from **139** and phosphorus pentasulfide in boiling dioxane (64M1335).

VI. Useful Applications of Condensed 4-Thiazolidinones

A. BIOLOGICAL PROPERTIES

Condensed 4-thiazolidinones have been shown to exhibit antifungal [69AJC2697, 69IJC767, 69IJC769; 70IJC10, 70IJC885; 72MI2; 78IJC(B)478; 81JIC1117; 84JIC89, 84JIC1050, 84JIC1053; 87IJC(B)693], antibacterial (69AJC2697, 69IJC767, 69IJC769, 70IJC10, 70IJC885; 72MI2; 84JIC1050, 84JIC1053; 85MI1), anticonvulsant [69JMC962; 70JMC1018; 76JAP74109398, 76USP3897446, 76USP3901879, 76USP3906001; 77JHC1093; 83IJC(B)785], anti-inflammatory (86JHC1359, 86USP4588812), herbicidal (85JIC147), central nervous system depressant (70USP3475424), antispasmodic (61ZOB1635; 69JMC962), carcinostatic (59JOC1410; 78MI1; 79MI1), reserpine antagonistic (73USP3732219), muscle relaxant (76JAP74109398, 76USP3897446, 76USP3901879, 76USP3906001), sedative (76JAP74109398, 76USP3897446, 76USP3901879, 76USP3906001), and antihypertensive (61-ZOB1635; 83AP569; 85AP502) properties. Arylidene derivatives of condensed 4-thiazolidinones are found to be better fungistatic agents than the parent condensed 4-thiazolidinones, whereas both possess almost the same antibacterial properties. Condensed 4-thiazolidinones are better antibacterial agents than their thiazole counterparts. Introduction of bromine enhances the fungicidal and bactericidal properties of condensed 4-thiazolidinones (72MI2).

B. DYESTUFFS

Condensed 4-thiazolidinones have found applications as intermediates in the synthesis of cyanine and merocyanine dyes used as photographic sensitizers for silver halide emulsion (50BRP634951; 51JCS734; 55BRP730489; 56BRP734792, 56BRP749193, 56JCS361; 57BRP743133, 57BRP749189, 57BRP749190; 58BRP785334, 58BRP785939; 65MI2; 66BEP668594; 67BRP1072384, 67EGP49396, 67GEP1235738, 67MI1; 68URP210658; 71KGS93; 75BRP1392499, 75JIC1193; 82USP4304908).

References

26JCS2531	H. W. Stephen and F. J. Wilson, *J. Chem. Soc.*, 2531 (1926).
28JCS1415	H. W. Stephen and F. J. Wilson, *J. Chem. Soc.*, 1415 (1928).
29MI1	F. J. Wilson, W. Baird, R. Burns, A. M. Munro, and H. W. Stephen, *J. R. Tech. Coll. Glasgow* **2**, 56 (1929) [*CA* **23**, 5164 (1929)].
36JCS1559	A. R. Todd and F. Bergel, *J. Chem. Soc.*, 1559 (1936).
42JA2706	T. B. Johnson and C. O. Edens, *J. Am. Chem. Soc.* **64**, 2706 (1942).
50BRP634951	J. D. Kendall and G. F. Duffin, Br. Pat. 634,951 (1950) [*CA* **44**, 9287 (1950)]
50BRP634952	J. D. Kendall and G. F. Duffin, Br. Pat. 634, 952 (1950) [*CA* **44**, 9287 (1950)].
50JCS1127	P. J. Heald and T. W. Walker, *J. Chem. Soc.*, 1127 (1950).
51JCS734	G. F. Duffin and J. D. Kendall, *J. Chem. Soc.*, 734 (1951).
54JIC787	M. S. Dhatt and K. S. Narang, *J. Indian Chem. Soc.* **31**, 787 (1954).
55BRP730489	J. D. Kendall and G. F. Duffin, Br. Pat. 730,489 (1955) [*CA* **49**, 15580 (1955)].
56BRP734792	J. D. Kendall and G. F. Duffin, Br. Pat. 734,792 (1956) [*CA* **50**, 1502 (1956)].
56BRP749193	J. D. Kendall and G. F. Duffin, Br. Pat. 749,193 (1956) [*CA* **50**, 16492 (1956)].
56JCS361	G. F. Duffin and J. D. Kendall, *J. Chem. Soc.*, 361 (1956).
56JOC24	J. A. van Allan, *J. Org. Chem.* **21**, 24 (1956).
56JOC193	J. A. van Allan, *J. Org. Chem.* **21**, 193 (1956).
56N(L)1042	J. Chatt, L. A. Duncanson, and L. M. Venanzi, *Nature (London)* **177**, 1042 (1956).
57BRP743133	J. D. Kendall, G. F. Duffin, and H. R. J. Waddington, Br. Pat. 743,133 (1957) [*CA* **51**, 899 (1957)].
57BRP749189	J. D. Kendall and G. F. Duffin, Br. Pat. 749,189 (1957) [*CA* **51**, 904 (1957)].
57BRP749190	J. D. Kendall and G. F. Duffin, Br. Pat. 749,190 (1957) [*CA* **51**, 902 (1957)].
58BRP785334	H. R. J. Waddington, G. F. Duffin, and J. D. Kendall, Br. Pat. 785,334 (1958) [*CA* **52**, 60308 (1958)].
58BRP785939	G. F. Duffin, D. J. Fry, and J. D. Kendall, Br. Pat. 785,939 (1958) [*CA* **52**, 10777 (1958)].
58JCS1974	D. J. Brown, *J. Chem. Soc.*, 1974 (1958).
58JOC897	A. L. Misra, *J. Org. Chem.* **23**, 897 (1958).
59JCS3332	D. J. Brown and R. J. Harrison, *J. Chem. Soc.*, 3332 (1959).
59JOC1410	R. C. Elderfield and R. N. Prasad, *J. Org. Chem.* **24**, 1410 (1959).
59MI1	O. F. Limar, *Farm. Zh. (Kiev.)* **14**, 12 (1959) [*CA* **54**, 22573 (1960)].
60JA2909	C. H. DePuy and P. P. R. Wells, *J. Am. Chem. Soc.* **82**, 2909 (1960).
60MI1	O. F. Limar, *Farmatsevt. Zh.* **15**, 4 (1960) [*CA* **56**, 13016 (1962)].
60USP2933497	R. M. Dodson, U.S. Pat. 2,933,497 (1960) [*CA* **54**, 24805 (1960)].
61ACS1575	J. Sandstrom, *Acta Chem Scand.* **15**, 1575 (1961) [*CA* **56**, 14294 (1962)]
61CRV463	F. C. Brown, *Chem. Rev.* **61**, 463 (1961).
61HC	W. L. Mosby, "Heterocyclic Systems with Bridgehead Nitrogen Atoms," Vol. 15, Parts I and 2. Wiley (Interscience), New York, 1961.

61JOC2715	E. C. Taylor, G. A. Berchtold, N. A. Goeckner, and F. G. Strochmann, *J. Org. Chem.* **26**, 2715 (1961).
61UKZ503	N. M. Turkevich and O. F. Lymar, *Ukr. Khim. Zh.* **27**, 503 (1961) [*CA* **56**, 7296 (1962)].
61ZOB394	E. I. Grinblat and I. Ya. Postovskii, *Zh. Obshch. Khim.* **31**, 394 (1961) [*CA* **55**, 22299 (1961)].
61ZOB1635	M. M. Turkevich and O. F. Lymar, *Zh. Obshch. Khim.* **31**, 1635 (1961) [*CA* **55**, 23503 (1961)].
61ZOB3267	P. M. Kochergin, *Zh. Obshch. Khim.* **31**, 3267 (1961) [*CA* **57**, 2208 (1962)].
62JA3514	H. C. Yao and P. Resmick, *J. Am. Chem. Soc.* **84**, 3514 (1962).
62JOC3701	J. C. Howard and G. Klein, *J. Org. Chem.* **27**, 3701 (1962).
62LA(657)113	T. Pyl, K. H. Wiensch, L. Buelling, and H. Beyer, *Justus Liebigs Ann. Chem.* **657**, 113 (1962) [*CA* **58**, 2443 (1963)].
63AG(E)565	R. Huisgen, *Angew. Chem., Int. Ed. Engl.* **2**, 565 (1963).
63MI1	K. Rothwell and R. L. Wain, *Ann. Appl. Biol.* **51**, 161 (1963) [*CA* **60**, 1041 (1964)].
63T2233	Z. Pelah, J. M. Wilson, M. Ohashi, H. Budzikiewicz, and C. Djerassi, *Tetrahedron* **19**, 2233 (1963).
63ZOB945	I. I. Chizhevskaya, L. I. Gapanovich, and L. V. Poznyak, *Zh. Obshch. Khim.* **33**, 945 (1963) [*CA* **59**, 8725 (1963)].
64JA107	J. B. Hendrickson, R. Rees, and J. F. Templeton, *J. Am. Chem. Soc.* **86**, 107 (1964).
64JOC865	J. J. D'Amico, R. H. Campbell, and E. C. Guinn, *J. Org. Chem.* **29**, 865 (1964).
64JOC1715	E. Campaigne and M. C. Wani, *J. Org. Chem.* **29**, 1715 (1964).
64JOC2959	H. C. Yao, *J. Org. Chem.* **29**, 2959 (1964).
64M1335	A. Asinger, H. Diem, and W. Schaefer, *Monatsh. Chem.* **95**, 1335 (1964) [*CA* **62**, 6478 (1965)].
64MI1	K. A. Zolotareva, I. P. Maslova, N. A. Glazunova, E. F. Burmistrov, and L. A. Pugacheva, *Sintez i Issled. Effektivn. Stabilizatorov dlya Polimern. Materialov, Sb., Voronezh* 5 (1964) [*CA* **65**, 18767 (1966)].
65JA1980	E. C. Taylor and R. W. Hendess, *J. Am. Chem. Soc.* **87**, 1980 (1965).
65MI1	B. A. Porai-Koshits, I. Y. Kvitko, and E. A. Shutkova, *Latv. PSR Zinat. Akad. Vestis, Kim. Ser.*, 587 (1965) [*CA* **64**, 8168 (1966)].
65MI2	E. J. Poppe, *Veroeff. Wiss. Photo-Lab., Wolfen* **10**, 115 (1965) [*CA* **65**, 9995 (1966)].
65ZOB1276	I. I. Chizhevskaya, L. I. Gapanovich, and L. V. Poznyak, *Zh. Obshch. Khim.* **35**, 1276 (1965) [*CA* **63**, 11539 (1965)].
66AP846	W. Schneider and E. Kaemmerer, *Arch. Pharm. (Weinheim, Ger.)* **299**, 846 (1966) [*CA* **66**, 371817 (1967)].
66BEP668594	VEB Filmfabrik Wolfen, Belg. Pat. 668,594 (1966) [*CA* **65**, 3213 (1966)].
66JA4112	M. P. Cava and N. M. Pollack, *J. Am. Chem. Soc.* **88**, 4112 (1966).
66URP175968	V. E. Bogachev and M. G. Fomenko, USSR Pat. 175,968 (1966) [*CA* **64**, 6671 (1966)].
66USP3225059	J. J. D'Amico, U.S. Pat. 3,225,059 (1966) [*CA* **64**, 8193 (1966)].
67BRP1072384	E. J. Poppe, Br. Pat. 1,072,384 (1967) [*CA* **67**, 69446 (1967)].
67CB3671	E. Winterfeldt and J. M. Nelke, *Chem. Ber.* **100**, 3671 (1967).
67CC935	A. Gomes and M. M. Joullie, *J. C. S. Chem. Commun.*, 935 (1967).
67CJC939	J. W. Lown and J. C. N. Ma, *Can. J. Chem.* **45**, 939 (1967).
67CJC953	J. W. Lown and J. C. N. Ma, *Can. J. Chem.* **45**, 953 (1967).

67CJC2903	A. E. Alper and A. Taurins, *Can. J. Chem.* **45**, 2903 (1967).
67EGP49396	E. J. Poppe, Ger. (East) Pat. 49,396 (1967) [*CA* **66**, 50702 (1967)].
67GEP1235738	E. J. Poppe, Ger. Pat. 1,235,738 (1967) [*CA* **66**, 110064 (1967)].
67IJC221	M. D. Nair, P. A. Malik, and S. R. Mehta, *Indian J. Chem.* **5**, 221 (1967).
67JHC577	H. F. Andrew and C. K. Bradsher, *J. Heterocycl. Chem.* **4**, 577 (1967).
67KGS93	P. M. Kochergin, A. M. Tsyganova, and L. M. Viktorova, *Khim. Geterotsikl. Soedin.*, 93 (1967) [*CA* **67**, 64298 (1967)].
67KGS894	I. A. Mazur and P. M. Kochergin, *Khim. Geterotsikl. Soedin*, 894 (1967) [*CA* **68**, 105109 (1968)].
67MI1	I. I. Chizhevskaya, L. I. Gapanovich, and R. S. Kharchenko, *Puti Sin. Izyskaniya Protivoopukholevykh Prep., Tr. Simp. Khim. Protivoopukholevykh, Veshchestv.*, 62 (1967) [*CA* **70**, 106421 (1969)].
68ACH191	G. Doleschall, G. Hornyak, L. Lang, K. Lempert, and K. Zauer, *Acta Chim. Acad. Sci. Hung.* **57**, 191 (1968) (*CA* **70**, 4063 (1969)].
68CPB2167	H. Ogura, T. Itoh, and Y. Shimada, *Chem. Pharm. Bull.* **16**, 2167 (1968) [*CA* **70**, 68261 (1969)].
68JCS(C)1777	R. Hull, *J. Chem. Soc. C*, 1777 (1968).
68JHC179	S. C. Bell and G. Conclin, *J. Heterocycl. Chem.* **5**, 179 (1968).
68JHC185	S. C. Bell and P. H. L. Wei, *J. Heterocycl. Chem.* **5**, 185 (1968).
68JHC609	E. F. Elslager, D. F. Worth, N. F. Haley, and S. C. Pericone, *J. Heterocycl. Chem.* **5**, 609 (1968).
68JIC710	H. S. Chaudhary and H. K. Pujari, *J. Indian Chem. Soc.* **45**, 710 (1968).
68KGS443	I. I. Chizhevskaya, N. N. Khovratovich, and Z. M. Garbovskaya, *Khim. Geterotsikl. Soedin.*, 443 (1968) [*CA* **69**, 86906 (1968)].
68KGS1008	I. I. Chizhevskaya, M. I. Zavadskaya, and N. N. Khovratovich, *Khim. Geterotsikl. Soedin.*, 1008 (1968) [*CA* **70**, 68257 (1969)].
68TH1	H. Singh, Ph.D. Thesis, Kurukshetra University, Kurukshetra (1968).
68URP210658	M. V. Diechneister, A. Z. Pinkhasova, E. B. Lifshits, and M. V. Krylova, USSR Pat. 210,658 (1968) [*CA* **69**, 28614 (1968)].
68ZOR179	I. I. Chizhevskaya and M. I. Zavadskaya, *Zh. Org. Khim.* **4**, 179 (1968) [*CA* **68**, 87242 (1968)].
69AJC2697	V. K. Chadha, H. S. Chaudhary, and H. K. Pujari, *Aust. J. Chem.* **22**, 2697 (1969).
69IJC735	S. V. Kessar and P. Jit, *Indian J. Chem.* **7**, 735 (1969).
69IJC767	H. S. Chaudhary and H. K. Pujari, *Indian J. Chem.* **7**, 767 (1969).
69IJC769	V. K. Chadha, H. S. Chaudhary, and H. K. Pujari, *Indian J. Chem.* **7**, 769 (1969).
69IJC881	A. Singh, A. S. Uppal, and K. S. Narang, *Indian J. Chem.* **7**, 881 (1969).
69JHC491	E. F. Elslager, D. F. Worth, and S. C. Perricone, *J. Heterocycl. Chem.* **6**, 491 (1969).
69JHC729	A. Gomes and M. M. Joullie, *J. Heterocycl. Chem.* **6**, 729 (1969).
69JHC797	H. Ogura, T. Itoh, and K. Kikuchi, *J. Heterocycl. Chem.* **6**, 797 (1969).
69JMC962	J. M. Singh. *J. Med. Chem.* **12**, 962 (1969).
69MI1	L. M. Jackman and S. Sternhell, "Applications of NMR Spectroscopy in Organic Chemistry," p. 202. Pergamon, Oxford, 1969.
69MI2	M. O. Lozinskii, A. F. Shiranyuk, and P. S. Pelkis, *Dopov. Akad. Nauk Ukr. RSR, Ser. B: Geol., Khim. Biol. Nauki* **31**, 1096 (1969) [*CA* **73**, 14767 (1970)].
69T4681	R. Cahill, R. C. Cookson, and T. A. Crabb, *Tetrahedron* **25**, 4681 (1969).
69URP230823	P. M. Kochergin and A. K. Krasovskii, USSR Pat. 230,823 (1969) [*CA* **70**, 68375 (1969)].

70CC890	A. F. Camerson, N. J. Hair, N. F. Elmore, and P. J. Taylor, *J. C. S. Chem. Commun.*, 890 (1970).
70IJC10	H. S. Chaudhary, C. S. Panda, and H. K. Pujari, *Indian J. Chem.* **8,** 10 (1970).
70IJC885	V. K. Chadha, H. S. Chaudhary, and H. K. Pujari, *Indian J. Chem.* **8,** 885 (1970).
70JHC1231	D. L. Trepanier and P. E. Krieger, *J. Heterocycl. Chem.* **7,** 1231 (1970).
70JIC579	B. D. Singh and D. N. Chaudhary, *J. Indian Chem. Soc.* **47,** 579 (1970).
70JIC1059	S. K. P. Sinha and D. N. Chaudhary, *J. Indian Chem. Soc.* **47,** 1059 (1970).
70JMC1018	J. M. Singh, *J. Med. Chem.* **13,** 1018 (1970).
70LA132	A. Mustafa, M. I. Ali, and A. Abou-State, *Justus Liebigs Ann. Chem.*, **740,** 132 (1970) [*CA* **74,** 13061 (1971)].
70M11	I. I. Chizhevskaya and M. I. Zavadskaya, *Vestsi. Akad. Navuk. Belarus, SSR, Ser. Khim. Navuk.*, 77 (1970) [*CA* **74,** 52890 (1971)].
70URP256774	M. O. Luzinskii, A. F. Shivanyuk, and P. S. Pelkis, USSR Pat. 256,774 (1970) [*CA* **72,** 132733 (1970)].
70USP3475424	P. H. L. Wei and S. C. Bell, U.S. Pat. 3,475,424 (1970) [*CA* **72,** 31850 (1970)].
70USP3475425	P. H. L. Wei and S. C. Bell, U.S. Pat. 3,475,425 (1970) [*CA* **72,** 31850 (1970)].
71ACR1	R. Hofman, *Acc. Chem. Res.* **4,** 1 (1971).
71CI(L)705	F. W. Short, B. C. Littleton, and J. L. Johnson, *Chem. Ind. (London)*, 705 (1971).
71IJC1216	V. K. Chadha, K. S. Sharma, and H. K. Pujari, *Indian J. Chem.* **9,** 1216 (1971).
71JCS(B)1648	G. V. Boyd and A. J. H. Summers, *J. Chem. Soc. B*, 1648 (1971).
71JCS(C)262	S. V. Kessar, M. Singh, V. K. Ahuja, and A. K. Lumb, *J. Chem. Soc. C*, 262 (1971).
71JCS(C)266	S. V. Kessar, P. Jit, K. P. Mundra, and A. K. Lumb, *J. Chem. Soc. C*, 266 (1971).
71JCS(C)3602	R. B. Blackshire and C. J. Sharpe, *J. Chem. Soc. C*, 3602 (1971).
71JHC621	D. L. Trepanier and P. E. Krieger, *J. Heterocycl. Chem.* **8,** 621 (1971).
71JHC1011	A. Shafiee and I. Lalezari, *J. Heterocycl. Chem.* **8,** 1011 (1971).
71JOC10	K. T. Potts and S. Husain, *J. Org. Chem.* **36,** 10 (1971).
71JOC1846	K. T. Potts and R. Armbruster, *J. Org. Chem.* **36,** 1846 (1971).
71JPR849	C. S. Bhandari, W. S. Mahnot, and N. S. Sogani, *J. Prakt. Chem.* **313,** 849 (1971) [*CA* **76,** 85346 (1972)].
71KGS93	I. I. Chizhevskaya and M. I. Zavadskaya, *Khim. Geterotsikl. Soedin.*, 93 (1971) [*CA* **77,** 164592 (1972)].
71KGS393	A. N. Krasovskii, P. M. Kochergin, and T. E. Kozlovskaya, *Khim. Geterotsikl. Soedin.* **7,** 393 (1971) [*CA* **76,** 14433 (1972)].
71KGS471	M. O. Lozinskii, A. F. Shivanyuk, and P. S. Pelkis, *Khim. Geterotsikl. Soedin.* **7,** 471 (1971) [*CA* **76,** 25184 (1972)].
71KGS822	A. N. Krasovskii, P. M. Kochergin, and A. B. Roman, *Khim. Geterotsikl. Soedin.* **7,** 822 (1971) [*CA* **76,** 25169 (1972)].
71KGS930	M. O. Lozinskii, A. F. Shivanyuk, and P. S. Pelkis, *Khim. Geterotiskl. Soedin.* **7,** 930 (1971) [*CA* **76,** 34200 (1972)].
72IJC274	J. Mohan and H. K. Pujari, *Indian J. Chem.* **10,** 274 (1972).
72IJC766	H. S. Chaudhary and H. K. Pujari, *Indian J. Chem.* **10,** 766 (1972).

72JPR785	A. Mustafa, M. I. Ali, M. A. Abou-State, and A-E. G. Hammam, *J. Prakt. Chem.* **314**, 785 (1972) [*CA* **78**, 72002 (1973)].
72KGS399	E. G. Knysh, A. N. Krasovskii, P. M. Kochergin, and P. M. Shabelnik, *Khim. Geterotsikl Soedin.* 399 (1972) [*CA* **77**, 88398 (1972)].
72MI1	M. Lacova and F. Volna, *Acta Fac. Rerum Nat. Univ. Comenianae, Chim.*, 1 (1972) [*CA* **79**, 18638 (1973)].
72MI2	J. Mohan, V. K. Chadha, H. S. Chaudhary, B. D. Sharma, H. K. Pujari, and L. N. Mohapatra, *Indian J. Expl. Biol.*, **10**, 37 (1972).
72ZOR1722	Yu. V. Svetkin, A. N. Minlibaeva, and A. G. Mansurova, *Zh. Org. Khim.* **8**, 1722 (1972) [*CA* **77**, 139872 (1972)].
73CPB270	H. Nagase, *Chem. Pharm. Bull. 21*, 270 (1973).
73GEP2160655	R. E. Manning, Ger. Pat. 2,160,655 (1973) [*CA* **79**, 66413 (1973)].
73IJC747	J. Mohan, V. K. Chadha, and H. K. Pujari, *Indian J. Chem.* **11**, 747 (1973).
73IJC1119	J. Mohan, V. K. Chadha, and H. K. Pujari, *Indian J. Chem.* **11**, 1119 (1973).
73KGS424	T. A. Kranistskaya, I. V. Smolanak, and A. L. Vais, *Khim. Geterotsikl. Soedin.*, 424 (1973) [*CA* **78**, 159509 (1973)].
73USP3732219	P. E. Krieger, U.S. Pat. 3,732,219 (1973) [*CA* **79**, 32109 (1973)].
74EGP105235	K. Peseke, Ger. (East) Pat. 105,235 (1974) (*CA* **81**, 169568 (1974)].
74FRP2164490	R. E. Manning, Fr. Pat. 2,164,490 (1974) [*CA* **80**, 27308 (1974)].
74IJC287	K. S. Dhaka, J. Mohan, V. K. Chadha, and H. K. Pujari, *Indian J. Chem.* **12**, 287 (1974).
74JOC167	D. L. Coffen, J. P. DeNoble, E. L. Evans, G. F. Field, R. I. Fryer, D. A. Katonak, B. J. Mandel, L. H. Sternbach, and W. J. Zally, *J. Org. Chem.* **39**, 167 (1974).
74JOC3619	K. T. Potts, J. Baum, E. Houghton, D. N. Roy, and U. P. Singh, *J. Org. Chem.* **39**, 3619 (1974).
74JOC3627	K. T. Potts, E. Houghton, and U. P. Singh, *J. Org. Chem.* **39**, 3627 (1974).
74JOC3631	K. T. Potts, J. Baum, and E. Houghton, *J. Org. Chem.* **39**, 3631 (1974).
74JPR147	M. I. Ali, M. A. Abou-State, and A. F. Ibrahim, *J. Prakt. Chem.* **316**, 147 (1974) [*CA* **81**, 49638 (1974)].
74JPR163	M. I. Ali, A. A. El-Sayed, and H. A. Hammouda, *J. Prakt. Chem.* **316**, 163 (1974).
74LA206	H. C. J. Ottenheijm, N. P. E. Vermenlen, and L. F. J. Breuer, *Liebigs Ann. Chem.*, 206 (1974) [*CA* **81**, 3813 (1974)].
74TH1	J. Mohan, Ph.D. Thesis, Kurukshetra University, Kurukshetra (1974).
75BRP1392499	D. J. Fry, G. E. Ficken, C. J. Palles, and A. W. Yates, Br. Pat. 1,392,499 (1975) [*CA* **83**, 149115 (1975)].
75FRP223371	Imperial Chemical Industries Ltd., Fr. Pat. 223,371 (1975) [*CA* **82**, 171031 (1975)].
75GEP2317109	R. T. Fox, J. R. Hadfield, P. Doyle, and B. Balasubramanyan, Ger. Pat. 2,317,109 (1975) [*CA* **82**, 31348 (1975)].
75IJC109	M. I. Ali, A. M. A-. Elfattah, H. A. Hammouda, and S. M. Hussain, *Indian J. Chem.* **13**, 109 (1975).
75IJC238	A. H. Harash, M. H. Elnagdi, M. Ezz-El-Din Sobby, and K. M. Foda, *Indian J. Chem.* **13**, 238 (1975).
75JAP7412699	Y. Kuwada, T. Souda, and K. Muguro, Jpn Pat. 74/12,699 (1975) [*CA* **83**, 28293 (1975)].

75JAP7495997	G. Hasegawa and A. Kotani, Jpn. Pat. 74/95,997 (1975) [*CA* **82**, 156299 (1975)].
75JAP7552065	G. Hasegawa and A. Kotani, Jpn. Pat. 75/52,065 (1975) (*CA* **83**, 206268 (1975)].
75JCS(P2)1294	B. T. Buzu, P. R. Olivate, R. Ritner, C. Trufen, H. Viertier, and B. Wladislaw, *J. C. S. Perkin 2*, 1294 (1975).
75JHC1207	S. C. Bell, C. Gochman, and P. H. L. Wei, *J. Heterocycl. Chem.* **12**, 1207 (1975).
75JIC1193	P. N. Dhal and A. Nayak, *J. Indian Chem. Soc.* **52**, 1193 (1975).
75KGS47	L. V. Zavyalova, N. K. Rozhkova, and K. L. Seitanidi, *Khim. Geterotsikl. Soedin.* **11**, 47 (1975). [*CA* **83**, 9941 (1975)].
75OMR617	U. Vogeli and W. von Philipsborn, *Org. Magn. Reson.* **7**, 617 (1975).
75USP3850948	D. L. Coffen and R. I. Fryer, U.S. Pat. 3,850,948 (1975) [*CA* **82**, 140211 (1975)].
76AP928	K. C. Liu, T. Y. Tuan, and B. J. Shih, *Arch. Pharm. (Weinheim Ger.)* **309**, 928 (1976) [*CA* **86**, 139986 (1977)].
76JAP74109398	Y. Kuwada, T. Souda, and K. Muguro, Jpn. Pat. 74 109,398 (1976) [*CA* **84**, 44177 (1976)].
76JAP74126828	Imperial Chem. Industries Ltd., Jpn. Pat. 74 126,828 (1976) [*CA* **85**, 8676 (1976)].
76JOC813	K. T. Potts, J. Baum, S. K. Datta, and E. Houghton, *J. Org. Chem.* **41**, 813 (1976).
76JOC818	K. T. Potts, J. Baum, and E. Houghton, *J. Org. Chem.* **41**, 818 (1976).
76JOC3863	C. A. Kingsbergy, D. Draney, A. Sopchik, W. Rissler, and D. Durham, *J. Org. Chem.* **41**, 3863 (1976).
76JPR12	M. I. Ali, A. B. Mostafa, and A. A. Soliman, *J. Prakt. Chem.* **318**, 12 (1976).
76JPR168	M. I. Ali, A. A. El-Sayed, and H. A. Hammouda, *J. Prakt. Chem.* **318**, 168 (1976).
76T615	L. A. Summers, *Tetrahedron* **32**, 615 (1976).
76USP3897446	J. B. Hunter, U.S. Pat. 3,897,446 (1976) [*CA* **84**, 44183 (1976)].
76USP3901879	D. L. Coffen and R. I. Fryer, U.S. Pat. 3,901,879 (1976) [*CA* **84**, 44191 (1976)].
76USP3906001	D. L. Coffen and R. I. Fryer, U.S. Pat. 3,906,001 (1976) [*CA* **84**, 31147 (1976)].
76ZN(B)111	A. F. A. Shalaby, H. A. Daboun, and M. A. Abdil-Aziz, *Z. Naturforsch., B: Anorg. Chem., Org. Chem.* **31B**, 111 (1976) [*CA* **84**, 121725 (1976)].
76ZN(B)380	N. A. Kassab, S. O. Abdalla, and H. A. R. Ead, *Z. Naturforsch., B: Anorg. Chem., Org. Chem.* **31B**, 380 (1976) [*CA* **85**, 32926 (1976)].
76ZN(B)1397	A. A. W. Soliman, *Z. Naturforsch., B: Anorg. Chem., Org. Chem.*, **31B**, 1397 (1976) [*CA* **86**, 55396 (1977)].
77IJC(B)46	A. Singh, K. S. Dhaka, H. S. Chaudhary, and H. K. Pujari, *Indian J. Chem., Sect. B* **15B**, 46 (1977).
77JCS(P2)1070	Chi-Kim Chan, J. C. N. Ma, and T. C. W. Mak, *J. C. S. Perkin 2*, 1070 (1977).
77JHC1093	S. P. Singh, S. S. Parmar, and B. R. Pandey, *J. Heterocycl. Chem.* **14**, 1093 (1977).
77MI1	K. C. Liu, J. Y. Tuan, and B. J. Shih, *J. Chin. Chem. Soc. (Taipei)* **24**, 65 (1977) [*CA* **87**, 135200 (1977)].

77MI2	V. Farcasan and I. Balazs, *Stud. Univ. Babes-Bolyai [Ser.] Chem.* **22,** 30 (1977) [*CA* **87,** 68238 (1977)].
77MI3	G. Kempter, H. J. Ziegner, G. Moser, and W. Natho, *Wiss. Z. Paedagog. Hochsch "Karl Liebknecht" Postdam* **21,** 5 (1977) [*CA* **89,** 163560 (1978)].
77S839	S. Kambe, K. Saito, H. Kishi, T. Hayashi, and A. Sakurai, *Synthesis,* 839 (1977).
77TL275	Y. Akasaki, M. Hatamas, and M. Fukuyama, *Tetrahedron Lett.,* 275 (1977).
77ZN(B)94	M. I. Ali, H. A. Hammouda, and A.-E. E. A-Elfattah, *Z. Naturforsch., B: Anorg. Chem., Org. Chem.* **32B,** 94 (1977).
78BCJ3091	H. A. Hammouda and N. Ishikawa, *Bull. Chem. Soc. Jpn.* **51,** 3091 (1978).
78HCA607	U. Vogeli, W. von Pilipsborn, K. Nagarajan, and M. D. Nair, *Helv. Chim. Acta* **61,** 607 (1978).
78IJC(B)329	R. P. Gupta and H. K. Pujari, *Indian J. Chem. Sect. B* **16B,** 329 (1978).
78IJC(B)478	A. Singh, R. N. Handa, and H. K. Pujari, *Indian J. Chem., Sect. B* **16B,** 478 (1978).
78IJC(B)537	R. P. Gupta, M. L. Sachdeva, R. N. Handa, and H. K. Pujari, *Indian J. Chem., Sect. B* **16B,** 537 (1978).
78JIC928	V. K. Singh and K. C. Joshi, *J. Indian Chem. Soc.* **55,** 928 (1978).
78JOC2697	K. T. Potts and D. R. Chaudhary, *J. Org. Chem.* **43,** 2697 (1978).
78MI1	M. I. Ali and A. H. Hammam, *J. Chem. Eng. Data* **23,** 351 (1978) [*CA* **89,** 146865 (1978)].
78MI2	H. Kishi and S. Kambe, *Mass Spectrom.* **26,** 259 (1978).
78OMR197	S. Braun, *Org. Magn. Reson.* **11,** 197 (1978).
78RRC820	G. Danila, *Rev. Roum. Chim.* **29,** 820 (1978) [*CA* **90,** 72086 (1979)].
78RRC1152	G. Danila, *Rev. Roum. Chim.* **29,** 1152 (1978).
78TH1	R. Pal, Ph.D. Thesis, Kurukshetra University, Kurukshetra (1978).
78TL2621	A. McKillop, G. C. A. Bellinger, P. N. Preston, A. Davidson, and T. J. King, *Tetrahedron Lett.,* 2621 (1978).
79AHC83	G. R. Newkome and A. Nayak, *Adv. Heterocycl. Chem.* **25,** 83 (1979).
79AP776	K. C. Liu, H.-H. Chen, L.-C. Lee, and J.-W. Chern, *Arch. Pharm. (Weinheim, Ger.)* **312,** 776 (1979) [*CA* **92,** 41875 (1980)].
79HC	J. V. Metzger, ed., "Thiazole and its Derivatives," Vol. 34, Parts 1-3. Wiley (Interscience), New York, 1979.
79IJC(17B)572	R. P. Gupta, R. N. Handa, and H. K. Pujari, *Indian J. Chem., Sect. B* **17B,** 572 (1979).
79IJC(18B)39	P. B. Talukdar, S. K. Sengupta, and A. K. Dutta, *Indian J. Chem., Sect. B* **18B,** 39 (1979).
79IJC(18B)479	M. D. Nair, K. Nagarajan, and J. A. Desai, *Indian J. Chem., Sect. B* **18B,** 479 (1979).
79JOC1816	J. J. Wade, *J. Org. Chem.* **44,** 1816 (1979).
79JOC3803	K. T. Potts and S. Kanemesa, *J. Org. Chem.* **44,** 3803 (1979).
79JOC3994	H. D. H. Showalter, M. T. Shipchandler, L. A. Mitscher, and E. W. Hagaman, *J. Org. Chem.* **44,** 3994 (1979).
79MI1	M. I. Ali, M.A-F. El-Kaschef, A. G. Hammam, and S. A. Khallaf, *J. Chem. Eng. Data* **24,** 377 (1979).
79MI2	H. Kishi and S. Kambe, *Mass. Spectrom.* **27,** 47 (1979).
79MI3	R. P. Gupta, M. L. Sachdeva, K. S. Dhaka, V. K. Chadha, and H. K. Pujari, *Ann. Soc. Sci. Bruxelles, Ser. I* **93,** 137 (1979).

79TL53	K. Nagarajan, M. D. Nair, and J. A. Desai, *Tetrahedron Lett.*, 53 (1979).
79UKZ1096	A. N. Krasovskii, N. A., Klyuev, A. B. Roman, I. I. Soroka, P. M. Kochergin, and A. B. Belikov, *Ukr. Khim. Zh.* **45**, 1096 (1979) [*CA* **92**, 94304 (1980)].
80EUP2978	D. C. H. Bigg, Eur. Pat. 2,978 (1980) [*CA* **92**, 163956 (1980)].
80H149	S. Bala, M. L. Sachdeva, R. N. Handa, and H. K. Pujari, *Heterocycles* **14**, 149 (1980).
80HC	G. Tennant, *in* "Benzimidazoles and congeneric Tricyclic Compounds" (P. N. Preston, ed.), Vol. 40, Part 2, pp. 96 and 170. Wiley (Interscience), New York, 1980.
80IJC(B)1035	G. D. Gupta, R. P. Gupta, and H. K. Pujari, *Indian J. Chem., Sect. B* **19B**, 1035 (1980).
80LA168	G. Bartels, R. Hinze, and D. Wullbrandt, *Liebigs Ann. Chem.*, 168 (1980) [*CA* **92**, 197893 (1980)].
80MI1	H. K. Pujari, *J. Chem. Sci.* **6**, 73 (1980).
80MI2	O. M. Krasovskvi, *Farm. Zh. (Kiev.)*, 38 (1980) [*CA* **93**, 168186 (1980)].
80OMR200	U. Vogeli, D. Herz, and W. von Pilipsborn, *Org. Magn. Reson.* **13**, 200 (1980).
81ACH197	T. Lorand, D. Szabo, A. Foldesi, and A. Neszmelyi, *Acta Chim. Acad. Sci. Hung.* **108**, 197 (1981) [*CA* **96**, 162638 (1982)].
81CPB1876	K. Tanaka, M. Ino, and Y. Murakami, *Chem. Pharm. Bull.* **29**, 1876 (1981).
81CRV175	S. P. Singh, S. S. Parmar, K. Raman, and V. I. Stenberg, *Chem. Rev.* **81**, 175 (1981).
81IJC(B)294	K. K. Jain and H. K. Pujari, *Indian J. Chem., Sect. B* **20B**, 294 (1981).
81IJC(B)296	G. D. Gupta and H. K. Pujari, *Indian J. Chem., Sect. B* **20B**, 296 (1981).
81IJC(B)538	P. B. Talukdar, S. K. Sengupta, and A. K. Datta, *Indian J. Chem., Sect. B* **20B**, 538 (1981).
81JCS(P1)415	R. M. Acheson and J. D. Wallis, *J. C. S. Perkin* 1, 415 (1981).
81JIC1117	A. Ali and R. K. Saksena, *J. Indian Chem. Soc.* **58**, 1117 (1981).
81MI1	A. G. Hammam and M. I. Ali, *J. Chem. Eng. Data* **26**, 99 (1981).
81MI2	A. G. Hammam and M. I. Ali, *J. Chem. Eng. Data* **26**, 101 (1981).
81MI3	M. I. Ali, A. G. Hammam, and N. M. Youssef, *J. Chem. Eng. Data* **26**, 214 (1981).
81MIP1	S. Groszkowski and M. Skopinska-Szmidt, Pol. Pat. 104,845 (1981) [*CA* **95**, 81018 (1981)].
81OMR316	R. M. Letcher and R. M. Acheson, *Org. Magn. Reson.* **16**, 316 (1981).
81S635	M. R. H. Elmoghayar, M. K. A. Ibraheim, A. H. H. Elghandour, and M. H. Elnagdi, *Synthesis*, 635 (1981).
81S981	M. Baudy-Floch and A. Robert, *Synthesis*, 981 (1981).
81ZN(B)501	A. F. A. Shalaby, M. A. Abdil-Aziz, and S. S. M. Boghdadi, *Z. Naturforsch, B: Anorg. Chem., Org. Chem.* **36B**, 501 (1981) [*CA* **95**, 97667 (1981)].
82CPB2714	K. Tanaka, M. Shimazaki, and Y. Murakami, *Chem. Pharm. Bull.* **30**, 2714 (1982).
82IJC(B)315	K. K. Jain and H. K. Pujari, *Indian J. Chem., Sect. B* **21B**, 315 (1982).
82IJC(B)517	P. B. Talukdar, S. K. Sengupta, and A. K. Datta, *Indian J. Chem., Sect. B* **21B**, 517 (1982).
82MI1	A. G. Hammam, *Egypt. J. Chem.* **25**, 471 (1982) [*CA* **99**, 212490 (1983)].
82NKK456	T. Nakamori, Y. Kogure, and T. Kasai, *Nippon Kagaku Kaishi*, 456 (1982) [*CA* **97**, 72297 (1982)].

82TH1	K. K. Jain, Ph.D. Thesis, Kurukshetra University, Kurukshetra (1982).
82USP4304908	M. D. Frishberg and J. J. Krutak, U.S. Pat. 4,304,908 (1982) [*CA* **96**, 144470 (1982)].
83AP394	H. A. Daboum and M. A. Abdil-Aziz, *Arch. Pharm. (Weinheim Ger.)* **316**, 394 (1983) [*CA* **99**, 22423 (1983)].
83AP569	K. C. Liu, J. W. Chern, M. H. Yen, and Y. O. Lin, *Arch. Pharm. (Weinheim Ger.)* **316**, 569 (1983).
83AP985	A. E. M. Khalifa, E. M. Zayed, G. H. Tammam, and A. A. A. Elbamani, *Arch. Pharm. (Weinheim Ger.)* **316**, 985 (1983) [*CA* **100**, 51516 (1984)].
83EGP156815	K. Peseke and C. Vogel, Ger. (East) Pat. 156,815 (1983) [*CA* **98**, 143409 (1983)].
83H1021	F. A. Khalifa, B. Y. Raid, and F. H. Hafez, *Heterocycles* **20**, 1021 (1983).
83H1549	M. R. Chaurasia and A. K. Sharma, *Heterocycles* **20**, 1549 (1983).
83IJC(B)243	P. B. Talukdar, S. K. Sengupta, and A. K. Datta, *Indian J. Chem., Sect. B* **22B**, 243 (1983).
83IJC(B)785	N. Soni, J. P. Barthwal, T. K. Gupta, T. N. Bhalla, S. S. Parmar, and K. P. Bhargava, *Indian J. Chem., Sect. B* **22B**, 785 (1983).
83JIC970	S. N. Dehuri and A. Nayak, *J. Indian Chem. Soc.* **60**, 970 (1983).
83MI1	H. A. Daboun and M. A. Abdil-Aziz, *Egypt. J. Chem.* **26**, 401 (1983) [*CA* **101**, 211081 (1984)].
83MI2	A. G. Hammam, A. S. Ali, and N. M. Yousif, *Egypt J. Chem.* **26**, 461 (1983) [*CA* **101**, 230459 (1984)].
83MI3	M. Szajda and E. Wryzykiewicz, *Pol. J. Chem.* **57**, 1027 (1983) [*CA* **102**, 45686 (1985)].
83ZN(B)781	G. E. H. Elgemeie, H. A. Elfahham, S. M. E. Hassan, and M. H. Elnagdi, *Z. Naturforsch., B: Anorg. Chem., Org. Chem.* **38B**, 781 (1983).
84CC213	K. T. Potts, K. Bordeaux, W. Kuehnlig, and R. Salsbury, *J. C. S. Chem. Commun.*, 213 (1984).
84IJC(B)316	P. B. Talukdar, S. K. Sengupta, and A. K. Datta, *Indian J. Chem., Sect. B* **23B**, 316 (1984).
84JCS(P1)2707	L. I. Giannola, G. Giammona, and S. Palazzo, *J. C. S. Perkin 1*, 2707 (1984).
84JHC1885	M. R. H. Elmoghayar, A. G. H. El-Agmey, M. Y. A-S. Nasr, and M. M. M. Sallam, *J. Heterocycl. Chem.* **21**, 1885 (1984).
84JIC89	V. M. Rao and V. MallaReddy, *J. Indian Chem. Soc.* **61**, 89 (1984).
84JIC1050	G. D. Gupta and H. K. Pujari, *J. Indian Chem. Soc.* **61**, 1050 (1984).
84JIC1053	K. Jain, K. K. Jain, V. K. Chadha, and R. N. Handa, *J. Indian Chem. Soc.* **61**, 1053 (1984).
84MI1	A. R. Katritzky and C. W. Rees, eds., "Comprehensive Heterocyclic Chemistry," Vols. 1–8. Pergamon, Oxford, 1984.
84MI2	S. O. Abd Allah, M. R. H. El-Moghayar, and H. A. Ead, *An. Quim., Ser. C.* **80**, 232 (1984) [*CA* **103**, 53992 (1985)].
84MI3	K. Peseke, R. R. Palacio, and C. Vogel, *Wiss. Z. Wilhelm-Pieck.-Univ. Rostock, Math.-Naturwiss. Reihe* **33**, 74 (1984) [*CA* **106**, 84575 (1987)].
84ZN(B)824	K. U. Sadek, A. A. Abtisam, A. E. E. Mourad, and M. H. Elnagdi, *Z. Naturforsch., B: Anorg. Chem., Org. Chem.* **39B**, 824 (1984) [*CA* **101**, 230397 (1984)].

85AP502	K.-C. Liu and L. Y. Hsu, *Arch. Pharm. (Weinheim Ger.)* **318**, 502 (1985) [*CA* **103**, 178233 (1985)].
85IJC(B)1224	B. R. Sharma and H. K. Pujari, *Indian J. Chem., Sect. B* **24B**, 1224 (1985).
85IJC(B)1227	V. Bindal and H. K. Pujari, *Indian J. Chem., Sect. B* **24B**, 1227 (1985).
85JIC147	H. Singh, L. D. S. Yadav, and A. K. Singh, *J. Indian Chem. Soc.* **62**, 147 (1985).
85M11	A. A. El-Barbary, *Proc. Pak. Acad. Sci.* **22**, 55 (1985) [*CA* **105**, 97412 (1986)].
85TH1	B. R. Sharma, Ph.D. Thesis, Kurukshetra University, Kurukshetra (1985).
85TH2	V. K. Bindal, Ph.D. Thesis, Kurukshetra University, Kurukshetra (1985).
85TH3	Sat Narayan, Ph.D. Thesis, Kurukshetra University Kurukshetra (1985).
86HC	P. N. Preston, "Condensed Imidazoles," Vol. 46, pp. 214, 233, 235, and 237. Wiley (Interscience), New York, 1986.
86IJC(B)264	B. R. Sharma, V. Bindal, and H. K. Pujari, *Indian J. Chem., Sect. B* **25B**, 264 (1986).
86IJC(B)267	S. Narayan, V. Kumar, and H. K. Pujari, *Indian J. Chem., Sect. B* **25B**, 267 (1986).
86IJC(B)354	V. Bindal, B. R. Sharma, and H. K. Pujari, *Indian J. Chem., Sect. B* **25B**, 354 (1986).
86IJC(B)776	A. P. Prasad, A. N. Rao, T. Ramalingam, and P. B. Sattur, *Indian J. Chem., Sect. B* **25B**, 776 (1986).
86IJC(B)807	V. Bindal, K. Jain, R. N. Handa, and H. K. Pujari, *Indian J. Chem., Sect. B* **25B**, 807 (1986).
86IJC(B)812	R. Dahiya and H. K. Pujari, *Indian J. Chem., Sect. B* **25B**, 812 (1986).
86JCS(P1)2095	L. I. Giannola, S. Palazzo, L. Lanartina, L. R. Sansevevino, and P. Sabatino, *J. C. S. Perkin 1*, 2095 (1986).
86JHC1359	V. St. Georgiev, G. A. Kennett, L. A. Radov, D. K. Kamp, and L. A. Trusso, *J. Heterocycl. Chem.* **23**, 1359 (1986).
86KGS1690	G. V. Bespalova, V. A. Sedavkina, E. V. Ponomareva, A. D. Shebaldova, and V. I. Labunskaya, *Khim. Geterotsikl. Soedin.*, 1690 (1986) [*CA* **107**, 217389 (1987)].
86M105	S. A. M. Osman, G. E. H. Elgemeie, G. A. M. Nawar, and M. H. Elnagdi, *Monatsh. Chem.* **117**, 105 (1986).
86PHA324	S. M. Rida, H. M. Salama, I. M. Labouta, and Y. S. A. Ghany, *Pharmazie* **41**, 324 (1986) [*CA* **106**, 119772 (1987)].
86USP4588812	S. A. Saeva and V. A. Georgiev, U.S. Pat. 4,588,812 (1986) [*CA* **105**, 226618 (1986)].
87IJC(B)216	M. K. A. Ibrahim, M. R. H. El-Moghayar, and M. A. F. Sharaf, *Indian J. Chem., Sect. B* **26B**, 216 (1987).
87IJC(B)437	J. Mohan and G. S. R. Anjaneyulu, *Indian J. Chem., Sect. B* **26B**, 473 (1987).
87IJC(B)532	V. Bindal and H. K. Pujari, *Indian J. Chem., Sect. B* **26B**, 532 (1987).
87IJC(B)556	M. S. Akhtar, M. Seth, and A. P. Bhaduri, *Indian J. Chem., Sect. B* **26B**, 556 (1987).
87IJC(B)693	S. Narayan and H. K. Pujari, *Indian J. Chem., Sect. B* **26B**, 693 (1987).
87IJC(B)739	S. Narayan, V. Bindal, B. R. Sharma, R. Dahiya, G. D. Gupta, K. K.

	Jain, R. N. Handa, and H. K. Pujari, *Indian J. Chem., Sect. B* **26B**, 739 (1987).
87UP1	S. Kumar and H. K. Pujari, unpublished results (1987).
87UP2	R. Dubey and H. K. Pujari, unpublished results (1987).
87UP3	R. Dahiya and H. K. Pujari, unpublished results (1987).
87UP4	S. Kumar, R. Dahiya, S. Narayan, V. Bindal, B. R. Sharma, K. Jain, G. D. Gupta, K. K. Jain, R. N. Handa, and H. K. Pujari, unpublished results (1987).
87UP5	H. K. Pujari, Y. Murakami, and T. Watanabe, unpublished results (1987).
88IJC(B)121	B. R. Sharma and H. K. Pujari, *Indian J. Chem., Sect. B* **27B**, 121 (1988).

Advances in Amination of Nitrogen Heterocycles

HELMUT VORBRÜGGEN

*Research Laboratories of Schering AG,
Berlin–Bergkamen, Federal Republic of Germany*

I. Introduction	118
II. Theory and General Mechanisms; Addition–Elimination Mechanism	118
III. Reactivity Factors in Aminations	119
A. Reactivity of Different Nitrogen Heterocycles	119
B. Reactivity of the Different Amines	120
C. Influence of Solvents	123
D. Effect of the Leaving Group	124
IV. Aminations via Leaving Groups Derived from Hydroxy-N-Heterocycles	125
A. Introduction	125
B. Chloro and Bromo Groups	127
1. Introduction	127
2. Preparation of Chloro- or Bromo-N-Heterocycles	127
3. Amination of Chloro- or Bromo-N-Heterocycles	135
C. Alkoxy or Aryloxy Groups	139
D. Silyloxy Groups	145
1. Introduction	145
2. Trimethylsilanol or Dimethylsilanol as a Leaving Group	146
3. One-Pot Reaction	149
4. Silylation–Amination of Nucleosides	150
5. Silylation–Amination of Hydroxy-N-Heterocycles	154
E. O-Sulfonates	159
F. O-Phosphoroamidates	162
G. 1,2,4-Triazoles	165
H. Quaternary Ammonium Salts	168
I. Sulfides, Sulfoxides, and Sulfones	169
V. Miscelleanous Aminations	173
A. Aminations via Nitrile, Trichloromethyl, and Carboxy Moieties as Leaving Groups	173
B. Transaminations and Dimroth Alkylations	176
C. Chichibabin Aminations	179
D. Amination by Vicarious Nucleophilic Substitution	182
VI. Comparison of Different Amination Methods	183
References	185

I. Introduction

Many natural products, drugs, or plant protection agents are amino-nitrogen heterocycles. Thus, methods for their efficient synthesis are of considerable practical and technical interest. This chapter will discuss primarily the advances made in the amination of nitrogen heterocycles since the comprehensive review by Shepherd and Fedrick on "Reactivity of Azine, Benzoazine, and Azinoazine Derivatives with Simple Nucleophiles" was published in this series in 1965 (65AHC145).

Because the mechanisms of nucleophilic substitution of N-heterocycles containing a leaving group as well as the reactivities of different types of N-heterocycles were dealt with in detail in the article by Shepherd and Fedrick, only the most important facts will be discussed here. Furthermore, because of the vast number of recent publications and patents describing the standard technology of conversion of hydroxy-N-heterocycles into the corresponding chloro derivatives, which are subsequently aminated, only the general problems connected with this method as well as some pertinent applications of this standard methodology will be discussed.

II. Theory and General Mechanisms; Addition–Elimination Mechanism

The rate of the attack of amines on nitrogen heterocycles containing a leaving group X, followed by subsequent elimination of X, is dependent on the nature of (a) the heterocycle, (b) the amine, (c) the solvent, and (d) the leaving group and will be dealt with in Section III. The aminations of N-heterocycles containing a leaving group X, as exemplified by **1** or **4**, lead to addition products **2** or **5** and are followed by subsequent eliminations of leaving group X to finally give amination products **3** or **6**. The formation of the addition–elimination (AE) intermediates **2** and **5** is favored by electron-withdrawing substituents at the heterocyclic system and additional nitrogens in the ring as well as the addition of protons and Lewis acids to the ring nitrogen in **1** or **4**. It is obvious that with X = H, intermediates **2** or **5** can be transformed into the aminated products only by oxidation, by vicarious nucleophilic amination as discussed in Sections V,C and V,D, or by elimination of metal hydride as in the Chichibabin reaction (Section V,C).

The addition of protons or Lewis acids to an α- or γ-ring nitrogen atom to facilitate the formation of AE intermediates **2** and **5**, and thus to accelerate the process of amination, was originally discovered by Banks (44JA1127) and subsequently investigated by a number of other authors (52JCS437; 54JCS1190; 46JCS1563; 73JHC511) and reviewed elsewhere (60AG294; 65AHC145). The structure of AE intermediates in heteroaromatic (72JA7927; 76TL4427; 79JOC2556) as well as aromatic [58JA6020; 79JA956; 85JCS(P2)87, 85JCS(P2)929, 85JOC649; 87JCS(P2)951, 87JCS(P2)987; 88JA3495, 88JA3512] nucleophilic substitution has been further examined, and evidence has been presented for single-electron transfer in some cases (88JA3495, 88JA3503, 88JA3512). Hydrochloric acid or H_2SO_4, preferably in the form of their ammonium salts—NH_4Cl, $(NH_4)_2SO_4$ or p-toluenesulfonic acid-dihydrate, camphorsulfonic acid, CF_3SO_3H or $(CH_3)_3SiOSO_2CF_3$—can be used as Lewis acids (84CB1523).

III. Reactivity Factors in Aminations

A. Reactivity of Different Nitrogen Heterocycles

As summarized by Shepherd and Fedrick (65AHC145), the aromaticity (and basicity) of heterocyclic systems increases with a decreasing number of nitrogen atoms. Consequently, the reactivity of nitrogen heterocycles with a leaving group X in α- or γ-position to a ring nitrogen atom decreases in the same order as illustrated by the following sequence (Scheme 1) (cf. 51JA4773; 54JOC1830; 66M11; 73CB3398).

SCHEME 1. Reactivities; X = Cl, Br, OSO$_2$R, OCH$_3$, OEt, and OSi(CH$_3$)$_3$.

Because of the annelated benzene ring, quinoline, isoquinoline, or phthalazine have enhanced reactivities compared to pyridine and pyridazine. Thus, 2-chloroquinoline reacts ~100 times faster with piperidine than 2-chloropyridine (66MI1; 73CB3398). Furthermore, an electron-attracting group such as the *o*- or *p*-nitro group, as in 2-chloro-3-nitropyridine, enhances the reactivity of the chlorine toward secondary amines. When compared to 2-chloropyridine, the enhancement is more than 10^6 (66MI1; 73CB3398). Because of this increase in aromaticity and basicity and consequent decrease in reactivity, the formation of the nonaromatic addition intermediates such as **2** and **5** becomes less and less favored and is most difficult in the case of 2-substituted pyridines, which therefore represent the ultimate test of each amination method.

Furthermore, the reactivity of a leaving group X in heterocyclic systems such as **7** (86M1305) and **8** (Section IV,D) is dramatically increased by conjugation with a carbonyl group. The reactivity of these systems with amines is further enhanced by addition of a proton or Lewis acid to the carbonyl group in **7** or **8** (cf. **150–166**).

B. REACTIVITY OF DIFFERENT AMINES

The more basic and sterically less hindered amines undergo a much more rapid amination than weakly basic amines, as indicated by the fol-

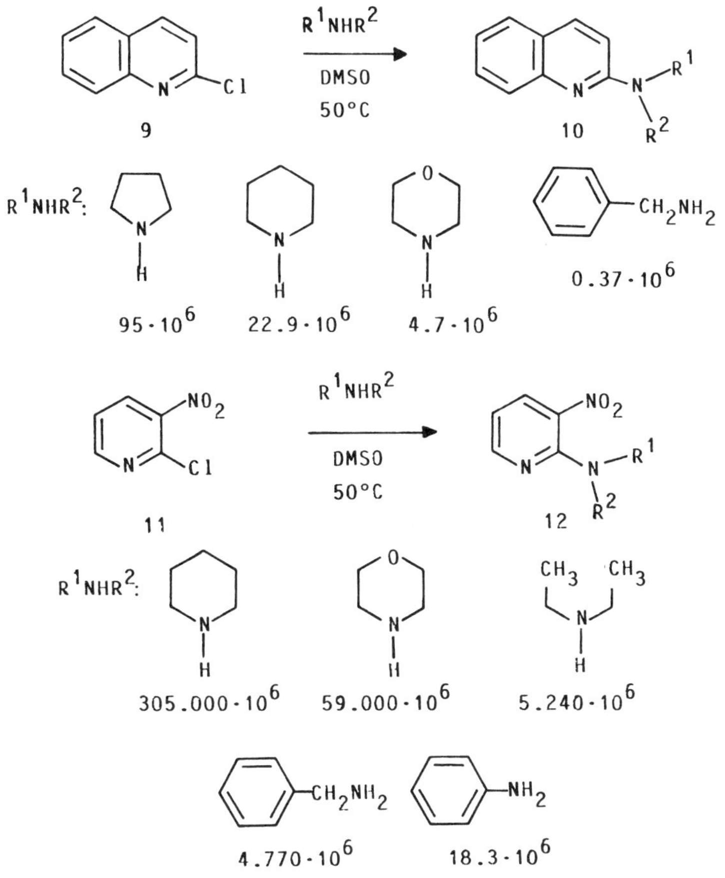

Sec. III.B]   AMINATION OF NITROGEN HETEROCYCLES   121

SCHEME 2

lowing rate constants (66MI1; 73CB3398) for the reaction of 2-chloroquinoline (9) and 2-chloro-3-nitropyridine (11) with amines to afford the corresponding aminated products 10 and 12 (Scheme 2). Significantly, the given order of reactivities is not influenced by changing the activating substituents, the leaving group, the solvent, or the reaction temperature. Piperidine, for example, usually reacts ~10^4 times faster than aniline.

The rate constant for aminolysis of a given 2-chloropyrimidine by an alkylamine is almost unaffected by increasing the chain length of by γ-branching of the chain. A β-branch has a small slowing influence, whereas one α-branch reduces the rate to 5%, and two α-branches reduce the rate to 0.1% of the corresponding n-alkylamine (69MI1). The reaction rates and activation energies of the reaction of piperidine with 2-bromopyridine and 2-bromoquinoline as well as with 2-chloropyrimidine and 2-chloroquinoline have been determined (51JA4773).

In addition to ammonia and primary and secondary amines, tertiary amines can also be reacted with chloro-nitrogen heterocycles to initially form quaternary salts (Section IV,H), which, on continued heating, are converted to the corresponding dialkylamino compounds and alkyl chlorides. Thus, 6-trifluoromethyluracil (13), on treatment with POCl$_3$ and N,N-dimethylaniline, affords not only 33% of the anticipated 2,4-dichloro-6-trifluoromethylpyrimidine (14), but also the 2-[N-methyl-N-phenylamino]-4-chloro-6-trifluoromethylpyrimidine (15) in 62% yield (84JHC1161). Analogously, 2,4-dichloro-6-trifluoromethylpyrimidine (14), on refluxing with excess triethylamine in toluene for 3 hr, gives 78% of 2-N,N-diethylamino-4-chloro-6-trifluoromethylpyrimidine (16) as well as 20% of 2-chloro-4-diethylamino-6-trifluoromethylpyrimidine (17) (87JHC205, 87JHC1243).

Cyclic tertiary amines, such as N-methylpyrrolidine, react with (1H,3H)-quinazoline-2,4-dione (18) in the presence of POCl$_3$ to give 2-(N-methyl-4-chlorobutylamino)-4-chloroquinazoline (19) in 67% yield (82CPB3471, 82H7, 82H15, 82H1879).

In the case of volatile amines, such as ammonia, methylamine, and dimethylamine, an autoclave must be used to permit reactions at higher temperatures; alternatively, silylated amines are applied (Section IV,D). Instead of these low-boiling and volatile amines, their corresponding acetates (cf. **128 → 129; 141 → 142**), formamides or phosphoroamidates can be used as solvents and reactants to permit reactions at higher temperature and normal pressure.

As demonstrated in Sections IV,B, IV,F, and IV,H, N-methylformamide, N,N-dimethylformamide (DMF) or hexamethylphosphoric triamide (HMPA) react readily with nitrogen heterocycles containing different leaving groups. A further possibility is the use of higher boiling hydrazine followed by subsequent cleavage of the N—N bond. Thus, 2-chloro-3-nitropyridine **(11)**, on heating with hydrazine in acetonitrile, affords 2-hydrazino-3-nitropyridine **(20)**, which, on Raney—nickel hydrogenation, furnishes 2,3-diaminopyridine **(21)** (85EUP159112). Analogously, 3-nitro-4-ethoxypyridine **(22)** gives rise to 3-nitro-4-hydrazinopyridine **(23)** in 89% yield, which can be hydrogenated to 3,4-diaminopyridine **(24)** in 90% yield. Ethyl carbazate, acethydrazide, benzhydrazide, or tosylhydrazide can be used instead of hydrazine (85EUP159112).

C. Influence of Solvents

The nucleophilic reaction between two uncharged partners is strongly dependent on the polarity of the solvent (69CRV1), as exemplified by the reaction between 4-nitrofluorobenzene and piperidine (63MI1) which is $\sim 10^4$ times faster in dimethylsulfoxide (DMSO), DMF, or N-methylpyrrolidone than in benzene and is furthermore base catalyzed by the piperidine (63MI1; 64CB3277). As mentioned in Section III,B, polar solvents such as N-methylformamide, DMF, N-methylpyrrolidone or HMPA can also serve as reactants in special cases to transform nitrogen heterocycles containing a leaving group into the corresponding amino compounds (Sections IV,F and IV,H).

In the case of silylation–amination (Section IV,D), usually no solvents are employed since silylation transforms the polar and often high-melting hydroxy–nitrogen heterocycles, as well as polar hydroxy amines, into nonpolar lipophilic silylated derivatives that mix readily without solvents (84CB1523).

D. Effect of the Leaving Group

The formation of reactive intermediates **2** and **5** (Section II) and their subsequent elimination of X^- to give the desired aminated products **3** and **6** is strikingly dependent on the nature of the leaving group X, as exemplified by the reaction of 2-substituted pyrimidines with amines (**25 → 26**) [69JCS(C)2720, 69MI1; 71JCS(C)1889; 79ACR198]. In Table I (69MI1),

TABLE I
Some Comparative Figures for Reactivity of 2-Substituted Pyrimidines

Pyrimidine	Approximate relative reactivity
2-SMe	20
2-OMe	80
2-Cl	3,000,000
4,6-diMe-2-2SMe	1
2-OMe-4,6-diMe	60
2-Cl-4,6-diMe	300,000
5-Br-2-SMe	80
5-Br-2-OMe	400
5-Br-2-Cl	60,000,000
2-SMe-5-NO_2	4,000,000
2-OMe-5-NO_2	100,000,000
2-Cl-5-NO_2	10,000,000,000,000

approximate comparative figures are given for the reactivity of such 2-substituted pyrimidines **25** toward aminolysis. Even the mildly electron-withdrawing 5-bromo substituent accelerates the aminolysis of 2-methoxypyrimidine by a factor of 5, whereas a 5-nitro substituent in **25** accelerates aminolysis of 2-chloropyrimidine by a factor of $\sim 10^7$!

Although the 2-methanesulfonyloxy or 2-trifluoromethanesulfonyloxy leaving groups are not included in Table I, substituted pyrimidines **25** or other nitrogen heterocycles containing a 2-trifluoromethanesulfonyloxy group can undergo nucleophilic substitutions several orders of magnitude faster than tosylates or bromides (79ACR198). The order of reactivity of leaving groups is, however, dependent on the system that undergoes nucleophilic substitution. Moreover, the reaction rate only reflects the nucleofugality of the leaving group if the leaving group is involved in the rate-determining step (79ACR198). Since the formation of AE intermediates **2** or **5** seems to exert an influence on the overall rate of aminations, the leaving groups X in **2** or **5** might not always be rate determining. The elucidation of the detailed reaction mechanism of these AE aminations would be highly desirable.

IV. Aminations via Leaving Groups Derived from Hydroxy-N-Heterocycles

A. Introduction

The thermal stability of hydroxy-N-heterocycles, which usually exist in the lactam form (Sections II–IV), has the consequence that the elimination of water during amination occurs only under forcing reaction conditions. In the absence of acidic catalysts, 2-(1H)-pyridinone (**27**) is aminated by ammonia, upon heating for $3\frac{1}{2}$ hr at 350°C and 100 bar, to give 2-amino-pyridine (**28**). The analogous amination of **27** with the weakly basic aniline furnishes, after 5 hr at 350°C, only 30% of 2-phenylamino-pyridine (**29**), whereas reaction of **27** with 2-aminopyridine gives rise to 20–30% 2,2'-dipyridylamine (**30**) (71GEP2032403).

In contrast to the unreactive 2-(1H)-pyridinone (**27**), the much more reactive 4-hydroxy-2-(1H)-pyridinone (**31**) (cf. **7** and **8** in Section III,A) is easily aminated by boiling it in excess benzylamine to give **32** in 89% yield (84S765). Some heterocycles, such as methyl-4-oxo-1,4-dihydroquinoline-2-carboxylate (**33**), are transformed by the very reactive chlorosulfonyl-isocyanate (CSI) or other reactive sulfonyl- or acylisocyanates in acetonitrile of 1,2-dichloroethane at room temperature with evolution of carbon dioxide to intermediates such as **34**, which are hydrolyzed by aqueous hydrochloric acid to methyl-4-aminoquinoline-2-carboxylate (**35**)

(84S1058). However, as pointed out in Section IV,B), CSI can also transform hydroxy-N-heterocycles into the corresponding chloro compounds (cf. **57 → 59**) (86SC543).

On comparing the different modes of activation of a given hydroxy-N-heterocycle, one must realize that leaving groups such as O-sulfonates or O-phosphates can react in two different ways, as depicted in the following case of 2-sulfonyloxypyridine (**36**).

On reaction with amines R^1NHR^2, only the AE (reaction sequence A) with C—O bond scission will lead to amination product **37** and sulfonic acid **38**, whereas the competing attack on the sulfonyl group (reaction sequence B) with S—O bond scission will lead to 2-(1H)-pyridinone **(27)** and a sulfonamide **39**, thus decreasing the yield of the desired amination product (cf. **209** → **210** in Section IV,E). This B-type of scission is also common for other leaving groups. Thus, derivatives of 2-(1H)-pyridinone **(27)**, such as bis(2-hydroxy-pyridyl)-carbonate, are used as activated derivates of carbon dioxide in the presence of 4-dimethylaminopyridine (DMAP) (880PP145).

It might be worthwhile to study the extent to which ortho-substituents in benzenesulfonates hinder the attack of amines on the sulfonyl group (attack B) and thus favor the desired amination (attack A) (cf. **212** → **213** in Section IV,E).

B. Chloro and Bromo Groups

1. Introduction

For many years, chloride, and to a lesser degree bromide, ions have been the most important leaving groups for preparative and large scale aminations of nitrogen heterocycles.

2. Preparation of Chloro- or Bromo-N-Heterocycles

A chlorine atom as a leaving group is usually introduced by treatment of hydroxy-N-heterocycles such as 4-(1H)-pyridinone **(40)** with either $POCl_3$, PCl_5, $SOCl_2$, SO_2Cl_2, or Vilsmeier reagent to give 4-chloropyridine **(41)** (47CA6245). These conversions of hydroxy-nitrogen heterocycles to the corresponding chloro heterocycles frequently give only moderate yields of the corresponding chlorine compounds because of side reactions of PCl_5, $POCl_3$ (62USP3021332; 81CPB1069; 86ABC495; 88OPP285), $SOCl_2$, or SO_2Cl_2 (36JPR188), especially when the heterocycles contain a methyl or alkyl substituent. Thus, 2-methyl-4-(1H)-quinolone (42), on heating in $POCl_3$, affords 4-chloro-2-trichloromethylquinoline (43) in 60% yield, 3,4-dichloro-2-trichloromethylquinoline (44) in 4% yield, and 3,4-dichloro-2-dichloromethylquinoline (45) in 2% yield (81CPB1069).

Apparently because of these side reactions, 2,4-dihydroxy-5,6,7-trimethylpyrido[2,3-d]pyrimidine (46) on reaction with excess $POCl_3$ followed by evaporation of POCl₃, aqueous workup, and subsequent treatment with alcoholic ammonia, gives less than 10% of the desired 2,4-diamino-5,6,7-trimethylpyrido[2,3-*d*]pyrimidine (**47**) (62USP3021332). Analogously, 2-methyl-4-hydroxypyrido[3,4-*d*]pyrimidine (**48**), on treatment with POCl₃ and *N*,*N*-dimethylaniline for 18 hr in boiling benzene, affords only a 45% yield of the corresponding unstable chloro compound **49**, which reacts readily with benzylamine in boiling ethanol to give 2-methyl-4-benzylaminopyrido[3,4-*d*]pyrimidine (**50**) in 96% yield (86ABC495). The silylation–amination of **48**, however, affords **50** in 96% yield in one reaction step (Section IV,D). All attempts to convert 2-alkyl- or 2-amino-5,8-dihydroxy-pyrimido[4,5-*d*]pyrazines into the corresponding 5,8-dichloro compounds by treatment with POCl₃/PCl₅ failed due to side reactions with the 2-alkyl or 2-amino groups (72CPB1513; 89T4485). These examples may suffice for reactions with POCl₃ or POCl₃/PCl₅.

Likewise, SOCl₂ chlorinates 6-methylnicotinic acid (**51**) to trichloronicotinoyl chloride (**52**) (36JPR188), whereas 4-methyl-nicotinic acid (**53**) is transformed by SOCl₂ into the dichlorothio lactone **54** (44JA1456;

69JOC247), as are 2-methylquinoline-3-carboxylic acids (78JHC687). Side reactions with SOCl₂ have been reviewed (81S661). As a consequence of these chlorinations of methyl or alkyl groups, substituted hydroxy-N-heterocycles often give higher yields when aminated using alternative methods.

In transformations of hydroxy-N-heterocycles with $SOCl_2$ into the corresponding chloro-N-heterocycles, the addition of methane-sulfonic acid, methanesulfonyl chloride, or 4-dimethylaminopyridine [87CI(L)694], as well as of phase-transfer catalysts such as benzyltriethylammonium chloride (87EUP225866) or tetraethylammonium chloride (86EUP203685), can increase the yield of the chlorinated products. In the case of maleic hydrazide (55), addition of methanesulfonic acid raises the yield of 3,6-dichloropyridazine (56) as well as that of chloropyridine-carboxylic acid chlorides [87CI(L)694]. Although the analogous conversion of 2-(1H)-pyridinone into 2-chloropyridine failed [87CI(L)694], this compound can be readily obtained on a preparative scale by chlorination of pyridine at temperatures of 270–300°C (39RTC709).

$$55 \xrightarrow[CH_3SO_3H]{SOCl_2} 56 \quad 90\%$$

Since the transformation of hydroxy-N-heterocycles into the corresponding chloro-heterocycles involves the prior acylation of the hydroxy-N-heterocycles by $POCl_3$ or $SOCl_2$ to intermediate chlorophosphates or sulfinic esters, the addition of nucleophilic catalysts such as DMAP [78AG(E)569] accelerates all these chlorination reactions as in the case of 4-(1H)-pyridinone [87CI(L)694] or 3'5'-di-O-benzoyl-2'-deoxyuridine (86MI1; 87MI2). However, DMAP can also react readily with the chloro compounds to form quaternary products. Other reagents such as diethylaniline hydrochloride have also been used to promote the transformation of 2', 3', 5'-tri-O-benzoyluridine with $POCl_3$ into the corresponding 4-chloro derivative in 45% yield (72CPB1050).

Whereas $SOCl_2$ on reaction with hydroxy-N-heterocycles, gives only SO_2 and HCl, and excess reagent can be readily distilled off, $POCl_3$ and PCl_5, after evaporation, leave phosphoric acids as a viscous concentrate, which can cause hydrolysis and loss of the chloro-heterocycles during aqueous workup. Instead of $SOCl_2$ or $POCl_3$, the Vilsmeier salt can be formed and used for chlorinations in situ by adding DMF to $SOCl_2$ or $POCl_3$ (65CCC2052; 88CJC61). However, an excess of thionyl chloride still seems to be necessary to achieve a good yield of the corresponding chloro compound (65CCC2052). Other reagents such as 4-chlorophenylphosphoryl chloride and other chlorophenylphosphonyl esters can probably be used also (Section V,G) since phenylphosphonic dichloride was used for analogous chlorinations (82JMC837; 85H2247, 85JMC1790; 86T2303; 88H1899).

In addition to the reagents previously discussed, chlorosulfonylisocyanate (CSI) **(58)** has been applied to convert 6-phenyl-3-(2H)-pyridazinone **(57)** into 6-phenyl-3-chloropyridazine **(59)** in 80% yield (86SC543).

Apart from the side reactions that occur during formation of chloro-N-heterocycles, the reactive chloro-heterocycles obtained, such as free 4-chloropyridine **(41)**, can dimerize or polymerize to give, after workup, compounds such as 4-(4'-pyridyl)-4-pyridinone **(60)** (54JOC1830;

59RTC593). Such 4-chloropyridines can be synthesized in one step on reaction of pyridines with thionyl chloride. Thus, picolinic acid hydrochloride (61), on prolonged heating with thionyl chloride and subsequent hydrolysis of the intermediate acid chloride (62), gives 4-chloropicolinic acid (63) in 50–55% yield (31CB21). The analogous reaction, with simultaneous irradiation with visible light, furnishes the corresponding 2,4-dichloropicolinic acid (32JPR36). Whereas the weakly basic picolinic acid, or more probably its acid chloride (62) will not dimerize, the reaction of pyridine (64) with thionyl chloride probably gives 4-chloropyridine (41) as an intermediate, which reacts with excess pyridine to give in high yield the chemically stable 4-pyridylpyridiniumchloride-hydrochloride (65). This useful compound, commercially available, is difficult to purify [31CB1049; 54ACS390, 54JCS1795; 56CB2921; 73OSC(5)977] and therefore used for further reactions in crude form (Section IV,H).

Pyridine or picolinic acid chloride (66; R = H, COCl) reacts presumably with thionyl chloride to form a σ-complex (67) that is attacked by a chloride ion to give 68. Elimination of HCl and SO, which disproportionate to sulfur and sulfur dioxide, ultimately furnishes the corresponding 4-chloropyridine hydrochloride (69). In the case of pyridine, the intermediate 4-chloropyridine (41) reacts with additional pyridine to form 65. The sometimes long initiation periods of these conversions (61 → 62 and 64 → 65) indicate they may in reality be complicated radical reactions (cf. 81S661).

The reaction of 2-aminopyridine (**28**) with sodium nitrite, bromine, and HBr gives 2-bromopyridine (**70**) in 87% yield (34JA231; 51JA4773). Bromine has also been introduced into quinolines by treating 2- or 4-methoxyquinolines with phosphorus tribromide in DMF to furnish 2-bromo- or 4-bromoquinoline in 78% and 68% yields, respectively

(77CL891), 2-Bromoquinoline is also obtained in 81% yield from 2-(1H)-quinolinone (carbostyril) with POBr$_3$ (51JA4773).

On treatment of heterocyclic N-oxides with POCl$_3$, PCl$_5$, or SO$_2$Cl$_2$, the corresponding chlorides are obtained with the chlorine atom in the α-, β-, or γ-position to the N-oxide nitrogen.

Nicotinamide-N-oxide (**71**) is converted by a mixture of phosphorus pentachloride and phosphorus oxychloride in 52% yield to 2-chloronicotinonitrile (**72**) (54JOC1633), whereas 2-phenylpyridine-N-oxide (**73**) reacts with sulfuryl chloride to afford a mixture of 43% of 2-phenyl-4-chloropyridine (**74**) and 35% of 2-phenyl-6-chloropyridine (**75**) (53CJC457).

If the N-oxides contain methyl groups, these can be chlorinated as discussed previously. Thus, 2-methyl-3-nitropyridine-N-oxide (**76**) is converted by phosphorus pentachloride-phosphorus oxychloride to 6% of **77** and 8% of **78** (78T3445). More complicated is the reaction of 5-nitroquinoline-N-oxide (**79**) with POCl$_3$ to give 35% of **80**, 20% of **81**, as

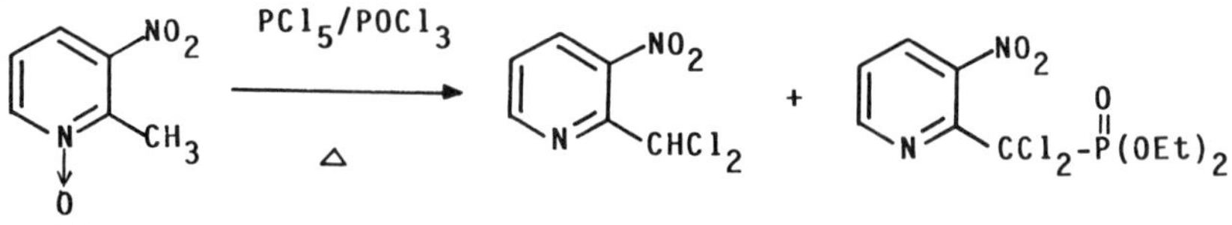

[Scheme: compound **76** (2-methyl-3-nitropyridine N-oxide) → PCl₅/POCl₃, Δ → **77** (2-dichloromethyl-3-nitropyridine) + **78** (with CCl₂-P(O)(OEt)₂ group)]

well as 10% of **82** (47JA303). For related reactions of quinoline-*N*-oxides, see also Bachman and Cooper (44JOC302).

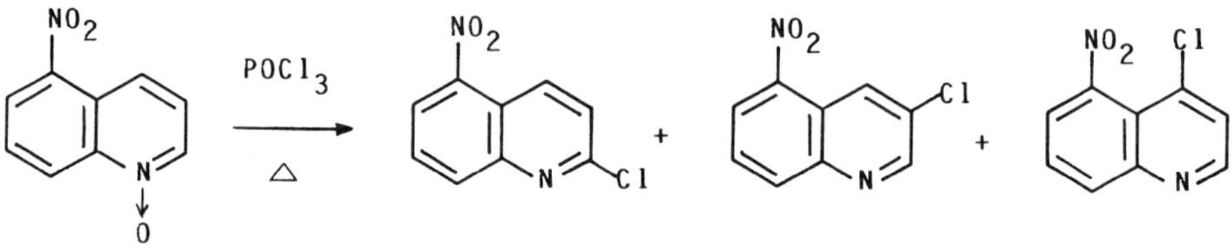

[Scheme: **79** (5-nitroquinoline N-oxide) → POCl₃, Δ → **80** (2-chloro) + **81** (3-chloro) + **82** (4-chloro) 5-nitroquinolines]

The isolation of chloro heterocycles is not always necessary. Chlorination with POCl₃ and amination can be combined in a one-pot reaction as exemplified by the reaction of 4-phenyl-1-(2*H*)-phthalazinone (**83**) with POCl₃ in the presence of excess neopentylamine in boiling xylene to furnish **84** in 77% yield (85EUP159652).

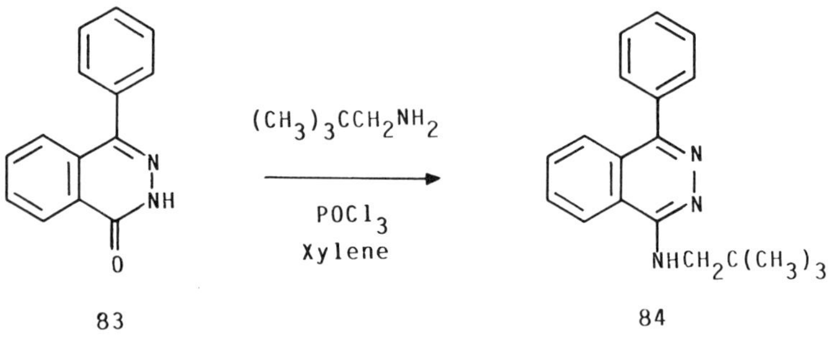

[Scheme: **83** (4-phenyl-1(2H)-phthalazinone) + (CH₃)₃CCH₂NH₂, POCl₃, xylene → **84** (1-neopentylamino-4-phenylphthalazine)]

Similarly, the reaction of hydroxy-nitrogen heterocycles with POCl₃ or P₂O₅ in the presence of secondary or tertiary amines at higher temperatures furnishes directly the corresponding dialkylamino-heterocycles as discussed in Sections III,B as well as in Sections IV,F and IV,H. These results indicate that the transformations of hydroxy-*N*-heterocycles into the corresponding chloro compounds must be carefully investigated and optimized to obtain uniformly high yields.

3. Amination of Chloro- or Bromo-N-Heterocycles

Because of the thousands of examples of aminations with chlorine or bromine as a leaving group, only some selected aminations of chloro- or bromo-heterocycles, in addition to those examples already discussed in Sections III,B and III,D will be mentioned here, with special emphasis on aminations of 2-halopyridines.

Because of the very low reactivity of 2-substituted pyridines, 2-chloropyridine does *not* react with the weakly basic aniline, whereas 2-bromopyridine **(70)** is aminated by aniline in the presence of zinc chloride/zinc dust to 2-phenylaminopyridine **(29)** in 85% yield (72CPB2678). Analogously, 1-ethoxycarbonylpiperazine gives **85** in 83% yield (53JOC1484) (Scheme 3). 2-Chloropyridine **(86)** (75GEP2341965), however, *does* react with 4-amino-1-benzylpiperidine in the presence of copper powder to afford 4-[N-(pyridyl)-amino]-1-benzylpiperidine **(87)** in 56% yield, as well as reacting slowly with pyrrolidine, morpholine, and benzylamine at 75°C in DMSO (66MI1). Furthermore, 2-chloropyridine **(86)** reacts at 130°C in DMSO with 4-acetamidopiperidine to give a 25% yield of 4-acetamido-1-(2-pyridyl)-piperidine **(88)** (80EUP45980).

The more reactive 2-chloro-3-fluoropyridine **(89)** is aminated readily by piperazine in boiling *n*-butanol to give ~60% of **90** (83JMC1696). The presence of a polar cyano group allows the 2-chloro-3-cyanopyridine **(72)** to be aminated already at 100°C by methylamine to 2-methylamino-3-cyanopyridine **(91)** in 63% yield (85SC1013). An additional 6-chloro group, as in **92**, leads, upon amination with ammonia to a mixture of the corresponding amino compounds **93** and **94**. With the weakly basic aniline in the

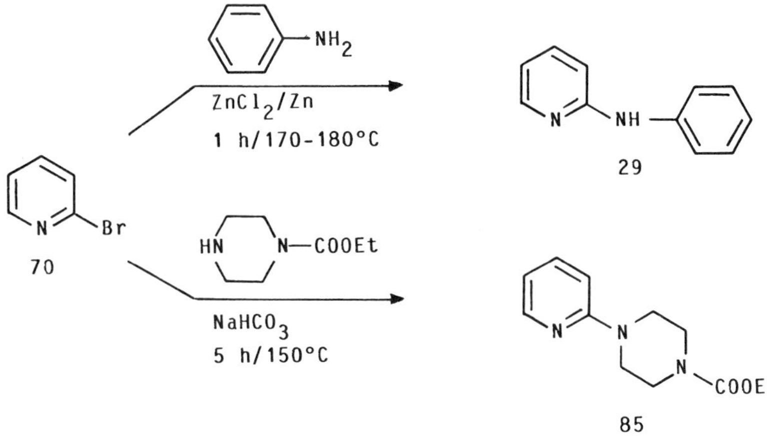

SCHEME 3

polar solvent N-methylpyrrolidone, **95** and some of compound **96** are furnished (77GEP2605467). The cyclization of enaminone **97** in tetrahydrofuran (THF) proceeds by addition of a solution of anhydrous tetrabutylammoniumfluoride (TBAF) in dioxane to give a 87% yield of the amination product **98** (88TL1931).

Concluding these reactions with halopyridines, 3-bromopyridine **(99)**, which is readily available by bromination of pyridine at 215°C (36CB1534), is converted by ammonia at the relatively low temperature of 140°C in the presence of cupric sulfate to 3-aminopyridine **(100)** (36CB1534,

36RTC122), whereas 2-bromopyridine **(70)** yields 2-aminopyridine only at 200°C (36RTC122).

2,3-Dichloroquinoxaline **(101)** is aminated by morpholine in the presence of excess cesium fluoride/18-crown-6 with simultaneous exchange of chlorine by fluorine to give **102** in 78% yield, whereas glycine ethylester-hydrochloride affords **103** in 61% yield (87H1215).

As already mentioned in Section III,B, the volatile amines can be replaced by the corresponding formamides or phosphoramides. Thus, 2-chloro-5-nitroquinoline (**80**) affords, on heating in DMF, the corresponding 2-dimethylamino-5-nitroquinoline (**104**) in 94% yield with evolution of carbon monoxide and hydrochloric acid. The less reactive 2-chloroquinoline gives 2-dimethylaminoquinoline in 76% yield (69CC38). 8-Amino-4-chloro-6-methoxyquinoline (**105**) furnishes analogously the 8-amino-4-dimethylamino-6-methoxyquinoline (**106**) in 65% yield (69CC38).

2-Chlorobenzimidazole (**107**), on heating with DMF, gives the corresponding 2-dimethylaminobenzimidazole (**108**) in 45% yield (66JHC107), and 2-chlorobenzothiazole affords 2-dimethylaminobenzothiazole in 90% yield (65JOC3618).

The electron-withdrawing nature of an N-oxide moiety results in an increased reactivity of any leaving group in an α- or γ-position to the N-oxide moiety. Thus, 4-chloropyridine-*N*-oxide (**109**), which is readily available from the reaction of 4-nitropyridine-*N*-oxide with acetylchloride (53JOC534), is aminated by dimethylamine or pyrrolidine to give the corresponding (56JCS2404; 57JCS4375; 85MI1) 4-dialkylaminopyridine-*N*-oxides (**110**). These 4-dialkylaminopyridine-*N*-oxides are of interest due to

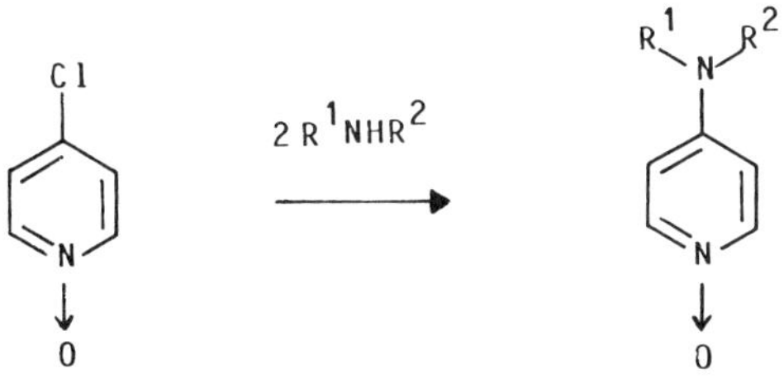

Sec. IV.C] AMINATION OF NITROGEN HETEROCYCLES

[Scheme: 109 (4-chloropyridine-N-oxide) + 2 R¹NHR² → 110 (4-(R¹R²N)pyridine-N-oxide)]

their high dipole moment of $\mu = 6.76$ D, as in the case of 4-dimethylamino pyridine-N-oxide **(110)** ($R^1 = R^2 = CH_3$) (57JCS1769), and their catalytic activities (86CS55) combined with a relatively low basicity of $pK_a = 3.88$ compared to 9.70 for DMAP [72JCS(P2)671].

Finally, the application of high-pressure methods (85S1) to the amination of chloro- and bromoquinolines and chloro- or bromopyridines (88H319) has been described. Thus, 2-chloropyridine **(86)** reacts at 60°C and 8 kbar pressure with piperidine to form 2-piperidinopyridine **(111)** in 72% yield. Analogously, 3-chloropyridine **(112)** reacts at 100°C with pyrrolidine to give a 10% yield of 3-pyrrolidinopyridine **(113)**, which is, however, obtained in 71% yield under the same reaction conditions from 3-bromo-pyridine **(99)** (88H319).

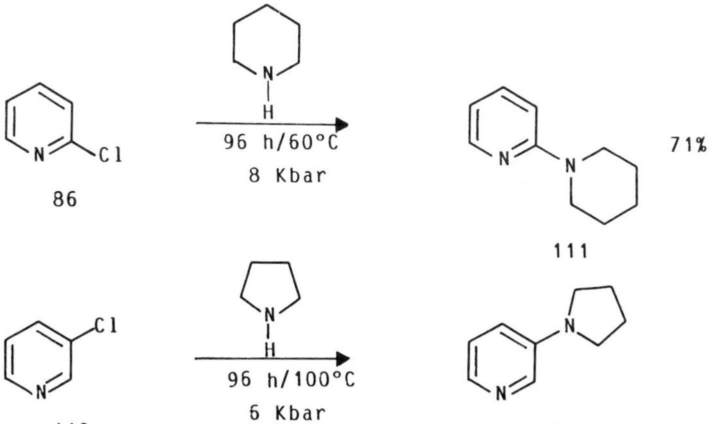

[Schemes: 86 + piperidine (96 h/60°C, 8 Kbar) → 111 (71%); 112 + pyrrolidine (96 h/100°C, 6 Kbar) → 113]

C. ALKOXY OR ARYLOXY GROUPS

In contrast to hydroxy-N-heterocycles, which usually occur in the lactam or vinylogous lactam form (Section IV,A), the alkoxy- or aryloxy-nitrogen heterocycles exist exclusively in the more reactive iminoether or vinylogous iminoether form and are therefore readily aminated with ammonia or primary or secondary amines. These O-alkyl or O-aryl derivatives, however, can rearrange, especially in the presence of basic amines at temperatures above 130°C, to the thermodynamically more stable N-alkyl or N-aryl derivatives, thereby decreasing the amination yields (69MI1; 72T5507; 86TL5997) (cf. the Chapman rearrangement).

The conversion of hydroxy-N-heterocycles into the corresponding alkoxy derivatives by alkylation with alkyl halides or diazo compounds usually leads to mixtures of O- and N-alkylated products. Thus, conversion of 1-phenyl-6-methyluracil (114) into the silver salt and subsequent treatment with isopropyl iodide in octane gives 50% of the desired 4-isopropyloxy derivative 115 as well as 1% of the N-isopropyl derivative 116, whereas the corresponding sodium salt gives primarily 116. The 4-isopropyloxy derivative 115, upon heating with aniline to 180–190°C, affords the 4-anilino derivative 117. Amination of 115 with butylamine or benzylamine gives the corresponding substituted cytosines 118 and 119 (70IZV1127; 72IZV2530).

Treatment of nucleosides, such as thymidine 120 in methanol, with an ether solution of diazomethane affords only a 6.5% yield of the desired O^4-methyl derivative 121 as well as 70.6% of N^3-methylthymidine (122) and 3.3% of O^2-methylthymidine (87MI4). Vilsmeier chlorination of uridine-2′,3′,5′-tri-O-acetate (123), followed by treatment with sodium ethylate in ethanol, gives an 84% overall yield of the crystalline O^4-ethyluridine (124). This derivative is modified in the ribose moiety and is subsequently aminated at 100°C with methanolic ammonia in an autoclave to the corresponding modified cytidine 125 (88CPB945).

2-Methoxy-5-nitropyridine (**126**) reacts readily with *N,N*-dimethylethylenediamine in boiling water to give 89% of **127** (85EUP136730). Analogously, 4-ethoxy-3-nitropyridine-hydrochloride (**128**), is heated at reflux for 8.5 hr with an aqueous solution of ammonium acetate to afford 87% of 3-nitro-4-aminopyridine (**129**) (85EUP149537).

In the absence of activating groups, more drastic conditions must be applied. Thus, 2-methoxy-5-phenylpyrimidine (**130**) has to be heated with ammonia to 195–200°C for 4 days to give only 23% of the desired 2-amino-5-phenylpyrimidine (**131**) as well as 27% of the rearrangement product 3,5-diphenylpyridine (**132**) (70AJC625). Addition of another ring nitrogen, as in 3-methoxy-1,2,4-triazine (**133**), results in a much higher reactivity of **133**, which is aminated by dimethylamine to the desired 3-dimethylamino-1,2,4-triazine (**134**) in practically quantitative yield (76CB1113).

The corresponding 1,3,5-triazines are even more reactive. Triallylcyanurate **(135)** reacts readily at room temperature with butylamine (74GEP2308560) or γ-aminopropyl triethoxysilane to the corresponding monoamino derivatives **136** and **137** in nearly quantitative yield (74GEP2308591). 4,5,8-Trimethoxy-6-aminoquinaldine **(138)** is aminated by ethanolic ammonia in an autoclave at 120°C to give **139** (78AFJ578).

Similarly, 4,7-dimethoxy-8-aminoquinoline reacts at 120°C with hydrazine-hydrate in 73% yield to give 4-hydrazino-7-methoxy-8-aminoquinoline (87MI1). 5-Methoxypyrimido-[5,4-*e*]-1,2,4-triazine is aminated by ammonia at −70°C to the 5-amino analogue (87JHC1657).

The kinetics of the amination of 2-phenoxy-1,3,5-triazine with piperidine was investigated. Instead of the acid autocatalysis (Section II), a cyclic mechanism is suggested in which a second piperidine, or much more effectively, valerolactam, acts as a catalyst for the amination (75AJ1851). Similar kinetic results are obtained with 3-phenoxyquinazoline (82G167) and 5-methyl-5-phenoxyquinazoline (85JPR865).

Since 1,4-dichloroisoquinoline **(140)** is difficult to aminate, it is converted by potassium phenoxide in 65% yield into 1-phenoxy-4-chloroisoquinoline **(141)**, which subsequently reacts with ammonium acetate at 190°C to give 1-amino-4-chloroisoquinoline **(142)** (78JHC1513).

2',3',5'-O-Protected derivatives of uridine and protected guanosine are activated by treatment with 2-mesitylenesulfonyl chloride, a tertiary base, and DMAP as the catalyst to the corresponding 4-O- or 6-O-mesitylenesulfonyl derivatives, which react readily with hydroxypyridines or O-nitrophenol to form the stable, reactive **143**, as well as the O-deprotected **144**. These are readily aminated, especially after quaternization of the pyridyloxy moiety with methyliodide, by ammonia, methylamine, dimethylamine, piperidine, and morpholine to the corresponding cytidines **(145)** or 2-aminoadenosines **(146)** [86ACS(B)806, 86ACS(B)826]. A recent study demonstrates that Michael addition of protected thymidine or 2'-deoxyguanosine with acrylonitrile or phenylvinylsulfone gives mainly, if not exclusively, N-alkylation (88MI1).

D. SILYLOXY GROUPS

1. Introduction

For the conversion of 2',3',5'-O-protected uridines into the corresponding cytidines, the 4-carbonyl group has to be activated by conversion with phosphorus pentasulfide into the 4-thiocarbonyl group or the even more reactive 4-methylthio group (Section IV,I.), or by Vilsmeier reagent into the 4-chloro derivative (Section IV,B). The 4-carbonyl group can furthermore be activated with sulfonyl halides or anhydrides (Section IV,E) as well as by introduction of the 4-(1,2,4-triazolyl) derivative to be subsequently reacted with ammonia or primary or secondary amines to give the corresponding cytidines. However, all these conversions of uridine into the corresponding cytidines imply at least three reaction steps:

(a) Protection such as acylation or silylation of the hydroxyl groups in the sugar moiety of the starting uridine or thymidine.
(b) Activation of the 4-position.
(c) Reaction with ammonia or primary or secondary amines with simultaneous or subsequent removal of the acyl groups.

In their first synthesis of cytidine, Todd and co-workers reacted the O-acetylated 4-O-ethyluridine **(147)** (obtained via the classic Hilbert–Johnson reaction of 2,4-diethoxypyrimidine with 2,3,5-tri-O-acetyl-1-chloro-ribose) with methanolic ammonia to afford, via AE intermediate

148, cytidine 149 as well as ethanol as leaving group (cf. Section IV,C) (47JCS1052).

4-O-Alkyl derivatives such as 147 cannot, however, be prepared from protected uridines of type 150 in reasonable yield, since N^3-alkylation is always a competing reaction (Section IV,C). Furthermore, at the higher temperatures necessary to convert 5-substituted 4-O-alkyl uridines with ammonia or amines into the corresponding cytidines, rearrangement of the 4-O-alkyl group occurs to yield the N^3-alkyl derivatives.

2. Trimethylsilanol or Dimethylsilanol as a Leaving Group

Since O-trimethylsilyl compounds are very similar to the corresponding O-alkyl compounds, the polar, hydrophilic uridine (150) is persilylated with hexamethyldisilazane (HMDS) with evolution of ammonia to the tetrasilylated-activated lipophilic intermediate 151, the structure of which is supported by mass spectral as well as UV data (75LA988). Thus, in one step, the alcoholic hydroxyl groups of the ribose moiety are protected, and the 4-position is activated. Because of the higher acidity of trimethylsilanol compared to ethanol, trimethylsilanol is a better leaving group than ethanol (59JA6145).

In a theoretically possible equilibrium between reactive intermediate **151** and the corresponding N^3-trimethylsilyl-compound **152**, the mobility (70ACR299) of the trimethylsilyl group, as well as the high affinity of silicon to oxygen, and the gain in aromatic energy in this and other hydroxy-N-heterocycles, always results in the reformation of **151**. The subsequent or simultaneous reaction of **151** with ammonia or amines R^1-NH-R^2 leads to the corresponding silylated cytidine **153** with the loss of trimethylsilanol.

The formation of the activated intermediate for the AE with R^1-NH-R^2 is accelerated by the addition of a proton or Lewis acid to the 2-carbonyl group in **151**. This is necessary for aminations of uridine **(150)** or thymidine **(162)** with weakly basic amines such as aniline, e.g., to **161e** (Section III,A). Silylated hydroxy-N-heterocycles lacking such an activating conjugated carbonyl group, however, usually react only with amines when a Lewis acid is present (Section III,A).

The O-trimethylsilyl group, which seems to resemble an O-tert-butyl group, is sterically much less demanding because of the longer O—Si bond compared to the O—C bond. In contrast to methanol or ethanol formed during these AE reactions, trimethylsilanol dimerizes on heating, especially in the presence of acids or bases (54JA3408), to form hexamethyldisiloxane and water.

To remove the water, which is formed on dimerization of trimethylsilanol, at least one additional equivalent of a silylating agent such as HMDS [boiling point = 126°C] must be used. Excess HMDS can then react with trimethylsilanol or water to give hexamethyldisiloxane and ammonia, but will also reconvert any "hydrolyzed" hydroxy-N-heterocycle into the corresponding activated O-silylated intermediates. Thus, sufficient HMDS should be used to silylate the hydroxy-N-heterocycle and all the alcoholic, phenolic, or other acidic hydroxy groups present in the heterocyclic or amine moiety. It should also be used to convert trimethylsilanol and water (or intruded humidity) from the reaction into hexamethyldisiloxane and ammonia (Eqs. 1–3).

$$2(CH_3)_3SiOH \rightarrow (CH_3)_3SiOSi(CH_3)_3 + H_2O \quad (1)$$

$$(CH_3)_3SiNHSi(CH_3)_3 + 2(CH_3)_3SiOH \rightarrow 2(CH_3)_3SiOSi(CH_3)_3 + NH_3 \quad (2)$$

$$(CH_3)_3SiNHSi(CH_3)_3 + H_2O \rightarrow (CH_3)_3SiOSi(CH_3)_3 + NH_3 \quad (3)$$

At normal pressure, the ammonia formed on silylation with HMDS evolves so rapidly it does not react with any silylated hydroxy-N-heterocycle to give the corresponding aminoheterocycle. Trimethylsilanol and hexamethyldisiloxane both exhibit a boiling point of 99°C, but form an azeotropic mixture boiling at 70–90°C. Since HMDS has a boiling point of 126°C, it is advisable to remove trimethylsilanol and hexamethyldisiloxane continuously with a small Vigreux distillation column to accelerate the reaction and follow its progress.

With low-boiling amines such as pyrrolidine (bp = 87°C) or diethylamine (bp = 55°C), the amines have to be transformed first into the higher-boiling N-trimethylsilylated pyrrolidine (bp = 138–140°C) or diethylamine (bp = 126°C) and then reacted with the hydroxy-N-heterocycle in the presence of HMDS (cf. reaction of **40** to **184** in this section). Because of the difference in relative electronegativities between nitrogen (3.16) and silicon (1.64) (76JA7869), the N-silylation of amines makes the nitrogen more basic for the nucleophilic attack on the lactam-carbonyl of heterocycles, yet also more bulky! Thus, these effects may negate each other depending on the sterical situation (cf. the reaction of **40** with N-silylated pyrrolidine to **184**).

Since HMDS boils at 126°C, and the least reactive silylated 2-(1*H*)-pyridinone **(27)** has to be heated to a much higher temperature to effect conversion, **27** is reacted with the stable crystalline and commercially available octamethylcyclotetrasilazane (OMCTS) **(157)** (melting point = 97°C, boiling point = 225°C) (57MI1) to afford activated dimeric intermediates such as **158**, which can be aminated at 180–200°C in the presence of

37

Lewis acids to give the desired 2-aminopyridines (37). The dimethyldisilanol (159), formed on amination, polymerizes to silicon oil (160) or cyclic oligomers, which can be readily removed by extraction with pentane or hexane.

If the heterocyclic systems such as uridine (150), thymidine (162), or the amine moieties contain alcoholic or phenolic hydroxyl groups, which are silylated and protected during this process of silylation–amination, these O-silyl groups have to be removed after completion of the reaction.

For analytical purposes, e.g., for thin-layer chromatography, the O-silyl groups are most rapidly removed by addition of TBAF (72JA6190). For preparative workup, the crude reaction mixture is transsilylated by boiling with excess methanol for several hours. These transsilylations are accelerated by continuous removal of the trimethylsilylated methanol (boiling point = 55°C) formed using a distillation column.

3. *One-Pot Reaction*

Since silylation and amination do not interfere with each other, they can be performed simultaneously. Thus, the hydroxy-N-heterocycle, a two- or threefold excess of the amine moiety, a slight excess of HMDS, as well as catalytical amounts of a Lewis acid are heated with simultanous distillation of trimethylsilanol and hexamethyldisiloxane as indicated previously.

a. *Solvents.* Apart from a few exceptions, such as the amination of uridine, thymidine (75LA988), or guanosine (76LA745) with ammonia or amines, where formamide, pyridine, or toluene were employed, solvents are usually not necessary. If pyridine is used as a solvent, it is important that the silylated hydroxy-N-heterocycle is approximately as basic as pyridine to permit protonation at the imino-silylether moiety to achieve amination. The polar hydroxy amines are converted by silylation into lipophilic silylated intermediates, which readily mix and react with the silylated hydroxy-N-heterocycle without solvents.

b. *Lewis Acids.* The most practical catalysis for the silylation–amination are $(NH_4)_2SO_4$, *p*-toluenesulfonic acid hydrate, camphorsulfonic acid, trifluoromethanesulfonic acid, and perfluorobutanesulfonic acid, which are probably transformed in situ into the corresponding bis(trimethylsilyl) sulfate, trimethylsilyl triflate, and trimethylsilyl perfluorobutanesulfonate. These acids, however, may also be present as amine salts. On employing amine hydrochlorides or traces of trimethylchlorosilane (TCS), NH_4Cl is formed during silylation with HMDS, which sublimes partially into the reflux condenser. Obviously, enough

NH₄Cl or amine hydrochloride remains in the reaction mixture to ensure a smooth silylation–amination.

$ZnCl_2$ and $SnCl_4$ sometimes function well, but are not considered as practical. Mercuric chloride, which was used in some of our silylation–aminations of purine nucleosides (76LA745), oxidizes the primary or secondary amines used as well as the reaction products. Furthermore, the toxic mercury-containing side products formed are difficult to remove (62CCC902, 62CCC906).

4. *Silylation–Amination of Nucleosides*

Nucleosides such as uridine (**150**) or thymidine (**162**) can easily be silylated–aminated in a one-step reaction to the corresponding cytidines (**161** and **163**). Uridine (**150**) and thymidine (**162**) are aminated by NH_3 and HMDS upon heating in an autoclave to the corresponding cytidines (**161a** and **163a**). Whereas uridine (**150**) can be readily heated to 160°C, the persilylated thymidine (**162**) decomposes to thymine or 5-methylcytosine and furanes at temperatures above 140°C (75LA988). The basic pyrrolidine reacts much faster with uridine **150** to afford **161b** than morpholine reacts to afford **161c**. As mentioned previously, the weakly basic aniline reacts only with uridine (**150**) in the presence of $(NH_4)_2SO_4$, after 14 hr at 140–145°C and subsequent treatment with methanolic ammonia, to give N^4-phenylcytidine (**161e**) (Section IV,G; reaction **238** → **239**).

Because of the steric hindrance of the 5-methyl group, thymidine (**162**) reacts considerably slower than uridine (**150**) to afford the corresponding cytidines **161** and **163**. Silylated diphenyllumazine-riboside (**164**) is analogously converted into **165** (73HCA1225).

The reaction was applied to convert 3′-azido-3′desoxythymidine **166** into the corresponding cytidine **167** in 51% yield (83JMC891), as well as 5-trifluoromethyl-2′-desoxyuridine to the corresponding cytidine (79GEP2838644). Furthermore, 2′,3′-didesoxyuridine was silylated–aminated with HMDS, NH_4Cl, and NH_3 for 92 hr at 160°C and 10 atm. to give 90.4% of 2′,3′-didesoxycytidine (87JAP1096192) (cf. the analogous aminations of 4-*O*-sulfonyl and 4-triazole groups in chapters IV,E, and G).

Persilylated intermediates such as **170** are also obtained during nucleoside synthesis. Thus, reaction of 3,5-bis-(trimethylsilyloxy)-1,2,4-triazine (**168**) with 1-*O*-acetyl-2,3,5-tri-*O*-benzoyl-β-D-ribofuranose (**169**) in the presence of 0.76 equivalents of $SnCl_4$ in 1,2-dichloroethane leads to **170** as intermediate, which reacts with a 10-fold excess of pyrrolidine to give O-benzoylated 6-azacytidine (**171**) in 57% yield (75LA988). Inosine (**172a**), guanosine (**172b**), or xanthosine (**172c**) have hitherto been trans-

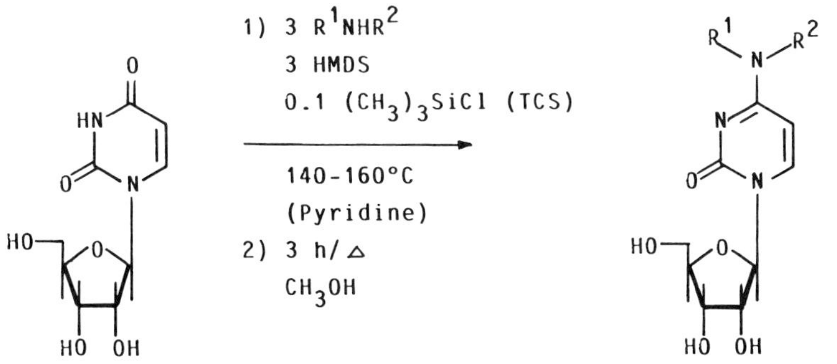

150 → **161**

1) 3 R^{1}NHR^2
   3 HMDS
   0.1 (CH$_3$)$_3$SiCl (TCS)
   140–160°C
   (Pyridine)
2) 3 h/△
   CH$_3$OH

a) R^1 = R^2 = H   48 h/160°C   80% (autoclave)
b) R^1 = R^2 = (CH$_2$)$_4$ 3 h/125°C   95%
c) R^1 = R^2 = (CH$_2$)$_2$O(CH$_2$)$_2$ 32 h/125°C   70%
d) R^1 = H; R^2 = (CH$_2$)$_2$N(CH$_3$)$_2$ 6 h/125°C   85%
e) R^1 = H; R^2 = C$_6$H$_5$ 14 h/140°C   86%

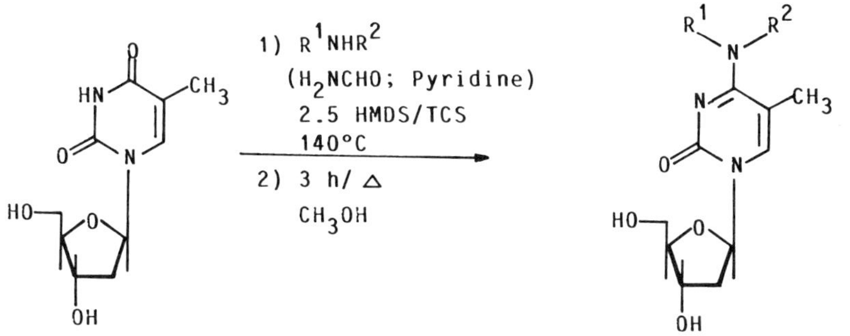

162 → **163**

1) R^{1}NHR^2
   (H$_2$NCHO; Pyridine)
   2.5 HMDS/TCS
   140°C
2) 3 h/△
   CH$_3$OH

a) R^1 = R^2 = H   80 h/140°C   79%
b) R^1 = H; R^2 = (CH$_2$)$_2$C$_6$H$_5$ 48 h/140°C   71%

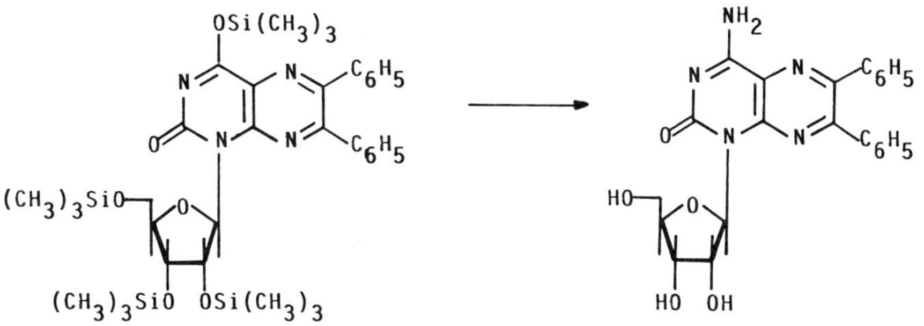

164 → **165**

formed via the 2′,3′,5′-tri-O-acetates with POCl$_3$ into the corresponding 6-chloro compounds, which were reacted with ammonia or amines, as such, or after mild removal of the O-acetyl groups with methanolic ammonia (cf. references in 76LA745).

We found that **172a** as well as **172b** and **172c** react readily with ammonia or primary or secondary amines and HMDS in the presence of acids such as (NH$_4$)$_2$SO$_4$ to afford the biologically potent and pharmacologically interesting N^6-substituted adenosines (**173**) in high yields (76LA745).

Other workers have reacted guanosine (81CJC3360), xanthosine [83IJC(B)48], and 1-β-D-ribofuranosyl-1-*H*-pyrazole[3,4-*d*]-pyrimidine-4-one (82JMC1040) with amines or ammonia in the presence of HMDS and ammonium sulfate to obtain the corresponding aminated nucleosides in high yields. Nucleotides such as the commercially available disodium salt of inosine-5′-phosphate (**174**) react analogously with HMDS and TCS to give the lipophilic persilylated intermediate **175** which, upon amination with excess benzylamine in the presence of (NH$_4$)$_2$SO$_4$, affords the crystalline **176** in 77.5% yield.

Sec. IV.D]   AMINATION OF NITROGEN HETEROCYCLES   153

		R^1NHR^2 =			
R = H	a)		β-phenethylamine	10 h/145°C	92%
	b)	"	dopamine · HCl	20 h/145°C	84%
	c)	"	HN⌒N-CH$_2$CH$_2$OH	46 h/145°C	72%
R = NH$_2$	a)	"	NH$_3$/toluene (CH$_3$)$_3$SiOSO$_2$CF$_3$	48 h/145°C	74%
	b)	"	tryptamine·HCl pyridine	49 h/135°C	42%
	c)	"	p-anisidine	72 h/145°C	32%
R = OH	a)	"	β-phenethylamine	23 h/145°C	80%

The analogous silylation–aminations of oligo- or polynucleotides containing uracil, thymine, or guanine bases have apparently not as yet been investigated. In analogy to the amination of the purine nucleosides, inosine, guanosine, or xanthosine, the purines hypoxanthine (**177a**), or guanine (**177b**) readily undergo silylation–amination with β-phenethylamine or pyrrolidine and HMDS to give the corresponding aminated purines (**178**) in high yields (76LA745).

a) R = H R^1 = H; R^2 = CH_2CH_2Ph
b) R = NH_2 R^1 = R^2 = $(CH_2)_4$

Although the two papers on the silylation–amination of pyrimidine (76-LA988) and purine nucleosides (76LA745) were unfortunately published in *Liebigs Annalen* in German, a few detailed descriptions, e.g., the silylation–amination of thymidine (**162**) into 5-methyl-2′-deoxycytidine (**163a**) and that of inosine (**172a**) into N^6-[2-(3,4-dihydroxyphenyl)-ethyl]-adenosine (**173b**) have been published in English (78M11, 78M12).

5. *Silylation–Amination of Hydroxy-N-Heterocycles*

Since the full paper on applications of the silylation–amination method to a representative series of hydroxy-N-heterocycles was published in English in 1984 (84CB1523), only a few examples from this paper will be discussed here. (1H,3H)-Quinazoline-2,4-dione (**18**) can be monoaminated at 130°C with equivalents of 1-(2-hydroxyethyl)-piperazine in 68% yield, **to 179** whereas with five equivalents of amine and OMCTS (**157**) at 180°C, 56% of the desired 2,4-bis-aminated product **180** is obtained along with 24% of **181** as a side product. Since this reaction and most of the other reactions described were not optimized, these yields can probably be improved considerably.

4-(1-*H*)-Quinolinone (**182**) is aminated with the dopamine-hydrochloride to afford 75% of the crystalline monohydrate **183**. Because of the rapid silylation of dopamine-hydrochloride during the reaction, the sensitive

Sec. IV.D]   AMINATION OF NITROGEN HETEROCYCLES   155

catechol moiety in dopamine is protected against oxydation (cf. the preparation of **173b**).

4-(1-*H*)-Pyridinone (**40**) reacts readily with N-silylated pyrrolidine (bp = 138–140°C) to give the supernucleophile 4-pyrrolidinopyridine (PPY) **184** [78AG(E)569] in 84% yield at normal pressure and in 92% yield in an autoclave with pyridine as solvent. The analogous reaction of **40** with benzylamine in the presence of OMCTS (**157**) and *p*-toluenesulfonic acid-hydrate gives rise to **185** in 76% yield.

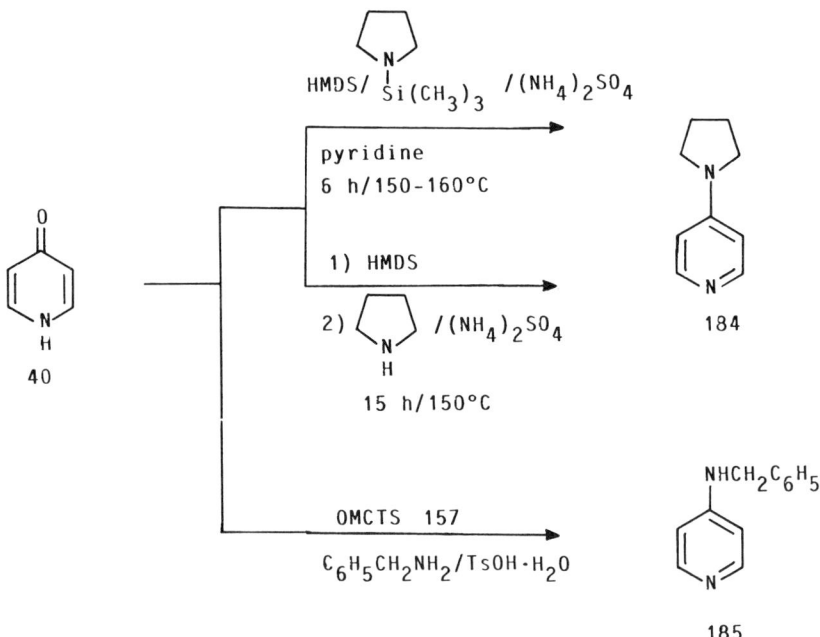

The least reactive hydroxy-N-heterocycle 2-(1*H*)-pyridinone (**27**) undergoes silylation–amination only at temperatures above 180°C. Thus, heating **27** with 2-phenethylamine, OMCTS (**157**), and perfluorobutanesulfonic acid for 24 hr at 200°C furnishes 2-(2-phenethylamino)-pyridine (**186**) in 71% yield. The analogous reaction of **27** with benzylamine, OMCTS (**157**), and perfluorobutanesulfonic acid gives, besides ~47% of the desired 2-benzylaminopyridine (**187**), some 2-aminopyridine (**28**) and dibenzyl- and tribenzylamine (**188–189**) formed by benzyl transfer from benzylamine. In aminations with amines containing aromatic ether moieties such as homoveratrylamine (**188**) at temperatures ≥180°C, the amino group can demethylate the ether moiety to form N-methylated amines **189** leading to mixtures of products.

Sec. IV.D] AMINATION OF NITROGEN HETEROCYCLES 157

Silylation–amination fails with 5-ring-hydroxy-N-heterocycles such as 2-benzoxazolone since, at temperatures of 130°C, the 2-trimethylsilyloxybenzoxazole (**190**) is in equilibrium with its open form **191** (73LA772), which reacts readily with benzylamine to eventually form N,N'-dibenzylurea ($C_6H_5CH_2$—NH—CO—NH—$CH_2C_6H_5$). The silylation–amination procedure has been applied to the amination of 6-oxadihydrouracil (**193**) with benzylamine to **194** (77JMC134). The 5-deazapteridine derivative **195** affords, with ammonia/HMDS in the presence of p-toluenesulfonic acid, the diaminated product **196** in 74% yield (86JMC709).

As already mentioned in Section IV,A, 2-methyl-4-hydroxypyrido-[3,4-*d*]pyrimidine (**48**) on silylation–amination with benzylamine, affords 97% of the corresponding N-benzyl derivative **50**, whereas the conventional two-step procedure via the 4-chloro compound **49** gives **50** (86ABC495) in only 40% yield.

The 5-aminoethyl-4-quinolone (**197**) is cyclized by silylation–amination with HMDS and *p*-toluenesulfonic acid to air-sensitive intermediate **198**, which is converted by methanolic hydrochloric acid in 51% overall yield into the natural product aaptamine (**199**) (87T4803). 2-Phenyl- and 2-amino-5,8-dihydroxypyrimido[4,5-*d*]pyridazines are silylated–aminated with HMDS, $(NH_4)_2SO_4$, and 2-amino-ethanol in collidine to give, depending on the reaction conditions, either the 5- or 8-mono- or the 5,8-bis-aminated products in moderate to good yields. On heating 2-amino-5,8-di(β-hydroxyethylamino)pyrimido[4,5-*d*]pyridazine, which cannot be chlorinated by $POCl_3/PCl_5$ (72CPB1513), under silylation–amination conditions for a longer period of time, the corresponding 2,5,8-tris(β-hydroxyethylamino)pyrido[4,5-*d*]pyridazine is apparently obtained by transamination (89T4485) (cf. Section V,B).

Analogously, 2,5-dioxo-8-ethyl-1,2,5,8-tetrahydropyrido[2,3-*d*]pyrimidine-6-carboxylic acid is silylated–aminated with piperazine, HMDS, and

Sec. IV.E] AMINATION OF NITROGEN HETEROCYCLES 159

p-toluenesulfonic acid-monohydrate, on heating for 6 hr at 160°C, to give 5,8-dihydro-8-ethyl-5-oxo-2-(piperazinyl)-pyrido [2,3-*d*]pyrimidine-6-carboxylic acid in high yield (86JAP3183582).

Hexamethyldisilazane and OMCTS **(157)** are produced on a technical scale and are therefore available at prices of about $15–20 per kilogram. Hence silylation–amination should also be considered for large scale aminations and especially for those cases where the hydroxy-N-heterocycle or the amine moiety contain further functional groups such as alcoholic or phenolic hydroxy groups. Finally, on using HMDS, the recovered hexamethyldisiloxane can be reconverted to TCS or HMDS.

E. O-SULFONATES

As discussed in Section III,D, sulfonyloxy groups and especially the triflate group are very good leaving groups. However, because of the competing attack of the employed amines on the sulfonyl group (cf. attack B in the reaction of **36** to **37** in Section IV,A), the starting hydroxy-N-heterocycle as well as the corresponding sulfonamide are always formed, leading thus to a decrease in the yield of the desired amino-N-heterocycle. A typical example is the conversion of 2-methylthio-5-methyl-pyrimidine-(3*H*)-4-one **(200)** by tosylchloride-triethylamine in DMF to the 4-*O*-tosylate **(201a)**, which reacts with dimethylamine in aqueous dioxane to give 80% of the desired aminated product **202** as well 10% of the recovered starting material **200** [79CF(S)125]. On replacing the tosyl group in **201a** by the more reactive methanesulfonyl group, as in **201b**, the analogous reaction with dimethylamine gives a 1 : 1 mixture of **202** and recovered **200**. A 1984 patent describes the conversion of 2-(1*H*)-pyrimidinone by methanesulfonylchloride-triethylamine to 2-methanesulfonyloxypyrimidine in 67% yield, as well as its subsequent efficient aminations with piperazines in acetone/Na$_2$CO$_3$ (84MIP1).

Likewise, heating of 2,5-diamino-1-oxy-4-tosyloxypyrimidine **(203)** in excess piperidine gives 60% of minoxidil **204** (87EUP270201).

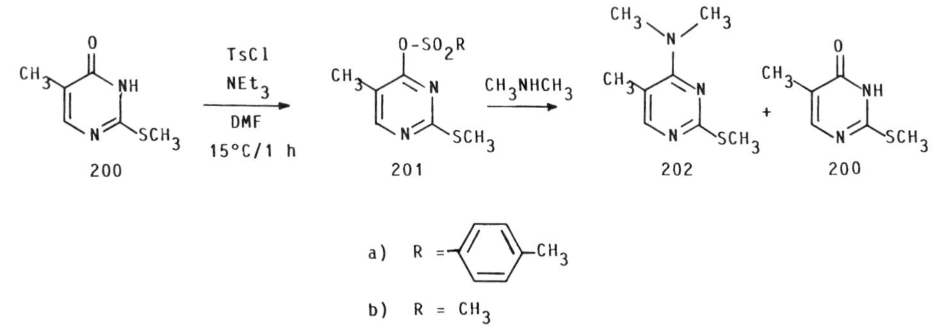

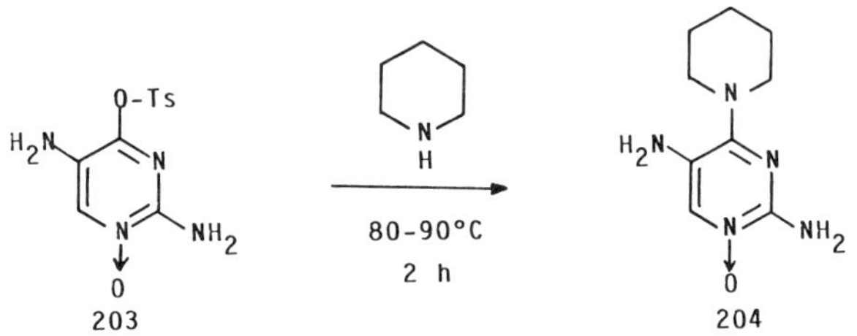

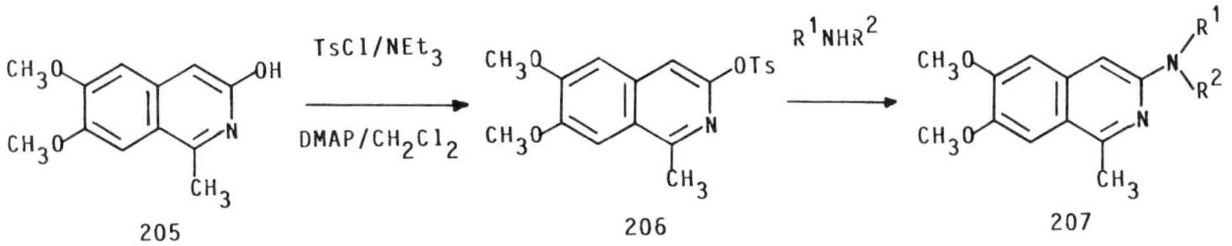

a) R = ⟨C₆H₄⟩–CH₃

a) R = –C₆H₄–CH₃
b) R = CH₃

The reactive 3-hydroxyisoquinoline (**205**) is tosylated by tosyl chloride in the presence of DMAP to **206**, which reacts readily with primary or secondary amines to give the corresponding aminated isoquinolines **207** in high yields (88T3391).

Whereas 2-tosyloxypyridine does not react with dimethyl- or diethylamine even under vigorous conditions, 2-methanesulfonyloxypyridine undergoes only S—O bond scission to give 2-(1H)-pyridinone and N,N-dimethylmethanesulfonamide [79JCR(S)125]. 5-Chloro-2-(1H)-pyridinone (**208**) can be transformed by treatment with sodium hydride in dioxane followed by subsequent reaction with trifluoromethane sulfonylchloride in THF to give in 69% yield 5-chloro-2-trifluoromethansulfonyloxy pyridine (**209**), which is aminated by piperazine-triethylamine in boiling acetonitrile in 17% yield to the corresponding 1-(5-chloro-2-pyridinyl)piperazine (**210**) (83JMC1696).

Protected 4-sulfonyloxypyrimidine nucleosides are readily available by treating protected nucleosides with arylsulfonyl chlorides and NaH in

Sec. IV.E] AMINATION OF NITROGEN HETEROCYCLES 161

THF (87TL2821), K_2CO_3 in methyl isobutyl ketone (86EUP204264), or triethylamine/4-dimethylaminopyridine [86ACS(B)806, 86ACS(B)826]. Because sulfonylated iminoethers conjugated with a carbonyl group are very reactive (Sections III,A and IV,C and D), they react readily under mild reaction conditions with ammonia or primary or secondary amines to afford the corresponding cytidines in high yields.

Thus, 3′,5′-di-O-(t-butyldimethylsilyl)thymidine **(211)** is sulfonylated by 2,4,6-triisopropylbenzenesulfonyl chloride (TIPS-Cl) to furnish the 4-O-TIPS derivative **212** in 93% yield. Amination with 2-(methylamino)ethanol gives rise to 72% of **213** (87TL2821).

Likewise, reaction of 2′,3′,5′-tri-O-acetyluridine with tosyl chloride/K_2CO_3 in methylisobutyl ketone for 4 hr at 80°C affords the corresponding 4-tosyloxy derivative in practically quantitative yield. Subsequent treatment with methanol/concentrated aqueous ammonia affords 91% cytidine and 8.1% uridine (86EUP204264).

$N^2, O^{2'}, O^{3'}, O^{5'}$-Tetrabenzoylguanosine **(214)** reacts readily with excess methanesulfonyl chloride/triethylamine in methylene chloride to give the crystalline 6-O-mesyl derivative **215** in 75% yield. This is aminated within 15 min at 20°C by methanolic dimethylamine in dioxane to give the crystalline 6-dimethylamino compound **216** in 85% yield (77CC447). The reaction of 2′,3′,5′-O-methylated or benzoylated derivatives containing a free 2-amino group gives essentially the same results (77CC447).

While O-protected guanosine with a free or acylated 2-amino group is selectively O^6-sulfonylated (77CC791; 80TL2265), the 3',5'-O-acetyl-2'-deoxyinosine, on reaction with TIPS-Cl, gives nearly equal amounts of the desired O^6-sulfonylated derivative, which is isolated in only 33% yield, and the corresponding undesired N^1-sulfonylated derivative (87JA1275).

F. O-PHOSPHOROAMIDATES

As a consequence of the low reactivity of hydroxy-N-heterocycles, Zavyalov and co-workers (Section IV,A) have reacted hydroxy-N-heterocycles with phosphoroamidates at higher temperatures. Thus, 6-methyluracil (**217**) is converted by the commercially available phenylphosphorodiamidate (PPDA) in 21% yield into 2,4-diamino-6-methylpyrimidine (**218**) and hypoxanthine **219** into adenine **220** in 25% yield (69IZV655). A condensed heterocycle such as 4-hydroxy-quinazoline (**221**), on heating for 30 min with PPDA to 235°C, gives 47% of 4-aminoquinazoline **222**. Analogously, 3-benzo[*f*]quinazolinone can be converted into 3-aminobenzo[*f*]quinazoline in 76% yield (70IZV953). The yields obtained are usually higher with N-substituted phosphoroamidates.

Thus, uracil **223** is converted by *N*-phenylphosphoric acid triamide in 46% yield into 2,4-bis-phenylaminopyrimidine (**224**), whereas HMPA aminates uracil **223** to the 2,4-bis-dimethylaminopyrimidine (**225**) in 78% yield (70IZV904). The amination of xanthine **226** by HMPA furnishes 79% of 2,6-bis-(dimethylamino)purine (**227**) (70IZV953).

Catalytic amounts of amine salts such as dimethylamine hydrochloride or $POCl_3$ seem to improve the yields (70IZV953; 72JHC1235). HMPA converts 4-(1*H*)-pyridinone (**40**) in 57% yield to the nucleophilic catalyst DMAP (**228**) and converts 2-(1*H*)-pyridinone (**27**) in 44% yield to 2-dimethylaminopyridine (**229**) (73S301).

For more reactions of 2-(1*H*)-quinolinone with HMPA to give 2-dimethylaminoquinoline in 79% yield see (74T875). Because of the carcin-

Sec. IV.F] AMINATION OF NITROGEN HETEROCYCLES 163

ogenic nature of HMPA, these aminations should be conducted with great care so as to avoid any contact with the HMPA vapors.

Zavyalov suggests that the phosphoroamidates react with hydroxy-N-heterocycles, via intermediates such as **230**, followed by subsequent elimination of phosphoroamidate to give the amination product **231** (69IZV655).

Instead of using any preformed phosphoroamidates, Pedersen *et al.* have used mixtures of amines and phosphorus pentoxide. Reaction of 4-(1*H*)-pyridinone (**40**) with pyrrolidine and P_2O_5 furnishes the nucleophilic catalyst PPY (**184**) in 39% yield (78S844) (see also Section IV,C). The analogous reaction of 2-(1*H*)-pyridinone (**27**) with pyrrolidine gives 2-pyrrolidinopyridine in 27% yield (**232**) and reaction with piperidine gives 2-piperidinopyridine (**233**) in 49% yield. 4-Methyl-(1*H*)-2-quinolinone (**234**) is aminated by a mixture of ammonium chloride, P_2O_5, and tributylamine to 2-amino-4-methylquinoline (**235**) in 68% yield (81CS240). For analogous aminations of **234** with alkylamines and a mixture of polyphosphoric acid and P_2O_5 see (81CS240); for aminations of hypoxanthines see (82S480); and for 7-*H*-pyrrolo[2,3-*d*]pyrimidine-2,4-diones see (88CS201).

Cyclization of aromatic or heteroaromatic ortho-acylamino esters (82-LA1012) or nitriles [87ACS(B)467] with amines and P_2O_5 leads to aminated, annellated nitrogen heterocycles in moderate yields. The aforementioned (Section IV,B) one-step reaction of 4-phenyl-1-(2H)-phthalazinone **(83)** with neopentylamine in the presence of $POCl_3$ in boiling xylene to give 1-neopentylamino-4-phenyl-phthalazine **(84)** might also involve the in situ formation of phosphoroamidates.

G. 1,2,4-Triazoles

During studies on oligonucleotide formation with excess 1-(mesitylene-2-sulfonyl)-1,2,4-triazole (MSNT) or tetrazole, Reese discovered that protected uridines are converted at room temperature into the chemically stable 1,2,4-triazole derivatives in high yields (80MI1). Reese

[82JCS(P1)1171] and Sung (81CC1089; 82JOC3623) found that these 1,2,4-triazole derivatives of uridine can be readily prepared at room temperature with excess 1,2,4-triazole and phosphoryl chloride and triethylamine or with O-chlorophenyldi-(1H-1,2,4-triazol-1-yl)phosphine oxide in acetonitrile. These stable, often crystalline 1,2,4-triazole derivatives react readily with aqueous ammonia and primary or secondary amines at room temperature to afford the corresponding cytidines in high yields.

The crystalline 1,2,4-triazole derivative **236**, on reaction with aqueous ammonia, gives 1-β-arabinofuranosylcytosine **(237)**. Heating the more reactive 3-nitro-1,2,4-triazole derivative **238** with the weakly basic aniline in pyridine, followed by subsequent hydrolysis with aqueous ammonia,

furnishes the corresponding free 1-β-arabinofuranosyl-N^4-phenylcytosine **(239)** [82JCS(P1)1171] (cf. Section V,C for a different preparation of N^4-arylcytidines). The reaction sequence can be used to convert the sensitive

2',3'-dideoxy-3'-azidouridines (**240**) into the corresponding cytidines (**241**) in high yields (83JMC1691; 86EUP217580) (for the analogous preparation of 2'-azidocarbocyclic cytidine as well as other cytidines, see 87JMC440; 88EUP261595, 88JMC268, 88JMC484, 88JMC1475).

Silylated 3-nitro-1,2,4-triazole derivative **242** is readily aminated by ethylenediamine to give dimeric cytidine **243**, as well by other ω-diamines, in 75% yield (86NJC643). The resulting cytidines can be attached to polymeric supports (87T3481, 87T3491) (for further applications to nucleotides compare 82RTC77, 83TL3171, 86JA2764). This methodology is very appropriate for the amination of protected, sensitive, and costly uridines, but implies several steps and the application of expensive reagents such as 1,2,4-triazole or 3-nitro-1,2,4-triazole and for these reasons, is rather unsuitable for large-scale preparation.

H. QUATERNARY AMMONIUM SALTS

Trialkylammonium derivatives, especially trimethylammonium derivatives of nitrogen heterocycles, have been used as activated intermediates because these groups are readily displaced by nucleophiles [71JCS(B)821]. The reactivity of trimethylammonium derivatives toward hydroxide ions has been measured for pyrimidine-2-yltrimethyl-ammonium chloride, which was found to be 700 times more reactive than 2-chloropyrimidine and ~ 5 times less reactive than 2-methanesulfonylpyrimidine [71JCS(B)1675].

As already discussed in Section IV,B (cf. reactions **13–20**), heating hydroxy-N-heterocycles with $POCl_3$ and tertiary amines leads, via the quaternary salts with elimination of alkyl chloride, to the dialkylamino-N-heterocycles. The heterocyclic chlorides formed can react in situ with the employed excess tertiary amines to the corresponding quaternary salts. As depicted in the reaction of 2,4-(1H,3H)-quinazoline-2,4-dione **(18)** with $POCl_3$ and N-methylpyrrolidine, the corresponding 2-(N-methyl-4-chlorobutylamino)-4-chloroquinazoline **(19)** can thus be obtained in 71% yield (82CPB1947, 82CPB3121). However, reaction of preformed chloropyridines, -pyrimidines, -quinolines, -isoquinolines, or -purines with trimethylamine to give the corresponding dimethylamino compounds via the quaternary salts apparently does not offer any advantage over the direct reaction of the chloro compounds with dimethylamine [71JCS(B)821; 71JCS(B)1675, 71JCS(B)2323; 72JCS(P1)1269]. The same arguments are also valid for the reactions of chloropyrimidines with triethylamine (75GEP2342881; 87JHC205).

Very useful is 4-pyridylpyridinium chloride-hydrochloride **(65)**, which is obtained by reaction of pyridine **64** with thionyl chloride as described in Section IV,B. 4-Pyridylpyridinium chloride-hydrochloride **(65)**, which can be readily prepared in large quantities, reacts on heating in DMF as solvent and reactant (cf. Section III,B and especially IV,B reactions **105–108**) to form the important nucleophilic catalyst DMAP **(228)** on a technical scale in ~60–70% yield. Similarly, N-formylpyrrolidine affords PPY **(184)** in 60% yield (77USP4140853).

Whereas ammonia or primary aliphatic or aromatic amines cleave the pyridinium ring in **65** with formation of 4-aminopyridine (58CB1266), secondary amines such as dimethylamine or pyrrolidine react with **65** to form DMAP **(228)** in ~40% yield and PPY **(184)** in 32% yield (84GEP3241429).

Heating **65** with N-methylformamide affords crystalline **244** (77USP4140853), which can be reacted with acrylic ester to give, after saponification, 3-[N-methyl-N-(4)-pyridylamino]propionic acid. This acid

Sec. IV.I] AMINATION OF NITROGEN HETEROCYCLES 169

can be readily attached to polyamines such as polyethyleneimine to give polymeric DMAP (79JA6020).

I. SULFIDES, SULFOXIDES, AND SULFONES

Mercapto or methylthio nitrogen heterocycles are either obtained by direct treatment of hydroxy-nitrogen heterocycles with reagents such as phosphorus pentasulfide or Lawesson reagent (78BSB223; 79T2433), or by treatment of activated intermediates such as chloro-nitrogen heterocycles with mercaptide anions as well as by total synthesis.

Methylation of 2-thiouracil with methyl iodide in the presence of sodium methylate gives 2-methylthio-4-hydroxypyrimidine (245), which is aminated by isopropylamine in toluene at 110–120°C to 2-isopropylamino-4-hydroxypyrimidine (246) in 84% yield (80GEP3009071).

Treatment of S-methylthiosemicarbazide (247) with glyoxal affords 3-methylthio-1,2,4-triazine (248), which reacts readily at 24°C with hydrazine to give 249. It also reacts with ethanolamine in boiling n-propanol to give 250 and also with ethyleneimine to give 251 (76JHC807). Also typical of this class of reactions is the transformation of 252 with 4-hydroxypiperidine to 253 (83EUP88593).

In the reaction of 2-chloro-4-methylthioquinazoline (254) with ethylamine in DMF at 80–85°C, the methylmercapto leaving group can react with the chlorine in 254 to afford, aside from 38% of 255 and 54% of 256, 3% of 2-methylthio-4-ethylaminoquinazoline (257), 2% of 2,4-bis-methylthioquinazoline (258), and 1% of 2,4-bis-ethylaminoquinazoline (259) (82H11).

According to the classic method of Fox (58JA1669), inosine-2',3',5'-tri-O-benzoate (260) is heated with P_2S_5 in pyridine or dioxane to give the corresponding 6-thio derivatives 261a, which can be reacted as such or, after transformation with methyl iodide into the more reactive S-methyl derivatives, 261b then gives corresponding adenosines 262. Amination of 2-amino-6-benzylthio-9-β-D-ribofuranosylpurine (263) with methyl- or dimethylamine furnishes the corresponding 6-amino derivative **264** in high yield (64CPB951).

Since kinetic studies with 2-methylthio-, 2-methylsulfinyl-, and 2-methylsulfonylpyrimidine demonstrated that the sulfinyl and the sulfonyl derivatives are more than 10^5 times more reactive toward aminolysis than the corresponding 2-methyl- or 2-phenylthiopyrimidines [67JCS(C)568; 69JCS(C)2720, 69MI1], sulfoxides and sulfones have been frequently prepared from the corresponding methylthio or phenylthio nitrogen heterocycles by oxidation with *m*-chloroperbenzoic acid. 2-Methylsulfinyl- or 2-methylsulfonylpyrimidine are approximately as active toward aminolysis with *n*-pentylamine in DMSO as 2-chloropyrimidine (69MI1) (see, however, reaction **270 → 271!**).

Thus, refluxing of 2-methylsulfonylpyrimidine (**265**) in excess isopropylamine for 1 hr gives a quantitative yield of 2-isopropylaminopyrimidine (**266**) (80BEP883751). Heating 3-methylsulfonyl-1,2,4-triazine (**267**) with butynylamine affords **268** in 64% yield (87TL379). (For further aminations of methylsulfonylpyrimidines see 81CPB948; 88JHC959.)

2,6-bis-(Methylsulfonyl)-pyrazine **(270)** is obtained in 53% yield by reaction of 2,6-dichloropyrazine **(269)** with sodium methylsulfinate in DMF. Amination of **270** with piperazine gives 6-methylsulfonyl-2-(piperazinyl)pyrazine **(271)** (78JMC536). 6-Methyl (or phenyl) sulfonyl-9-β-D-ribofuranosylpurine reacts analogously with primary amines (75JOC658; see also 87TL2469).

V. Miscellaneous Aminations

A. Amination via Nitrile, Trichloromethyl, and Carboxy Moieties as Leaving Groups

The replacement of nitrile groups by amino groups is exemplified by the reaction of 2,3-dicyano-pyrazine **(272)** with an aqueous solution of methylamine to give 3-methylamino-2-cyanopyrazine **(273)** (86JHC1299; see also

75CPB2044). However, much more interesting are two further technical syntheses of the important nucleophilic catalyst DMAP **(228)** and related 4-aminopyridines. Treatment of 4-cyanopyridine **(274)**, produced on a technical scale by ammonoxidation of γ-picoline, with 2-vinylpyridine **(275)** in the presence of hydrochloric acid gives the crystalline quaternary salt **276**. This salt is readily aminated by an aqueous solution of dimethylamine at room temperature with liberation of cyanide. The resulting aminated quaternary salt is finally cleaved by 40% NaOH to DMAP **(228)** and 2-vinylpyridine **(275)**, which is recycled after separation of **228** and **275** by distillation (79USP4158093).

Reaction of quaternary salt **226** with 4-methylpiperidine gives rise to 4-(4-methylpiperidino)pyridine **(277)**, which is a liquid, in contrast to the crystalline DMAP **(228)** and PPY **(184)** (83EUP74837). The reaction of 4-cyanopyridine **(274)** with acrylamide in the presence of HCl furnishes the crystalline quaternary salt **278**, in quantitative yield, which is readily aminated with dimethylamine. Subsequent cleavage of the aminated quaternary salt by heating with strong alkali solution gives monomeric or polymeric sodium acrylate, as well as a nearly quantitative yield of DMAP (228), which can simply be extracted from the aqueous phase (86EUP192003).

Like nitrile groups, trichloromethyl groups as in tris(trichloromethyl)-S-triazine (279) can be replaced stepwise by N-(2-hydroxyethyl)-piperazine to 280 and subsequently by ethylamine to 281 (67GEP1670564; 85KGS1557). For the analogous conversion of 2-trifluoroquinoline by sodium amide to 2-aminoquinoline see Kobayashi et al. (75CPB2044).

The Curtius degradation of carboxylic groups to amino groups is standard chemistry as in the degradation of 4-chloro-2-pyridinecarbocylic acid to 2-amino-4-chloropyridine via the acid azide (81JMC39) and the corresponding isocyanate, or of 5-ethyl-2-pyridinecarboxylic acid (**282**) to the carbamate (**283**) by reaction with diphenylphosphoryl azide and benzylalcohol/triethylamine (78GEP2822544).

The conversion of 2,5-pyridinedicarboxylic acid N oxide (**284**) in acetonitrile/acetic anhydride to the corresponding 6-amino-nicotinic acid (**288**) seems to be a Ritter reaction of intermediate cation **285** with acetonitrile, followed by rearrangement of intermediate **286** to **287**, which is then saponified by potassium hydroxide to give 49% of 6-aminonicotinic acid (**288**) and 8% of 6-hydroxynicotinic acid (**289**) (83EUP90173).

B. Transaminations and Dimroth Alkylations

Since amino groups are rather poor leaving groups, acid-catalyzed transaminations, which proceed probably via the typical AE mechanisms (Section II), demand rather vigorous conditions. Thus, heating 6-amino-2,4-dimethylpyrimidine hydrochloride (290) for 20 hr at 170°C with butylamine affords, via 291, compound 292a in 91% yield. The more weakly basic aniline gives 292b in only 53% yield. Under the same reaction conditions, 2-methyl-6-amino-[3H]-pyrimidine-4-one is transaminated with benzylamine and aniline in moderate yields (60JA3971). Adenine, on heating for 8 hr to 170°C with excess benzylamine/benzylamine hydrochloride, gives rise to 6-benzylaminopurine in 54% yield (73HJC133).

On refluxing 2,4,7-triamino-6-methylsulfonylpteridine (293) with excess benzylamine, the transamination product 2,4,7-tribenzylaminopteridine (294) is obtained in 44% yield without replacement of the 6-methylsulfonyl group. Other amines such as β-phenethylamine react analogously. Interestingly, heating 293 with thiophenol effects selective replacement of the methylsulfonyl group (73JHC133).

Persilylated adenosine 295, on heating with excess phenethylamine for several days at 170°C in the presence of mercuric chloride, affords only 20% of N^6-β-phenethyladenosine (296) (76LA745). N-Tosylation of O-protected 2'-deoxycytidine to 297 with excess tosylchloride-pyridine, and amination with n-butylamine readily gives the protected N^4-butyl-2'-deoxycytidine (298) and p-toluenesulfonamide (87MI3). The transamination of cytidine 149 with hydrazine and bisulfite takes place via intermediate 299 to give N^4-aminocytidine (300) in 29% yield (87CPB3884). The transamination of 2-amino-benzoxazole with benzylamine proceeds readily in toluene on heating for 8 hr to 160°C in the presence of CH_3SO_3H to give 99% of 2-benzylamino-benzoxazole (86MIP1). 5-Phenacyl-2-aminooxazole, however, is partially rearranged by dimethylamine to give, besides 6% of the expected 5-phenacyl-2-dimethylaminooxazole, 67% of 2-dimethylamino-4-phenyl-5-hydroxypyrimidine (87JHC1485).

It is obvious that some of the trans-amination products discussed in the preceding paragraphs are also accessable via reductive alkylation, as exemplified by the classic reductive alkylation of 2-amino-pyridine (28) with benzaldehyde to 2-benzylaminopyridine (187) (31CB2839) or by reductive alkylation of 3′,5′-cyclic-adenosine-phosphate (cAMP) (301) to N'-alkylated derivative 302 (88CPB2212).

Another classic transformation of amino-nitrogen-heterocycles is the Dimroth rearrangement, which has been reviewed (68MI1; 69ZC241) and further investigated [cf. 74JCS(P1)372; 79AJC1585]. In the Dimroth rearrangement of adenosine 303, the basic ^1N-nitrogen is alkylated to intermediate 304, which adds water to 305 and is followed by ring opening to 306 and ring closure (cf. the ^{15}N-label) to finally give N^6-alkylated adenosines 307 in up to 50% yield (69JMC1056; 73JOC2247).

C. Chichibabin Aminations

The Chichibabin reaction is exemplified by the amination of 2-methylpyridine (**308**) by sodium amide in toluene via intermediate **309** to yield 6-amino-2-methylpyridine (**310**). This reaction has been applied to pyridines, quinolines, isoquinolines, and naphthyridine as well as to benzimidazole, and was reviewed in great detail (66AHC229; 78RCR1042; 88AHC1).

Recently, the Chichibabin amination of 3-methylpyridine (**311**) with sodium amide and ammonia has been performed in the gas phase to give a 3.7:1 mixture of 2-amino-5-methyl pyridine (**312**) and 2-amino-3-methylpyridine (**313**) in good yield (84USP4386209). Heating 3-

methylpyridine (311) with sodium butylamide to 105°C furnishes 61% of 2-*n*-butylamino-5-methylpyridine (314) and ~10% of 2-*n*-butylamino-3-methylpyridine (315) (84USP4405790). However, all attempts to introduce dialkylamino groups via the Chichibabin amination have as yet failed.

The reaction of strongly basic amide anions, $R^1\text{-}\overline{N}\text{-}R^2$, with chloro- or bromopyridines, -pyrimidines, and other heterocycles can lead to ring opening and subsequent ring closure to result in substituted amino-N-heterocycles (addition of nucleophile, ring opening, ring closure = ANRORC).

Thus, reaction of 6-bromo-4-phenylpyrimidine (316) with lithium isopropylamide in isopropylamine proceeds via 317 to the ring-opened product 318 to eventually give 6-isopropylamino-4-phenylpyrimidine (319) in 70% yield. Since these particular aminations have been reviewed (78ACR462), they will not be discussed further. Nor will nucleophilic aminations proceeding via aryne-type intermediates (65AG557, 65AHC121) be a topic of this review.

Whereas metal hydride is eliminated from the σ-complex intermediates (cf. 309) in Chichibabin aminations, which results in the subsequent reaction of the metal hydride with ammonia or amines and the evolution of hydrogen, the σ-complex intermediates can also be oxidized to the corresponding amino-N-heterocycles. Thus, 5-azacinnoline (320) can be aminated with primary and secondary aliphatic or benzyl amines in the presence of potassium hydroxide/potassium ferricyanide to give the corresponding amino derivatives 321 in up to 65% yield. The corresponding amino group ($R^1 = R^2 = H$) can be readily introduced by hydroxylamine ·hydrochloride and KOH (77KGS1554).

Aromatic amines such as aniline or *p*-chloroaniline can be introduced analogously in the presence of KOH and air in up to 50% yield (78KGS809). Preparatively much more useful, however, is the oxidation of Chichibabin-σ-complexes by potassium permanganate in liquid ammonia. This topic has been reviewed by van der Plas (86MI12; 87KGS1011).

Reaction of quinoline **322** gives the kinetically controlled 2-aminoquinoline **(323)** at −60°C. At +10°C, however, the thermodynamically favored 4-aminoquinoline **324** is obtained (85JHC353). The amination of 5-substituted pyrimidines furnishes up to 95% of the corresponding 4-aminopyrimidines. 5-Nitropyrimidine **325**, which is decomposed by potassium amide, is converted by liquid ammonia/potassium permanganate into 2-amino-5-nitropyrimidine **(326)** (83JOC1354).

Analogously, pyrazine, pyridazine, 1,2,4- as well as 1,3,5-triazines, naphthyridines, and pteridines can be aminated. (For other applications, see 87JHC1377, 87JHC1657, 87JOC5643; 88JHC831.) However, only

1,2,4,5-tetrazines such as **327** have as yet been aminated to the corresponding alkyl-amino derivative **328** by primary amines such as ethylamine in liquid ammonia in the presence of potassium permanganate. If these new oxidative aminations can be performed on a larger scale *without* any explosion hazard, they might prove to be very useful preparatively.

D. AMINATION BY VICARIOUS NUCLEOPHILIC SUBSTITUTION

Katritzky *et al.* applied the new synthetic methodology of vicarious nucleophilic substitution of aromatic systems, as developed by Makosza (87ACR282), to the amination of nitrobenzenes. His group reacted nitrobenzenes **329** with 4-amino-1,2,4-triazoles (**330**) in DMSO in the presence of *t*-butoxide to give *p*-aminonitro compounds **332** via intermediate **331** in good yields. N-Substituted 4-amino-1,2,4-triazoles afford the corresponding *p*-alkylamino-nitrobenzenes. Analogously, 1-nitronaphthaline yields mainly a mixture of 2-amino- and 4-amino-1-nitronaphthaline, whereas 2-nitronaphthalene and 2-nitrothiophene give rise to 1-amino-2-nitronaphthalene and 3-amino-2-nitrothiophene, respectively, in good yields (86JOC5039; 88JOC3978).

As yet, however, no reactive nitro-N-heterocycle, such as 5-nitropyrimidine **(325)**, 1-alkyl-5-nitropyrimidine-2-one, or 4-nitropyridine-*N*-oxide [or N-heterocycles containing other acceptor groups such as 2-methylsulfonylpyrimidine **(265)** or 3-methylsulfonyl-1,2,4-triazine **(267)**] have been aminated using this interesting new procedure.

VI. Comparison of Different Amination Methods

On approaching the problem of aminating a certain N-heterocycle, three principal questions must be answered:

(1) What are the properties of the heterocyclic systems or amines to be employed? Do they contain groups sensitive to heat or reactive groups such as alcoholic or phenolic hydroxyls that have to be protected?

(2) How reactive is the particularly heterocyclic system (cf. Section III,A) and amine (cf. Section III,B), and how is the heterocyclic system to be activated for the amination?

(3) On what scale will the amination be performed—in g, kg or ton quantities?

Concerning heat sensitive groups (question 1), it is, for example, not advisable to heat azide moieties (cf. reaction **166** → **167** in Section IV,D) above temperatures of 120–140°C or heat 2'-deoxy nucleosides higher than 140°C (cf. reaction **162** → **163** in Section IV,D). Most importantly, reactive groups such as alcohol or phenol groups in a hydroxy-N-heterocycles, as well in the amine moiety, usually have to be protected, e.g., by acylation or silylation, *before* activating the hydroxy-*N*-heterocycle for subsequent amination. These protective groups must be removed *after* amination. Thus, at least three reaction steps are usually necessary if such reactive groups are present in the heterocyclic moiety (e.g., in nucleosides). Only silylation–amination permits protection–activation–amination in one step (cf. Section IV,D).

On comparing the reactivities of N-heterocycles with different leaving groups (question 2) in Section III and IV,A, it is obvious that the least reactive 2-(1*H*)-pyridinone **(27)** can only be aminated with great difficulty. The most reactive 2-sulfonyloxypyridines do undergo the competing S—O bond cleavage (cf. Sections IV,A and IV,E and reaction **209** → **210**) often giving only low yields of the desired 2-aminopyridines as well as recovered hydroxy-N-heterocycles. The silylation–amination sequence (Section IV,D) requires heating at 200°C for 24 hr as exemplified by the transformation of 2-(1H)-pyridinone (**27**) to give the corresponding 2-aminopyridines in 50–70% yield. Due to benzyl-transfer at $T > 180°C$, benzylamine gives only moderate yields of 2-benzylaminopyridine on silylation–amination. Furthermore, on longer heating at these temperatures, amines such as homoveratrylamine containing aromatic O-methyl groups will be partially demethylated by the amino group (cf. reaction **188** → **189** in Section IV,D).

A possible alternative is the reaction of 2-halopyridines with amines in polar solvents such as N-methylpyrrolidone in the presence of cupric ions or under high pressure, as discussed in Section IV,B.

The price of the starting N-heterocycle and the amine as well as the number of steps (question 3) will dictate the choice of amination method and determine the price of the resulting amination product. The large-scale preparation of the important nucleophilic catalysts, DMAP (**228**) or PPY (**184**), may serve as examples. The conversion of 4-(1H)-pyridone (**40**) into DMAP (**228**) by treatment with HMPA (cf. Section IV,F) or conversion into PPY (**184**) by silylation–amination with pyrrolidine (cf. Section IV,D) would produce **228** or **184** at prohibitive costs because of the relatively high price of 4-(1H)-pyridone (**40**) and HMPA and the carcinogenicity of the latter.

The conversion of pyridine by thionyl chloride to 4-pyridyl-pyridinium chloride hydrochloride (**65**), followed by heating in DMF as solvent and reactant to afford DMAP (**228**) and pyridine, was the first commercial process to give DMAP (**228**) in ton amounts at reasonable prices. Amination of 4-pyridylpyridinium chloride hydrochloride with pyrrolidine also furnishes PPY (**184**) directly (cf. Section IV,H). Alternatively, the inexpensive 4-cyanopyridine (**274**) is quaternized by either 2-vinylpyridine-HCl or acrylamide-HCl and readily aminated with dimethylamine or pyrrolidine to give, after alkaline treatment, DMAP (**228**) or PPY (**184**) (cf. Section V,A).

The available equipment might also influence the choice of the amination method. Thus, amination with volatile amines such as ammonia, methyl- or ethylamine via silylation–amination (cf. Section IV,D) on a laboratory scale demands the use of a small stainless-steel autoclave, whereas activation (e.g., of nucleosides by O-sulfonylation or O-phosphorylation-1,2,4-triazolide formation) proceeds readily at room temperature and normal pressure (cf. Section IV,G). The availability of high-pressure equipment will facilitate the difficult aminations of halopyridines (cf. reactions **86** → **111** and **112** → **113** in Section IV,B).

Finally, it might be useful to compare several amination methods as in the aforementioned amination of 2-methyl-4-hydroxypyrido-[3,4-d]pyrimidine **(48)**. Treatment with $POCl_3/N,N$-dimethylaniline and subsequent reaction with benzylamine gives 2-methyl-4-benzylaminopyrido[3,4-d]pyrimidine **(50)** in 45% overall yield (cf. Section IV,B), whereas one-step silylation–amination of **48** with benzylamine/HMDS affords **50** in 92% yield (cf. Section IV,D).

ACKNOWLEDGMENTS

I want to thank Dr. M. G. Pettett, for carefully reading and correcting the English text, and my secretary, Mr. W. Becker, for repeatedly typing and drawing as well as proofreading the manuscript.

References

31CB21	R. Graf, *Chem. Ber.* **64**, 21 (1931).
31CB1049	E. Königs and H. Greiner, *Chem. Ber.* **64**, 1049 (1931).
31CB2839	A. E. Tschitschibabin and I. L. Knunjanz, *Chem. Ber.* **64**, 2839 (1931).
32JPR36	R. Graf, *J. Prakt. Chem.* **133**, 36 (1932).
34JA231	L. C. Craig, *J. Am. Chem. Soc.* **34**, 231 (1934).
36CB1534	H. Maier-Bode, *Chem. Ber.* **69**, 1534 (1936).
36JPR188	R. Graf and F. Zettl, *J. Prakt. Chem.* **147**, 188 (1936).
36RTC122	H. J. den Hertog and J. P. Wibaut, *Recl. Trav. Chim. Pays-Bas* **55**, 122 (1936).
36RTC709	J. P. Wibaut and J. R. Nicolai, *Recl. Trav. Chim. Pays-Bas* **58**, 709 (1939).
44JA1127	C. K. Banks, *J. Am. Chem. Soc.* **66**, 1127 (1944).
44JA1456	J. L. Webb and A. H. Corwin, *J. Am. Chem. Soc.* **66**, 1456 (1944).
44JOC302	G. B. Bachman and D. E. Cooper, *J. Org. Chem.* **9**, 302 (1944).
47CA6245	M. V. Rubtsov and V. T. Klimko, *CA* **41**, 6245 (1947).
47JA303	R. W. Gouley, G. W. Moersch, and H. S. Mosher, *J. Am. Chem. Soc.* **69**, 303 (1947).
47JCS1052	G. A. Howard, B. Lythgoe, and A. R. Todd, *J. Chem. Soc.*, 1052 (1947).
51JA4773	T. E. Young and E. D. Amstutz, *J. Am. Chem. Soc.* **73**, 4773 (1951).
52JCS437	R. R. Bishop, E. A. S. Cavell, and N. B. Chapman, *J. Chem. Soc.*, 437 (1952).
53CJC457	H. Gilman and J. T. Edward, *Can. J. Chem.* **31**, 457 (1953).
53JOC534	E. Ochiai, *J. Org. Chem.* **18**, 534 (1953).
53JOC1484	K. L. Howard, H. W. Stewart, E. A. Conroy, and J. J. Denton, *J. Org. Chem.* **18**, 1484 (1953).
54ACS390	B. Bak and D. Cristensen, *Acta Chem. Scand.* **8**, 390 (1954).
54JA3408	W. T. Grubb, *J. Am. Chem. Soc.* **76**, 3408 (1954).
54JCS1190	N. B. Chapman and C. W. Rees, *J. Chem. Soc.*, 1190 (1954).
54JCS1795	K. Bowden and P. N. Green, *J. Chem. Soc.*, 1795 (1954).
54JOC1633	E. C. Taylor, Jr. and A. J. Crovetti, *J. Org. Chem.* **19**, 1633 (1954).
54JOC1830	K. R. Brower, J. W. Way, W. P. Samuels, and E. D. Amstutz, *J. Org. Chem.* **19**, 1830 (1954).
56CB2921	D. Jerchel, H. Fischer, and K. Thomas, *Chem. Ber.* **89**, 2921 (1956).

56JCS1563	N. B. Chapman and D. Q. Russell-Hill, *J. Chem. Soc.*, 1563 (1956).
56JCS2404	A. R. Katritzky, *J. Chem. Soc.*, 2404 (1956).
57JCS1769	A. R. Katritzky, E. W. Randall, and L. E. Sutton, *J. Chem. Soc.*, 1769 (1957).
57JCS4375	J. N. Gardner and A. R. Katritzky, *J. Chem. Soc.* 4375 (1957).
57MI1	R. C. Osthoff and S. W. Kantor, *Inorg. Synth.* **5**, 61 (1957) [*CA* **52**, 945a (1958)].
58CB1266	D. Jerchel and L. Jakob, *Chem. Ber.* **91**, 1266 (1958).
58JA1669	J. J. Fox, I. Wempen, A. Hampton, and I. L. Doerr, *J. Am. Chem. Soc.* **80**, 1669 (1958).
58JA6020	J. F. Bunnett and J. J. Randall, *J. Am. Chem. Soc.* **80**, 6020 (1958).
59JA6145	R. West and R. H. Baney, *J. Am. Chem. Soc.* **81**, 6145 (1959).
59RTC593	J. P. Wibaut and F. W. Broekman, *Recl. Trav. Chim. Pays-Bas* **78**, 593 (1959).
60AG294	J. Sauer and R. Huisgen, *Angew. Chem.* **72**, 294 (1960).
60JA3971	C. W. Whitehead and J. J. Traverso, *J. Am. Chem. Soc.* **82**, 3971 (1960).
62CCC902	J. J. K. Novak and F. Šorm, *Collect. Czech. Chem. Commun.* **27**, 902 (1962).
62CCC906	J. Skoda, I. Bartosek, and F. Šorm, *Collect. Czech. Chem. Commun.* **27**, 906 (1962).
62USP3021332	G. H. Hitchings and R. K. Robins, U.S. Pat. 3,021,332 (1962) [*CA* **57**, 839i (1962)].
63MI1	H. Suhr, *Ber. Bunsenges. Phys. Chem.* **67**, 893 (1963).
64CB3277	H. Suhr, *Chem. Ber.* **97**, 3277 (1964).
64CPB951	T. Naito, K. Ueno, and F. Ishikawa, *Chem. Pharm. Bull.* **12**, 951 (1964).
65AG557	T. Kauffmann, *Angew. Chem.* **77**, 557 (1965).
65AHC121	H. J. den Hertog and H. C. van der Plas, *Adv. Heterocycl. Chem.* **4**, 121 (1965).
65AHC145	R. G. Shepherd and J. L. Fedrick, *Adv. Heterocycl. Chem.* **4**, 145 (1965).
65CCC2052	J. Zemlicka and F. Šorm, *Collect. Czech. Chem. Commun.* **30**, 2052 (1965).
65JOC3618	J. J. D'Amico, S. T. Webster, R. H. Campbell, and C. E. Twine, *J. Org. Chem.* **30**, 3618 (1965).
66AHC229	R. A. Abramovitch and J. G. Saha, *Adv. Heterocycl. Chem.* **6**, 229 (1966).
66JHC107	L. Joseph and A. H. Albert, *J. Heterocycl. Chem.* **3**, 107 (1966).
66MI1	H. Suhr and H. Grube, *Ber. Bunsenges. Phys. Chem.* **70**, 544 (1966).
67GEP1670564	DEGUSSA, Ger. Pat. 1,670,564 (1967).
67JCS(C)568	D. J. Brown and P. W. Ford, *J. Chem. Soc. C*, 568 (1967).
68MI1	D. J. Brown, in "Mechanism Molecular Migration" (B. S. Thyagarajan, ed.), Vol. 1, p. 209. Wiley (Interscience), New York, 1968.
69CC38	N. D. Heindel and P. D. Kennewell, *J. C. S. Chem. Commun.*, 38 (1969).
69CRV1	A. J. Parker, *Chem. Rev.* **69**, 1 (1969).
69IZV655	E. A. Arutyunyan, V. I. Gunar, E. P. Gracheva, and S. I. Zav'yalov, *Izv. Akad. Nauk SSSR, Ser. Khim.*, 655 (1969).
69JCS(C)2720	D. J. Brown and P. W. Ford, *J. Chem. Soc. C*, 2720 (1969).
69JMC1056	M. H. Fleysher, A. Bloch, M. T. Hakala, and C. A. Nichol, *J. Med. Chem.* **12**, 1056 (1969).
69JOC247	E. Wenkert, F. Haglid, and S. L. Mueller, *J. Org. Chem.* **34**, 247 (1969).
69MI1	G. B. Barlin and D. J. Brown, in "Topics in Heterocyclic Chemistry" (R. N. Castle, ed.), p. 122. Wiley (Interscience), New York, 1969.
69ZC241	M. Wahren, *Z. Chem.* **9**, 241 (1969).

70ACR299	J. F. Klebe, *Acc. Chem. Res.* **3**, 299 (1970).
70AJC625	D. J. Brown and B. T. England, *Aust. J. Chem.* **23**, 625 (1970).
70IZV904	E. A. Arutyunyan, V. I. Gunar, and S. I. Zav'yalov, *Izv. Akad. Nauk SSSR, Ser. Khim.*, 904 (1970).
70IZV953	E. A. Arutyunyan, V. I. Gunar, and S. I. Zav'yalov, *Izv. Akad. Nauk SSSR, Ser. Khim.* 953 (1970).
70IZV1127	Z. A. Martirosyan, V. I. Gunar, and S. I. Zav'yalov, *Izv. Akad. Nauk SSSR, Ser. Khim.* 1127 (1970).
71GEP2032403	P. Aviron-Violet and A. Blind, Ger. Pat. 2,032,403 (1971) [*CA* **74**, 76337p (1971)].
71JCS(B)821	G. B. Barlin and A. C. Young, *J. Chem. Soc. B*, 821 (1971).
71JCS(B)1675	G. B. Barlin and A. C. Young, *J. Chem. Soc. B*, 1675 (1971).
71JCS(B)2323	G. B. Barlin and A. C. Young, *J. Chem. Soc. B*, 2323 (1971).
71JCS(C)1889	B. W. Arantz and D. J. Brown, *J. Chem. Soc. C*, 1889 (1971).
72CPB1050	M. Kaneko and B. Shimizu, *Chem. Pharm. Bull.* **20**, 1050 (1972).
72CPB1513	S. Yurugi, M. Hieda, T. Fushimi, Y. Kawamatsu, H. Sugihara, and M. Tomimoto, *Chem. Pharm. Bull.* **20**, 1513 (1972).
72CPB2678	Y. Ito, Y. Hamada, and M. Hirota, *Chem. Pharm. Bull.* **20**, 2678 (1972).
72IZV2530	S. I. Zav'yalov, V. I. Gunar, Z. A. Martirosyan, and L. F. Ovechkina, *Izv. Akad. Nauk SSSR, Ser. Khim.*, 2530 (1972).
72JA6190	E. J. Corey and A. Venkateswarlu, *J. Am. Chem. Soc.* **94**, 6190 (1972).
72JA7927	C. L. Liotta and A. Abidaud, *J. Am. Chem. Soc.* **94**, 7927 (1972).
72JCS(P1)1269	G. B. Barlin and A. C. Young, *J. C. S. Perkin 1*, 1269 (1972).
72JCS(P2)671	P. Forsythe, R. Frampton, C. D. Johnson, and A. R. Katritzky, *J. C. S. Perkin 2*, 671 (1972).
72JHC1235	A. Rosowsky and N. Papathanasopoulos, *J. Heterocycl. Chem.* **9**, 1235 (1972).
72T5507	P. Beak, T. S. Woods, and D. S. Mueller, *Tetrahedron* **28**, 5507 (1972).
73CB3398	H. Clauss and H. Suhr, *Chem. Ber.* **106**, 3398 (1973).
73HCA1225	K. Harzer and W. Pfleiderer, *Helv. Chim. Acta* **56**, 1225 (1973).
73JHC133	W. D. Johnston, H. S. Broadbent, and W. W. Parish, *J. Heterocycl. Chem.* **10**, 133 (1973).
73JHC511	J.-D. Bourzat and E. Bisagni, *J. Heterocycl. Chem.* **10**, 511 (1973).
76JOC2247	M. H. Wilson and J. A. McCloskey, *J. Org. Chem.* **38**, 2247 (1973).
73LA772	H. R. Kricheldorf, *Liebigs Ann. Chem.*, 772 (1973).
73OSC(5)977	R. F. Evans, H. C. Brown, and H. C. van der Plas, *Org. Synth., Coll. Vol.* **5**, 977 (1973).
73S301	H. Vorbrüggen, *Synthesis*, 301 (1973).
74GEP2308560	W. Kleeberg, H. Ahne, and R. Wiedenmann, Ger. Pat. 2,308,560 (1974) [*CA* **82**, 59008e (1975)].
74GEP2308591	W. Kleeberg, H. Ahne, and R. Wiedenmann, Ger. Pat. 2,308,591 (1974) [*CA* **82**, 73803s (1975)].
74JCS(P1)372	D. J. Brown and K. Ienega, *J. C. S. Perkin 1*, 372 (1974).
74T875	E. B. Pedersen and S.-O. Lawesson, *Tetrahedron* **30**, 875 (1974).
75CPB2044	Y. Kobayashi, I. Kumadaki, Y. Hanzawa, and M. Mimura, *Chem. Pharm. Bull.* **23**, 2044 (1975).
75GEP2341965	A. Langbein, H. Merz, G. Walther, and K. Stockhaus, Ger. Pat. 2,341,965 (1975) [*CA* **82**, 156121u (1975)].
75GEP2342881	J.-M. Claverie, G. Louiseau, G. Mattioda, R. Millischer, and F. Percheron, Ger. Pat. 2,342,881 (1975) [*CA* **82**, 156364a (1975)].
75JA1851	G. Illuminati, F. La Torre, G. Liggieri, G. Sleiter, and F. Stegel, *J. Am. Chem. Soc.* **97**, 1851 (1975).

75JOC658	R. Wetzel and F. Eckstein, *J. Org. Chem.* **40**, 658 (1975).
75LA988	H. Vorbrüggen, K. Krolikiewicz, and U. Niedballa, *Liebigs Ann. Chem.*, 988 (1975).
76CB1113	H. Neunhoeffer and B. Lehmann, *Chem. Ber.* **109**, 1113 (1976).
76JA7869	G. Simons, M. E. Fandler, E. R. Talaty, *J. Am. Chem. Soc.* **98**, 7869 (1976).
76JHC807	B. T. Keen, D. K. Krass, and W. W. Paudler, *J. Heterocycl. Chem.* **13**, 807 (1976).
76LA745	H. Vorbrüggen and K. Krolikiewicz, *Liebigs Ann. Chem.*, 745 (1976).
76TL4427	C. L. Liotta and A. Abidaud, *Tetrahedron Lett.* **17**, 4427 (1976).
77CC447	P. K. Bridson, W. T. Markiewicz, and C. B. Reese, *J. C. S. Chem. Commun.*, 447 (1977).
77CC791	P. K. Bridson, W. T. Markiewicz, and C. B. Reese, *J. C. S. Chem. Commun.*, 791 (1977).
77CL891	T. Yajima and K. Munakata, *Chem. Lett.*, 891, 1977.
77GEP2605467	G. Lamm, Ger. Pat. 2,605,467 (1977) [*CA* **87**, 152039d (1977)].
77JMC134	P. T. Berkowitz, R. A. Long, P. Dea, R. K. Robins, and T. R. Matthews, *J. Med. Chem.* **20**, 134 (1977).
77KGS1554	M. F. Budyka, P. B. Terent'ev, and A. N. Kost, *Khim. Geterotsikl. Soedin.* **11**, 1554 (1976).
77USP4140853	H. Vorbrüggen, U. S. Pat. 4,140,853 (1977) [*CA* **86**, 55293d (1977)].
78ACR462	H. C. van der Plas, *Acc. Chem. Res.* **11**, 462 (1978).
78AF578	P. Nickel, H.-R. Seidel, L. Preissinger, H. Barnickel, E. Fink, and O. Dann, *Arzneim.-Forsch.* **28**, 578 (1978).
78AG(E)569	G. Höfle, W. Steglich, and H. Vorbrüggen, *Angew. Chem., Int. Ed. Engl.* **17**, 569 (1978).
78BSB223	B. S. Pedersen, S. Scheibye, N. H. Nilsson, and S.-O. Lawesson, *Bull. Soc. Chim. Belg.* **87**, 223 (1978).
78GEP2822544	P. F. Juby, Ger. Pat. 2,822,544 (1978) [*CA* **90**, 103998u (1979)].
78JHC687	A. Walser and T. Flynn, *J. Heterocycl. Chem.* **15**, 687 (1978).
78JHC1513	A. Nuvole and G. A. Pinna, *J. Heterocycl. Chem.* **15**, 1513 (1978).
78HMC536	W. C. Lumma, Jr., R. D. Hartman, W. S. Saari, E. L. Engelhardt, R. Hirschmann, B. V. Clineschmidt, M. L. Torchiana, and C. A. Stone, *J. Med. Chem.* **21**, 536 (1978).
78KGS809	M. F. Budyka, P. B. Terent'ev, and A. N. Kost, *Khim. Geterotsikl. Soedin.*, 809 (1978).
78MI1	H. Vorbrüggen and K. Krolikiewicz, *in* "Nucleic Acid Chemistry" (L. B. Townsend and R. S. Tipson, eds.), p. 227. Wiley, New York, 1978.
78MI2	H. Vorbrüggen and K. Krolikiewicz, *in* "Nucleic Acid Chemistry" (L. B. Townsend and R. S. Tipson, eds.), p. 533. Wiley, New York, 1978.
78RCR1042	A. F. Pozharskii, A. M. Simonov, and V. N. Doron'kin, *Russ. Chem. Rev.* **47**, 1042 (1978).
78S844	E. B. Pedersen and D. Carlsen, *Synthesis*, 844 (1978).
78T3445	T. Kato, N. Katagiri, and A. Wagai, *Tetrahedron* **34**, 3445 (1978).
79ACR198	C. J. M. Stirling, *Acc. Chem. Res.* **12**, 198 (1979).
79AJC1585	D. J. Brown and T. Nagamatsu, *Aust. J. Chem.* **32**, 1585 (1979).
79GEP2838644	S. B. Greer, E. G. Stump, Jr., and T. Psarras, Ger. Pat. 2,838,644 [*CA* **92**, 192254u (1980)].
79JA956	C. A. Fyfe, S. W. H. Damji, and A. Koll, *J. Am. Chem. Soc.* **101**, 956 (1979).

79JA6020	M. A. Hierl, E. P. Gamson, and I. M. Klotz, *J. Am. Chem. Soc.* **101**, 6020 (1979).	
79JCR(S)125	S. R. James, *J. Chem. Res., Synop.*, 125 (1979).	
79JOC2556	S. Sekiguchi, I. Ohtsuka, and K. Okada, *J. Org. Chem.* **44**, 2556 (1979).	
79T2433	B. S. Pedersen and S.-O. Lawesson, *Tetrahedron* **35**, 2433 (1979).	
79USP4158093	T. D. Bailey and C. K. McGill, U. S. Pat. 4,158,093 (1979) [*CA* **91**, 123636y (1979)].	
80BEP883751	C. Aspisi and C. Demosthene, Belg. Pat. 883,751 (1980) [*CA* **94**, 192364g (1981)].	
80EUP45980	D. Nisato and P. Carminati, Eur. Pat. 45,980 (1980).	
80GEP3009071	A. Esanu, Ger. Pat. 3,009,071 (1980) [*CA* **94**, 103410d (1981)].	
80MI1	C. B. Reese and A. Ubasawa, *Nucleic Acids Res., Symp. Ser.* **7**, 5 (1980).	
80TL2265	C. B. Reese and A. Ubasawa, *Tetrahedron Lett.* **21**, 2265 (1980).	
81CC1089	W. L. Sung, *J. C. S. Chem. Commun.*, 1089 (1981).	
81CJC3360	M. J. Robins, F. Hansske, and S. E. Bernier, *Can. J. Chem.* **59**, 3360 (1981).	
81CPB948	T. Sekiya, H. Hiranuma, M. Uchide, S. Hata, and S.-I. Yamada, *Chem. Pharm. Bull.* **29**, 948 (1981).	
81CPB1069	T. Kato, N. Katagiri, and A. Wagai, *Chem. Pharm. Bull.* **29**, 1069 (1981).	
81CS240	E. B. Pedersen and D. Carlsen, *Chem. Scr.* **18**, 240 (1981).	
81JMC39	J. G. Lombardino, *J. Med. Chem.* **24**, 39 (1981).	
81S661	K. Oka, *Synthesis*, 661 (1981).	
82CPB1947	H. Miki, *Chem. Pharm. Bull.* **30**, 1947 (1982).	
82CPB3121	H. Miki, *Chem. Pharm. Bull.* **30**, 3121 (1982).	
82CPB3471	H. Miki and F. Kasahara, *Chem. Pharm. Bull.* **30**, 3471 (1982).	
82G167	D. Corvi, A. Mitidieri-Costanza, and G. Sleiter, *Gazz. Chim. Ital.* **112**, 167 (1982).	
82H7	H. Miki, *Heterocycles* **19**, 7 (1982).	
82H11	H. Miki and J. Yamada, *Heterocycles* **19**, 11 (1982).	
82H15	H. Miki, *Heterocycles* **19**, 15 (1982).	
82H1879	H. Miki and F. Kasahara, *Heterocycles* **19**, 1879 (1982).	
82JCS(P1)1171	K. J. Divakar and C. B. Reese, *J. C. S. Perkin 1*, 1171 (1982).	
82JMC837	G. Y. Lesher, B. Singh, and Z. E. Mielens, *J. Med. Chem.* **25**, 837 (1982).	
82JMC1040	J. L. Rideout, T. A. Krenitsky, G. W. Koszalka, N. K. Cohn, E. Y. Chao, G. B. Elion, V. S. Latter, and R. B. Williams, *J. Med. Chem.* **25**, 1040 (1982).	
82JOC3623	W. L. Sung, *J. Org. Chem.* **47**, 3623 (1982).	
82LA1012	O. R. Andresen and E. B. Pedersen, *Liebigs Ann. Chem.*, 1012 (1982).	
82RTC77	G. A. van der Marel, G. Wille, H. Westerink, A. H.-J. Wang, A. Rich, J. R. Mellema, C. Altona, and J. H. van Boom, *Recl. Trav. Chim. Pays-Bas* **101**, 77 (1982).	
82S480	N. S. Girgis and E. B. Pedersen, *Synthesis*, 480 (1982).	
83EUP74837	E. F. V. Scriven and L. M. Huckstep, Eur. Pat. 74,837 (1983) [*CA* **99**, 70568j (1983)].	
83EUP88593	D. T. Wong and W. B. Lacefield, Eur. Pat. 88,593 (1983) [*CA* **100**, 103396w (1984)].	
83EUP90173	D. Quarroz, Eur. Pat. 90,173 (1983) [*CA* **100**, 68178c (1984)].	
83IJC(B)48	D. S. Bhakuni and P. K. Gupta, *Indian J. Chem., Sect. B* **22B**, 48 (1983).	
83JMC891	T. A. Krenitsky, G. A. Freeman, S. R. Shaver, L. M. Beacham, III, S. Hurlbert, N. K. Cohn, L. P. Elwell, and J. W. T. Selway, *J. Med. Chem.* **26**, 891 (1983).	

83JMC1691	T.-S. Lin, Y.-S. Gao, and W. R. Mancini, *J. Med. Chem.* **26**, 1691 (1983).
83JMC1696	W. S. Saari, W. Halczenko, S. W. King, J. R. Huff, J. P. Guare, Jr., C. A. Hunt, W. C. Randall, P. S. Anderson, V. J. Lotti, D. A. Taylor, and B. V. Clineschmidt, *J. Med. Chem.* **26**, 1696 (1983).
83JOC1354	H. C. van der Plas, V. N. Charushin, and B. van Veldhuizen, *J. Org. Chem.* **48**, 1354 (1983).
83TL3171	B. C. Froehler and M. D. Matteucci, *Tetrahedron Lett.* **24**, 3171 (1983).
84CB1523	H. Vorbrüggen and K. Krolikiewicz, *Chem. Ber.* **117**, 1523 (1984).
84GEP3241429	H. Häuser and G. Walter, Ger. Pat. 3,241,429 (1984) [*CA* **101**, 72621e (1984)].
84JHC1161	H. Gershon and A. T. Grefig, *J. Heterocycl. Chem.* **21**, 1161 (1984).
84MIP1	L. O. Sandefur, W. Slusarek, B. D. Wilson, and C. A. Maggiulli, PCT Pat. WO 84/00751 (1984) [*CA* **101**, 23492e (1984)].
84S765	N. C. Hung and E. Bisagni, *Synthesis*, 765 (1984).
84S1058	R. G. McR. Wright, *Synthesis*, 1058 (1984).
84USP4386209	C. K. McGill and J. J. Sutor, U. S. Pat. 4,386,209 (1984) [*CA* **100**, 6343u (1984)].
84USP4405790	C. K. McGill and T. D. Bailey, U. S. Pat. 4,405,790 (1984) [*CA* **100**, 22584d (1984)].
85EUP136730	W. Orth and W. Fickert, Eur. Pat. 136,730 (1985) [*CA* **103**, 22472c (1985)].
85EUP149537	J. M. Greene, A. J. Pike, E. R. Lavagnino, and E. C. Taylor, Eur. Pat. 149,537 (1985) [*CA* **103**, 215190t (1985)].
85EUP159112	J. B. Campbell, E. R. Lavagnino, and A. J. Pike, Eur. Pat. 159,112 (1985) [*CA* **104**, 148751g (1986)].
85EUP159652	Y. Morinaka, K. Iseki, T. Kanayama, T. Watanabe, and H. Nishi, Eur. Pat. 159,652 (1985) [*CA* **104**, 75039u (1986)].
85H2247	R. S. Hosmane and B. B. Lim, *Heterocycles* **23**, 2247 (1985).
85JCS(P2)87	Y. Hasegawa, *J. C. S. Perkin 2*, 87 (1985).
85JCS(P2)929	Z. Zhu, C. Sun, and T. Han, *J. C. S. Perkin 2*, 929 (1985).
85JHC353	H. Tondys, H. C. van der Plas, and M. Wozniak, *J. Heterocycl. Chem.* **22**, 353 (1985).
85JMC1790	G. A. Youngdale and T. F. Oglia, *J. Med. Chem.* **28**, 1790 (1985).
85JOC649	Y. Hasegawa, *J. Org. Chem.* **50**, 649 (1985).
85JPR865	A. Mitidieri-Costanza and C. Sleiter, *J. Prakt. Chem.* **327**, 865 (1985).
85KGS1557	W. I. Kjelarjew, A. Dibi, and A. F. Lunin, *Khim. Geterotsikl. Soedin.*, 1557 (1985).
85MI1	V. A. Efimov and O. G. Chakhmakhcheva, *Nucleosides & Nucleotides* **4**, 265 (1985).
85S1	K. Matsumoto, A. Sera, and T. Uchida, *Synthesis*, 1 (1985).
85SC1013	G. M. Coppola, *Synth. Commun.* **15**, 1013 (1985).
86ABC495	S. Nishikawa, Z. Kumazawa, N. Kashimura, S. Maki, and Y. Nishikimi, *Agric. Biol. Chem.* **50**, 495 (1986).
86ACS(B)806	X.-X. Zhou, C. J. Welch, and J. Chattopadhyaya, *Acta Chem. Scand., Ser. B* **B40**, 806 (1986).
86ACS(B)826	N. Nyilas and J. Chattopadhyaya, *Acta Chem. Scand., Ser. B* **B40**, 826 (1986).
86CS55	V. A. Efimov and O. G. Chakhmakhcheva, *Chem. Scr.* **26**, 55 (1986).
86EUP192003	L. J. Nummy, Eur. Pat. 192,003 (1986) [*CA* **106**, 4887z (1987)].
86EUP203685	M. R. Harnden and R. L. Jarvest, Eur. Pat. 203,685 (1986).
86EUP204264	M. Kawada, K. Matsumoto, and M. Tsurushima, Eur. Pat. 204,264 (1986) [*CA* **106**, 214307m (1987)].

86EUP217580	J. L. Rideout, D. W. Barry, S. N. Lehrman, M. H. St. Clair, P. A. Furman, G. A. Freeman, T. P. Zimmerman, G. Wolberg, P. M. De Miranda, S. R. Shaver, G. White, L. E. Kirk, III, D. H. King, and R. H. Clemons, Eur. Pat. 217,580 (1986) [*CA* **107**, 40276d (1987)].
86JA2764	T. R. Webb and M. D. Matteucci, *J. Am. Chem. Soc.* **108**, 2764 (1986).
86JAP3183582	Farmacim Eng. Spa, Jap. Pat. 3,183,582 (1986).
86JHC1299	K. Tsuzuki and M. Tada, *J. Heterocycl. Chem.* **23**, 1299 (1986).
86JMC709	T.-L. Su, J.-T. Huang, J. H. Burchenal, K. A. Watanabe, and J. J. Fox, *J. Med. Chem.* **29**, 709 (1986).
86JOC5039	A. R. Katritzky and K. S. Laurenzo, *J. Org. Chem.* **51**, 5039 (1986).
86M1305	W. Stadlbauer, *Monatsh. Chem.* **117**, 1305 (1986).
86MI1	L. J. Petrauskiene, S. J. Klimasauskas, V. V. Butkus, and A. A. Janulaitis, *Bioorg. Khim.* **12**, 1597 (1986).
86MI2	H. C. van der Plas and M. Wozniak, *Croat. Chem. Acta* **59**, 33 (1986).
86MIP1	F. Brunner, L. Due, R. Tabacchi, and K. J. Boosey, Swiss Pat. CH667,649 [*CA* 110,192821c (1989)].
86NJC643	A. F. Maggio, M. Lucas, J. L. Barascut, A. Pompon, and J. L. Imbach, *Nouv. J. Chim.* **10**, 643 (1986).
86SC543	T. N. Srinivasan, K. Rama Rao, and P. B. Sattur, *Synth. Commun.* **16**, 543 (1986).
86T2303	N. Chi Hung and E. Bisagni *Tetrahedron* **42**, 2303 (1986).
86TL5997	P. Barraclough, R. Iyer, J. C. Lindon, and J. M. Williams, *Tetrahedron Lett.* **27**, 5997 (1986).
87ACR282	M. Makosza and J. Winiarski, *Acc. Chem. Res.* **20**, 282 (1987).
87ACS(B)467	K. M. H. Hilmy, J. Mogensen, and E. B. Pedersen, *Acta Chem. Scand., Ser. B* **B41**, 467 (1987).
87CI(L)694	G. L. Goe, C. A. Huss, J. G. Keay, and E. F. V. Scriven, *Chem. Ind. (London)*, 694 (1987).
87CPB3884	K. Negishi, M. Kawakami, K. Kayasuga, J. Odo, and H. Hayatsu, *Chem. Pharm. Bull.* **35**, 3884 (1987).
87EUP225866	L. H. Schlager, Eur. Pat. 225,866 (1987) [*CA* **107**, 77834k (1987)].
87EUP270201	J. Lamsa, Eur. Pat. 270,201 (1987).
87H1215	K. Makino and H. Yoshioka, *Heterocycles* **26**, 1215 (1987).
87JA1275	X. Gao and R. A. Jones, *J. A. Chem. Soc.* **109**, 1275 (1987).
87JCS(P2)951	N. S. Nudelman, P. M. E. Mancini, R. D. Martinez, and L. R. Vottero, *J. C. S. Perkin 2*, 951 (1987).
87JCS(P2)987	S. M. Chiacchiera, J. O. Singh, J. D. Anunziata, and J. J. Silber, *J. C. S. Perkin 2*, 987 (1987).
87JHC205	H. Gershon, A. Grefig, and D. D. Clarke, *J. Heterocycl. Chem.* **24**, 205 (1987).
87JHC1243	H. Gershon, A. T. Grefig, and D. D. Clarke, *J. Heterocycl. Chem.* **24**, 1243 (1987).
87JHC1377	D. J. Buurman and H. C. van der Plas, *J. Heterocycl. Chem.* **24**, 1377 (1987).
87JHC1657	A. C. Brouwer and H. C. van der Plas, *J. Heterocycl. Chem.* **24**, 1657 (1987).
87JHC2485	F. Y. Walker and V. G. Kraus, *J. Heterocycl. Chem.* **24**, 1485 (1987).
87JMC440	T.-S. Lin, M. S. Chen, C. McLaren, Y.-S. Gao, I. Ghazzouli, and W. H. Prusoff, *J. Med. Chem.* **30**, 440 (1987).
87JOC5643	M. Wozniak, A. Baranski, K. Nowak, and H. C. van der Plas, *J. Org. Chem.* **52**, 5643 (1987).
87KGS1011	H. C. van der Plas, *Khim. Geterotsikl. Soedin.* 1011 (1987).

87MI1	P. Helissey, H. Parrot-Lopez, J. Renault, and S. Cros, *Eur. J. Med. Chem.* **22**, 366 (1987).
87MI2	V. Butkus, S. Klimasauskas, L. Petrauskiene, Z. Maneliene, A. Janulaitis, L. E. Minchenkova, and A. K. Schyolkina, *Nucleic Acids Res.* **15**, 8467 (1987).
87MI3	W. T. Markiewicz, R. Kierzek, and B. Hernes, *Nucleosides & Nucleotides* **6**, 269 (1987).
87MI4	R. Saffhill, *Nucleosides & Nucleotids* **6**, 679 (1987).
87T3481	S. Pochet, T. Huynh-Dinh, and J. Igolen, *Tetrahedron* **43**, 3481 (1987).
87T3491	S. R. Sarfati, S. Pochet, C. Guerreiro, A. Namane, T. Huynh-Dinh, and J. Igolen, *Tetrahedron* **43**, 3491 (1987).
87T4803	R. G. Andrew and R. A. Raphael, *Tetrahedron* **43**, 4803 (1987).
87TL379	E. C. Taylor and J. L. Pont, *Tetrahedron Lett.* **28**, 379 (1987).
87TL2469	M. D. Matteucci and T. R. Webb, *Tetrahedron Lett.* **28**, 2469 (1987).
87TL2821	N. Bischofberger, *Tetrahedron Lett.* **28**, 2821 (1987).
88AHC1	C. K. McGill and A. Rappa, *Adv. Heterocycl. Chem.* **44**, 1 (1988).
88CJC61	L. J. J. Hronowski and W. A. Szarek, *Can. J. Chem.* **66**, 61 (1988).
88CPB945	A. Matsuda, H. Itoh, K. Takenuki, T. Sasaki, and T. Ueda, *Chem. Pharm. Bull.* **36**, 945 (1988).
88CPB2212	S. Kataoka, J. Isono, N. Yamaji, M. Kato, T. Kawada, and S. Imai, *Chem. Pharm. Bull.* **36**, 2212 (1988).
88CS201	A. Jorgensen, H. H. Moharram, and E. B. Pedersen, *Chem. Scr.* **28**, 201 (1988).
88EUP261595	T.-S. Lin and W. H. Prusoff, Eur. Pat. 261,595 (1988).
88H319	S. Hashimoto, S. Otani, T. Okamoto, and K. Matsumoto, *Heterocycles* **27**, 319 (1988).
88H1899	D. Donati, S. Fusi, F. Ponticelli, and P. Tedeschi, *Heterocycles* **27**, 1899 (1988).
88JA3495	R. Bacaloglu, C. A. Bunton, G. Cerichelli, and F. Ortega, *J. Am. Chem. Soc.* **110**, 3495 (1988).
88JA3503	R. Bacaloglu, C. A. Bunton, and F. Ortega, *J. Am. Chem. Soc.* **110**, 3503 (1988).
88JA3512	R. Bacaloglu, C. A. Bunton, and F. Ortega, *J. Am. Chem. Soc.* **110**, 3512 (1988).
88JHC831	A. T. M. Marcelis, H. Tondijs, and H. C. van der Plas, *J. Heterocycl. Chem.* **25**, 831 (1988).
88JHC959	S. Kohra, Y. Tominaga, and A. Hosomi, *J. Heterocycl. Chem.* **25**, 959 (1988).
88JMC268	A. S. Jones, J. R. Sayers, R. T. Walker, and E. De Clercq, *J. Med. Chem.* **31**, 268 (1988).
88JMC484	T.-S. Lin, X.-H. Zhang, Z.-H. Wang, and W. H. Prusoff, *J. Med. Chem.* **31**, 484 (1988).
88JMC1475	T. R. Webb, H. Mitsuya, and S. Broder, *J. Med. Chem.* **31**, 1475 (1988).
88JOC3978	A. R. Katritzky and K. S. Laurenzo, *J. Org. Chem.* **53**, 3978 (1988).
88MI1	M. Mag and J. W. Engels, *Nucleic Acids Res.* **16**, 3525 (1988).
88OPP145	S. Kim, *Org. Prep. Proced. Int.* **20**, 145 (1988).
88OPP285	G. Cristalli, P. Franchetti, M. Grifantini, E. Nasini, and S. Vittori, *Org. Prep. Proced. Int.* **20**, 285 (1988).
88T3391	J. S. Hinkle and O. W. Lever, Jr., *Tetrahedron* **44**, 3391 (1988).
88TL1931	D. Bouzard, P. Di Cesare, M. Essiz, J. P. Jacquet, and P. Remuzon, *Tetrahedron Lett.* **29**, 1931 (1988).
89T4485	K. J. Szabo, J. Csaszar, and A. Toro *Tetrahedron* **45**, 4485 (1989).

Saturated Bicyclic 6/5 Ring Fused Systems with Bridgehead Nitrogen and a Single Additional Heteroatom

TREVOR A. CRABB

Department of Chemistry, Portsmouth Polytechnic, Portsmouth, Hampshire, England

DAVID JACKSON

The Frewen Library, Portsmouth Polytechnic, Portsmouth, Hampshire, England

ASMITA V. PATEL

School of Pharmacy, Portsmouth Polytechnic, Portsmouth, Hampshire, England

I. Introduction	194
Scope of Review	194
II. Perhydropyrazolo[1,2-*a*]pyridazines	196
A. Synthesis and Reactions	196
B. Stereochemistry	198
III. Perhydropyrazolo[1,5-*a*]pyridines	198
Synthesis	198
IV. Perhydro-isoxazolo[2,3-*a*]pyridines	200
A. Synthesis and Reactions	200
B. Stereochemistry	205
V. Perhydropyrrolo[1,2-*b*]pyridazines	207
Synthesis and Reactions	207
VI. Perhydro-imidazo[1,5-*a*]pyridines	208
A. Synthesis	208
B. Stereochemistry	212
VII. Perhydro-oxazolo[3,4-*a*]pyridines	214
A. Synthesis and Reactions	214
B. Stereochemistry	219
VIII. Perhydrothiazolo[3,4-*a*]pyridines	220
A. Synthesis and Reactions	220
B. Stereochemistry	222
IX. Perhydropyrrolo[1,2-*c*]pyrimidines	223
Synthesis and Stereochemistry	223
X. Perhydropyrrolo[1,2-*c*][1,3]oxazines	224
A. Synthesis and Reactions	224
B. Stereochemistry	226

XI.	Perhydro-imidazo[1,2-a]pyridines	226
	Synthesis and Reactions	226
XII.	Perhydro-oxazolo[3,2-a]pyridines	229
	A. Synthesis and Reactions	229
	B. Stereochemistry	238
XIII.	Perhydrothiazolo[3,2-a]pyridines	238
	A. Synthesis	238
	B. Stereochemistry	241
XIV.	Perhydropyrrolo[1,2-a]pyrazines	242
	A. Synthesis	242
	B. Stereochemistry	248
XV.	Perhydropyrrolo[2,1-c][1,4]oxazines	249
	A. Synthesis and Reactions	249
	B. Stereochemistry	254
XVI.	Perhydropyrrolo[2,1-c]thiazines	254
	Synthesis	254
XVII.	Perhydropyrrolo[1,2-a]pyrimidines	255
	A. Synthesis	255
	B. Stereochemistry	257
XVIII.	Perhydropyrrolo[2,1-b][1,3]oxazines	258
	Synthesis	258
XIX.	Perhydropyrrolo[2,1-b][1,3]thiazines	259
	Synthesis	259
XX.	Conformational Analysis of Saturated 6/5 Ring Systems with Bridgehead Nitrogen and a Single Additional Heteroatom	259
	References	267

I. Introduction

Scope of Review

Saturated bicyclic 6/5-ring fused systems with a bridgehead nitrogen atom are found in many compounds of biological or medicinal interest. This review covers the literature from 1967 on compounds containing such a system with a single additional heteroatom (nitrogen, oxygen, or sulfur in either the six- or five-membered ring). Compounds in which the system is embedded in a more complex ring or cage structure have normally been excluded from consideration.

The ring systems reviewed include the perhydropyrazolo[1,2-a]-pyridazines (**1**), perhydropyrazolo[1,5-a]pyridines (**2**), perhydro-isoxazolo[2,3-a]pyridines (**3**), perhydropyrrolo[1,2-b]pyridazines (**4**), perhydro-imidazo[1,5-a]pyridines (**5**), perhydro-oxazolo[3,4-a]pyridines (**6**), perhydrothiazolo[3,4-a]pyridines (**7**), perhydropyrrolo[1,2-c]pyrimidines (**8**), perhydropyrrolo[1,2-c][1,3]oxazines (**9**), perhydro-imidazo[1,2-a]pyridines (**10**), perhydro-oxazolo[3,2-a]pyridines (**11**), perhydrothiazolo[3,2-*a*]pyridines **(12)**, perhydropyrrolo[1,2-*a*]pyrazines **(13)**, perhydropyrrolo[2,1-*c*][1,4]oxazines **(14)**, perhydropyrrolo[2,1-*c*]- [1,4]thiazines **(15)**, perhydropyrrolo[1,2-*a*]pyrimidines **(16)**, perhydropyrrolo[2,1-*b*][1,3]oxazines **(17)**, and perhydropyrrolo[2,1-*b*][1,3]thiazines **(18)**. The nature and extent of the available material has largely dictated the presentation of each section, but the synthesis, reactions, and stereochemistry of the systems have been covered.

In the previous review on "Conformational Equilibria in Nitrogen Containing Saturated Six-Membered Rings," both the general features of conformational equilibria in the six-membered nitrogen heterocycles and the physical methods used in investigating conformation and conformational equilibria in such systems were described in detail [84AHC(36)1]. Much of that discussion is directly relevant to the material covered in the present review and, accordingly, cross-referencing to the original review will be made as appropriate.

II. Perhydropyrazolo[1,2-*a*]pyridazines

A. Synthesis and Reactions

Perhydropyrazolo[1,2-*a*]pyridazine-1-one (**19**) and the 3-phenyl substituted derivative (**20**) are readily obtained by the reaction between hexahydropyridazine and ethyl acrylate (81JA461) and ethyl cinnamate (80TL3493), respectively. Use of epichlorohydrin provides a route to 2-hydroxyperhydropyrazolo[1,2-*a*]pyridazine (82EUP47620; 83BRP2105707).

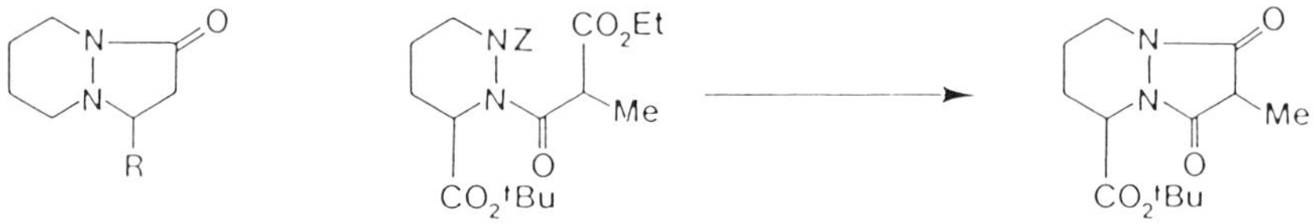

(19: R=H)
(20: R=Ph)

(21: Z= benzyloxycarbonyl)

(22)

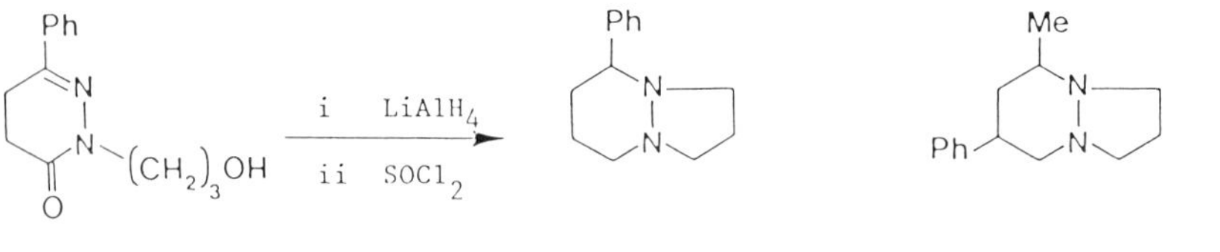

(23)   (24)   (25)

Reaction between α-carbethoxypropionyl chloride and *t*-butyl-1-benzyloxy carbonyl hexahydropyridazine-3-carboxylate followed by catalytic hydrogenolysis of the protecting group in the product (**21**) and cyclization gave *t*-butyl hexahydro-2-methyl-1,3-dioxo-1-*H*-pyrazolo[1,2-*a*]-pyridazine-5-carboxylate (**22**) [80EUP42100; 82MI1; 84JCS(P1)155]. An alternative ring-closure reaction on the hexahydropyridazine ring is illustrated by the sequence starting from 2-(3-hydroxypropyl)-6-phenyl-4,5-dihydropyridazin(2H)-3-one (23), obtained from 3-benzoylpropionic acid and 3-hydrazinopropanol, and leading to the 5-phenyl substituted derivative (24) (69JOC2720; 70FRP7076M, 70USP3497513; 72USP3663536). By a similar route from α-phenyllevulinic acid, 5-methyl-7-phenylperhydropyrazolo[1,2-a]pyridazine (25) is prepared (69USP3420831). Numerous variations on this theme are available, leading, for example, to compounds of biological activity and proceeding via ring formation on pyrazolidine derivatives (66FRP1441519; 68BSF4210; 71FES1017; 83BRP2128984; 83EUP94095).

The in situ Diels–Alder reaction of 4,4-diethylpyrazoline-3,5-dione (from 4,4-diethylpyrazolidine-3,5-dione (26) and lead tetracetate) with 1,3-dienes provides a convenient route to 5,8-dihydropyrazolo[1,2-a]-pyridazine-1,3-(2H)-diones of type 27, which may be converted to perhydropyrazolo[1,2-a]pyridazin-1,3-diones (28) by catalytic hydrogenation over Pd/C [69JOC3181, 69LA(724)150]. Conversion of 28 to 29 by lithium aluminum hydride reduction has been described (72JHC41).

Irradiation of diazo compound 30 in dry ether and ethanol gave t-butylethyl hexahydro-2-oxo-1,2-diazeto[1,2-a]pyridazine-4,1-dicarboxylate (31). The X-ray structure of the diazo compound shows that the migrating nitrogen atom is the more planar of the two [85TL3167; 87JCS(P1)885]. Ring enlargement of the methiodide (32) to 33 is accomplished by treatment with sodium methoxide in methanol (69JOC2720).

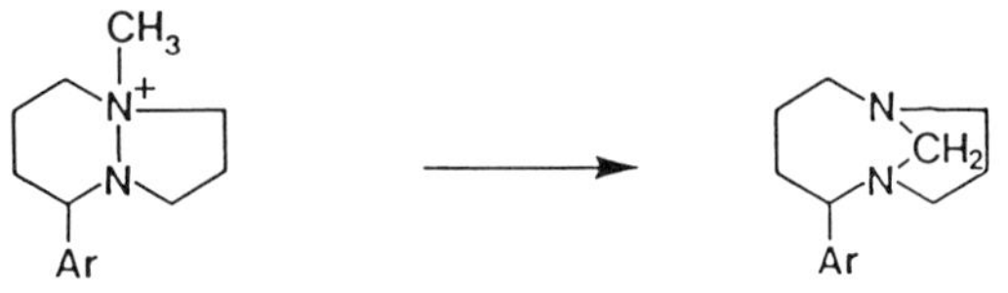

(32) → (33)

B. Stereochemistry

In the photoelectron spectra of hydrazines, the separation between the two lone-pair-dominated ionization potentials is sensitive to the lone-pair–lone-pair dihedral angle (84JOC1891). In this connection, the photoelectron spectrum of perhydropyrazolo[1,2-*a*]pyridazine was consistent only with the presence of the trans-conformer **(34)** (74JA6982, 74JA6987;

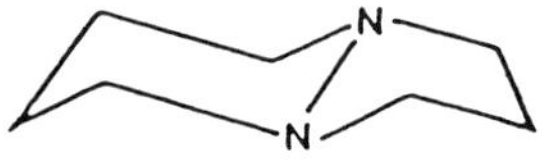

(34)

75CB1557). This is supported by low temperature, cyclic voltammetry (76JA5269; 77JA1130), and vibrational studies [76SA(A)157]. Low temperature ^{13}C-NMR spectroscopy shows the trans-fused form to be considerably lower in enthalpy than the cis-fused form (76JA7007).

Perhydropyrazolo[1,2-*a*]pyridazine deviates from the straight-line plot of δ ^{15}N for hydrazines against δ ^{13}C for the hydrocarbon analogue of the hydrazine with N replaced by CH. This has been taken as an indication of flattening at the nitrogen in the favored trans-conformer (80JOC3609).

X-ray data on a reduced pyrazolo[1,2-*a*]pyridazine are provided with structure **35** in which N-5 is more planar than N-9; the sum of the angles around N is 353.6° at N-5 and 343.9° at N-4 [87JCS(P1)885]. The corresponding bond angles in thiol **36** are 355.6° and 345.4°, respectively, and the carboxylic acid group adopts the axial position [84JCS(P1)155].

III. Perhydropyrazolo[1,5-*a*]pyridines

Synthesis

Literature reports of syntheses of perhydropyrazolo[1,5-*a*]pyridines are limited. Perhydropyrazolo[1,5-*a*]pyridine has been prepared from

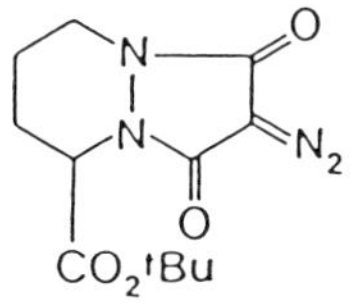

Bond distances (Å)

N —— C-1	1.353
C-1 — C-2	1.440
C-2 — C-3	1.437
C-3 —— N	1.376
N —— C-5	1.459
C-5 — C-6	1.521
C-6 — C-7	1.509
C-7 — C-8	1.523
C-8 —— N	1.445

(35)

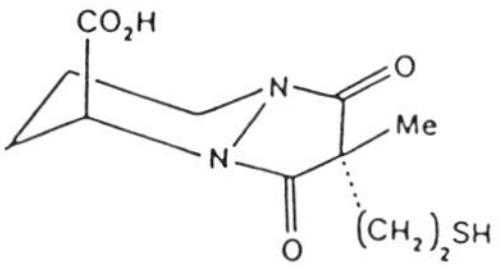

1-nitrosopiperidine-2-acetic acid via reduction and cyclization to 2-oxo-perhydropyrazolo[1,5-*a*]pyridine (37). Consecutive reduction of 37 by lithium aluminum hydride and hydrogenation over platinum gave perhydropyrazolo[1,5-*a*]pyridine (38) in poor yield. However, 1-substituted perhydropyrazolo[1,5-*a*]pyridines (40) were readily obtained by reduction of the amides (39) with lithium aluminum hydride (69CHE629). A different approach involves the reaction between the Δ^2-pyrazole (41) and diketene in chloroform to give compounds of type 42 (66FRP1441937).

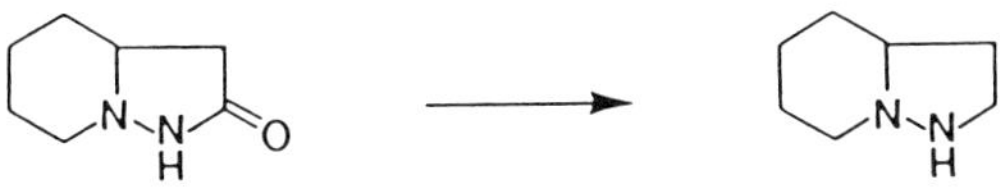

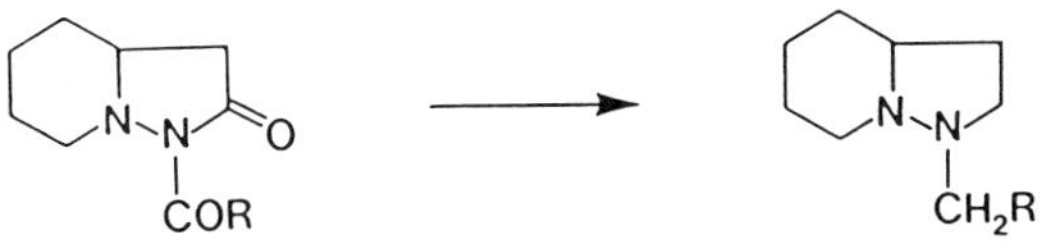

(41) → (42)

IV. Perhydro-isoxazolo[2,3-a]pyridines

A. Synthesis and Reactions

Large numbers of perhydro-isoxazolo[2,3-*a*]pyridines have been synthesized via nitrones. This, together with a survey of syntheses involving nitrones, has been well reviewed (84MI1). Accordingly, only work in this area published in 1983 and after is included in this review.

Many quinolizidine alkaloids have been obtained by way of substituted perhydro-isoxazolo[2,3-*a*]pyridines. Reaction between 3,4-dimethoxybenzaldehyde and the phosphorane derived from allyltriphenylphosphonium bromide gave an *E*- and *Z*-isomer mixture of 1-(3,4-dimethoxyphenyl)butadiene, which reacted with 2,3,4,5-tetrahydropyridine 1-oxide in boiling toluene to give the corresponding isomers of the isoxazolidines (43–46). The formation of the trans- (H-2, H-3a) isomers was favored in

(43: α-H, R=3,4-(MeO)$_2$Ph)   (45: α-H, R=3,4-(MeO)$_2$Ph)
(44: β-H, R=3,4-(MeO)$_2$Ph)   (46: β-H, R=3,4-(MeO)$_2$Ph)

each case. Initial treatment of the isoxazolidine (43) with hydrogen chloride to give the side chain chloro derivatives, followed by hydrogenation over Pd/C catalyst in ethanol, gave mainly lasubine-I (47, 44% yield) with 2-epi-lasubine II (48, 14% yield). *Z*-isomers 45 and 46, on similar treatment, gave lasubine-I as the sole product (83CC1143; 84JOC1909).

Isoxazolidines 51 and 52, obtained from the homoallylic alcohols 49 and 50, were transformed to simple phenylquinolizidines (53–56) by initial

(47: R=3,4-(MeO)₂Ph)   (48: R=3,4-(MeO)₂Ph)

(49: R=PhCH₂)
(50: R=MeO(CH₂)₂OCH₂)

(51: R=PhCH₂)
(52: R=MeO(CH₂)₂OCH₂)

(53: R=PhCH₂)
(54: R=MeO(CH₂)₂OCH₂)

(55: R=PhCH₂)
(56: R=MeO(CH₂)₂OCH₂)

treatment with methanesulfonyl chloride in pyridine followed by reduction with zinc in acetic acid (84CPB3892). 1,3-Cycloaddition between 2,3,4,5-tetrahydropyridine 1-oxide and 3,4-dimethoxystyrene gave a trans- **(57)** and cis- **(58)** isomer mixture of perhydro-isoxazolo[2,3-*a*]pyridines in the ratio 20:1. Treatment of **57** with zinc in acetic acid gave **59**, which was converted to cryptopleurine **(60)** via a synthetic sequence beginning with acylation with *p*-methoxyphenylacetyl chloride (84JOC2412).

A novel rearrangement of suitably 2-substituted side-chain perhydro-isoxazolo[2,3-*a*]pyridines has been shown to afford indolizine and quinolizine derivatives (86TL1727). Thus, 2-spirocyclopropylperhydro-isoxazolo[2,3-*a*]pyridine **(61)**, obtained as a mixture with the 3-spiro isomer from the reaction between 3,4,5,6-tetrahydropyridine-1-oxide and methylene cyclopropane, undergoes thermal rearrangement (400°C, 0.2 mmHg) to the quinolizidin-2-one **(62)**.

(57: R=3,4-(MeO)$_2$Ph)   (58: R=3,4-(MeO)$_2$Ph)

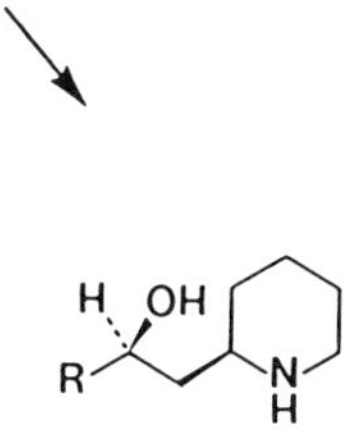

(59: R=3,4-(MeO)$_2$Ph)

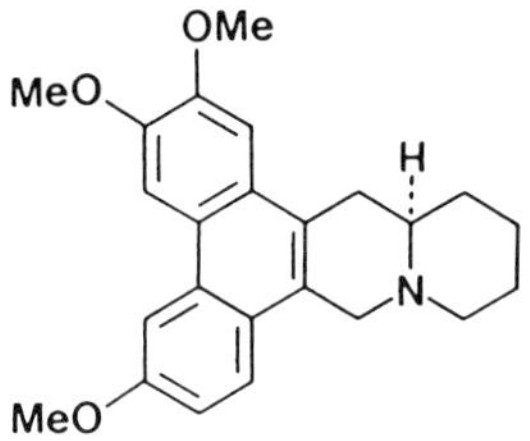

(60)

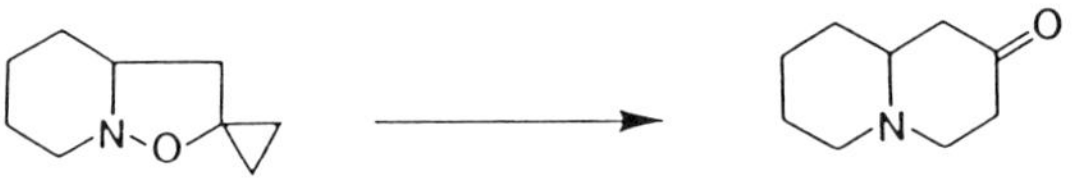

(61)   (62)

A reinvestigation of the stereoselectivity of the reaction between styrene and 2,3,4,5-tetrahydropyridin-1-oxide in chloroform shows a predominance of isomer **63** over **64** (ratio 98:2). Changing the solvent to toluene made no significant difference to the result. The reductive cleavage of the N—O bond of the derived N-methylated isoxazolidine **(65)** with W2 Raney nickel in methanol gives allosedamine **(67)** in 90% yield, whereas lithium aluminum hydride reduction of **65** affords 80% allosedamine and 20% sedamine **(68)** (87BSB57).

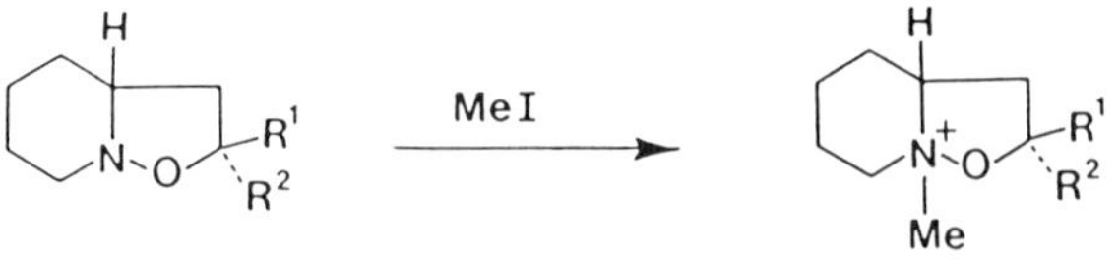

(63: R^1=Ph, R^2=H)   (65: R^1=Ph, R^2=H)
(64: R^1=H, R^2=Ph)   (66: R^1=H, R^2=Ph)

(67: $R^1=H, R^2=OH$)
(68: $R^1=OH, R^2=H$)

The perhydro-isoxazolo[2,3-*a*]pyridine system offers a route to 2,6-disubstituted piperidines. For example, the reaction between styrene and the substituted 3,4,5,6-tetrahydropyridine-1-oxide **(69)** gives the isoxazolidine **(70)** which, on benzylation, gives **71**. This was subjected to reductive cleavage of the N—O bond with lithium aluminum hydride, followed by mesylation of the aminodiol produced. Treatment of the mesylate with lithium triethylborohydride gave the trans-2,6-disubstituted piperidine **(72)** (86TL1489). An alternative route to a 2,6-disubstituted piperidine **(74)** involves the reductive cleavage of the N—O bond of the isoxazolidine **(73)** (from 2-methyl-2,3,4,5-tetrahydropyridin-1-oxide and undec-1-ene) by zinc in acetic acid (86CC1287).

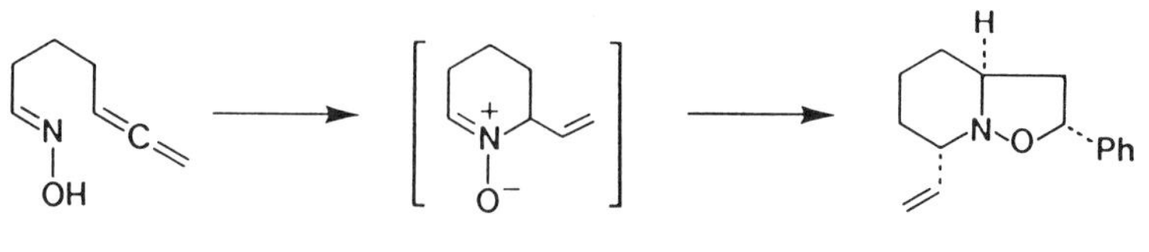

(75) (76) (77)

The preparation of vinyl nitrones by catalytic cyclization of allenic oximes has been demonstrated. Oxime **75**, prepared from 5,6-heptadienal, was cyclized to nitrone **76** in the presence of silver(I) tetrafluoroborate. This was trapped by, for example, styrene to give **77**. Such a compound again opens up a route to 2,6-disubstituted piperidines (85TL6249).

Heating the 2-exomethylene-substituted perhydro-isoxazolo[2,3-*a*]-pyridine **(78)** in dimethylformamide (DMF) at 80°C has been shown to result in rearrangement to the indolizidinone **(79)** (87TL755). The stereochemistry involved in the reaction was emphasized by results on other substrates. By comparison, similar treatment of the related 3-sulfonyl-2-methylene-substituted derivative **(80)** results in a 1,3-hydrogen shift to give **81** (86TL2683).

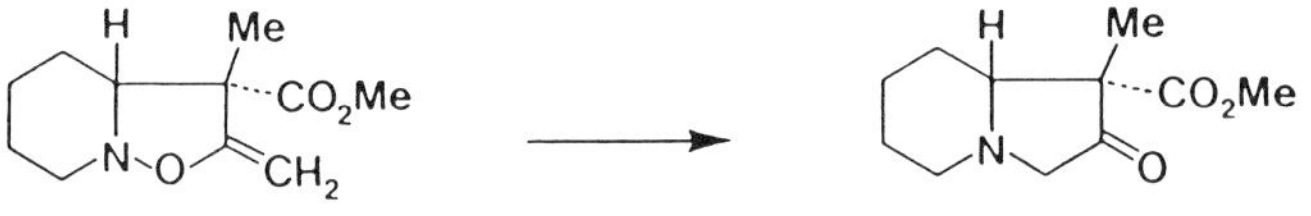

(78) (79)

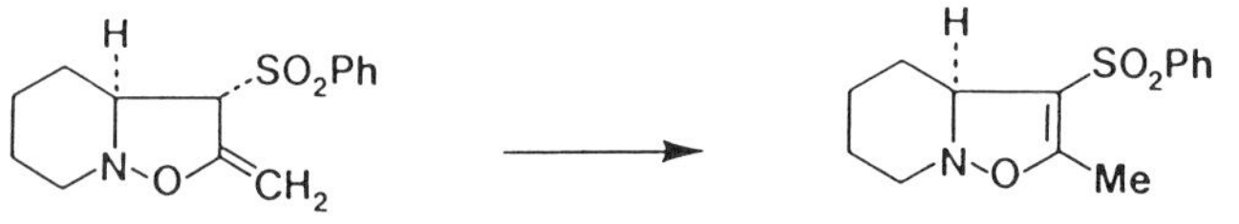

(80) (81)

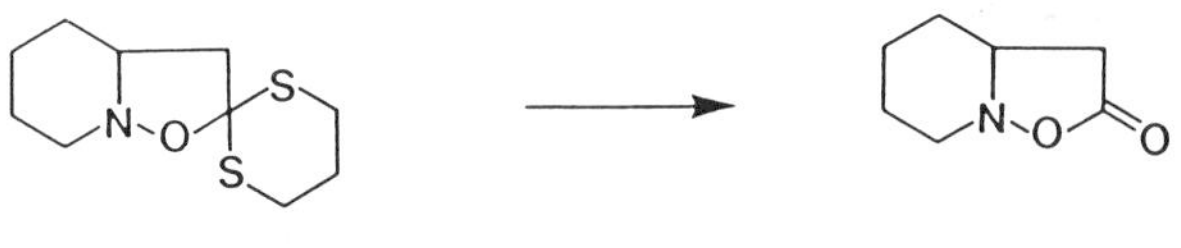

(82) (83)

The reaction between 2,3,4,5-tetrahydropyridine 1-oxide and 2-methylene-1,3-dithiane in boiling tetrahydrofuran gives the spiro compound **82**. Hydrolysis of this gave the isoxazolidinone **(83)** (87H755).

B. Stereochemistry

The structures of α- and β-2,4,7-trimethylperhydro-isoxazolo[2,3-a]-pyridine-2,7-dicarboxaldehyde dioxime (**84** and **85**), obtained by thermal oligomerization of methacrylaldehyde oxime, have been assigned by X-ray analysis. The bond lengths (Å) are given with the structural formulae (80BCJ3240; 83BCJ487). The ^{13}C-NMR shifts for the two isomers given in **86** and **87** confirm the stereochemistry. The ^{13}C-NMR spectra of other derivatives **88** and **89** show the preference for the cis-fused conformer as a result of conformational biasing by the piperidine ring substituent. The spectrum of **90** ⇌ **91** shows the preference for the trans-conformer (**90**) and, relative to analogous 6/5 systems with nitrogen at a bridgehead, a high energy barrier to ring inversion (81M1017).

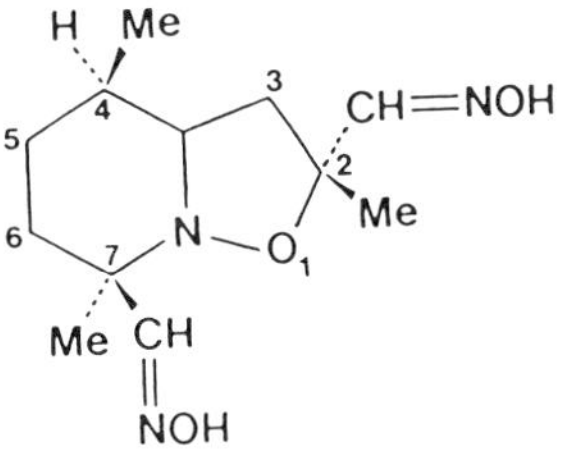

Bond distances (Å)

O — C-2	1.467
C-2 — C-3	1.548
C-3 — C-3a	1.526
C-3a — C-4	1.541
C-4 — C-5	1.540
C-5 — C-6	1.519
C-6 — C-7	1.529
C-7 — N	1.490
N — C-3a	1.484

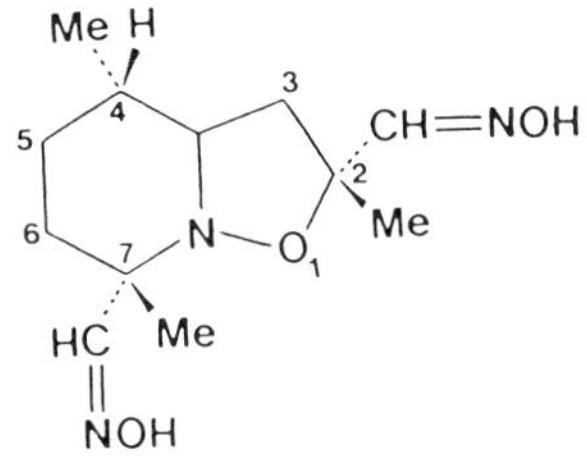

(84)

Bond distances (Å)

O — C-2	1.478
C-2 — C-3	1.517
C-3 — C-3a	1.520
C-3a — C-4	1.543
C-4 — C-5	1.523
C-5 — C-6	1.544
C-6 — C-7	1.545
C-7 — N	1.484
N — C-3a	1.471

(85)

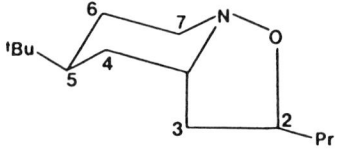

Chemical Shifts (δ)

C-2	77.62
C-3	30.16
C-3a	65.99
C-4	36.44
C-5	35.28
C-6	44.41
C-7	60.75
Me	27.42
Me	18.79

(86)

Chemical Shifts (δ)

C-2	77.16
C-3	30.78
C-3a	59.13
C-4	30.47
C-5	27.15
C-6	41.75
C-7	61.41
Me	12.59
Me	10.86

(87)

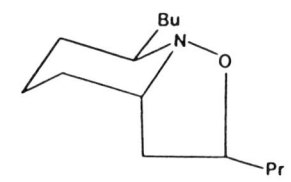

Chemical shifts (δ)

C-2	76.9
C-3	38.0
C-3a	60.1
C-4	35.6
C-5	39.95
C-6	26.4
C-7	50.2

(88)

Chemical shifts (δ)

C-2	76.1
C-3	37.9
C-3a	60.1
C-4	33.8
C-5	18.9
C-6	35.8
C-7	58.1

(89)

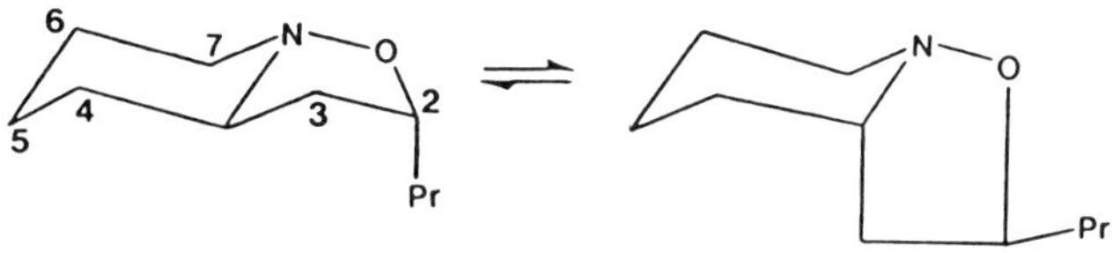

Sec. V] BICYCLIC SIX- AND FIVE-MEMBERED RINGS

Chemical shifts (δ)		Chemical shifts (δ)	
C-2	75.1	C-2	76.6
C-3	40.3	C-3	38.6
C-3a	66.1	C-3a	59.6
C-4	37.7	C-4	36.5
C-5	24.3	C-5	19.3
C-6	24.9	C-6	24.4
C-7	55.0	C-7	50.2

(90)   (91)

V. Perhydropyrrolo[1,2-b]pyridazines

Synthesis and Reactions

The synthesis of perhydropyrrolo[1,2-b]pyridazines using pent-4-enoic acid hydrazide (92) and a variety of halocarbonyl compounds (93; R^1, R^2 = H, alkyl, and/or aryl) has been investigated. Cyclization of the hydrazones (94), formed through the transient azoalkenes (95) to give 96, was promoted by treatment with sodium carbonate under high dilution. Catalytic reduction of 96 to perhydropyrrolo[1,2-b]pyridazine-7-one (97) over Adams catalyst is stereoselective. Cleavage of the N—N bond in 97 to give 98 was only effected by initial N-acetylation followed by reaction with sodium in liquid ammonia (87TL1573).

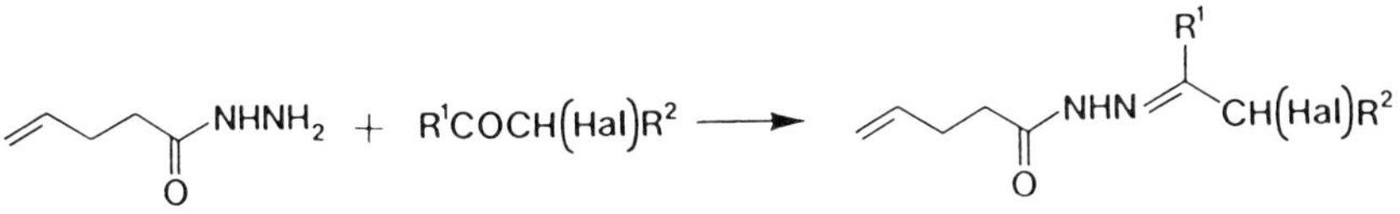

(92)   (93)   (94)

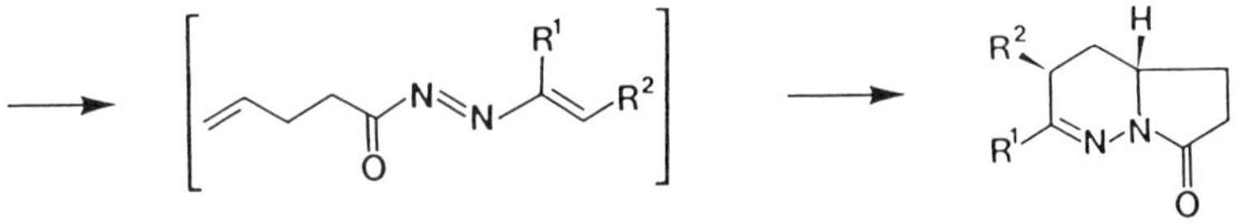

(95) (96) 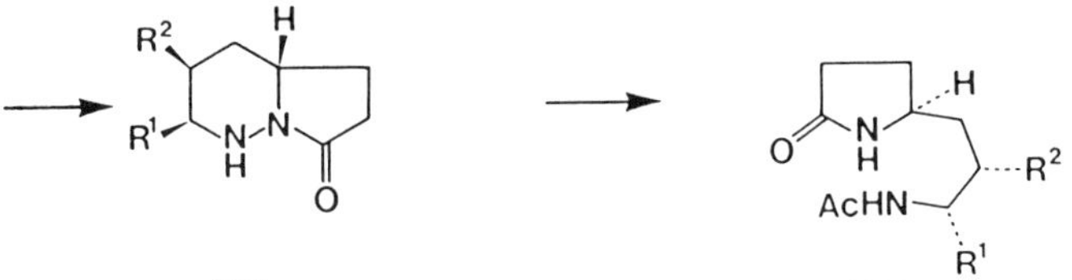 (97) (98)

VI. Perhydro-imidazo[1,5-a]pyridines

A. Synthesis

The parent system (99) has been obtained by lithium aluminum hydride reduction of perhydro-imidazo[1,5-*a*]pyridine-1,3-dione (from ethyl 2-piperidine carboxylate and cyanic acid) and by ring closure of 2-aminomethylpiperidine with formaldehyde. 3-Substituted derivatives have also been described. The 3-phenyl- and 3-*t*-butyl derivatives (but not the propyl and benzyl derivatives) show ring–chain tautomerism (70JHC355).

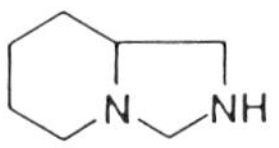

(99)

(100) (101)

Irradiation of *N*-(*N*-piperidylmethyl)glutarimide (100) in acetonitrile gives the diastereoisomers of 101 [83JCS(P1)2857]. The reaction between

5,5-dimethylhydantoin and a series of olefinic alcohols under basic conditions has been used to produce imidazolidin-2-ones. The general sequence of reactions is illustrated by the preparation of the isomeric 7-(formyloxy)-1,1,7-trimethylperhydro-imidazo[1,5-a]pyridines (**104** and **105**) from the hydantoin (**102**). The hydantoin was reduced with LiAlH$_4$ to give the 4-hydroxy-imidazolidin-2-one (**103**), which was intramolecularly cyclized under acidic conditions (formic acid) to give **104** and **105** in an approximate 1 : 1 ratio.

The NMR data of compounds **106** and **107**, obtained by hydrolysis of **104** and **105**, are shown with the structures. In the case of the axial hydroxy-substituted isomer **106**, downfield shifts of the syn-axial 5-H and 8a-H are noted. In agreement, ^{13}C-NMR data show upfield shifts of C-5 and C-8a (84JOC3812).

Perhydro-imidazo[1,5-a]pyridines of type **109** are obtained from 1-p-chlorophenyl-2-piperidyl ketone (**108**) and phenyl isocyanate. The 3-thioxo analogues (**110**) are prepared using phenyl isothiocyanate (70USP3494926).

The substituted perhydro-imidazo[1,5-a]pyridine (**113**) was prepared in order to test for inhibition of enzymes thymidylate synthetase and dihydrofolate reductase. Pyridine-2-carboxaldehyde was condensed with ethyl-p-aminobenzoate to give the Schiff base (**111**) which, on sodium borohydride reduction followed by catalytic hydrogenation, resulted in **112**. Condensation of **112** with 5-formyluracil gave **113** (70JMC276).

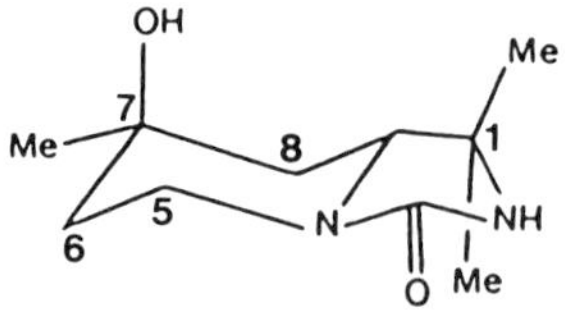

Chemical Shift (δ)

C-1	53.5
C-3	159.4
C-5	37.0
C-6	36.1
C-7	66.6
C-8	36.6
C-8a	59.6
H-5a	2.80
H-5eq	3.40
H-8a	3.22

(106)

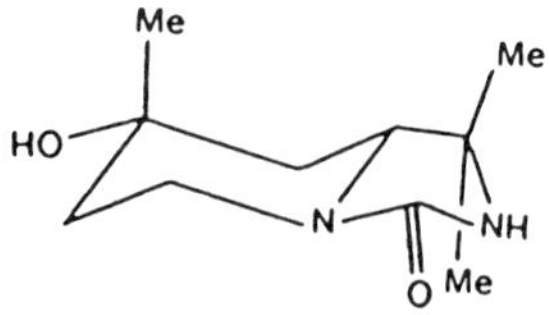

Chemical Shift (δ)

C-1	54.6
C-3	159.9
C-5	39.5
C-6	38.4
C-7	69.9
C-8	38.8
C-8a	62.5
H-5ax	2.68
H-5eq	3.85
H-8a	3.09

(107)

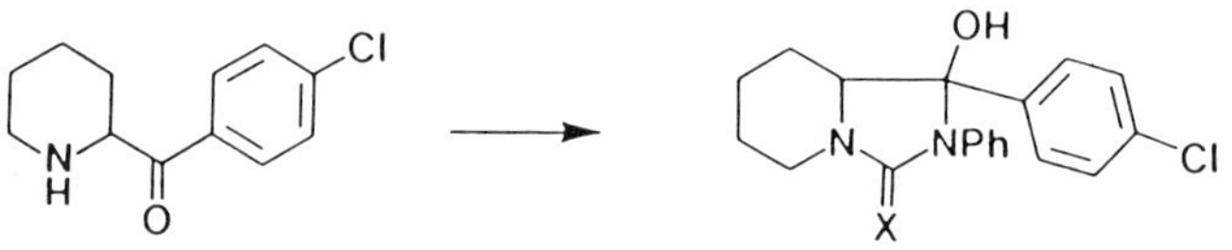

(108)

(109: X=O)
(110: X=S)

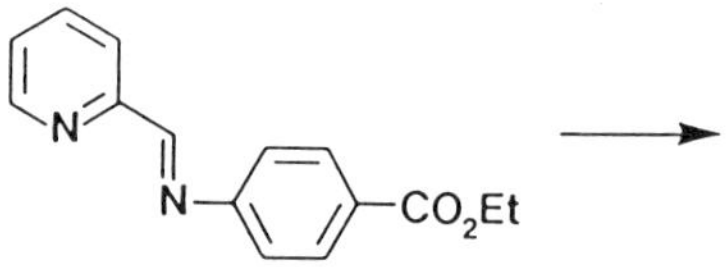

(111)

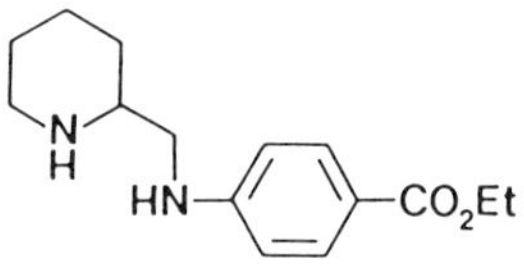

(112)

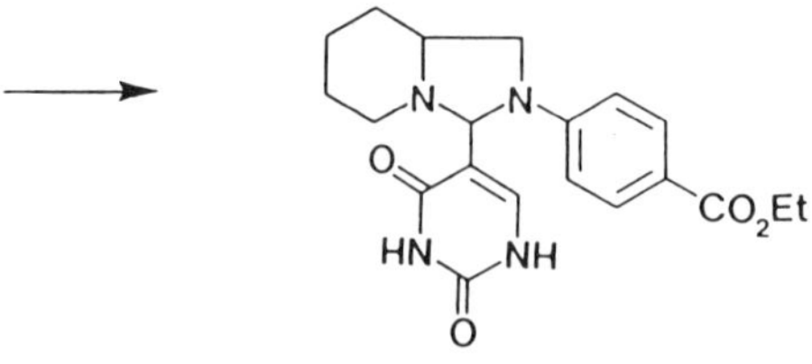

(113)

The patent literature abounds with methods of preparation of perhydroimidazo[1,5-a]pyridine-1,3-diones (**114**) and 3-thioxo-analogues, particularly in connection with their fungicidal and herbidical activity. These involve reactions between an isocyanate or isothiocyanate and pipecolinic acid, followed by acid cyclization of the product (78BRP1503244) by stirring pipecolinic acid with aryl isocyanates at room temperature in the presence of sodium hydroxide [85JAP(K)97980; 86JAP(K)27985] or by using the ethyl ester of pipecolinic acid, which reacts with the aryl isocyanate in toluene solution in the presence of triethylamine, to give the required derivative of **114** [87JAP(K)158280]. Other examples of related syntheses are available (68RTC11; 70CB2394, 70JHC355; 76BRP1503244, 76USP3958976; 79USP4138242; 83EUP70389; 84EUP104532).

Reaction between diethyl piperidyl-1,2-carboxylate and hydrazine gives amino derivative **115** (71AP216), and the synthesis of other N-substituted derivatives with pesticidal properties is illustrated by the preparation of **118**. The hydantoin (**116**) is treated with formaldehyde followed by hydrogen bromide to give **117**, which reacts with S-ammonium-O,O-dimethylphosphorothioate to give **118** (71BRP1337269).

Alternative routes to other derivatives involve heating the 2-benzoxazolinone (**120**) with ethyl pipecolinate to give **119** (69JPS1282),

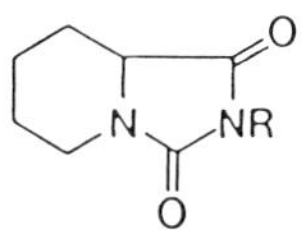

(114)
(115: R=NH$_2$)
(116: R=H)
(117: R=CH$_2$Br)
(118: R=CH$_2$SPO(OMe)$_2$)
(119: R=o-HOPh)

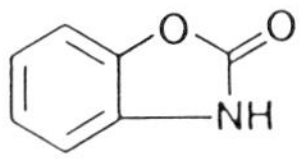

(120)

(121) (122)

cyclization of the 5-phenyl-5-(4-hydroxybutyl) hydantoin tosylate **(121)** to 8a-phenylperhydro-imidazo[1,5-*a*]pyridine-1,3-dione **(122)** in the presence of sodium hydroxide (70JOC3818), and treatment of benzylthiourea with dibromobutyryl chloride in the presence of benzyltrimethylammonium chloride [86JCS(P1)1733].

B. Stereochemistry

2-Alkylperhydro-imidazo[1,5-*a*]pyridines adopt conformational equilibria in CDCl$_3$ solution containing 98% trans-fused conformers **(123)** at 25°C (68T6327; 87MI2). The two diastereoisomeric 1,2-dimethylperhydroimidazo[1,5-*a*]pyridines **(124** and **125)** adopt the conformations shown, and comparison of the geminal coupling constants of the C-3 methylene protons [84AHC(36)1] in these compounds with that for **126** indicates an equilibrium for the latter containing 83% of the conformer with a trans-arrangement of nitrogen lone pairs (87MI2). These results will be discussed in Section XX of this review.

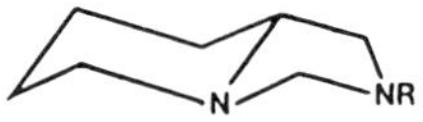

(123)

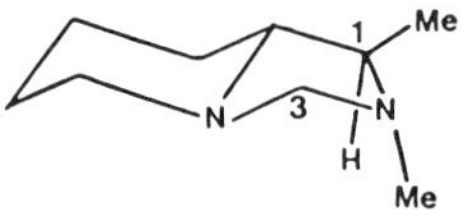

J3eq,3ax −5.25Hz
(124)

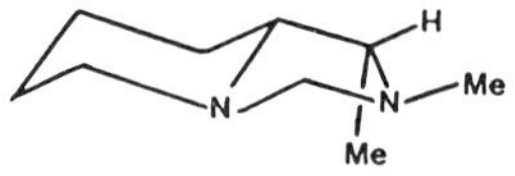

J3eq,3ax −3.75Hz
(125)

83%   J3eq,3ax −5.0Hz   17%

(126)

Whereas **127** adopts the trans-fused conformation ($J_{3eq,3ax} = -4.8$ Hz), the equilibrium for the C-5 epimer **(128)** shows a shift towards the cis-conformer as evidenced by the change in $J_{3eq,3ax}$ to -6.5 Hz [87MI2]. The diastereoisomer pair of **129** shows similar conformational preferences to those of **127** and **128** (69JHC301).

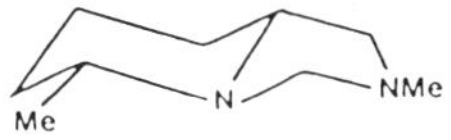

(127)

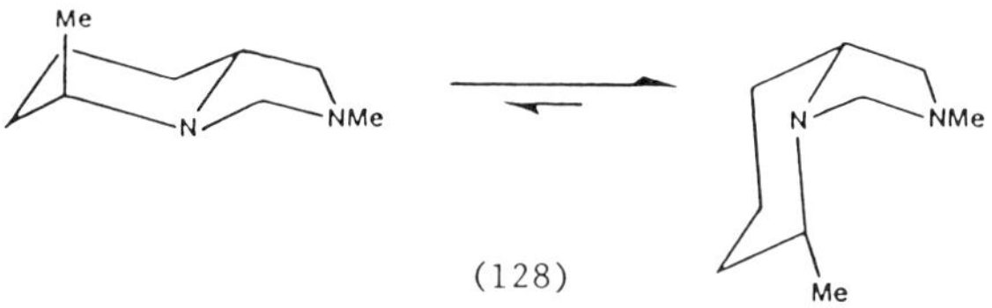

(128)

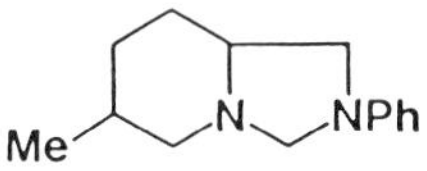

(129)

The parent perhydro-imidazol[1,5-*a*]pyridine (**130**) and the two rotamers of the corresponding N-acetyl derivatives (**131** and **132**) all favor

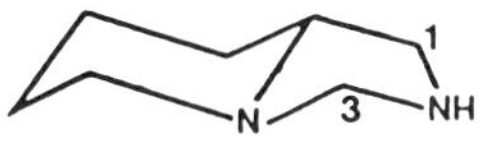

δ 3eq-H   3.95
δ 3eq-H   3.00
$J_{3eq,3ax}$ -6.5Hz

(130)

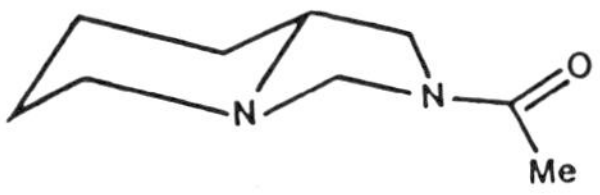

3eq-H   3.93
3ax-H   3.13
$J_{3eq,3ax}$ -4.4Hz

(131)

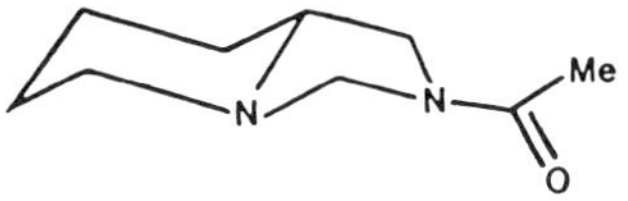

δ 3eq-H   3.95
δ 3ax-H   3.47
$J_{3eq,3ax}$ -6.4Hz

(132)

the trans-fused ring conformers. The latter pair of compounds shows distinctive differences in the values of the J_{gem} of the C-3 methylene protons (73OMR397). J_{gem} values again permit assignments of conformation to the perhydro-imidazo[1,5-*a*]pyridine-1-ones (**133** and **134**) [72JCS(P2)1920].

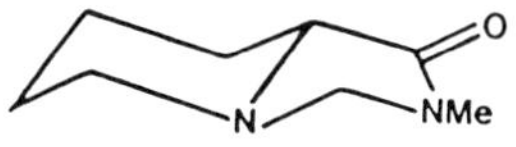

δ 3eq-H 4.06
         J3eq,3ax −4.9Hz
δ 3eq-H 3.81

(133)

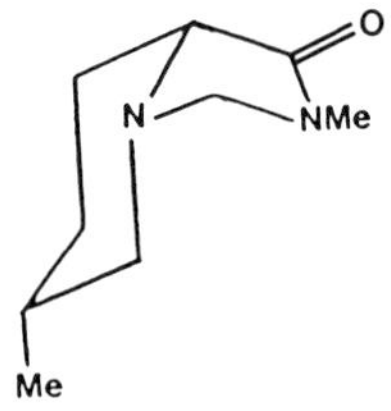

δ 3eq-H 4.13
         J3eq,3ax −7.1Hz
δ 3eq-H 3.83

(134)

VII. Perhydro-oxazolo[3,4-*a*]pyridines

A. Synthesis and Reactions

Perhydro-oxazolo[3,4-*a*]pyridines have long been prepared by condensation between the appropriate piperidyl-2-carbinol and an aldehyde or ketone. For example, the antidepressant spiropiperidine derivative (**135**) is obtained by the reaction between α-phenyl-2-piperidine methanol and *N*-benzyl-4-piperidone (81USP4260623). Perhydro-oxazolo[3,4-*a*]pyridin-3-one (**136**) is obtained by phosgenation of 2-(hydroxymethyl)piperidinium *p*-toluene sulfonate followed by cyclization of the resultant chloroformate salt by treatment with triethylamine in dichloromethane. This ring system

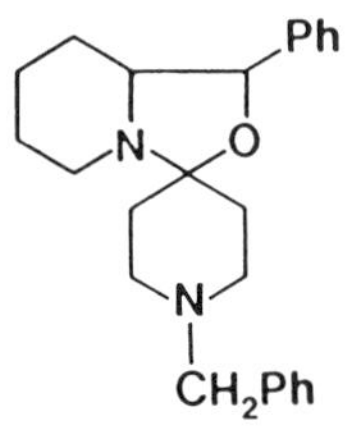

(135) 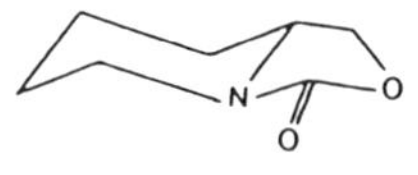   (136)

has been found to be resistant to a variety of attempts at polymerization (80JOC5325).

Ring closure of erythro-α-phenyl-2-piperidinemethanol **(137)** with bromocyanogen proceeds to the 2-imino-oxazolidine **(138)** with retention of its configuration. The alternative route (HOCN followed by $SOCl_2$) to **139** is characterized by inversion (73AP284). Related studies have been published (78MI2; 80MI1). Heating a benzene solution of 4-phenylthio-oxazolidin-2-one **(140)** in the presence of tri-n-butyltin hydride and a trace of azobisisobutyronitrile gave the single isomer **141** of 6-methyl-7-phenyl-perhydro-oxazolo[3,4-a]pyridin-3-one. It is suggested that diastereoselectivity in this reaction is due to A-type strain in the benzyl radical intermediate (86CC1717).

The synthesis of 5-methoxyperhydro-oxazolo[3,4-a]pyridin-3,8-dione **(143)** is summarized below. The final step **(142)** → **(143)** proceeds stereospecifically with production of the axial methoxy-substituted product (86TL5085) (Scheme 1).

A stereoselective synthesis of (±) conhydrine via an oxazolidin-3-one proceeds by anodic α-methoxylation of N-carbomethoxypiperidine **(144)** to give **145**. Structure **145** is converted to oxazolone **146** by the series of

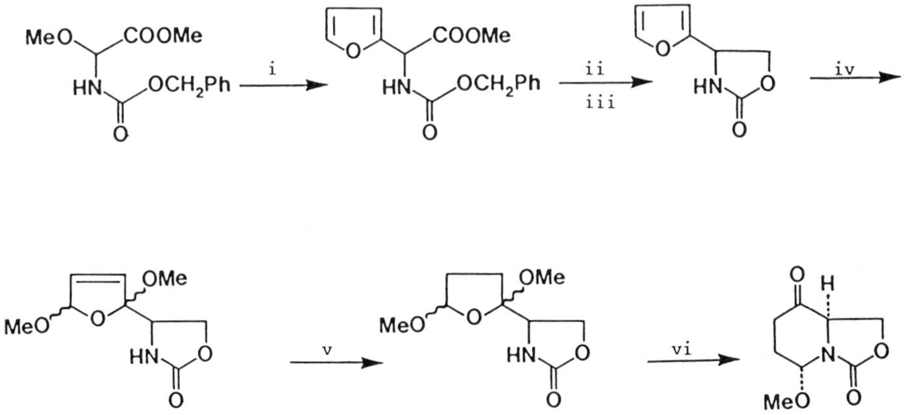

(142) (143)

SCHEME 1. Reagents: i, Furan, BF_3OEt_2; ii, $LiBH_4$; iii, NaH; iv, Br_2-MeOH then NH_3; v, H_2-Rh(Al_2O_3); vi, CF_3SO_3H.

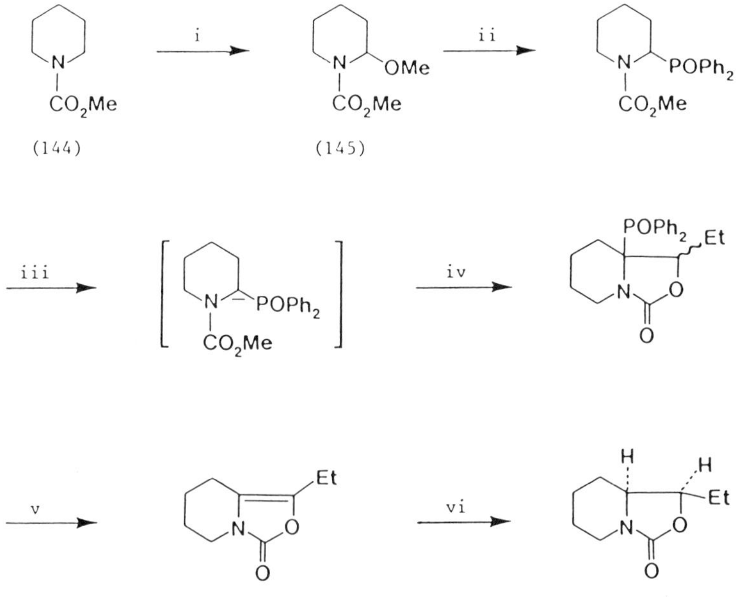

(144) (145) (146) (147)

SCHEME 2. Reagents: i, −2e MeOH; ii, Ph_2PCl; iii, LDA; iv, C_2H_5CHO; v, Δ; vi, H_2-Pt.

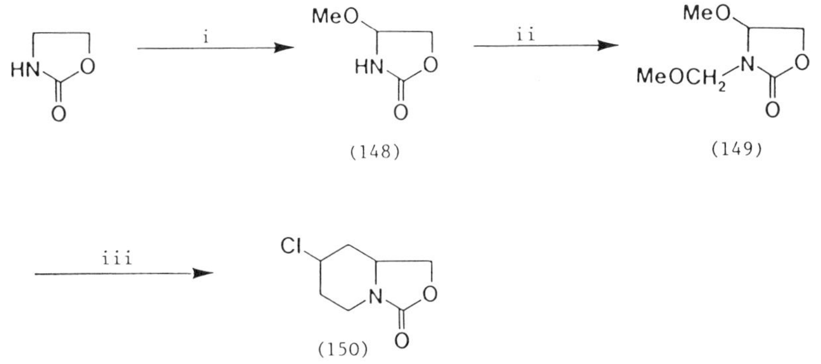

SCHEME 3. Reagents: i, −2e MeOH; ii, ClCH$_2$OMe; iii, TiCl$_4$.

reactions shown in Scheme 2. The hydrogenation of **146** proceeds stereoselectively to give the erythro-isomer **(147)** which may readily be converted to (±) conhydrine (83TL4577) (Scheme 2). Anodic α-monomethoxylation also plays a part in the synthesis of **150** (Scheme 3). The anodically α-methoxylated oxazolidin-2-one **(148)** is methoxymethylated to **149**. A [3 + 3]-type annelation occurs between this compound and allyltrimethylsilane in the presence of titanium tetrachloride to give **150** (85JOC3243). A variation on this theme is exemplified by the synthesis of **151** (86H621) (Scheme 4).

Treatment of perhydro-oxazolo[3,4-*a*]pyridine with acetophenone in ethanolic hydrochloric acid gives β-(2-hydroxymethylpiperidino) pro-

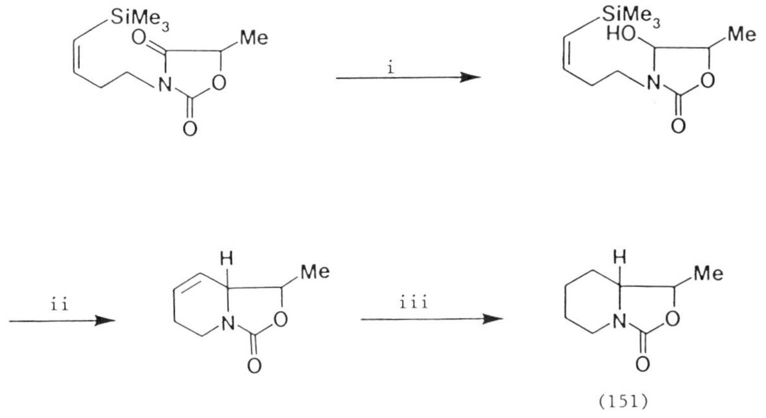

SCHEME 4. Reagents: i, NaBH$_4$-MeOH; ii, TiCl$_4$; iii, H$_2$-Pt.

piophenone hydrochloride **(152)** (76MI1). Boiling an ethanolic solution of the oxazolidine **(153)** under reflux with 25% (by weight) of 5% Pd/C or Raney nickel forms **154**, whereas the 1,1-dimethyl analogue **(155)** undergoes a much slower reaction to form **156**. A mechanism for reaction via ion **157** has been proposed (67TL1039).

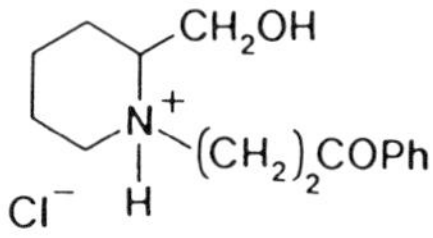

(152)

(153: R=H, PhCH$_2$, p-HOPh)

(154)

(155: R=Ph, PhCH$_2$)

(156)

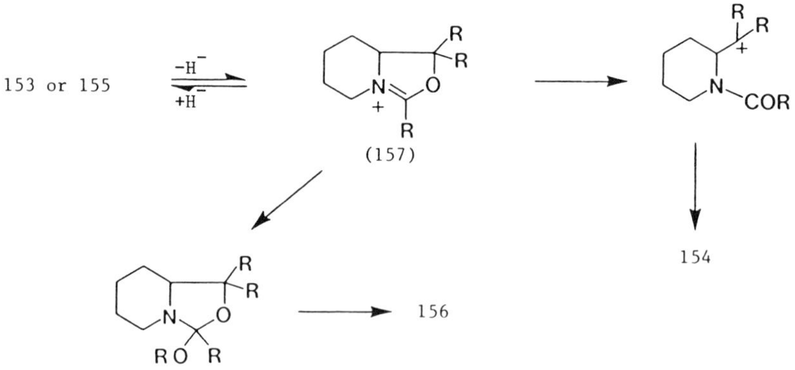

153 or 155

(157)

154

156

B. STEREOCHEMISTRY

The perhydro-oxazolo[3,4-*a*]pyridine system has been studied extensively by NMR spectroscopy and IR measurements in the 2800–2600 cm^{-1} region of the spectrum [see 84AHC(36)1: Sections II,B,E] [66JHC418; 68T1997; 74CC210; 75JCS(P2)51; 80OMR63, 80OMR159; 82OMR242; 83OMR203; 88MI1]. Low temperature ^{1}H- and ^{13}C-NMR measurements show the parent system to exist in CDCl$_3$–CFCl$_3$ solution at 183 K as an equilibrium between 73% trans-conformer **(158)** and 27% O-inside cis-conformer **(159)** (88MI1). The trans-conformer is characterized by a large chemical shift difference between the NCH$_2$O protons (Δae = 0.82 ppm) and J_{gem} of 0 Hz, whereas the cis-conformer shows corresponding values of Δae = 0.05 ppm and a J_{gem} of 5.8 Hz (68T1997; 88MI1). The low temperature ^{15}N-NMR spectrum ($-111°$C) gives ^{15}N shifts of 67.7 and 67.1 for **158** and **159**, respectively (relative to NH$_3$) (83OMR203).

cis-(H-1,H-8a)-1-Alkylperhydro-oxazolo[3,4-*a*]pyridines adopt equilibria heavily biased towards the trans-fused conformer **(160)**, whereas the epimers adopt 18% O-inside cis-fused **(162)** ⇌ 82% trans-fused **(161)**

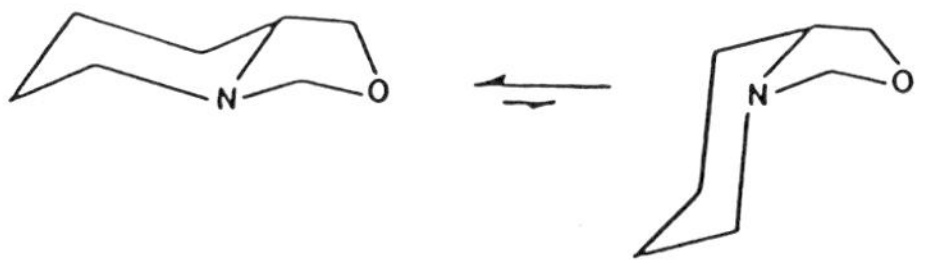

(158)   (159)

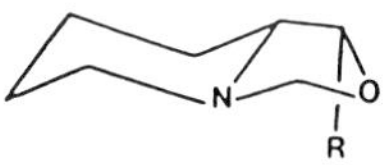

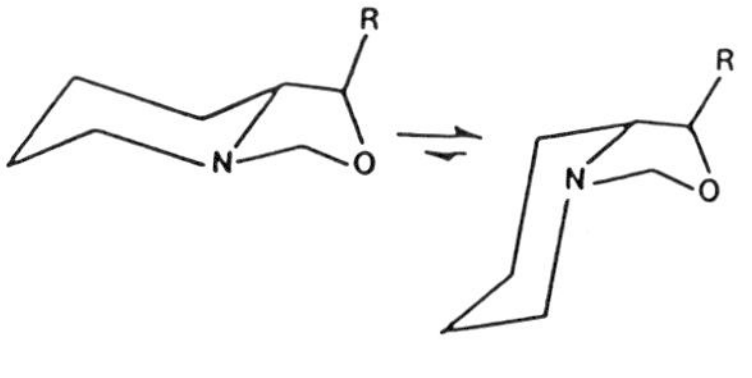

(160)   (161)   (162)

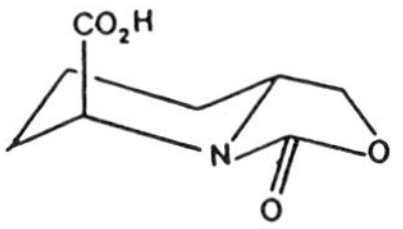

(163)

conformational equilibria at 183 K (66JHC418; 80OMR63; 88MI1). The configurations of some 1-arylperhydro-oxazolo[3,4-*a*]pyridines (**160–162**; R = Ar) have been assigned by analogy to those used for the 1-alkyl analogues (74JMC519). The 1,5-, 1,6-, and 1,8-dimethylperhydro-oxazolo[3,4-*a*]pyridines show equilibria positions representing a balance between nonbonded interactions brought about by the two substituents (82OMR242). This is discussed in Section XX.

X-ray analysis shows the pseudoaxial orientation of the carboxy group in **163** (81JA7573), and circular dichroism (CD) measurements on 1-phenyl-3-iminoperhydro-oxazolo[3,4-*a*]pyridine are available (78MI1).

VIII. Perhydrothiazolo[3,4-*a*]pyridines

A. Synthesis and Reactions

Methyl substituted perhydrothiazolo[3,4-*a*]pyridines (**164**) have been obtained by condensation between the appropriately substituted 2-mercaptomethylpiperidines and formaldehyde (68T2485). Derivatives of type **166** may be prepared from piperidyl-2-carbinols by treatment with a substituted isothiocyanate to give a thiocarboxamide (**165**), which undergoes reaction with thionyl chloride to yield **166** (70USP3505340). Other variations on this theme are available (76IJC(B)770), including conversion of a piperidyl-2-carbinol to the 2-piperidylmethyl chloride with thionyl chloride and treatment of this with carbon disulfide to give **167** (70USP3496185). Hydrolysis of the isopropylidene group of 2-thioxothia-

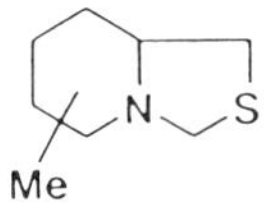

(164)

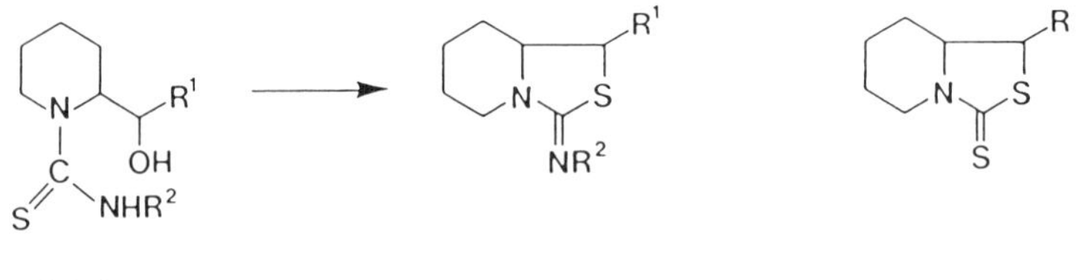

(165)   (166)   (167)

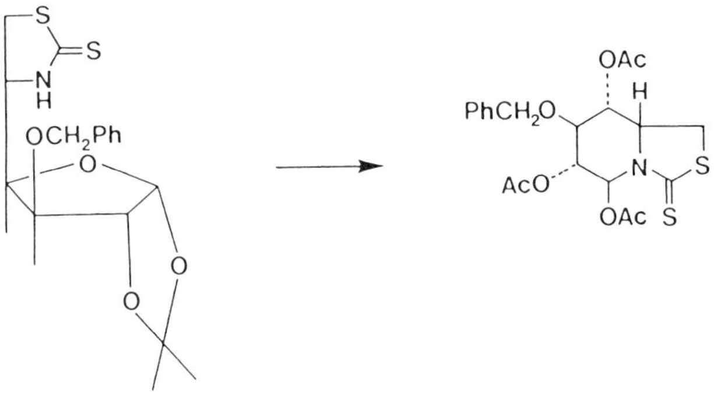

(168) → (169)

zolidine (168) followed by acetylation gave the 3-thioxoperhydrothiazolo[3,4-a]pyridine (169) (74CL519; 75BCJ610).

The N-alkylated 2,4-thiazolidinediones of type 170 provide a ready route to a variety of perhydrothiazolo[3,4-a]pyridines. Thus, sodium borohydride reduction of 170 gives the corresponding N-alkylated-4-hydroxythiazolidin-2-one (171) which, on treatment with formic acid, provides the required ring system 172. Cyclization onto the acetylenic bond in 173 gives entry into another type of functionalized product 174. Such reactions of thiazolidinediones enable stereoselective synthesis of piperidine derivatives. For example, lithium aluminum hydride reduction of 176, obtained by cyclization of 175, gives 177 which, on treatment with Raney nickel, provides the substituted piperidine (178) (79TL1347; 82T3255; 85T2007, 85T2861).

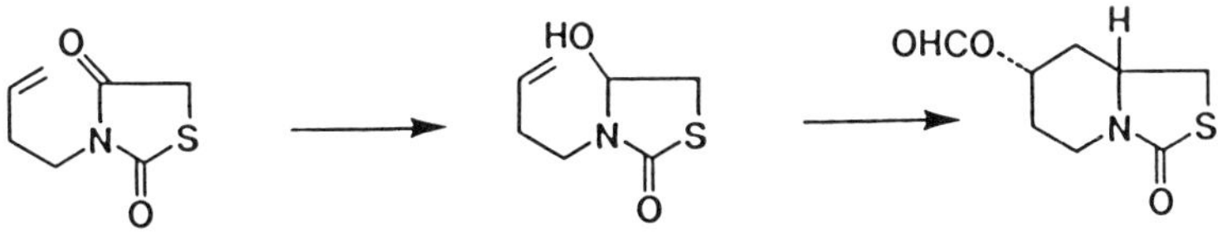

(170)   (171)   (172)

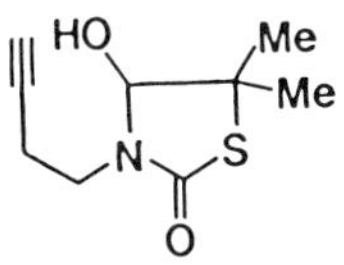

(173) 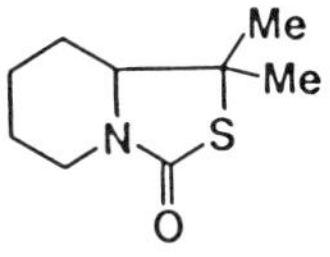   (174)

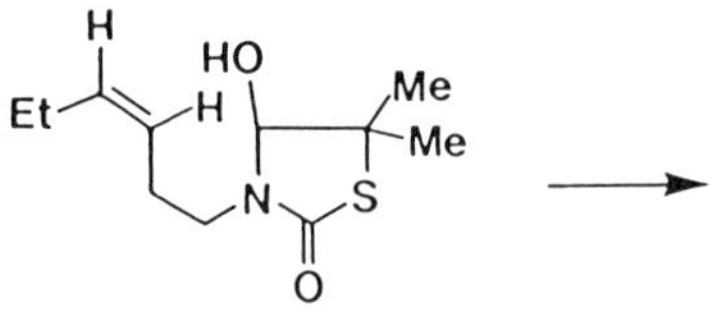

(175) 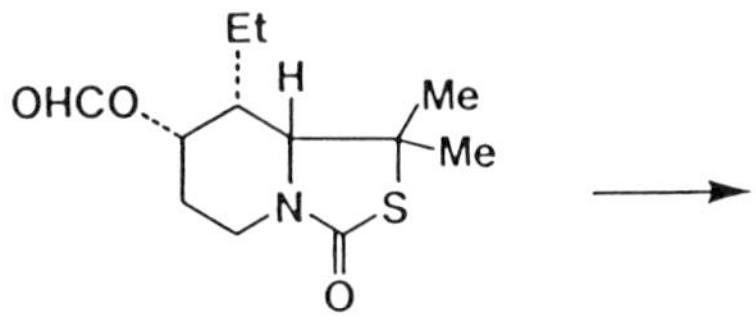 (176) 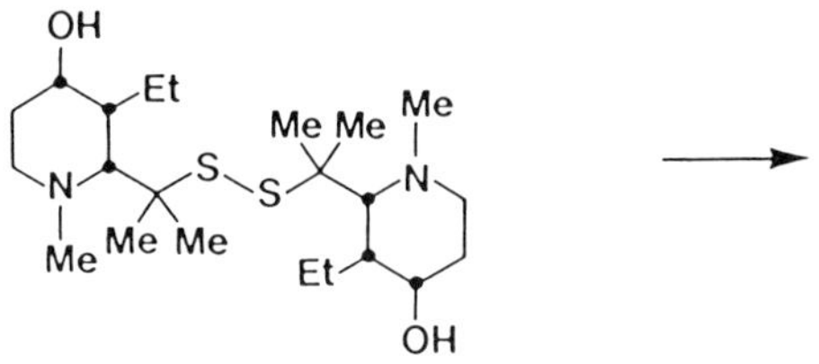 (177) 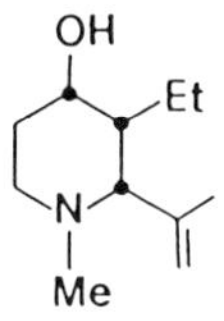 (178)

B. Stereochemistry

Perhydrothiazolo[3,4-a]pyridine adopts a 60% trans-fused **(179)** ⇌ 40% cis-fused **(180)** equilibrium in $CDCl_3$–$CFCl_3$ at 193 K. The cis-(1-H, 8a-H)-1-methyl derivative favors exclusively the trans-conformer **(181)**, whereas the epimer adopts an 87% trans-fused **(182)** ⇌ 13% cis-fused **(183)** conformational equilibrium at 193 K. The pair of 6-ethyl substituted derivatives adopts the trans- and cis-fused conformations **(184)** and **(185)**. These assignments were based on ^{1}H- and ^{13}C-NMR spectral data [68T2485; 88JCS(P1)1173].

(179) 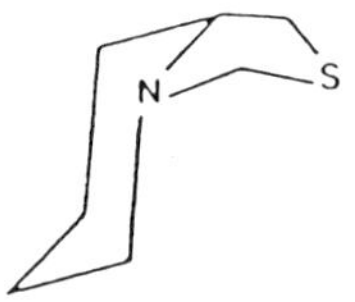 (180) 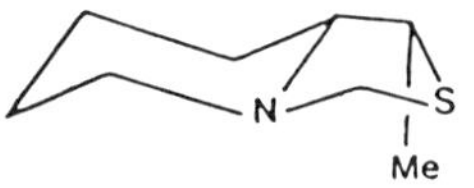 (181) 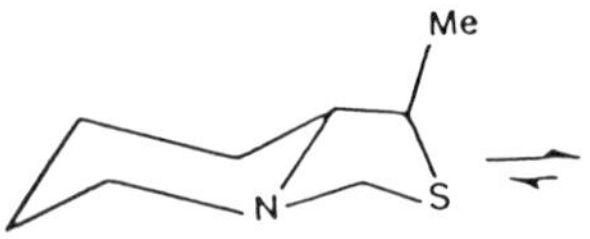 (182) 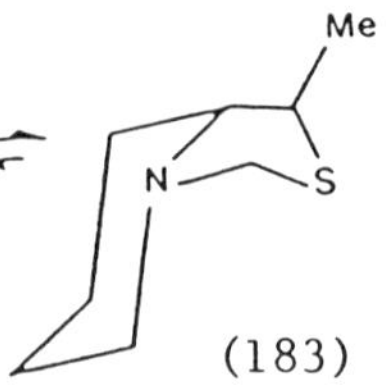 (183)

(184) 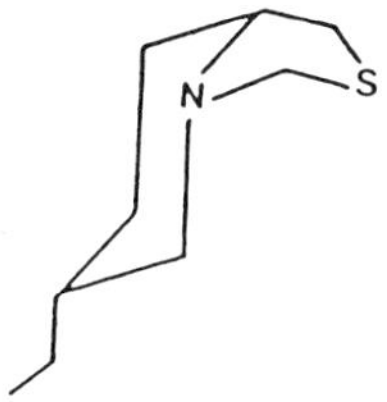 (185)

IX. Perhydropyrrolo[1,2-c]pyrimidines

SYNTHESIS AND STEREOCHEMISTRY

Although the 2,3,4,6-tetrahydro-1,1'-methylene-2,2'-pyrromethen-5[1-H]-one **(190)** is not strictly a perhydropyrrolo[1,2-c]pyrimidine, it carries the essential features of this ring system. Synthesis was effected by treatment of pyrrole-2-carboxaldehyde **(186)** with diiodomethane in the presence of a base to give **187** which was oxidized to **188** by m-chloroperbenzoic acid. Intramolecular condensation of **188** gave **189** which was

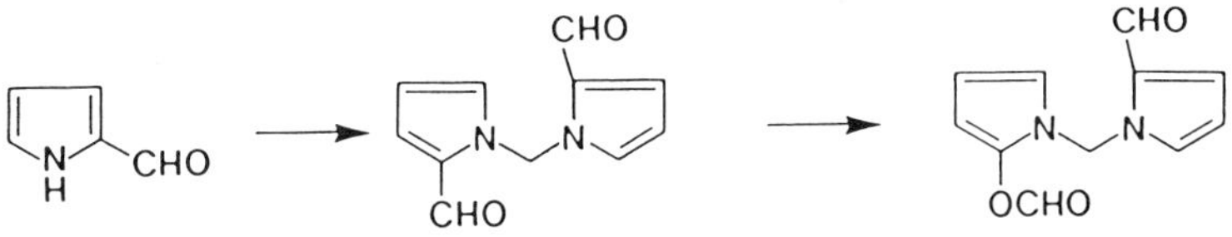

(186) (187) (188)

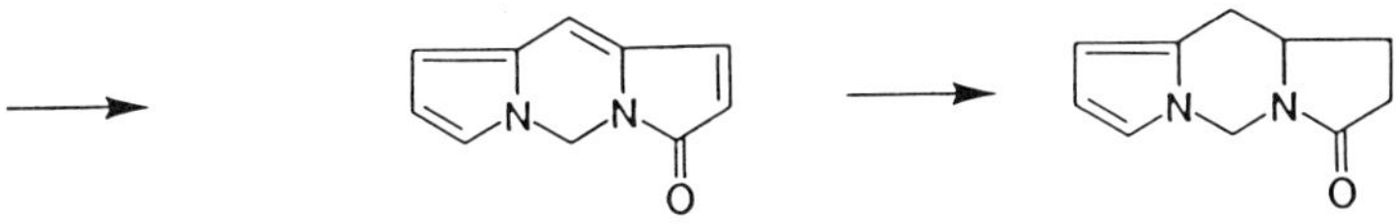

(189) (190)

converted to **190** by catalytic hydrogenation. The ^{1}H- and ^{13}C-NMR chemical shifts (CDCl$_3$) of **190** are provided along with the proposed conformation in **191**. The ^{1}H vicinal coupling constants between the H-3 and H-4 protons lead to the assignment of an envelope conformation to the pyrrolidone ring (86RTC360).

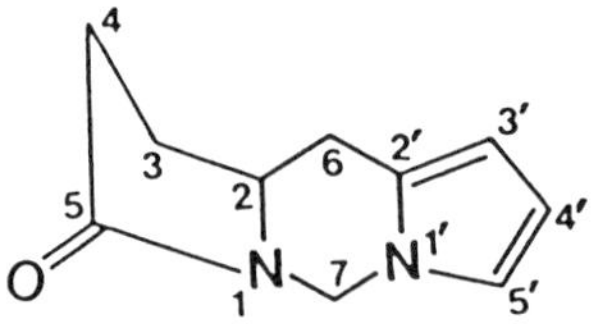

Chemical Shifts (δ)

C-2'	125.2		
C-3'	105.0	H-3'	5.88
C-4'	108.8	H-4'	6.15
C-5'	116.8	H-5'	6.57
C-2	53.1	H-2	3.91
C-3	24.2	H-3	1.80, 2.25
C-4	29.5	H-4	2.40 (J=8.5, 9.9)
			2.45 (J=4.2, 9.2)
C-5	173.5		
C-6	30.0	H-6	2.65, 3.00
C-7	54.0	H-7	4.94, 5.80 (J=-11.0)

(191)

X. Perhydropyrrolo[1,2-c][1,3]oxazines

A. Synthesis and Reactions

The (+) form of the alkaloid gephyrotoxin **(198)** was synthesized through a series of reactions starting with L-pyroglutamic acid **(192)**, which was converted to the nitrile **(193)**. Treatment of this with P_2S_5,

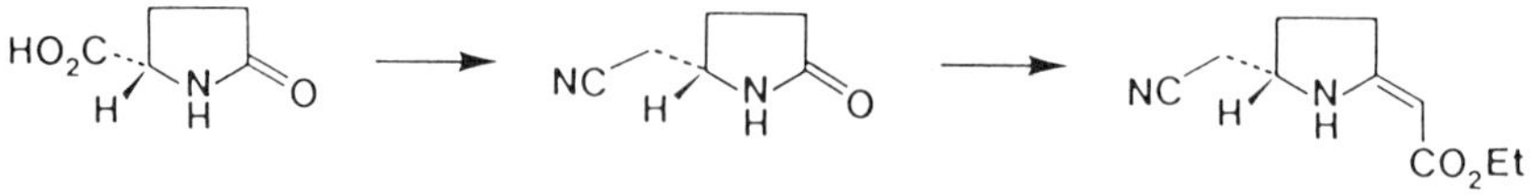

(192)   (193)   (194)

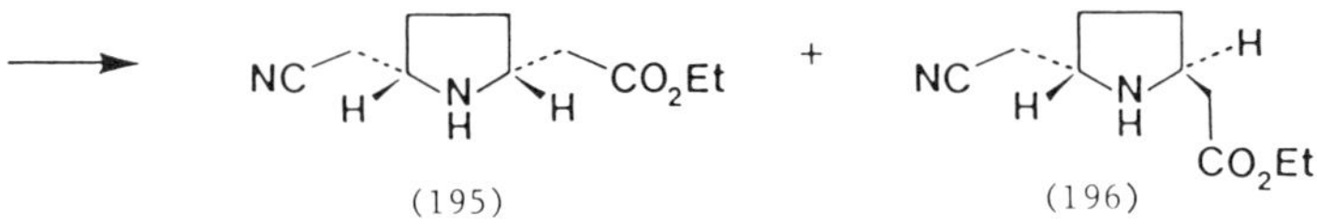

(195)   (196)

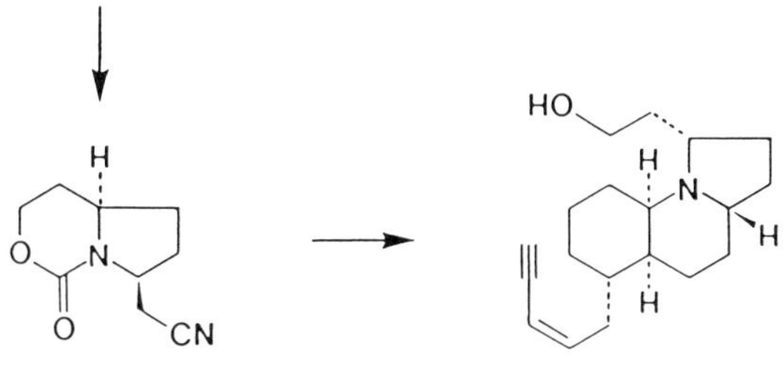

(197)   (198)

followed by Eschenmoser sulfide contraction and deacylation reactions gave **194**. The cis- and trans-isomers **195** and **196** were then obtained by reduction of **194**. The cis-isomer was converted to the perhydropyrrolo [1,2-c][1,3]oxazine **(197)** through three reaction steps; this compound served as a stepping stone to (+) gephyrotoxin (81TL4197).

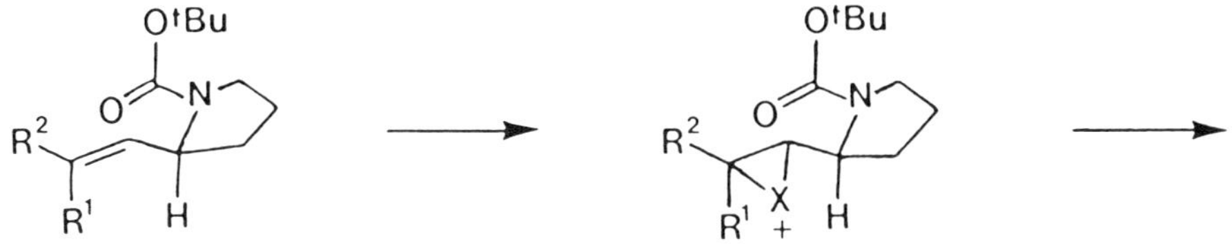

199)
200: R^1=H, R^2=Ph)
201: R^1=H, R^2=p-NO$_2$Ph)
202: R^1=p-NO$_2$Ph, R^2=H)

(203)

204: R^2=Ph, R^1=H, R^3=I)
205: R^2=p-NO$_2$Ph, R^1=H, R^3=Br)
206: R^1=p-NO$_2$Ph, R^2=H, R^3=Br)

(207: R^1=p-NO$_2$Ph, R^2=H, R^3=Br)

The stereo- and regioselective cyclizations of the allylic urethanes **(199)** to oxazolidin-2-ones and/or perhydropyrrolo[1,2-c][1,3]oxazin-1-ones by iodonium and bromonium dicollidine perchlorates in dichloromethane via intermediate **203** have been studied. In this same manner, **204** was generated from **200**. In the cyclization of **201** in the presence of bromonium dicollidine perchlorate, the only product recovered was **205**, while a similar reaction with isomer **202** resulted in a mixture of **206** and **207** (1 : 5 ratio, respectively) (83JA654).

B. Stereochemistry

Perhydropyrrolo[1,2-c][1,3]oxazine can exist in solution as an equilibrium between the trans-fused **(208)**, the O-inside cis-fused **(209)**, and the O-outside cis-fused **(210)** conformers. Comparison of the ^{1}H-NMR geminal coupling constants [84AHC(36)1; Section II,B,2] for the NCH$_2$O pro-

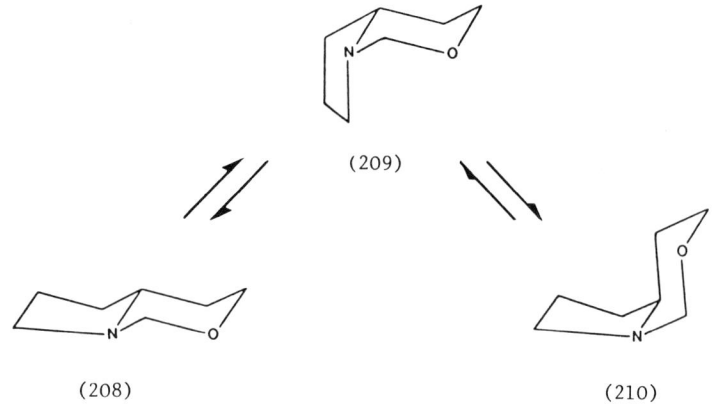

tons (−10.2 Hz) with those in fixed stereochemical systems shows the preference for the O-inside cis-fused conformer **(209)** (82OMR113).

XI. Perhydro-imidazo[1,2-*a*]pyridines

Synthesis and Reactions

The parent perhydro-imidazo[1,2-*a*]pyridine system **(211)** has been obtained by condensation between 5-chloropentanal and 1,2-diaminoethane in dichloromethane in the presence of excess powdered K$_2$CO$_3$ (82TL4181).

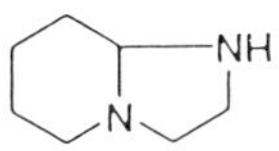

(211) 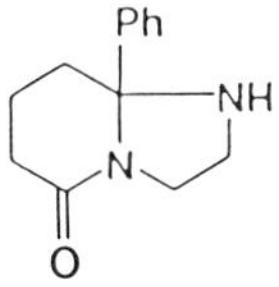 (212) 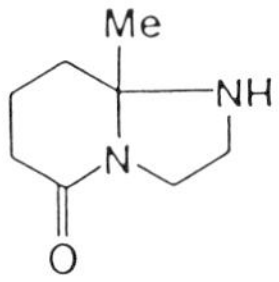 (213) 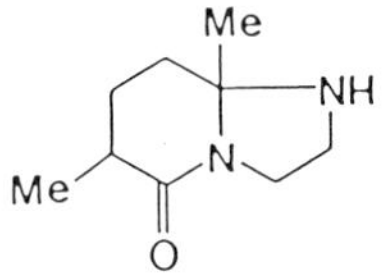 (214)

The synthesis of a variety of perhydro-imidazo[1,2-a]pyridin-5-ones is illustrated by the preparation of **212** by treatment of γ-benzoylbutyric acid with 1,2-diaminoethane in the presence of p-toluenesulfonic acid in boiling xylene (69USP3454585). Compounds **213** (70GEP1802468) and **214** (72AKZ438; 82EUP65724) were similarly obtained by the reaction between the appropriate ketoester and 1,2-diaminoethane.

The Strecker reaction of glutaraldehyde with 1,2-diaminoethane gave three products: **215**, **216**, and the perhydro-imidazo[1,2-a]pyridine **(217)**. The yields of **217** were found to be dependent on the concentration of glutaraldehyde and the reaction time. The 1H-NMR spectrum of **217** shows the absorption of the 5-H proton at δ 3.94 as a near triplet ($J = 3.6$ Hz), indicating the axial CN group in conformation **218** (86H2835).

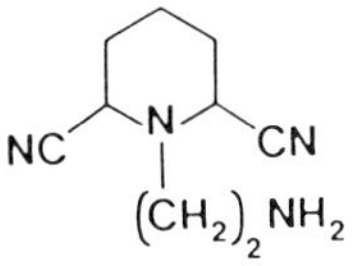

(215) 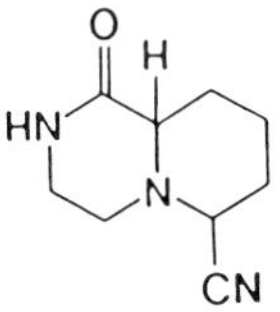 (216) 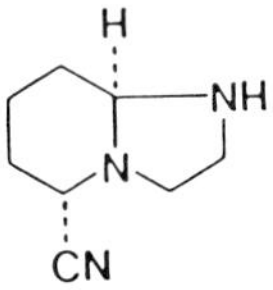 (217)

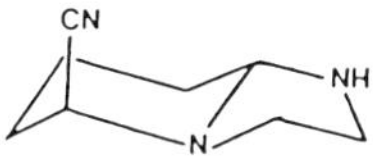

(218)

In the preparation of 2-phenoxymethylperhydro-imidazo[1,2-*a*]pyridine **(221)**, the ring opening of 1,2-epoxy-3-phenoxypropane **(219)** with 2-aminopyridine followed by catalytic reduction of the hydrochloride of the resultant amino alcohol over Rh/C gave **220**. Treatment of **220** with thionyl chloride followed by base produced the required **221** (70IJC707).

Photo-oxidation of the imidazole **(222)** in methanol gives the perhydro-imidazo[1,2-*a*]pyridine **(223)** in 12% yield [81T(S9)191].

Cleavage of perhydro-imidazo[1,2-*a*]pyridine **(211)** with diisobutyl-aluminum hydride produces 1,4-diazacyclononane (82TL4181). Addition of phenylmagnesium bromide to 2-phenyl-6,7,8,8a-tetrahydro-3*H*,5*H*-imidazo[1,2-*a*]pyridin-1-oxide **(224)** resulted in the formation of 2,2-diphenyl-perhydro-imidazo[1,2-*a*]pyridin-1-ol **(225)** (83AP47).

XII. Perhydro-oxazolo[3,2-*a*]pyridines

A. Synthesis and Reactions

Very diverse routes have been followed in the preparation of perhydro-oxazolo[3,2-*a*]pyridines, particularly those incorporating amide functions. Systems of type **226** have been obtained by the reaction between substituted Δ^2-oxazolines with diketene (66FRP1441937) and of **228**, from the oxazoline **(227)** with azidoketene (77MI1).

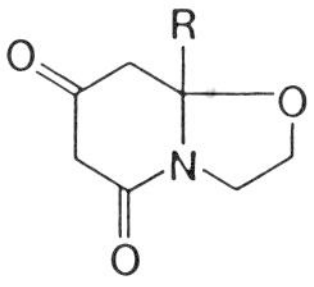

(226) 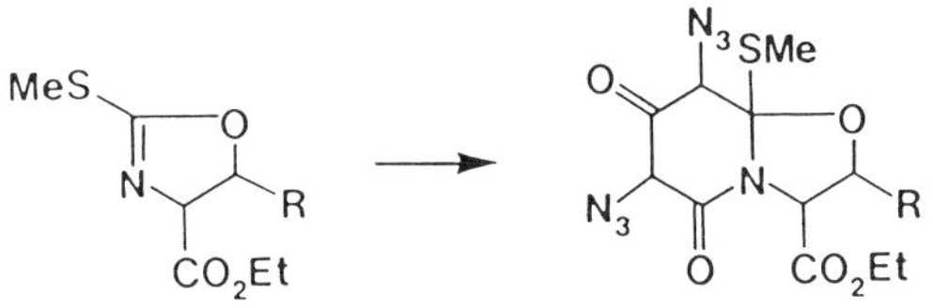   (227: R=p-NO$_2$Ph)   (228: R=p-NO$_2$Ph)

Also, in a very straightforward manner, condensation between 4-benzoylbutyric acid and ethanolamine gives the lactam **(229)** (69JOC165), and the oxidation of 2-piperidinoethanol **(230)** with potassium hexacyanoferrate(III) in aqueous $2M$ potassium hydroxide solution gives perhydro-oxazolo[3,2-*a*]pyridine **(232)** via the iminium ion **(231)** [71JCS(B)1745].

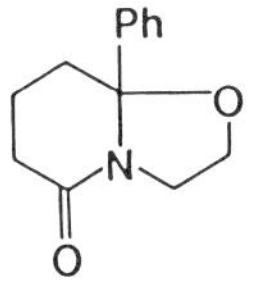

(229)

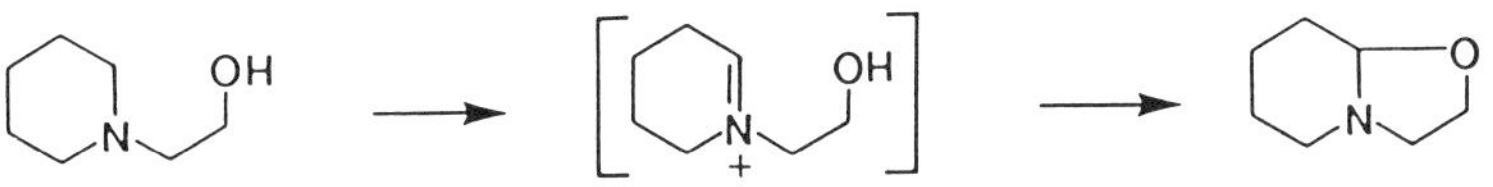

(230)   (231)   (232)

Results of related studies on the cyclization of aminoalcohols by hydride abstraction and mercury(II)-ethylenediaminetetraacetic acid (EDTA) have been published (83PHA512). Methiodide **233** obtained by a route involving a similar Hg(II)-EDTA reaction, on heating with sodium ethoxide in ethanol, gave the medium ring ortho ester **(234)** (79TL809). 1,5-Diketones have also provided a starting point for the synthesis of the perhydro-oxazolo[3,2-*a*]pyridine system. For example, **236** was obtained from the addition of ethanolamine to diketone **235** in DMF (79CHE85). The hydroxyethylamination reaction with diketone **237** over Ni/Ru catalysts gave **238**, which was converted to **239** by hydrogenation (83CHE1090).

The major products from the photocyclizations of 1-(oxophenylacetyl)- and *N*-(2-oxopropionyl)piperidine **(240, 241)** include the perhydro-oxazolo[3,2-*a*]pyridin-3-ones **(242–245)** [69TL371; 76ACS(B)383; 79JA5343].

During a search for a route to cyclic enamides, treatment of **246** with 10% Pd/C in tetrahydrofuran at 120°C for 6 hr in a sealed tube was found to

(240: R=Ph)
(241: R=Me)

(242: R^1=Ph, R^2=H)
(243: R^1=H, R^2=Ph)
(244: R^1=Me, R^2=H)
(245: R^1=H, R^2=Me)

give **247**, **248**, and **249** (75:17:18). The same reaction conducted in the presence of triethylamine gave **247** only (87H917).

(246)

(247)

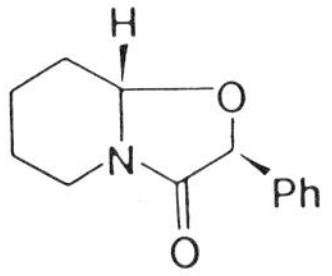

(248)

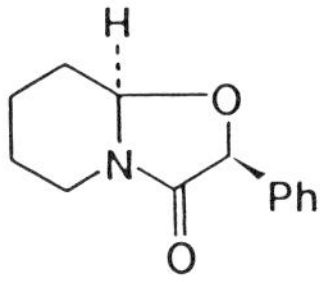

(249)

Steps taken to prepare derivatives of the cyclol **250**, e.g., **255** and **256** involved acylation of ethyl 2-oxonipecotate with benzyloxyacetyl chloride and 2-(benzyloxy) propionyl chloride to give **251** and **252** which, on hydrogenolysis over palladium, gave **253** and **254**, respectively. The respective cyclols were then obtained by heating **253** and **254** in aqueous solution (67JGU1623). Similarly, cyclol **258** was prepared by acylation of valerolactam with acid chloride **257**, followed by hydrogenolysis of the product over Pd/C (76JHC781).

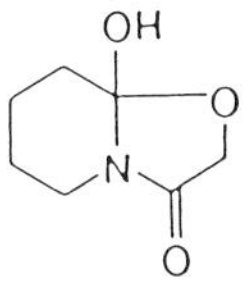

(250)

(251: R=H)
(252: R=Me)

(253: R=H)
(254: R=Me)

(255: R=H)
(256: R=Me)

(257)

(258)

5-Cyanoperhydro-oxazolo[3,2-*a*]pyridines have been claimed to possess bactericidal, fungicidal, and nematicidal properties. The synthesis of the parent compound is effected by the reaction between glutaraldehyde dicyanohydrin and ethanolamine (68USP3375249). Compounds of this type, e.g., **259** (obtained as the single chiral product by condensation between (+)-norephedrine and glutaraldehyde dicyanohydrin), also provide valuable chiral synthons since they open up the possibility of stereoselective reactions at C-2 (α-aminonitrile) and C-5 (α-aminoether). For example, alkylation [lithium diisoproplamide (LDA), PrBr] of **259** gave **260** which was reduced by sodium borohydride in ethanol stereoselectively to **261** (9:1 mixture of 2*S*:2*R* diastereoisomers). Removal of the N-substituent by heating with 70% sulfuric acid gave *S*-(+)-coniine **(262)**. *R*-(−)-Coniine was obtained by first complexing the cyano group of **259** with AgBF$_4$ in tetrahydrofuran before its reaction with propyl-magnesium bromide to give the 5*R*-epimer of **263**. The 5*R*-epimer was then converted to *R*-(−)-coniine, as in the conversion of **261** → **262**. Treatment of **260** with AgBF$_4$, followed by reaction with Zn(BH$_4$)$_2$, resulted in removal of the cyano group to produce **263**. Compound **263** reacted with methylmagnesium iodide to give **264** which, on removal of the N-substituent, gave 2*S*,6*R*-(+)-dihydropinidine **(265)** (83JA7754).

Reaction between the anion of **266** and propionaldehyde gave **267** as the single product. Reduction of **267** by sodium borohydride to **268** was stereospecific, and hydrogenolysis of **268** provided (+)-β-conhydrine **(269)**

(85TL3803). Treatment of the chiral perhydro-oxazolo[3,2-a]pyridine **(270)** with trimethylsilyl cyanide in the presence of zinc bromide resulted in the formation of **271** which, on alkylation (LDA, undecylbromide), gave **272**. Sodium borohydride reduction of **272** gave **273** as the major isomer, which was then converted to 2S,6S-(+)-solenopsin A **(274)** by hydrogenolysis (86JOC4475). (−)-Gephyrotoxin-223AB **(278)** has been obtained by similar transformations. Thus, alkylation of the anion of **266** with 3-bromo-1-pentanal ethylene ketal led to the formation of **275** as the single product.

(266) (267) (268) (269) (270) (271) (272) (273) (274)

Application of the previous reactions resulted in conversion to the *cis*-2,6-disubstituted piperidine (**276**), which gave the indolizidine (**277**) on treatment with HCl and KCN in a biphasic medium. Treatment of **277** with butylmagnesium bromide gave the required (−)-gephyrotoxin-233 AB (**278**) along with a smaller amount of the S-epimer (85TL1515). A similar reaction sequence starting with **266** results in the formation of (−)-monomorine (**279**) (85JOC670).

Perhydro-oxazolo[3,2-*a*]pyridine derivatives have also been used in the synthesis of cyclohexenones. For example, amino diol **280** was treated with 4-acetylbutanoic acid to obtain mainly the bicyclic lactams (**281** and **282**). Compound **282** may be alkylated to give the perhydro-oxazolo[3,2-*a*]pyridin-5-one derivatives **283–285**, and Red-Al reduction of these give the respective derivatives **286–288**. Ring opening of **286–288** by heating under reflux with aqueous tetrabutylammonium dihydrogen phosphate in ethanol was followed by aldol cyclization of the intermediate **289** to give the 4,4-dialkylcyclohexenones (**290**, **291** and **292**) (86JOC1936).

An efficient use of bicyclic lactams as chiral precursors to 4-substituted cyclopentenones and cyclohexenones has been illustrated by the synthesis of (+)-mesembrine (**302**). The required bicyclic lactams **294–297** were prepared by the reaction between **293** and (S)-valinol. The mixture of isomers was transformed to **298** by metalation (BuLi) followed by treatment with allyl bromide. This was converted to **299** by the steps shown. Reduction of **299** with LiAl(OEt)H$_3$ in dimethoxyethane/toluene at −20°C gave **300**, which was heated with tetrabutylammonium dihydrogen phosphate in ethanol to obtain **301**. Cyclization of **301** was promoted in the presence of base to give (+)-mesembrine (**302**). A cyclic system **303** similar to **299** was treated with Red-Al in toluene and then heated with tetrabutylammonium dihydrogen phosphate in ethanol to give **304**. The conversion to (R)-(−)-4-methyl-4-(acetoxyethyl)-2-cyclohexenone (**305**) was achieved by treating **304** with triethylamine and acetic anhydride in tetrahydrofuran (85JA7776).

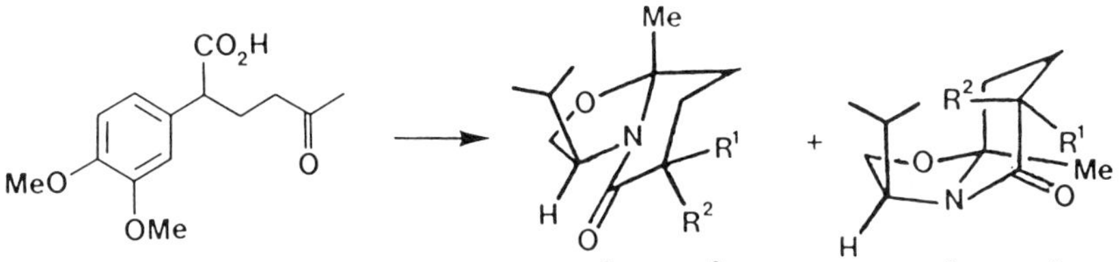

Sec. XII.A] BICYCLIC SIX- AND FIVE-MEMBERED RINGS 237

(293)

(294: R^1=Ar, R^2=H)
(295: R^1=H, R^2=Ar)

(296: R^1=Ar, R^2=H)
(297: R^1=H, R^2=Ar)

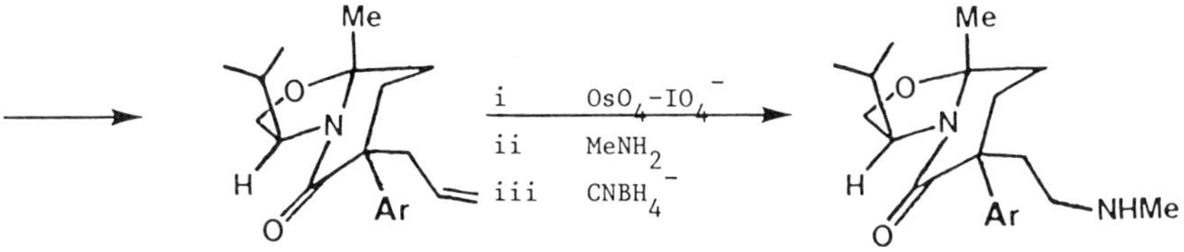

(298)

i OsO_4-IO_4^-
ii $MeNH_2$
iii $CNBH_4^-$

(299)

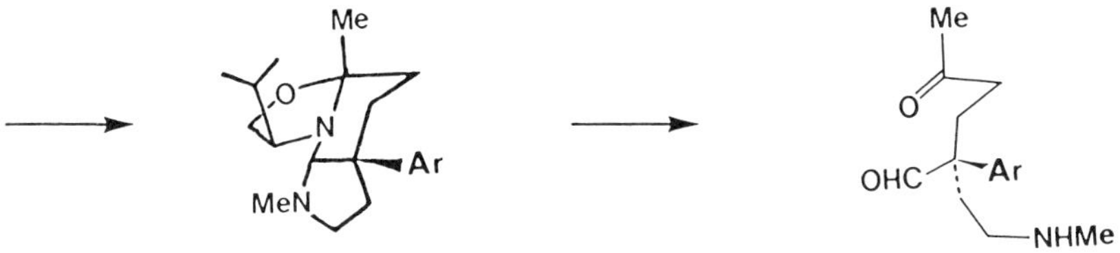

(300)

(301)

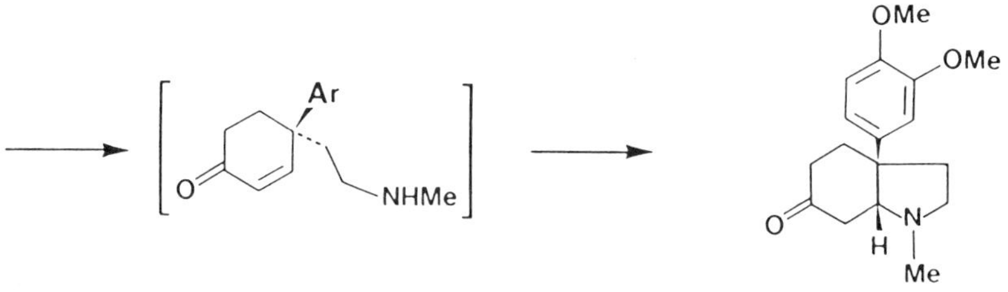

(302)

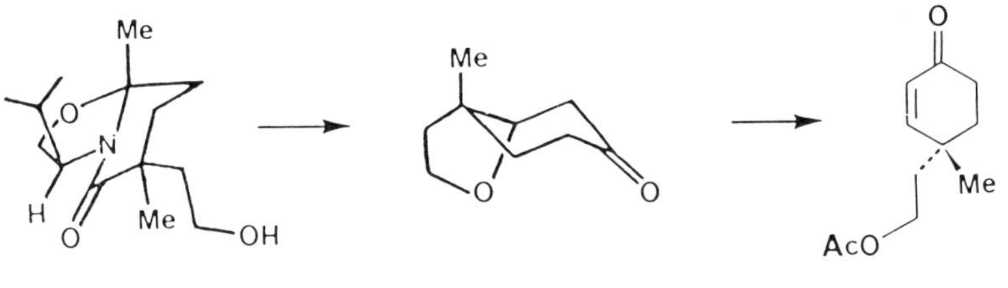

(303)

(304)

(305)

B. Stereochemistry

The stereochemistry of an isomer of 6,7,8,8a-tetrahydro-2-phenyl-5*H*-oxazolo[3,2-*a*]pyridin-3[2*H*]one has been established by X-ray analysis, and the piperidine ring has been shown to adopt a chair conformation. The published bond lengths are given in **306** (77CSC553).

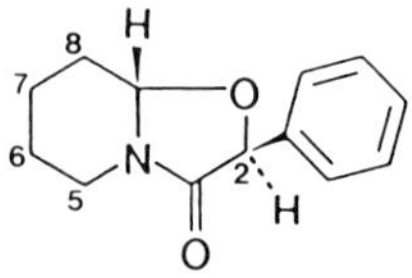

Bond lengths (Å)

O — C-2	1.422
C-2 — C-3	1.506
C-3 — N	1.332
N — C-5	1.452
C-5 — C-6	1.504
C-6 — C-7	1.510
C-7 — C-8	1.516
C-8 — C-8a	1.496

(306)

XIII. Perhydrothiazolo[3,2-*a*]pyridines

A. Synthesis

Perhydrothiazolo[3,2-*a*]pyridine (**308**) has been synthesized by condensation of aziridine with α-hydroxytetrahydropyran and hydrogen sulfide to give 2-(4-hydroxybutyl)thiazolidine (**307**). Treatment of **307** with thionyl chloride followed by sodium hydroxide gave **308** [75CR(C)953].

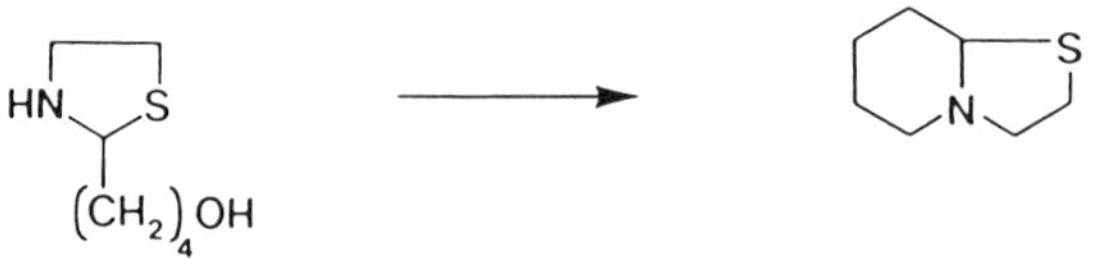

(307)   (308)

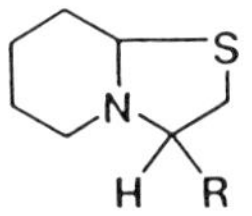

(309: R=Me)
(310: R=Et)

Two other methods of preparing **308** involve dehydration of **307** on alumina in the liquid phase, and the reaction between **307** and triphenylphosphine, carbon tetrachloride, and triethylamine in acetonitrile. The latter method gave good yields of pure products. Using this method and starting with 2-methylaziridine gave **309** and **310** as cis/trans mixtures in the ratios 76:24 and 65:35, respectively (80S387).

A regioselective free radical initiated reaction between 1,2,5,6-tetrahydropyridine and thiiranes has been shown to produce β-aminothiols **(311)**, which cyclize on boiling under reflux with di-t-butyl peroxide in chlorobenzene to give perhydrothiazolo[3,2-a]pyridine derivatives **(312)** (82TL5039).

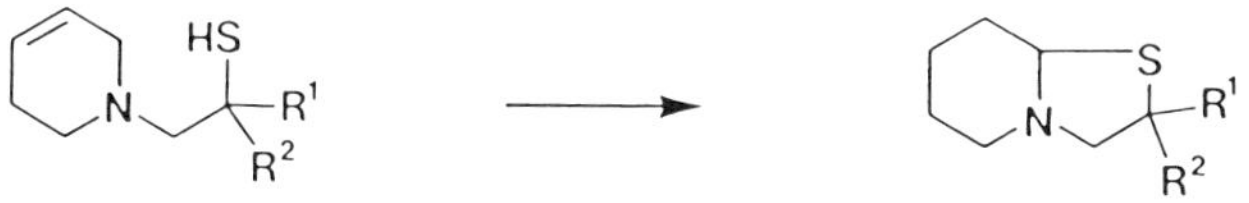

(311)　　→　　(312)

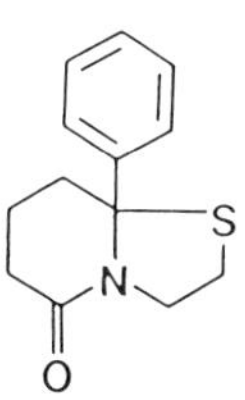

(313)

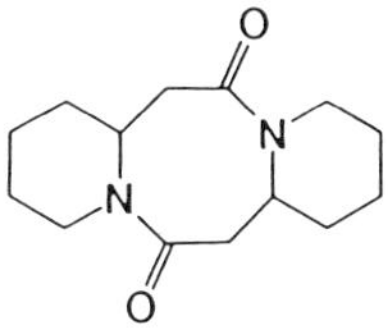

(314)　　→　　(315)　　(316)

The preparation of 8a-phenylperhydrothiazolo[3,2-*a*]pyridin-5-one **(313)** by the reaction between γ-benzoylbutyric acid and mercaptoethylamine illustrates the synthesis of a number of similar compounds (67USP3334091).

Imine trimers such as **314** have been used as a means of activating the α-position of alicyclic amines as sites for introducing functional groups. In one reaction sequence, the perhydrothiazolo[3,2-*a*]pyridin-3-one **(315)** was synthesized by heating the imine trimer **(314)** with ethyl thioglycolate. Oxidation of **315** with *m*-chloroperbenzoic acid gave the corresponding sulfoxide which, on heating at 150°C, resulted in dimerization to **316** with release of sulfur dioxide (86CPB105).

Analogues of penicillin G containing the perhydrothiazolo[3,2-*a*]-pyridine nucleus have been prepared. For example, γ-methyl hydrogen *N*-phthaloylglutamate **(317)** was converted via the Rosenmund reaction on the corresponding acid chloride to the aldehyde **(318)**. Compound **318**, on condensation with L-cysteine in aqueous ethanol, gave **319** which, on hydrazinolysis followed by phenylacetylation, gave **320** [69JCS(C)408].

(317)

(318)

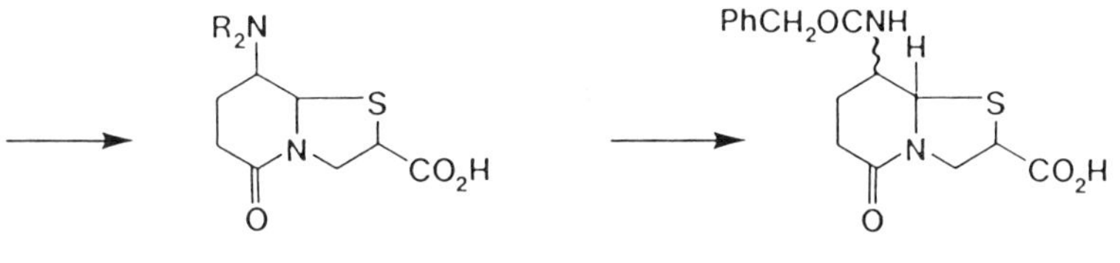

(319: R_2=phthaloyl)

(320)

The synthesis of **325** involved the reaction between *N*-phthalyl-L-glutamic anhydride **(321)** and thiophenol in the presence of dicyclohexylamine in dioxane to give **322**, which was converted by a sequence of steps to the aldehyde **(323)**. Treatment of **323** with L-cysteine hydrochloride in aqueous ethanol and sodium acetate gave **324** which, on heating at 170°C, underwent ring closure to give **325** (84MI2; 85TL647).

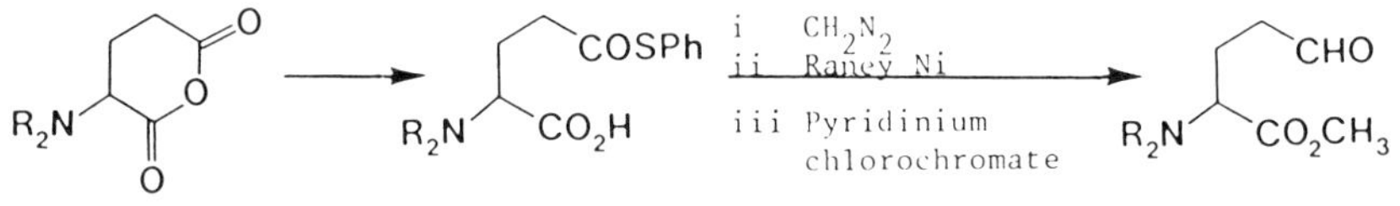

Sec. XIII.B]   BICYCLIC SIX- AND FIVE-MEMBERED RINGS   241

(321: R$_2$=phthaloyl)   (322)   (323)

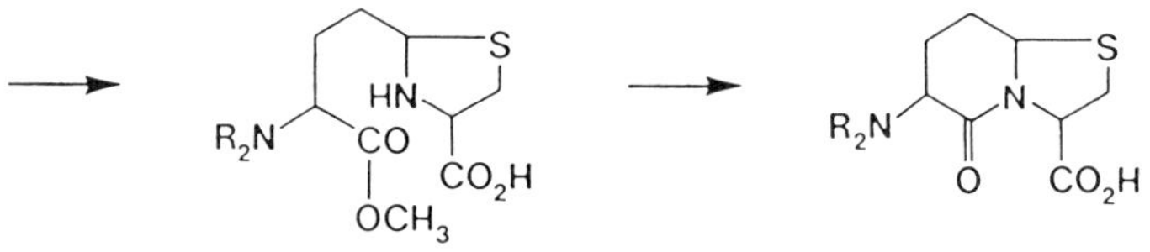

(324)   (325)

B. STEREOCHEMISTRY

Perhydrothiazolo[3,2-*a*]pyridine may exist in solution as an equilibrium mixture of the trans-fused (328) and two cis-fused conformers (326, 327). NMR spectroscopy shows the absence of 327 and the adoption of an equilibrium between 326 and 328 favoring 326 ($\Delta G°_{298}$ = 0.7 kJ mol^{-1}). The two conformers (326 and 328) were detected by low temperature ^{13}C-NMR spectroscopy, and the chemical shifts are shown in 329 and 330. The 3,3-dimethyl compound adopts the trans-conformation (331) (81OMR103). Other systems have also been examined (83CHE622).

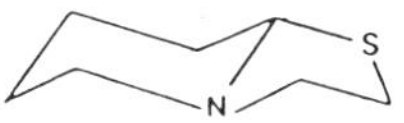

(328)

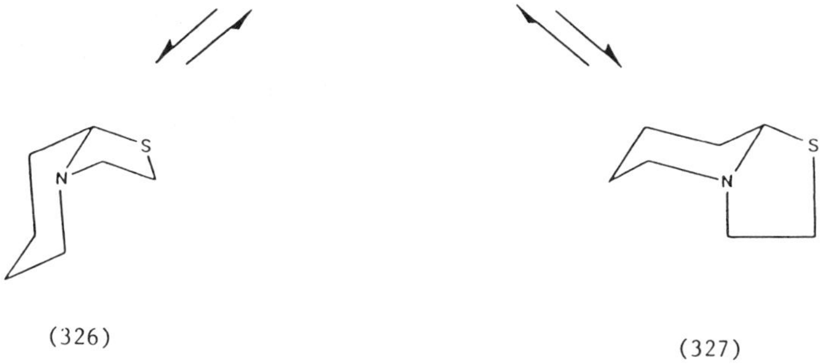

(326)   (327)

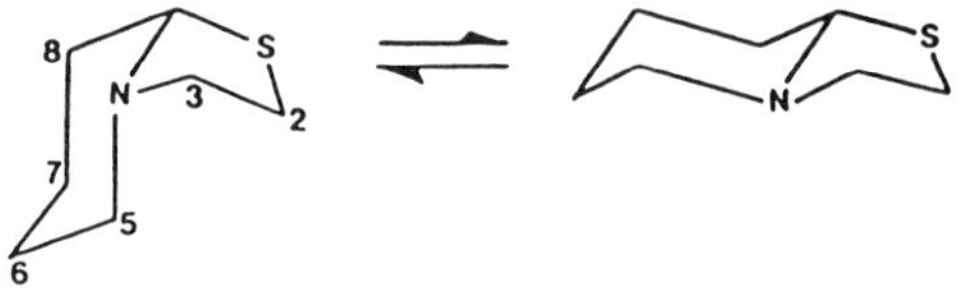

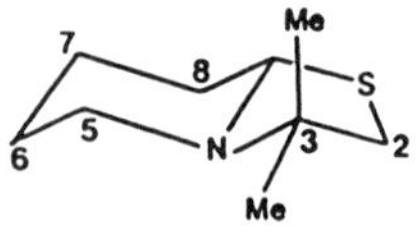

Chemical shifts (δ)		Chemical shifts (δ)		Chemical shifts (δ)	
C-2	31.59	C-2	28.36	C-2	42.33
C-3	60.04	C-3	57.76	C-3	62.84
C-5	45.71	C-5	52.65	C-5	45.34
C-6	25.08	C-6	25.50	C-6	25.01
C-7	18.11	C-7	24.11	C-7	24.56
C-8	26.14	C-8	31.30	C-8	33.26
C-8a	72.12	C-8a	69.11	C-8a	65.88

(329)   (330)   (331)

XIV. Perhydropyrrolo[1,2-*a*]pyrazines

A. Synthesis

Perhydropyrrolo[1,2-*a*]pyrazine (**334**) may be prepared from the dialkylacetal of 2,5-dialkoxytetrahydrofurfural (**332**) by treatment with ethylenediamine to give the 3,4-dihydropyrrolo[1,2-*a*]pyrazine (**333**). Compound **333** is reduced to the required system **334** (79GEP2844549; 80BRP2025936). Alternative syntheses include the lithium aluminum hydride reduction of the diketo derivative (**335**) obtained by, for example, cyclization of 1-glycylproline (84KFZ1445; 86JMC1814), and also similar reductions of **337** (obtained from treatment of **336** with hydrazoic acid; 68JOC2379), **338** (from ethyl prolinate and aziridine; 68JOC2379) and **339** [from 2-aminomethylpyrrolidine and diethyloxalate; 83IJC(B)644].

Related pyrazinones may also be obtained by, for example, the photolysis of succinimide **340**, to give **341**, (82H2057; 86CPB3142, 86LA859), and

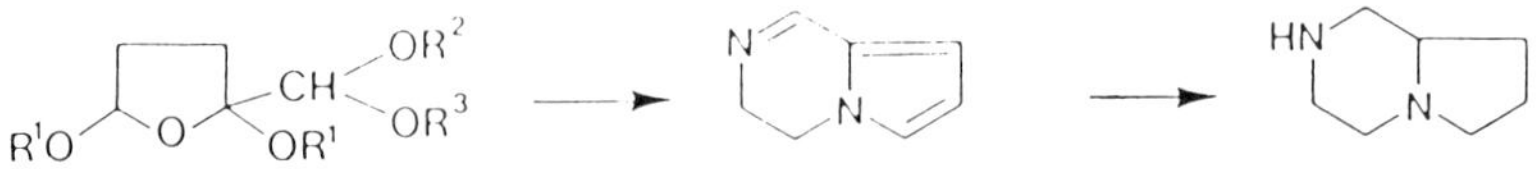

(332)   (333)   (334)

(335) (336) (337) (338) (339) (340) (341) (342)

some tricyclic derivatives of **337**, such as **342**, have been prepared from 2,2'-dipyrrolidine (86MI1).

Numerous perhydropyrrolo[1,2-*a*]pyrazines carrying a range of substituents have been synthesized in the search for compounds of medicinal interest. The following examples provide illustrations. The reaction between 2-aminoperhydropyrrolo[1,2-*a*]pyrazine **(343)** with 3-formylrifamycin S (or SV) gives hydrazones **(344)** with antituberculotic activity

(343) (344)

(85EUP135837, 85USP4551450). Antihypertensive compounds of type **348** are obtained from 2-benzylaminomethylpyrrolidine **(345)** which, on treatment with ethyl 2,3-dibromopropionate, produces **346**. Hydrogenolysis of **346** followed by hydrolysis yields **347** which, on condensation with 3-benzoylthio-2-methylpropionyl chloride, gives **348** (83USP4400511). Other substituted derivatives, such as antiarrythmic agent **349**, have been described (82EUP54593, 82FRP2492374).

Reaction between **334** and 5(β-chloropropionyl)-10,11-dihydro-5H-dibenzo[b,f]azepine leads to **350** (82FRP2493314). Condensation of **334** with the dibenzo[b,e][1,4]diazepine **(351)** in the presence of titanium chloride gives **352** (81EGP151166; 84PHA812). (In structures **349**, **350**, and **352**, R = perhydropyrrolo[1,2-a]pyrazin-2-yl.) Similar reactions involving **334** provide other systems of biological interest such as **353** (79FRP2396757) and **354** (81KFZ55).

(353: R=perhydropyrrolo[1,2-a]pyrazin-2-yl)   (354)

In an alternative approach, antihypertensives of type **356** were obtained by treatment of substituted perhydropyrrolo[1,2-*a*]pyrazines with 4-vinylpyridine, giving **355** which, on catalytic reduction followed by acylation with the appropriate derivative, gives **356** (83USP4414389). The

(355: R^1=4-pyridyl)   (356)

synthesis of the antihypertensive perhydropyrrolo[1,2-*a*]pyrazine **(358)** proceeds via condensation between 2-amino-6-chloropyrazine and ethyl prolinate to give **357**, which is reduced to the corresponding amine. Ring

(357)   (358)

(359)   (360: R^1=p-tolyl, R^2=H)   (362: R=p-tolyl)
        (361: R^1=p-tolyl, R^2=BrCH$_2$CO)

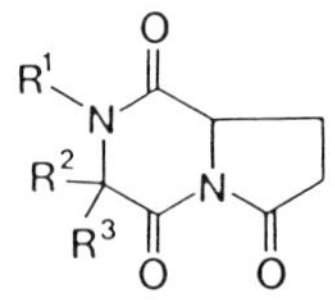

(363)

closure of the amine with oxalyl chloride gives the piperazinedione, which is converted to **358** by diborane reduction (82USP4339579).

The preparation of a number of 1,4-diketopyrrolo[1,2-*a*]pyrazine derivatives with central nervous system (CNS) depressant properties is illustrated by the preparation of compound **362**. 2,5-Dicarbethoxypyrrolidine **(359)** is treated with the appropriate amine to give **360**, which reacts with 2-bromoacetylbromide to produce **361**. Ring closure of **361** to **362** was promoted by heating **361** with sodamide (73FES463; 75BRP1409185; 84FES718). The synthesis of related CNS stimulants and depressants of type **363** have also been described (71USP3563992).

Among various diazabicyclic systems with anticonvulsant, depressant, and analgesic properties, 2-(2-phenethyl)-3-phenylperhydropyrrolo[1,2-*a*]pyrazine **(365)** was synthesized by the reaction between 1-(2-chloro-2-phenethyl)-2-chloromethylpyrrolidine **(364)** and 2-phenethylamine

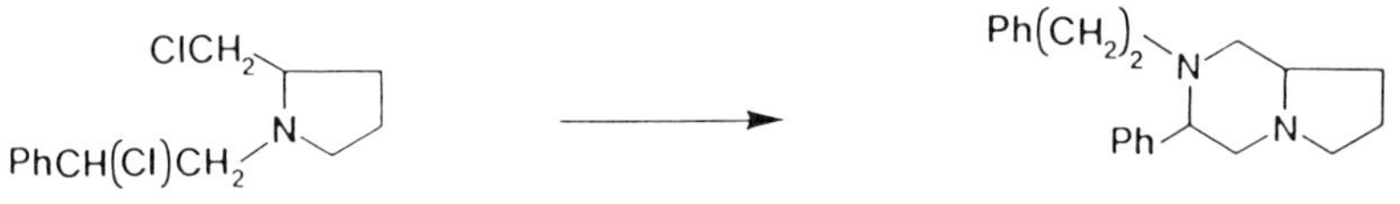

(364)   (365)

(70USP3531485). Ring closure of the dehydrodipeptide **(366)** with 1,5-diazabicyclo[5.4.0]undec-5-ene (DBU) in boiling benzene gives **367**, which was further transformed to compounds such as **368** and **369** (77CB921). Similarly, ring closure of *N*-methyl-1-pyruvoyl-(2*S*)-pyrrolidin-2-carboxamide **(370)** gives **371**, which is isomeric with **369** (74CB2804, 74CB2816; 75CB2907, 75CB2917).

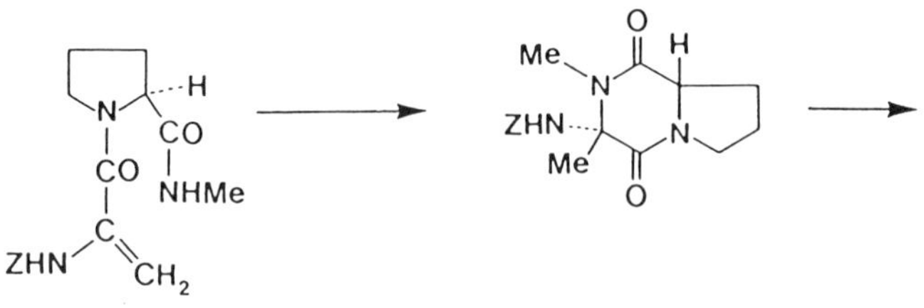

(366)   (367)

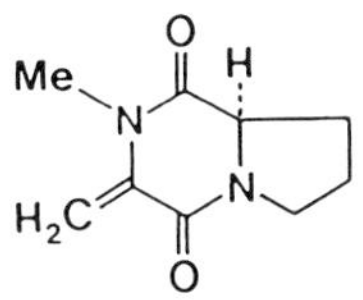

(368)

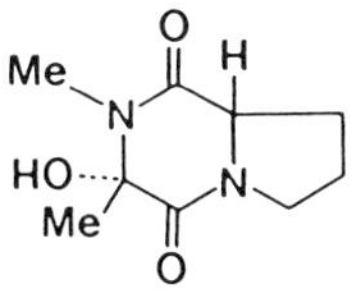

(369)

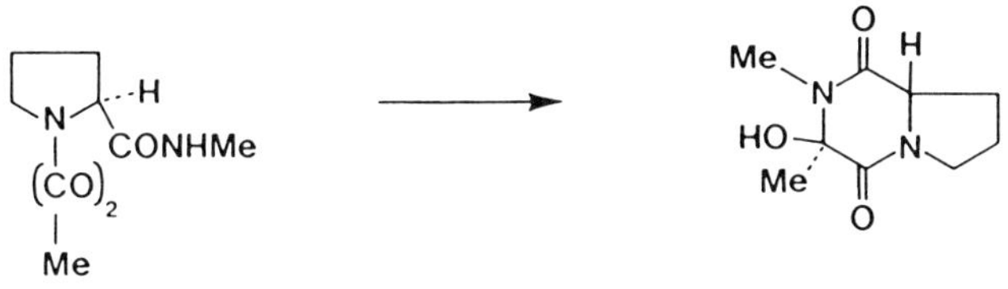

(370) → (371)

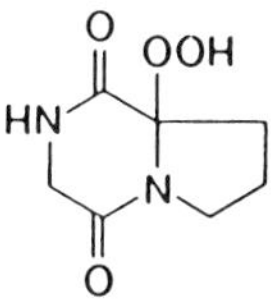

(372)

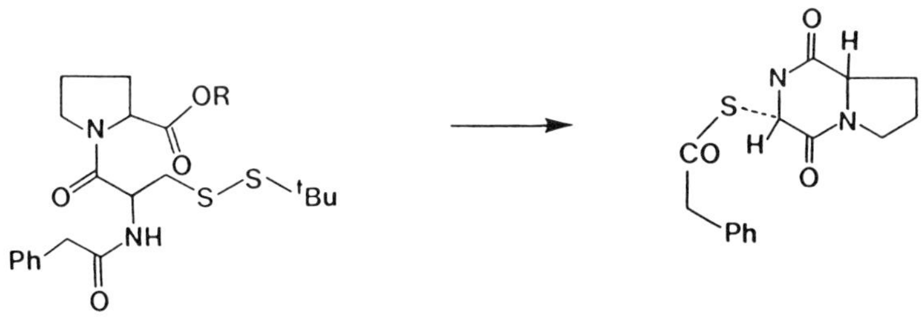

(373: R=p-NO$_2$Ph) → (374)

Photochemical oxidation of cyclo-DL-Pro-Gly gives peroxide **372** (78CB361). Deprotection of the thiol group of **373** gives the perhydropyrrolo[1,2-*a*]pyrazin-1,4-dione **(374)** (85TL5481).

B. STEREOCHEMISTRY

The dioxopiperazine ring in N-(phenylacetyl-L-alanyl)-cyclo-(-L-Phe-D-Pro) **(375)** adopts a boat conformation with the PhCH$_2$ side chain in the axial position and the pyrrolidine ring in a β-envelope conformation. The differences in N—C(O) bond lengths are shown in structure **375** [82CSC(A)1487; 84MI3]. Similar conformations are adopted by N-(pyruvoyl)-cyclo-(-L-Phe-D-Pro-) (85MI1) and pyroergotamine **(376)** (85HCA724). X-ray data on the N-hydroxy-dioxopiperazine **(377)** are provided with the structure (86MI2).

Computational studies on proline containing cyclic dipeptides have been performed (76JA5358), and lanthanide NMR shifts have been used to

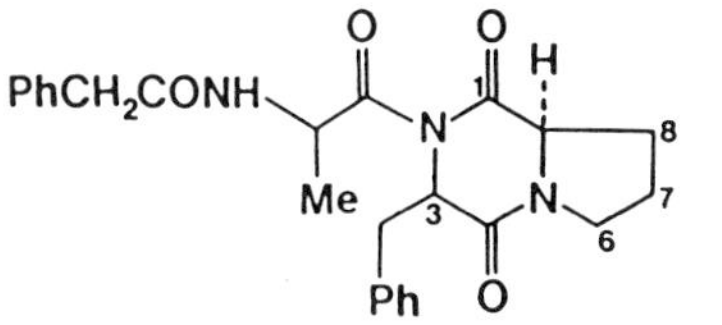

Bond Lengths (Å)

C-1	N(2)	1.396
N(2)	CO	1.416
N(2)	C-3	1.483
C-4	N(5)	1.314
N(5)	C-8a	1.464
N(5)	C-6	1.467

(375)

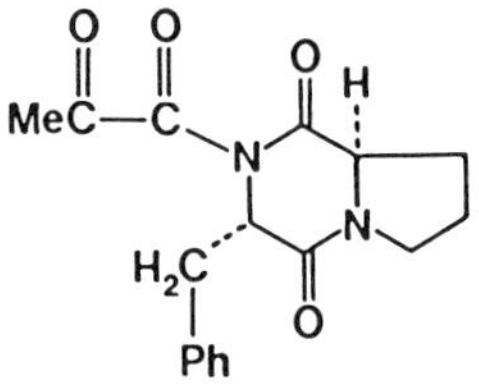

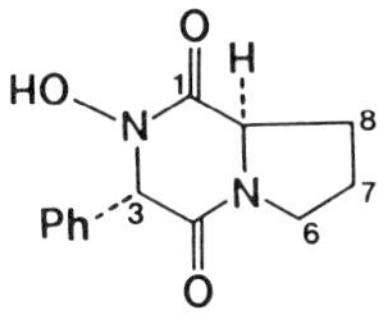

Bond lengths (Å)

C-8a	C-1	1.499
C-1	N	1.329
N	C-3	1.461
C-3	C-4	1.510
C-4	N	1.322
N	C-8a	1.467
N	C-6	1.471
C-6	C-7	1.528
C-7	C-8	1.528
C-8	C-8a	1.518

(377)

study side-chain conformations in these (76JA5365). The circular dichroism (CD) spectrum of cyclo-(-Pro-D-Val) is available (85MI2).

The $J(^{13}CH_3\text{—}N\text{—}C\text{—}H_{A(B)})$ values in **378** have been obtained in order to establish a Karplus-type relationship with the dihedral angle [80CR(C)291]. Detailed ^{1}H-NMR studies using a generalized Karplus relationship have been used in assigning conformations to, for example, cyclo-(-Pro-D-Ala) (85BSB187). ^{13}C-NMR shifts for **379** are given with the structure (82CCC3312).

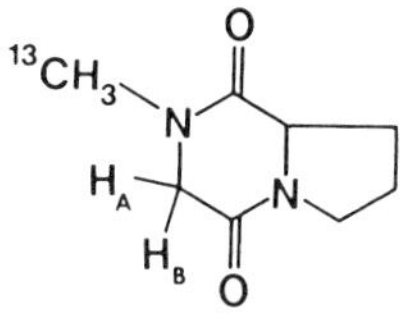

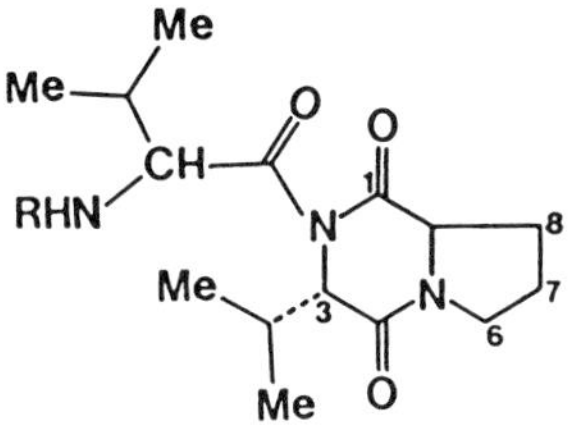

Chemical shifts (δ)

C-1	169.9
C-3	64.3
C-4	165.3
C-6	45.5
C-7	22.8
C-8	23.3
C-8a	58.2

(378)

(379: R=ergine moiety)

XV. Perhydropyrrolo[2,1-*c*][1,4]oxazines

A. SYNTHESIS AND REACTIONS

Most perhydropyrrolo[2,1-*c*][1,4]oxazines reported during the period covered by this review have been been derived from proline. The simplest of these, **382**, is obtained by ring closure of **381** which is obtained by the

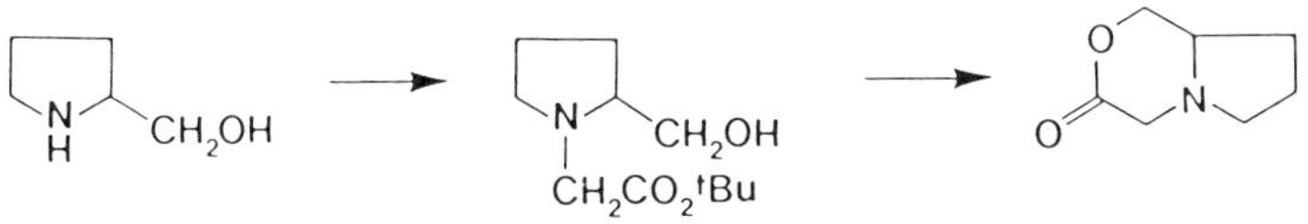

(380)   (381)   (382)

reaction btween prolinol **(380)** and *t*-butyl bromoacetate (74CC395). In a similar manner, L-prolinol in converted to **383** by treatment with (*R*)-2-bromo-3-phenylpropionic acid and dicyclohexylcarbodiimide in dichloromethane, which is then cyclized to **384** by a sodium hydride treatment. This compound provides a route to Boc-Pro(ψ)[CH$_2$O]Ph—OH **(385)** by acid hydrolysis followed by amine protection (87JOC418). Similar routes to modified oxytocin and vasopressin dipeptides have been described (86USP4596819).

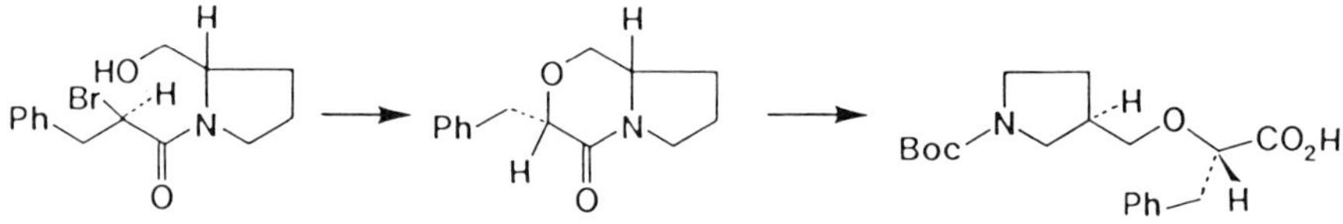

(383)   (384)   (385)

Routes to perhydropyrrolo[2,1-*c*][1,4]oxazin-1,4-diones are varied. Thus, **386** is obtained from prolinol by double carbonylation in the presence of PdCl$_2$(MeCN)$_2$/CuI catalyst under CO and O$_2$ in an autoclave at room temperature (87CC125); **388** is obtained from **387** with a diastereoisomeric excess of 65% using 2,3-dichloro-5,6-dicyanobenzoquinone

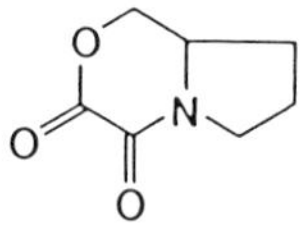

(386)

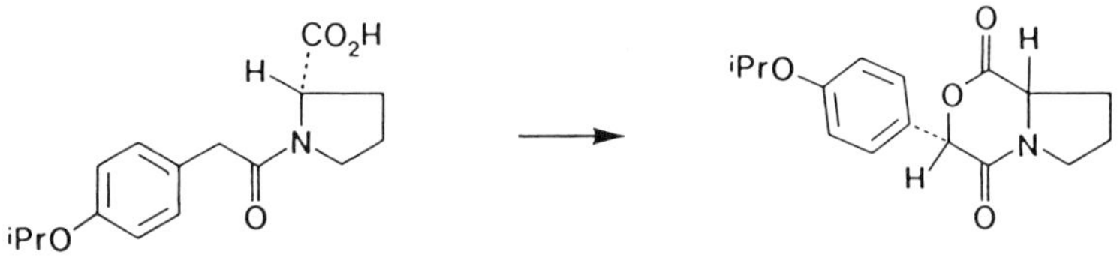

(387)   (388)

(389)   (390)

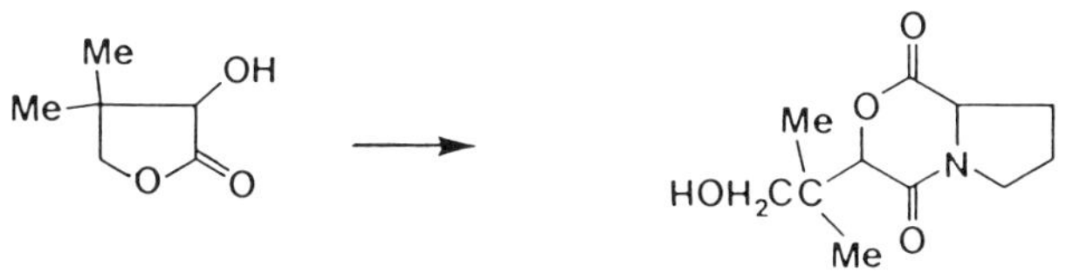

(391) (392)

(86CC741); **390** is obtained by an oxo-Wittig-type rearrangement of carbobenzyloxy-L-proline **(389)** [76CI(L)693]; and cyclo (*N*-D-pantoyl-L-proline) **(392)** is obtained as a byproduct from the reaction between D-pantolactone **(391)** and L-proline (79MI1, 79MI2).

3,3-Dimethylperhydropyrrolo[2,1-*c*][1,4]oxazin-1-one **(393)** has been obtained by reacting isobutene oxide and proline (83AP339). 8-Hydroxy-8-phenylperhydropyrrolo[2,1-*c*][1,4]oxazin-6-one **(396)** has been obtained by photocyclization of *N*-benzoylacetamidomorpholine **(394)** via δ-hydrogen abstraction to the 1,5-diradical intermediate **(395)** by the ketone carbonyl group [74CC743; 76JCS(P1)2054].

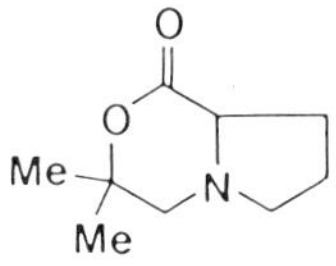

(393)

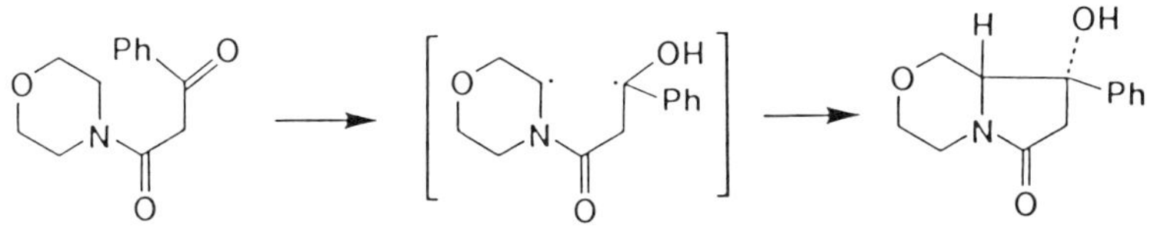

(394) (395) (396)

3-Substituted perhydropyrrolo[2,1-*c*][1,4]oxazin-1,4-diones (e.g., **400**) have been used in the synthesis of optically active 2-hydroxy-2-methylalkanoic acids as shown in the conversion of **397** to the S-isomer **(402)**. The acid chloride **(398)** derived from 2-methylenehexanoic acid **(397)** reacted under Schotten–Baumann conditions with L-proline to give amide **399** which, on treatment with *N*-bromosuccinimide, gave **400**. Reduction of **400** with tributyltin hydride gave 3,3-dialkylperhydropyrrolo-[2,1-*c*][1,4]oxazin-1,4-dione **(401)**, which was hydrolyzed by 48% aqueous hydrogen bromide to (+)*S*-2-hydroxy-2-methylhexanoic acid **(402)** (86EUP198348, 86EUP198352; 87TL2801).

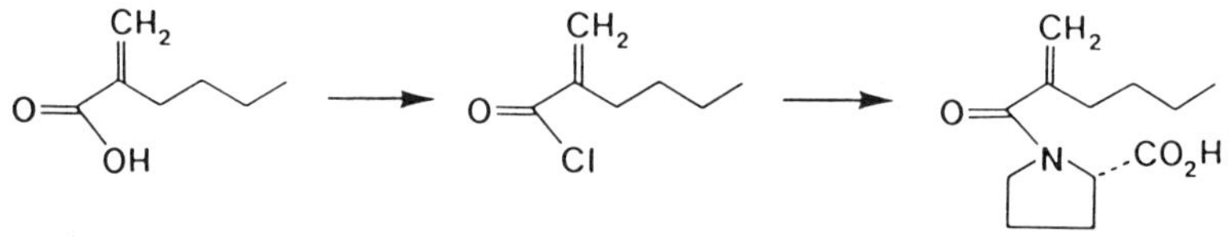

(397) (398) (399)

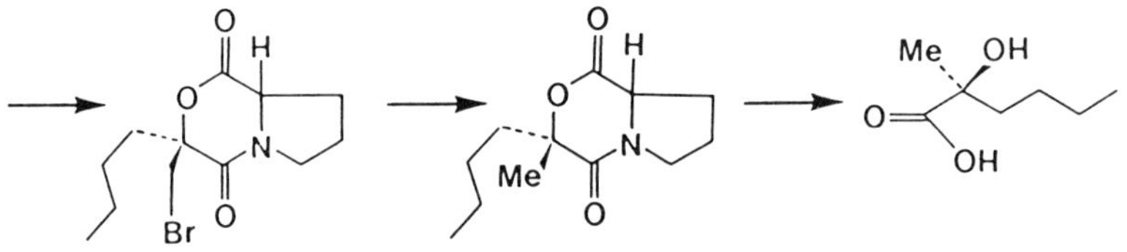

(400) (401) (402)

A related route to α-hydroxyalkanoic acids is illustrated by the synthesis of R(−)-2-hydroxy-2-methylbutyric acid (407) from (S)(−)-tigloylproline (403). The initial step of halolactonization of 403 with N-bromosuccinimide in DMF proceeds stereospecifically to give 405 (94.5%) and 407 (5.5%). Debromination of 405 to 409 with tributyltin hydride in benzene, followed by hydrolysis, gave the desired compound 410 (77TL1005; 79T2337, 79T2345).

In a similar way, the 3-substituted perhydropyrrolo[2,1-c][1,4]oxazin-1,4-diones (406, 408) provided a route to optically active α,β-epoxyaldehydes from α,β-unsaturated acids. Condensation of α-methylcinnamic

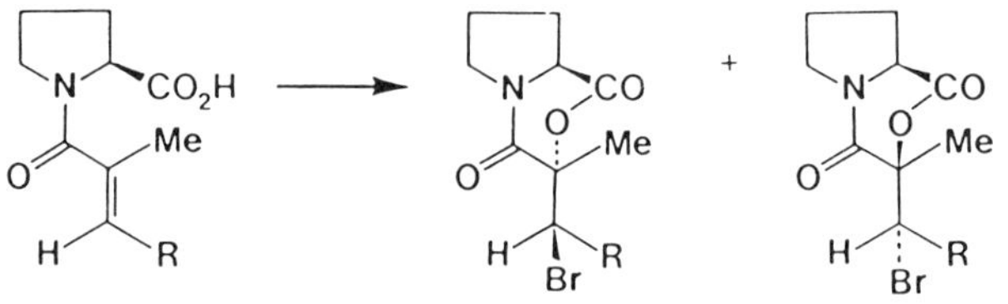

(403: R=Me)   (405: R=Me)   (407: R=Me)
(404: R=Ph)   (406: R=Ph)   (408: R=Ph)

(409)   (410)

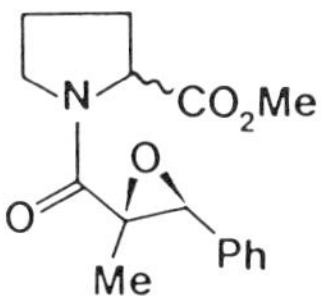

(411)

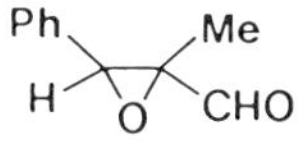

(412)

acid with proline gave **404** which, on bromolactonization, gave predominantly **408** along with a small amount of **406**. This mixture was treated with sodium methoxide in methanol and ring opened to the epimeric epoxy esters **(411)**. Reductive cleavage of the proline moiety gave (2*R*),(3*S*)-2-methyl-3-phenylepoxypropanal **(412)** (80TL2733; 81T2797).

A different usage of such oxazin-1,4-diones in synthesis is exemplified by the preparation of the enantiomerically pure acyloin **(417)**. The *S*-proline derivative **(413)** was treated with an allylsilane in the presence of a Lewis acid to give a mixture of optically active tertiary homoallyl alcohols

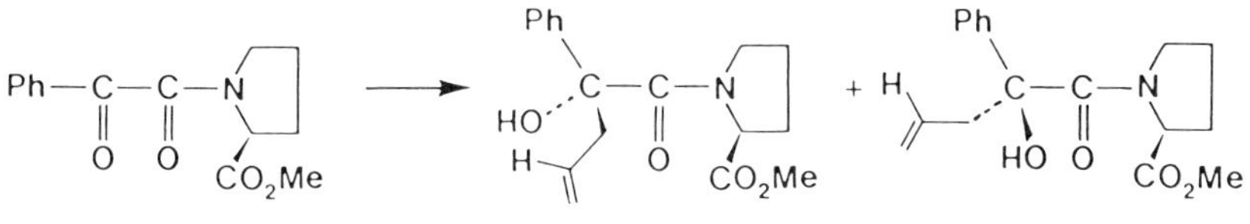

(413)

(414)

(415)

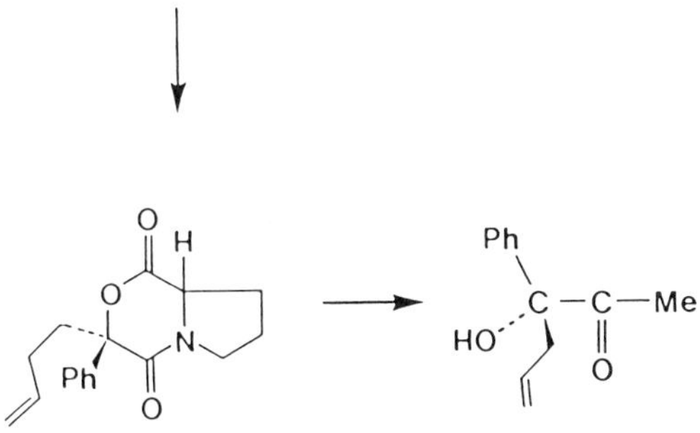

(416)

(417)

(414, 415). Diastereomer **414** was readily cyclized to 3,3-disubstituted perhydropyrrolo[2,1-*c*][1,4]oxazin-1,4-dione **(416)**, while **415** remained unchanged under the conditions employed. Acyloin **417** was then obtained by treatment of **416** with methyl lithium in tetrahydrofuran (86JOC3290).

B. Stereochemistry

The ^{1}H-NMR coupling constants provided with the structures show perhydropyrrolo[2,1-c][1,4]oxazin-3-ones **418** and **419** to exist in the conformations shown (74CC395).

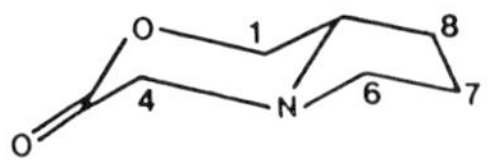

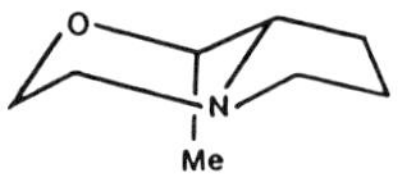

Chemical shift (δ)

H-1ax   4.14   (J=10, 10Hz)

H-1eq   (J=10, 4.5Hz)

(418)

Chemical shift (δ)

H-1eq   4.62   (J=6.5Hz)

(419)

XVI. Perhydropyrrolo[2,1-c][1,4]thiazines

Synthesis

Derivatives of perhydropyrrolo[2,1-c][1,4]thiazine have been prepared in connection with the development of angiotension-converting enzyme inhibitors. For example, condensation between 2-(acetylthio)propionic acid **(420)** and *t*-butyl prolinate in the presence of dicyclohexylcarbodiimide gave *t*-butyl N[2-(acetylthio)propanoyl]prolinate which, after successive deprotection [anisole/trifluoroacetic acid; (ii) NH$_3$/MeOH], produced **421**. Cyclization of **421** using ethyl chloroformate and triethylamine gave rise to the required 3-methylperhydropyrrolo[2,1-c][1,4]thiazin-1,4-dione **(422)** (80BRP2036743, 80USP4209617; 86JMC784). In an alternative sequence, treatment of a mercaptopropionic acid ester with a N-protected proline gave **423**. Again, deprotection followed by cyclization gave **422** (80EUP17390).

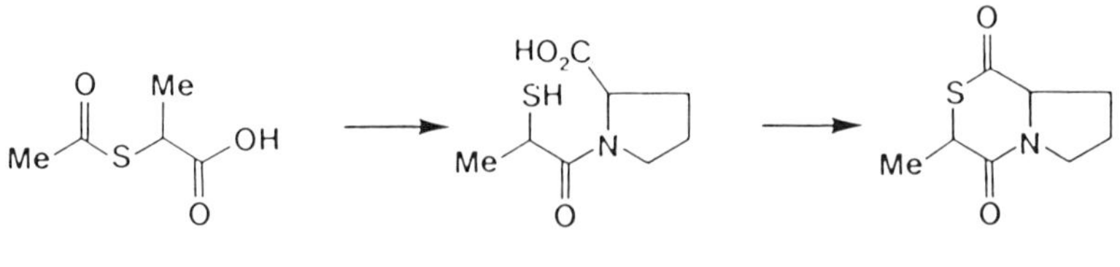

(420)   (421)   (422)

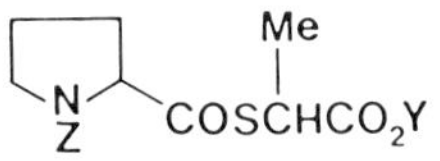

(423)

The search for dopamine antagonists has led to the synthesis of compounds of type **427**. The diester **425**, prepared by treatment of methyl (3-thiomorpholino) acetate **(424)** with ethyl bromoacetate on Dieckmann cyclization, gave perhydropyrrolo[2,1-*c*][1,4]thiazin-6-one **(426)**. This was converted to the oxime, which was then reduced and acylated to **427** (82EUP57536).

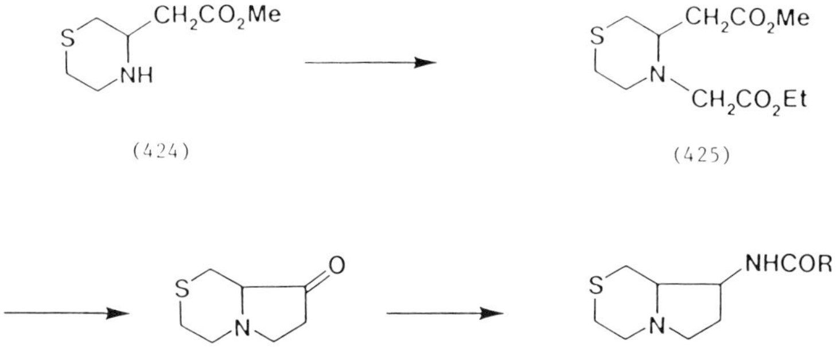

(424)   (425)

(426)   (427)

XVII. Perhydropyrrolo[1,2-*a*]pyrimidines

A. Synthesis

The parent compound, perhydropyrrolo[1,2-*a*]pyrimidine **(428)**, is obtained by the reaction between 4-chlorobutanal and 1,3-diaminopropane in the presence of potassium carbinate (83TL1559). A study of ring-chain tautomerism in such systems shows that in neutral aqueous solutions **428** predominates [84ACS(B)526]. 6-Propyl and 6-heptyl derivatives **(430)** were prepared by reduction of **429** with LiAlH$_4$. Hydroxyethylation of **430** with 50% excess ethylene oxide gives rise to **431** (76URP527433).

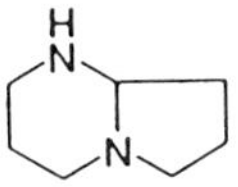

(428)

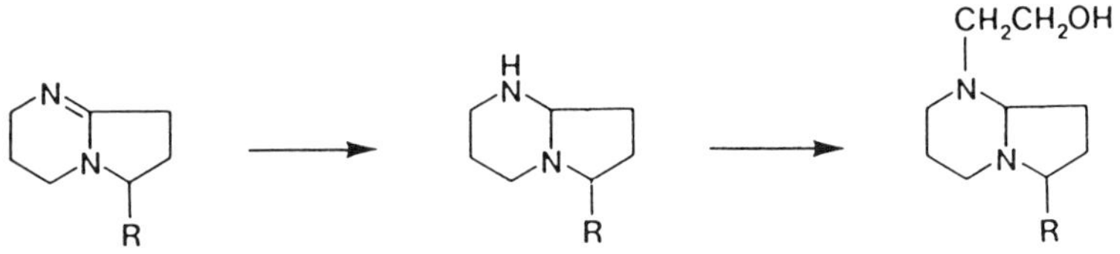

(429) (430) (431)

Studies aimed at the total synthesis of (S)-(−)-celacinnine have involved the preparation of an optically active nine-membered azalactam **(435)** via pyrrolo[1,2-*a*]pyrimidine intermediates. The (S)-(−)-β-Phenyl-β-alanine methyl ester **(432)** was condensed with 2-methoxypyrroline **(433)** by heating at 130°C to give **434**. Ring expansion of **434** was accomplished by treatment with three equivalents of NaBH$_3$CN in acetic acid to give S-(−)-**435** (31%) along with 32% of S-(−)-2-phenylperhydropyrrolo[1,2-*a*]pyrimidin-4-one **(436)**. The latter was also obtained by reduction of **434** with NaBH$_4$ in methanol (87H85). Similarly, reaction between 2-methoxypyrroline **(433)** and the ethyl ester of β-alanine gave **437** which, on catalytic hydrogenation, afforded **438** (69BSF3139).

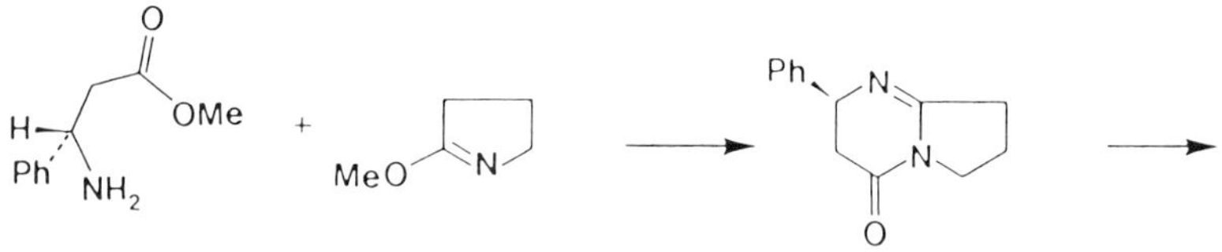

(432) (433) (434)

(435) (436)

(437) (438)

8a-Phenylperhydropyrrolo[1,2-*a*]pyrimidin-6-one **(440)** has been synthesized by treating 3-benzoylpropionic acid **(439)** with 1,3-diaminopropane in the presence of a catalytic amount of *p*-toluenesulfonic acid (70USP3526626). Other related derivatives have been described

PhCO(CH$_2$)$_2$CO$_2$H ⟶

(439)   (440)

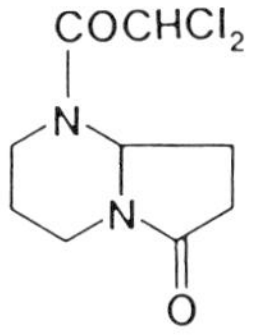

(441)

(67USP3334099; 70USP3526660), and compounds of type **441** have been reported as agrochemicals (81EUP31042; 86EUP190105; 87GEP3600903).

B. Stereochemistry

The ^{1}H-NMR coupling constants and chemical shifts of 8a-methylperhydropyrrolo[1,2-*a*]pyrimidin-6-one, shown with the structure, are in accordance with conformation **(442)**. Data on 1,8a-dimethyl- and 8a-phenyl-substituted derivatives of the same system are also available (71AP774).

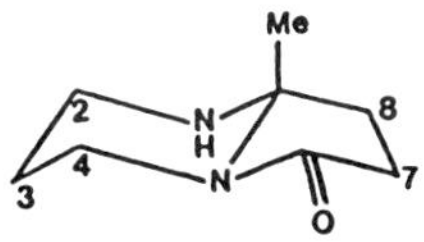

Chemical Shifts (δ)

H-2	3.0, 3.12
H-3ax	1.6 (J=12)
H-3eq	1.4 (J=4)
H-4ax	3.08 (J=2, 13.5)
H-4eq	4.12 (J=13.5)
H-7	2.43
H-8	1.8, 2.18

(442)

XVIII. Perhydropyrrolo[2,1-*b*][1,3]oxazines

Synthesis

The general procedure for preparing a variety of perhydropyrrolo-[2,1-*b*][1,3]oxazin-6-ones with sedative properties is illustrated by the preparation of 8a-*p*-tolylperhydropyrrolo[2,1-*b*][1,3]oxazin-6-one **(444)**. 3-(*p*-Methylbenzoyl)-propionic acid **(443)** was converted to *N*-(3-hydroxypropyl)-3-(4-methylbenzoyl)propionamide by treatment with ethyl chloroformate and triethylamine followed by 3-aminopropanol. The amide was then ring closed to **444** upon heating in xylene in the presence of *p*-toluenesulfonic acid (68USP3414585). The preparations of related compounds were conducted in a similar manner (76JMC436). In a variation on this synthesis, 3-benzoyl-propionic acid was replaced by 4-chlorobutyrophenone (79AF983).

RCO(CH$_2$)$_2$CO$_2$H ⟶

(443: R=p-tolyl)   (444: R=p-tolyl)

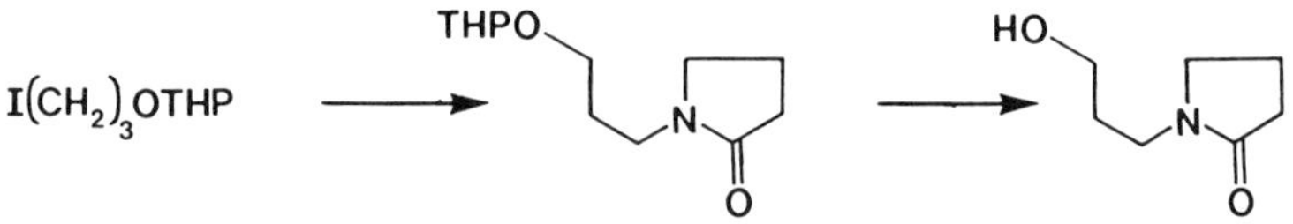

I(CH$_2$)$_3$OTHP ⟶   ⟶

(445)   (446)   (447)

⟶

(448)

A different route to these types of compounds is illustrated by the synthesis of perhydropyrrolo[2,1-*b*][1,3]oxazin-6-one **(448)**. The tetrahydropyranyl ether **(445)**, derived from trimethylene iodohydrin, was boiled under reflux with 2-pyrrolidinone in THF in the presence of sodium hydride to give **446**. Deprotection of **446** gave **447**, and oxidation of this resulted in cyclization to **448** (80H1089).

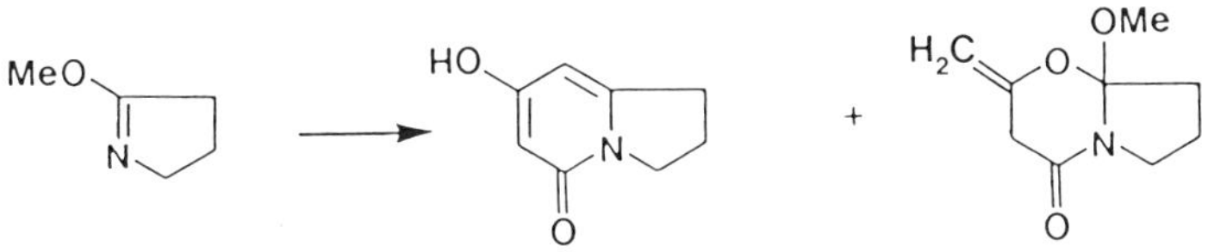

(449)   (450)   (451)

The direct reaction between diketene and 2-methoxy-1-pyrroline **(449)** gave a mixture of **450** (11%) and 8a-methoxy-2-methyleneperhydropyrrolo[2,1-*b*][1,3]oxazin-4-one (**451**, 73%) (75H927).

XIX. Perhydropyrrolo[2,1-*b*][1,3]thiazines

SYNTHESIS

Among the compounds evaluated for their ability to reverse electroconvulsive shock-induced amnesia in mice, perhydropyrrolo[2,1-*b*][1,3]thiazin-4,6-dione **(454)** was synthesized. A mixture of 5-ethoxypyrrolidone **(452)** and 3-mercaptopropanoic acid was heated at 60°C for 16 hr to give **453**, which ring closed to **454** on heating with acetic anhydride in acetic acid (87JMC498).

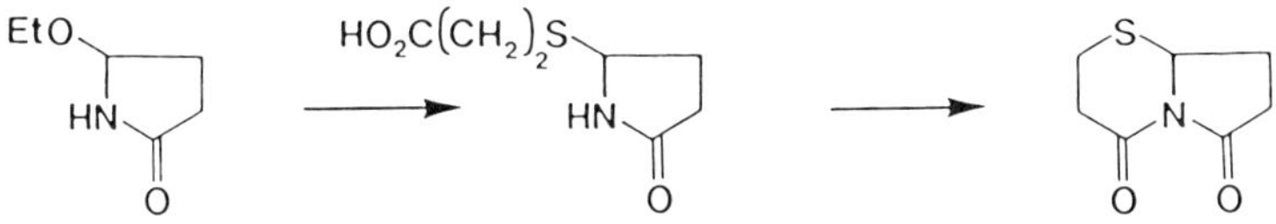

(452)   (453)   (454)

XX. Conformational Analysis of Saturated 6/5 Ring Systems with Bridgehead Nitrogen and a Single Additional Heteroatom

All 18 systems (Sections II–XIX) described in this review may exist in solution as an equilibrium mixture of one trans-fused and two cis-fused conformers interconvertible by nitrogen inversion and ring inversion. For the parent compound, indolizidine **(455)**, this is illustrated by the equilibrium **456** ⇌ **457** ⇌ **458**. IR studies (68TL6191) on trans-(H-8,H-8a)-8-hydroxyindolizidine suggested an equilibrium for indolizidine, itself with $\Delta G°_{298} = 2.4$ kcal mol^{-1}, favoring the trans-conformer **456** (98%). The 360 MHz NMR spectrum of **455** confirmed the trans-conformation with a chair piperidine ring [84JCS(PI)1]. In addition, studies on the ^{13}C-NMR spectra of the 6-ethylindolizidines showed no detectable amounts of cis-fused conformers (87MI1). The equilibrium for cis-(H-6,H-8a)-6-ethylindolizidine (**459** ⇌ **460**), relative to that for indolizidine, should be biased towards the cis-conformer (**460**) by the *gauche*-butane and *gauche*-propylamine-type interactions resulting from the presence of the axial ethyl group in **459**, which may be estimated as 1.5 kcal mol^{-1} (75CC844) or 1.8 kcal mol^{-1} (76TL3765). Taking these values and the literature value of $\Delta G°_{298} = 2.4$ kcal mol^{-1} for the indolizidine equilibrium gives $\Delta G°_{298}$ values of 0.9 kcal mol^{-1} (82% of **459**) or 0.6 kcal mol^{-1} (73% of **459**) for **459** ⇌ **460**.

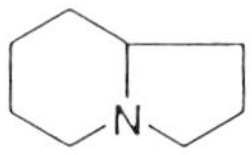

(455)

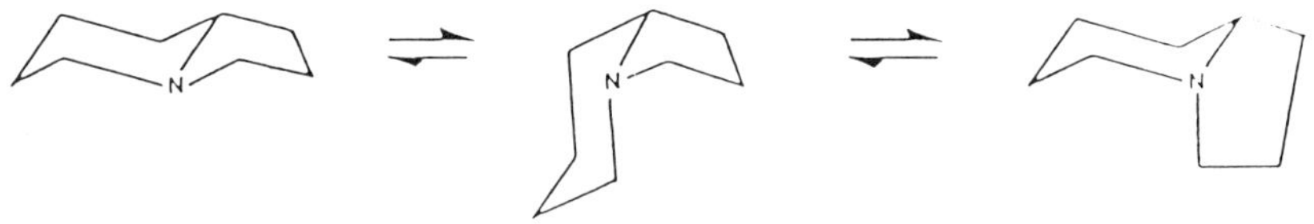

(456)   (457)   (458)

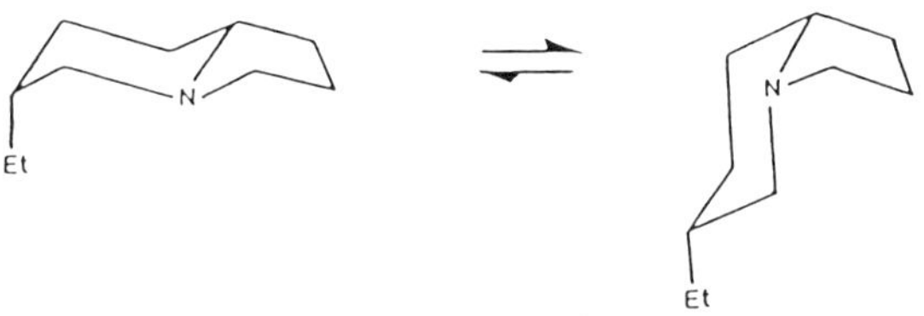

(459)   (460)

Such equilibria were not indicated by the ^{13}C-NMR spectra, and it would seem that the equilibrium **459** ⇌ **460** contains no less than 90% of **459** ($\Delta G°_{298} = 1.3$ kcal mol^{-1}). Taking the lower limit of 1.5 kcal mol^{-1} for the *gauche*-butane and *gauche*-propylamine interactions in **459** gives a minimum estimate for the indolizidine equilibrium of $\Delta G°_{298} = 2.8$ kcal mol^{-1} (>99.1% of **456**) (87MI1).

Only the trans-conformer (**461**) of perhydropyrazolo[1,2-*a*]pyrazine has been detected by ^{13}C-NMR spectroscopy (Section II), a result that parallels the strong conformational preference for the trans-conformer [$\Delta G°_{224}$ (*trans* ⇌ *cis*) = 2.4 kcal mol^{-1}) shown by the six-membered ring analogue **462** [84AHC(36)1; Section III,C,2].

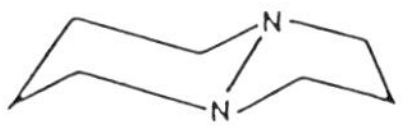

(461) 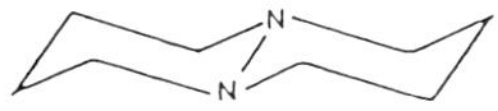   (462)

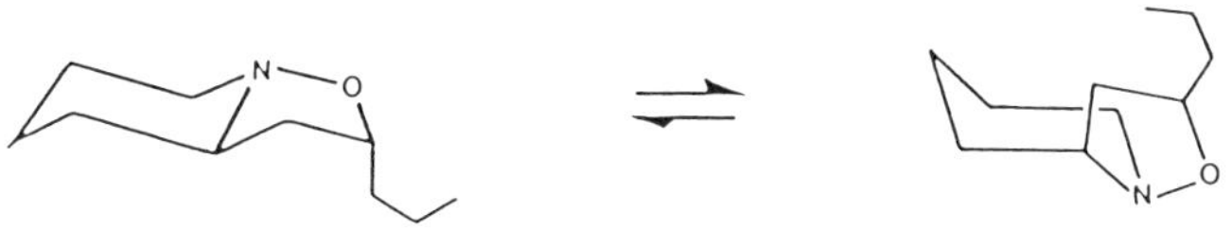

(463)   (464)

Turning to the perhydro-isoxazolo[2,3-*a*]pyridines (Section IV), the presence of the oxygen atom α to the bridgehead nitrogen is expected to raise the barrier to nitrogen inversion [80AG(E)521]. In line with this, a coalescence temperature of 35–40°C was observed for the **463** ⇌ **464** equilibrium (Section IV,B).

The conformational analysis of the perhydro-imidazo[1,5-*a*]pyridines (**465**) (Section VI), the perhydro-oxazolo[3,4-*a*]pyridines **466** (Section VII), and the perhydrothiazolo[3,4-*a*]pyridines (**467**) (Section VIII) may best be considered together since the conformational preferences of these systems are dominated by the generalized anomeric effect [83MI1;

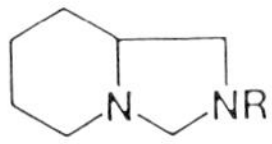

(465) 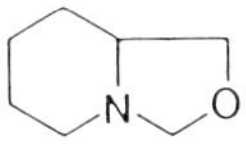   (466) 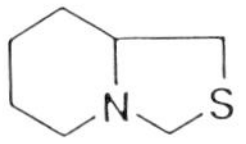   (467)

84AHC(36)1] and also involve differences in nonbonded interactions and ring fusion strain. Thus, comparison of the equilibrium positions for indolizidine with that for **466** shows a relative stabilization of the O-inside cis-conformer (**469**) (Scheme 5) of the 1,3-heterocycle. This stabilization is due to the favorable anomeric effect in **469** and the smaller oxygen/C-5 methylene interaction in **469** relative to the C-2/C-5 methylene interaction in the corresponding conformer **(457)** of indolizidine.

Examination of other systems (Scheme 5) shows the importance of the third major factor, ring fusion strain, in influencing the position of conformation equilibria in these types of systems. The change in position of conformational equilibrium from 73% trans-conformation in perhydro-oxazolo[3,4-*a*]pyridine **(468 ⇌ 469)** to 14% trans-conformation in perhydro-oxazolo[3,4-*d*][1,4]oxazine **(472 ⇌ 473)** [74JCS(P2)1419] may at first sight be attributed to a reduction in the magnitude of the indicated *gauche*-butane-type interaction in the cis-fused conformation, since Me/ heteroatom interactions are less severe than CH_2/CH_2 interactions. The swing back to 84% trans-conformer in the perhydro-oxazolo[3,4-*d*]- [1,4]thiazine equilibrium **(474 ⇌ 475)**, however, is inconsistent with this

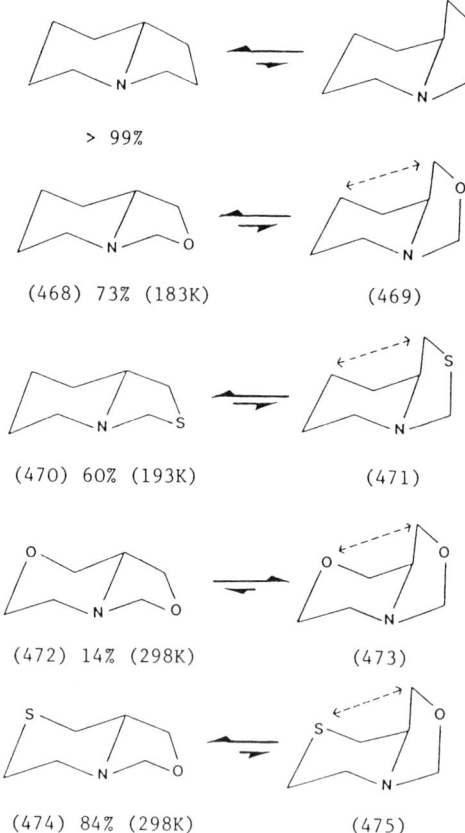

(468) 73% (183K)   (469)

(470) 60% (193K)   (471)

(472) 14% (298K)   (473)

(474) 84% (298K)   (475)

SCHEME 5. Conformational equilibria in 6/5 systems.

[76JCS(P2)203]. Since the C—S bond length (1.82 Å) is larger than the C—O length (1.43 Å), the indicated nonbonded interaction between the CH_2 and sulfur atom in **475** should be less than that between the CH_2 and oxygen atom in **473**. Thus, a shift towards the cis-conformer in the 1,4-thiazine system is expected. In order to explain the observed opposite direction of the shift, ring fusion strain in trans-fused 6/5 systems was proposed [76(JCS(P2)203]. Such trans-fusions can only be accomplished by distortion of the geometry present in the strain-free ring components, which results in an unfavorable puckering of the six-membered ring. The problem is much less severe in cis-fused six-membered to five-membered ring systems, since the ring fusion strain is accommodated by a slight flattening of the chair six-membered ring against a relatively soft potential barrier (65MI1). Thus, in fused 6/5 systems, the expectation is for a relative destablization of the trans-conformer by the ring fusion strain.

Following such an argument, the shifts in conformational equilibria (Scheme 5) may be rationalized by accepting greater ring-fusion strain in the trans-fused 1,4-oxazine than in the 1,4-thiazine as a result of the relative inability of the smaller oxygen-containing ring to incorporate the puckering caused by trans-fusion. The change in equilibrium position, on replacement of the oxygen atom in 466 by sulfur as in **467**, must represent a balance between a decrease in ring-fusion strain in **466** and an increase in the anomeric stabilization of the *cis*-fused perhydrothiazolo[3,4-*a*]pyridine **(471)** relative to the oxygen analogue **469**. Thus, perhydro-oxazolo-[3,4-*a*]pyridine, because of the short C—O bond length (1.43 Å), may be relatively more strained than *trans*-hydrindane (65MI1), whereas the longer C—S bond (1.82 Å) in perhydrothiazolo[3,4-*a*]pyridine will reduce such strain.

The increase in the anomeric stabilization of the S-inside cis-fused conformer of perhydrothiazolo[3,4-*a*]pyridine **(471)** is expected from a comparison of the equilibria positions of perhydropyrido[1,2-*c*]-[1,3]oxazine **(476 ⇌ 477)** (90% **476** at 298 K) and perhydropyrido[1,2-*c*]-[1,3]thiazine **(478 ⇌ 479)** (64% **478** at 198 K) [84AHC(36)1; Sections III,D,1 and 2]. In these systems, the trans-conformers are strain-free, so the position of equilibria reflects largely the change in anomeric stabilization.

(476) ⇌ (477)

(478) ⇌ (479)

In contrast to the equilibrium for **466** and **467**, the equilibrium in 2-methylperhydroimidazolo[1,5-*a*]pyridine (**465**; R = Me) shifts back towards the trans-conformer (> 98%). This shift must be due to stabilization of the trans-fused conformer by the generalized anomeric effect brought about by N-2 inversion, rather than by N-4 inversion as in **466** and **467**. This is supported by the position of equilibrium (83% **480** ⇌ 17% **481**) in **465** (R = Me) (87MI2).

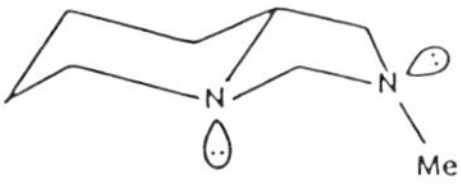

(480) 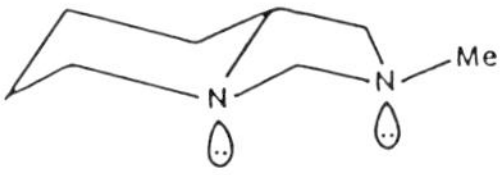   (481)

The conformational equilibria for the various alkyl substituted derivatives of **466** and **467** may also be rationalized in terms of nonbonded interactions and the generalized anomeric effect. Thus, the ring A alkyl substituted derivatives show heavy bias towards a particular conformation. Those isomers in which the substituent is equatorial in the trans-conformation, such as **482** and **483**, show no detectable amounts of

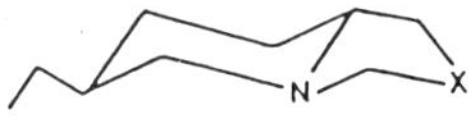

(482: X=O)
(483: X=S)

the cis-conformers. The epimer of **482** adopts an equilibrium containing ~3% trans-conformer (**484**) in equilibrium with 97% O-inside cis-fused conformer (**485**) at 190 K in CDCl$_3$ (88MI1). In this case, conformer **484** is disfavored by nonbonded interactions arising from the axial ethyl substituent and the generalized anomeric effect. Similar equilibrium positions have been observed for the sulfur analogues [88JCS(P1)1173].

Whereas the cis-(H-1,H-8a) 1-alkyl substituted derivates of **466** and **467** adopt exclusively the trans-conformers (**486, 487**) because of unfavorable

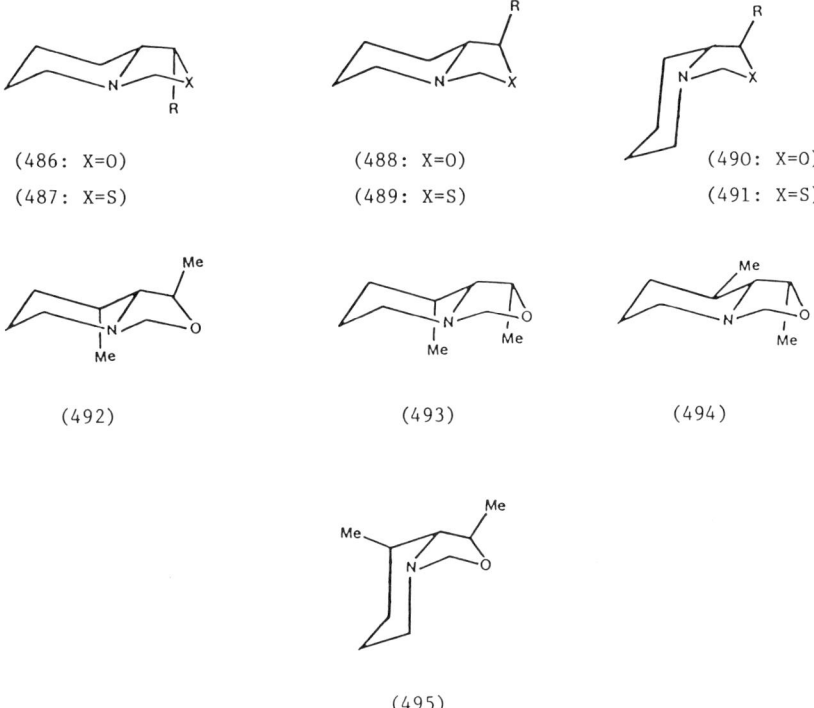

nonbonded interactions in the alternative cis-conformers, the epimers adopt 82% trans **488** ⇌ 18% cis **490** equilibria at 183 K in CDCl$_3$/CFCl$_3$ (88MI1) and 87% trans **489** ⇌ 13% cis **491** in CDCl$_3$ at 193K [88JCS(P1)1173].

Derivatives of **466** carrying alkyl substituents in each ring have also been studied (82OMR242). For example, the preferred conformation of the 1,8-dimethyl derivatives are shown in **492–495**. One isomer adopts the cis-conformation **(495)** and the rest adopt the trans-conformation, with **493** providing evidence of conformational distortion of the ring system.

The stereochemistry of protonated **466** shows important differences to that of the free base. Thus, protonation of a solution of **466** in ether by hydrogen chloride gas gave a mixture containing 28% trans- and 72% cis-salts **(496, 497)**. The ^{1}H-NMR spectrum of the cis-salt **(496)** in CDCl$_3$ showed an equilibrium **498** ⇌ **499** containing 24% of the O-outside cis-conformer **499**, whereas the corresponding O-outside cis-conformer of the free base has not been detected. This must be due to the stabilization of **499**

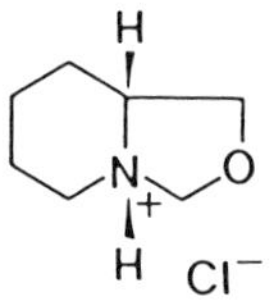

(496)

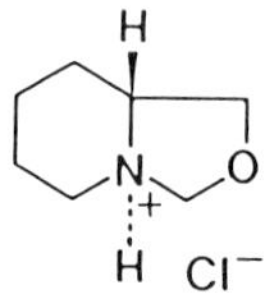

(497)

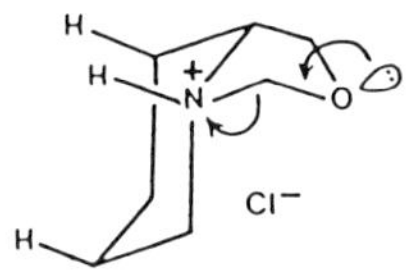

(498)

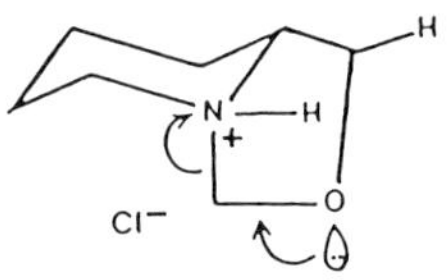

(499)

(relative to the situation in the free base conformers) by the generalized anomeric effect and the smaller number of nonbonded interactions relative to those in the O-inside cis-conformer introduced by protonation (88MI2).

The conformational preference of perhydropyrrolo[1,2-c][1,3]oxazine (Section X) for the O-inside cis-conformation **(501)** must be due to its stabilization relative to the trans-conformation **(500)** and the O-outside cis-conformation **(502)** by the generalized anomeric effect and the presence of ring fusion strain (see previous discussion) in the trans-conformer **(500)**. The nonbonded interaction between the oxygen atom

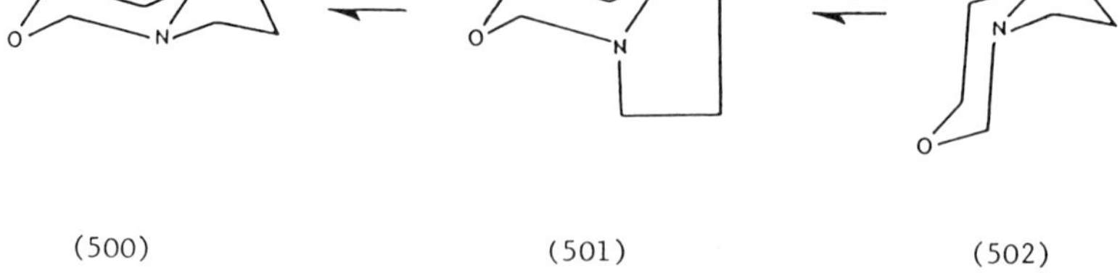

(500)   (501)   (502)

and the C-6 methylene in **501** must also be less than that between the C-4 and C-7 methylenes in **502**. The difference in position of conformational equilibria for this compound and indolizidine **(455)** parallels the difference between the equilibria for *N*-methyltetrahydro-1,3-oxazine and *N*-methylpiperidine [84AHC(36)1; Sections III,A,1 and III,D,1].

Perhydrothiazolo[3,2-*a*]pyridine (Section XIII) has been shown to adopt an equilibrium between 43% of the trans-fused conformer **(503)** and 57% of the cis-fused conformer **(504)** in CS$_2$ solution at 308 K (81OMR103), again showing the importance of the geometry of the 1,3-heterocyclic moiety on conformational equilibria positions.

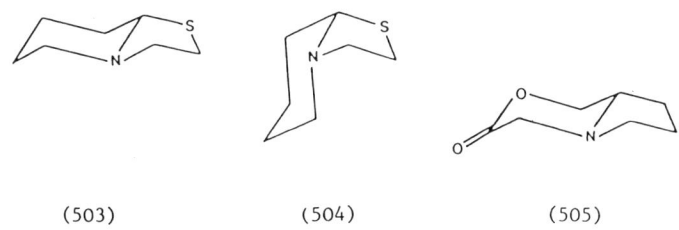

(503)   (504)   (505)

In systems described in Sections XIV–XVI, the 1,4-arrangement of heteroatoms is not expected to affect the relative importance of the trans-conformation shown by indolizidine. In line with this, the perhydropyrrolo[2,1-*c*][1,4]oxazin-3-one has been shown to adopt the trans-conformation **(505)** (74CC395) (see Section XV).

References

65MI1	E. L. Eliel, N. L. Allinger, S. H. Angyal, and G. A. Morrison, Conformational Analysis. Wiley (Interscience), London, 1965.
66FRP1441519	A. Etienne, A. Le Berre, and J. Godin, Fr. Pat. 1,441,519 (1966) [*CA* **66**, 10950 (1967)].
66FRP1441937	Farbwerke Hoechst A-G, Fr. Pat. 1,441,937 (1966) [*CA* **66**, 55492 (1967)].
66JHC418	T. A. Crabb and R. F. Newton, *J. Heterocycl. Chem.* **3**, 418 (1966).
67JGU1623	V. K. Antonov, L. I. Andreeva, A. G. Lyakisheva, and M. M. Shemyakin, *J. Gen. Chem. USSR (Engl. Transl.)* **37**, 1623 (1967).
67TL1039	D. Ghiringhelli and L. Bernardi, *Tetrahedron Lett.*, 1039 (1967).
67USP3334091	W. J. Houlihan, U.S. Pat. 3,334,091 (1967) [*CA* **69**, 86988 (1967)].
67USP3334099	W. J. Houlihan, U.S. Pat. 3,334,099 (1967) [*CA* **69**, 96769 (1967)].
68BSF4210	J. Godin and A. Le Berre, *Bull Soc. Chim. Fr.*, 4210 (1968).
68JOC2379	L. A. Paquette and M. K. Scott, *J. Org. Chem.* **33**, 2379 (1968).
68RTC11	H. C. Beyermen, L. Maat, A. Sinnema, and A. Van Veen, *Recl. Trav. Chim. Pays-Bas* **87**, 11 (1968).

68T1997	T. A. Crabb and R. F. Newton, *Tetrahedron* **24**, 1997 (1968).
68T2485	T. A. Crabb and R. F. Newton, *Tetrahedron* **24**, 2485 (1968).
68T6327	T. A. Crabb and R. F. Newton, *Tetrahedron* **24**, 6327 (1968).
68TL6191	H. S. Aaron and C. Ferguson, *Tetrahedron Lett.*, 6191 (1968).
68USP3375249	H. E. Johnson, U.S. Pat. 3,375,249 (1968) [*CA* **69**, 59222 (1968)].
68USP3414585	W. J. Houlihan, U.S. Pat. 3,414,585 (1968) [*CA* **70**, 68382 (1968)].
69BSF3139	A. Le Berre and C. Renault, *Bull. Soc. Chim. Fr.* **9**, 3139 (1969).
69CHE629	E. E. Mikhlina, N. A. Komarova, and M. V. Rubtsov, *Chem. Heterocycl. Compd. (Engl. Trans.)*, 629 (1969).
69JCS(C)408	D. Todd, R. J. Cornell, R. T. Wester, and T. A. Wittstruck, *J.Chem. Soc. C*, 408 (1969).
69JHC301	T. A. Crabb and R. F. Newton, *J. Heterocycl. Chem.* **6**, 301 (1969).
69JOC165	P. Aeberli and W. J. Houlihan, *J. Org. Chem.* **34**, 165 (1969).
69JOC2720	P. Aeberli and W. J. Houlihan, *J. Org. Chem.* **34**, 2720 (1969).
69JOC3181	B. T. Gillis and R. A. Izydore, *J. Org. Chem.* **34**, 3181 (1969).
69JPS1282	J. Sam, E. B. McLaurin, and R. M. Shafik, *J. Pharm. Sci.* **58**, 1282 (1969).
69LA(724)150	H. Stetter and P. Woernle, *Justus Liebigs Ann. Chem.* **724**, 150 (1969).
69TL371	B. Åkermark, N. G. Johanssen, and B. Sjöberg, *Tetrahedron Lett.*, 371 (1969).
69USP3420831	W. J. Houlihan, U.S. Pat. 3,420,831 (1969) [*CA* **70**, 68402 (1969)].
69USP3454585	W. J. Houlihan, U.S. Pat. 3,454,585 (1969) [*CA* **71**, 61387 (1969)].
70CB2394	L. Capuano, M. Welter, and R. Zander, *Chem. Ber.* **103**, 2394 (1970).
70FRP7076M	W. J. Houlihan, Fr. Pat. 7076M (1970) [*CA* **74**, 53873 (1970)].
70GEP1802468	H. Wollweber, Ger. Pat. 1,802,468 (1970) [*CA* **73**, 25543 (1970)].
70IJC707	D. K. Banerjee, T. K. Das Gupta, S. Mukerjee, and B. N. Mitra, *Indian J. Chem.* **8**, 707 (1970).
70JHC355	H. J. Beim and A. R. Day, *J. Heterocycl. Chem.* **7**, 355 (1970).
70JMC276	M. P. Mertes and A. J. Lin, *J. Med. Chem.* **13**, 276 (1970).
70JOC3818	E. E. Smissman, P. L. Chien, and R. A. Robinson, *J. Org. Chem.* **35**, 3818 (1970).
70USP3494926	A. J. Frey and R. E. Manning, U.S. Pat. 3,494,926 (1970) [*CA* **72**, 90474 (1970)].
70USP3496185	B. Loev, U.S. Pat 3,496,185 (1970) [*CA* **72**, 111462 (1970)].
70USP3497513	W. J. Houlihan, U.S. Pat. 3,497,513 (1970) [*CA* **73**, 35390 (1970)].
70USP3505340	A. J. Frey and R. E. Manning, U.S. Pat 3,505,340 (1970) [*CA* **73**, 3910 (1970)].
70USP3526626	W. J. Houlihan, U.S. Pat. 3,526,626 (1970) [*CA* **73**, 98976 (1970)].
70USP3526660	W. J. Houlihan, U.S. Pat. 3526660 (1970) [*CA* **73**, 120339 (1970)].
70USP3531485	M. E. Freed, U.S. Pat. 3,531,485 (1970) [*CA* **73**, 120672 (1970)].
71AP216	K. Winterfeld and G. Nair, *Arch. Pharm. (Weinheim, Ger.)* **304**, 216 (1971).
71AP774	H. Wollweber, J. Kurz, and W. Naegele, *Arch. Pharm. (Weinheim, Ger.)* **304**, 774 (1971).
71BRP1337269	Esso Research and Engineering Co., Br. Pat. 1,337,269 (1971) [*CA* **75**, 110318 (1971)].
71FES1017	E. Testa and L. Fontanella, *Farmaco, Ed. Sci.* **26**, 1017 (1971).
71JCS(B)1745	J. R. L. Smith and C. A. Audeh, *J. Chem. Soc. B*, 1745 (1971).
71USP3563992	M. R. Harnden, U.S. Pat. 3,563,992 (1971) [*CA* **75**, 20443 (1971)].
72AK2438	A. G. Terzyan, E. G. Vaganyan-Chilingaryan, and G. T. Tatevosyan, *Arm. Khim. Zh.* **25**, 438 (1972).

72JCS(P2)1920	T. A. Crabb and R. F. Newton, *J.C.S. Perkin 2*, 1920 (1972).
72JHC41	B. T. Gillis and R. A. Izydore, *J. Heterocycl. Chem.* **9**, 41 (1972).
72USP3663536	W. J. Houlihan, U.S. Pat. 3,663,536 (1972) [*CA* **77**, 62039 (1972)].
73AP284	H. Wollweber and R. Hiltmann, *Arch. Pharm. (Weinheim, Ger.)* **306**, 284 (1973).
73FES463	L. Fontanella, E. Occelli, and A. Perazzi, *Farmaco, Ed. Sci.* **28**, 463 (1973).
73OMR397	T. A. Crabb, P. J. Chivers, and R. F. Newton, *Org. Magn. Reson.* **5**, 397 (1973).
74CB2804	J. Haeusler and U. Schmidt, *Chem. Ber.* **107**, 2804 (1974).
74CB2816	U. Schmidt, A. Perco, and E. Oehler, *Chem. Ber.* **107**, 2816 (1974).
74CC210	Y. Takeuchi, P. J. Chivers, and T. A. Crabb, *J.C.S. Chem. Commun.*, 210 (1974).
74CC395	A. K. Ganguly, S. Szmulewicz, O. Z. Sarre, D. Greves, J. Morton, and J. McGlotten, *J.C.S. Chem. Commun.*, 395 (1974).
74CC743	T. Hasegawa and H. Aoyama, *J.C.S. Chem. Commun.*, 743 (1974).
74CL519	M. Iwakawa and J. Yoshimura, *Chem. Lett.*, 519 (1974).
74JA6982	S. F. Nelsen and J. M. Buschek, *J. Am. Chem. Soc.* **96**, 6982 (1974).
74JA6987	S. F. Nelsen and J. M. Buschek, *J. Am. Chem. Soc.* **96**, 6987 (1974).
74JCS(P2)1419	T. A. Crabb and M. J. Hall, *J.C.S. Perkin 2*, 1419 (1974).
74JMC519	M. P. LaMontagne, A. Markovac, and P. Blumbergs, *J. Med. Chem.* **17**, 519 (1974).
75BCJ610	M. Iwakawa and J. Yoshimura, *Bull. Chem. Soc. Jpn.* **48**, 610 (1975).
75BRP1409185	L. Fontanella and L. Mariani, Br. Pat. 1,409,185 (1975) [*CA* **81**, 37576 (1975)].
75CB1557	P. Rademacher and H. Koopman, *Chem. Ber.* **108**, 1557 (1975).
75CB2907	E. Oehler and U. Schmidt, *Chem. Ber.* **108**, 2907 (1975).
75CB2917	H. Poisel and U. Schmidt, *Chem. Ber.* **108**, 2917 (1975).
75CC844	M. J. T. Robinson, *J.C.S. Chem. Commun.*, 844 (1975).
75CR(C)953	G. Ricart, D. Couturier, and C. Glacet, *C.R. Hebd. Seances Acad. Sci., Ser. C* **280**, 953 (1975).
75H927	T. Kato, Y. Yamamoto, and M. Kondo, *Heterocycles* **3**, 927 (1975).
75JCS(P2)51	Y. Takeuchi, P. J. Chivers, and T. A. Crabb, *J.C.S. Perkin 2*, 51 (1975).
76ACS(B)383	N. G. Johanssen, B. Akermark, and B. Sjöberg, *Acta Chem. Scand., Ser. B* **B30**, 383 (1976).
76BRP1503244	O. Wakabayashi, K. Matsuya, H. Ohta, T. Jikihara, and H. Watanabe, Br. Pat. 1,503,244 (1976) [*CA* **85**, 160100 (1976)].
76CI(L)693	P. A. Crooks, R. H. B. Galt, and Z. S. Matusiak, *Chem. Ind. (London)*, 693 (1976).
76IJC(B)770	V. P. Arya, V. Honkan, and S. J. Shenoy, *Indian J. Chem., Sect. B* **14B**, 770 (1976).
76JA5269	S. F. Nelsen, V. Peacock, and G. R. Weisman, *J. Am. Chem. Soc.* **98**, 5269 (1976).
76JA5358	V. Madison, P. E. Young, and E. R. Blout, *J. Am. Chem. Soc.* **98**, 5358 (1976).
76JA5365	P. E. Young, V. Madison, and E. R. Blout, *J. Am. Chem. Soc.* **98**, 5365 (1976).
76JA7007	G. R. Weisman and S. F. Nelson, *J. Am. Chem. Soc.* **98**, 7007 (1976).
76JCS(P1)2054	T. Hasegawa, H. Aoyama, and Y. Omote, *J.C.S. Perkin 1*, 2054 (1976).
76JCS(P2)203	T. A. Crabb and M. J. Hall, *J.C.S. Perkin 2*, 203 (1976).
76JHC781	D. E. Portlock, A. C. Ghosh, W. E. Schwarzel, R. R. Kurtz, E. Boger,

	H. C. Dalzell, and R. K. Razdan, *J. Heterocycl. Chem.* **13,** 781 (1976).
76JMC436	P. Aeberli, J. H. Gogerty, W. J. Houlihan, and L. C. Iorio, *J. Med. Chem.* **19,** 436 (1976).
76MI1	H. Ulbrich, H. Priewe, and E. Schroeder, *Eur. J. Med. Chem.—Chim. Ther.* **11,** 343 (1976).
76SA(A)157	H. P. Koopman and P. Rademacher, *Spectrochim. Acta, Part A* **32A,** 157 (1976).
76TL3765	E. L. Eliel, D. Kandasamy, and W. R. Kenan, *Tetrahedron Lett.*, 3765 (1976).
76URP527433	V. A. Sedavkina and N. A. Morozova, U.S.S.R. Pat. 527,433 [*CA* **86,** 29869 (1976)].
76USP3958976	S. J. Goddard, U.S. Pat. 3,958,976 (1976) [*CA* **85,** 123923 (1976)].
77CB921	E. Oehler and U. Schmidt, *Chem. Ber.* **110,** 921 (1977).
77CSC553	G. Malmros and A. Wagner, *Cryst. Struct. Commun.* **6,** 553 (1977).
77JA1130	S. F. Nelsen, L. Echegoyen, E. L. Clennan, D. H. Evans, and D. A. Corrigan, *J. Am. Chem. Soc.* **99,** 1130 (1977).
77MI1	T. Kobayashi, Y. Iwano, and K. Hirai, *Sankyo Kenkyusho Nempo* **29,** 138 (1977).
77TL1005	S. Terashima and S. S. Jew, *Tetrahedron Lett.*, 1005 (1977).
78BRP1503244	Mitsubishi Chemical Industries Co., Ltd., Br. Pat. 1,503,244 (1978) [*CA* **93,** 46672 (1978)].
78CB361	J. Haeusler, R. Jahn, and U. Schmidt, *Chem. Ber.* **111,** 361 (1978).
78MI1	U. Pohl and H. Wollweber, *Eur. J. MEd. Chem.—Chim. Ther.* **13,** 127 (1978).
78MI2	H. Wollweber, U. Pohl, and K. Stoepel, *Eur. J. Med. Chem.—Chim. Ther.* **13,** 141 (1978).
79AF983	P. Kourounakis, W. H. Hunter, and P. G. Morris, *Arzneim.-Forsch.* **29,** 983 (1979).
79CHE85	G. V. Pavel and M. N. Tilichenko, *Chem. Heterocycl. Compd. (Engl. Transl.)*, 85 (1979).
79FRP2396757	Société d'Etudes Scientifiques et Industrielles de l'Ile-de-France, Fr. Pat. 2,396,757 (1979) [*CA* **92,** 163975 (1979)].
79GEP2844549	A. P. Skoldnikov, A. M. Likhosherstov, V. P. Peresada, N. V. Kaverina, M. I. Schmaryan, and A. G. Petukhov, Ger. Pat. 2,844,549 (1979) [*CA* **91,** 108003 (1979)].
79JA5343	H. Aoyama, T. Hasegawa, and Y. Omote, *J. Am. Chem. Soc.* **101,** 5343 (1979).
79MI1	T. D. Marieva, V. M. Kopelevich, and V. I. Gunar, *Chem. Nat. Compd. (Engl. Transl.)*, 93 (1979).
79MI2	T. D. Marieva, V. M. Kopelevich, V. V. Mishchenko, A. K. Starostina, L. Y. Yuzefovich, Z. K. Torosyan, and V. I. Gunar, *Chem. Nat. Compd. (Engl. Transl.)*, 328 (1979).
79T2337	S. S. Jew, S. Terashima, and K. Koga, *Tetrahedron* **35,** 2337 (1979).
79T2345	S. S. Jew, S. Terashima, and K. Koga, *Tetrahedron* **35,** 2345 (1979).
79TL809	S. Yoshifuju, K. Tanaka, and Y. Arata, *Tetrahedron Lett.*, 809 (1979).
79TL1347	J. A. M. Hamersma, H. E. Schoemaker, and W. N. Speckamp, *Tetrahedron Lett.*, 1347 (1979).
79USP4138242	S. J. Goddard, U.S. Pat. 4,138,242 (1979) [*CA* **91,** 39478 (1979)].
80AG(E)521	A. R. Katritzky, R. C. Patel, and F. G. Riddell, *Angew. Chem., Int. Ed. Engl.* **20,** 521 (1980).

80BCJ3240	T. Ota, S. Masuda, and M. Kido, *Bull. Chem. Soc. Jpn.* **53**, 3240 (1980).
80BRP2025936	A. P. Skoldnikov, A. M. Likhosherstov, V. P. Peresada, M. I. Schmaryan, and A. G. Petukhov, Br. Pat. 2,025,936 (1980) [*CA* **93**, 186406 (1980)].
80BRP2036743	M. A. Ondetti, Br. Pat. 2,036,743 (1980) [*CA* **94**, 156944 (1980)].
80CR(C)291	M. T. Cung, G. Boussard, B. Vitoux, and M. Marraud, *C.R. Hebd. Seances Acad. Sci., Ser. C* **290**, 291 (1980).
80EUP17390	D. H. Kim, Eur. Pat. Appl. 17390 (1980) [*CA* **94**, 121618 (1980)].
80EUP42100	C. H. Hassall, G. Lawton, and C. J. Moody, Eur. Pat. Appl. 42100 (1980) [*CA* **96**, 199715 (1980)].
80H1089	M. Okita, T. Wakamatsu, M. Mori, and Y. Ban, *Heterocycles* **14**, 1089 (1980).
80JOC3609	S. F. Nelsen and W. C. Hollinsed, *J. Org. Chem.* **45**, 3609 (1980).
80JOC5325	H. K. Hall and A. El-Shekeil, *J. Org. Chem.* **45**, 5325 (1980).
80MI1	H. Wollweber, R. Hiltmann, K. Stoepel, and H. G. Kroneberg, *Eur. J. Med. Chem.—Chim. Ther.* **15**, 111 (1980).
80OMR63	T. A. Crabb and P. A. Jupp, *Org. Magn. Reson.* **13**, 63 (1980).
80OMR159	T. A. Crabb and P. A. Jupp, *Org. Magn. Reson.* **13**, 159 (1980).
80S387	D. Barbry, D. Couturier, and G. Ricart, *Synthesis,* 387 (1980).
80TL2733	S. Terashima, M. Hayashi, and K. Koga, *Tetrahedron Lett.* **21**, 2733 (1980).
80TL3493	H. H. Wasserman, R. P. Robinson, and H. Matsuyama, *Tetrahedron Lett.* **21**, 3493 (1980).
80USP4209617	R. L. White, U.S. Pat 4,209,617 (1980) [*CA* **94**, 15753 (1980)].
81EGP151166	C. Rueger, A. Rostock, H. Roehnert, A. P. Skoldnikov, K. S. Raevskii, A. M. Likhosherstov, and A. W. Stavrovskaya, Ger. (East) Pat. 151,166 (1981) [*CA* **96**, 162757 (1981)].
81EUP31042	W. Rohr, H. Hansen, P. Plath, and B. Wuerzer, Eur. Pat. Appl. 31042 (1981) [*CA* **95**, 150697 (1981).
81JA461	H. H. Wasserman and H. Matsuyama, *J. Am. Chem. Soc.* **103**, 461 (1981).
81JA7573	B. Nader, T. R. Bailey, R. W. Franck, and S. M. Weinreb, *J. Am. Chem. Soc.* **103**, 7573 (1981).
81KFZ55	A. M. Likhosherstov, Z. P. Senova, A. V. Stavrovskaya, N. V. Kaverina, and A. P. Skoldnikov, *Khim. Farm. Zh.* **15**, 55 (1981).
81M1017	E. Goessinger, *Monatsh. Chem.* **112**, 1017 (1981).
81OMR103	D. Barbry, G. Ricart, and D. Couturier, *Org. Magn. Reson.* **17**, 103 (1981).
81T2797	M. Hayashi, S. Terashima, and K. Koga, *Tetrahedron* **37**, 2797 (1981).
81T(S9)191	H. H. Wasserman, M. S. Wolff, K. Stiller, I. Saito, and J. E. Pickett, *Tetrahedron, Suppl.* **9**, 191 (1981).
81TL4197	R. Fujimoto and Y. Kishi, *Tetrahedron Lett.* **22**, 4197 (1981).
81USP4260623	H. A. De Wald, U.S. Pat. 4,260,623 (1981) [*CA* **95**, 62173 (1981).
82CCC3312	J. Stuchlik, A. Krajicek, L. Cvac, J. Spacil, P. Sedmera, M. Flieger, J. Vokoun, and Z. Rehacek, *Collect. Czech. Chem. Commun.* **47**, 3312 (1982).
82CSC(A)1487	G. Lucente, F. Pinne, G. Zanotti, S. Cerrini, W. Fedeli, and F. Mazza, *Cryst. Struct. Commun.* **11**(4,Pt A), 1487 (1982).
82EUP47620	M. S. Hadley and F. D. King, Eur. Pat. Appl. 47620 (1982) [*CA* **97**, 23814 (1982)].

82EUP54593	C. Bernhart, J. P. Gagnol and P. Gautier, Eur. Pat. Appl. 54593 (1982) [*CA* **97**, 72250 (1982).
82EUP57536	F. D. King and M. S. Hadley, Eur. Pat. Appl. 57536 (1982) [*CA* **98**, 143462 (1982)].
82EUP65724	P. Plath, W. Hoffmann, V. Schwendemann, and B. Wuerzer, Eur. Pat. Appl. 65724 (1982) [*CA* **98**, 126150 (1982)].
82FRP2492374	H. Demarne, C. Bernhart, and M. Serre, Fr. Pat. 2,492,374 (1982) [*CA* **97**, 127492 (1982)].
82FRP2493314	N. V. Kaverina, V. V. Lyskovtsev, Z. P. Senova, A. N. Gritsenko, Z. I. Ermakova, A. P. Skoldnikov, E. Carstens, H. Wunderlich, and S. Andreas, Fr. Pat. 2,493,314 (1982) [*CA* **97**, 162853 (1982)].
82H2057	M. Machida, S. Oyadomari, H. Takechi, K. Ohno, and Y. Kanaoka, *Heterocycles* **19**, 2057 (1982).
82MI1	C. H. Hassall, A. Krohn, C. J. Moody, and W. A. Thomas, *FEBS Lett.* **147**, 175 (1982).
82OMR113	T. A. Crabb, R. F. Newton, and J. Rouse, *Org. Magn. Reson.* **20**, 113 (1982).
82OMR242	T. A. Crabb and G. C. Heywood, *Org. Magn. Reson.* **20**, 242 (1982).
82T3255	J. A. M. Hamersma and W. N. Speckamp, *Tetrahedron* **38**, 3255 (1982).
82TL4181	R. W. Alder, P. Eastment, R. E. Moss, R. B. Sessions, and M. A. Stringfellow, *Tetrahedron Lett.* **23**, 4181 (1982).
82TL5039	M. P. Crozet, M. Kaafarani, W. Kassar, and J. M. Surzur, *Tetrahedron Lett.* **23**, 5039 (1982).
82USP4339579	M. E. Freed, U.S. Pat. 4,339,579 (1982) [*CA* **97**, 182458 (1982)].
83AP47	H. Moehrle and B. Schmidt, *Arch. Pharm. (Weinheim, Ger.)* **316**, 47 (1983).
83AP339	J. Lehmann, *Arch. Pharm. (Weinheim, Ger.)*, **316**, 339 (1983).
83BCJ487	T. Ota, S. Masuda, H. Tanaka, Y. Inazawa, and M. Kido, *Bull. Chem. Soc. Jpn.* **56**, 487 (1983).
83BRP2105707	M. S. Hadley and F. D. King, Br. Pat. 2,105,707 (1983) [*CA* **99**, 70747 (1983)].
83BRP2128984	M. R. Attwood, C. H. Hassall, R. W. Lambert, G. Lawton, and S. Redshaw, Br. Pat. 2,128,984 (1983) [*CA* **100**, 139158 (1983)].
83CC1143	H. Iida, M. Tanaka, and C. Kibayashi, *J.C.S. Chem. Commun.*, 1143 (1983).
83CHE622	A. A. Potekhin, V. V. Sokolov, K. A. Oglobin, and S. M. Esakov, *Chem. Heterocycl. Compd. (Engl. Transl.)*, 622 (1983).
83CHE1090	T. G. Nikolaeva, P. V. Reshetov, A. P. Kriven'ko, and V. G. Kharchenko, *Chem. Heterocycl. Compd. (Engl. Transl.)*, 1090 (1983).
83EUP70389	E. Nagano, S. Hashimoto, R. Yoshido, H. Matsumoto, and K. Kamoshita, Eur. Pat. Appl. 70389 (1983) [*CA* **99**, 5630 (1983)].
83EUP94095	M. R. Attwood, C. H. Hassall, R. W. Lambert, G. Lawton, and S. Redshaw, Eur. Pat. Appl. 94095 (1983) [*CA* **100**, 139158 (1983)].
83IJC(B)664	U. K. Shukla, J. M. Khanna, S. Sharma, N. Anand, R. K. Chatterjee, and A. B. Sen, *Indian J. Chem., Sect B* **22B**, 664 (1983).
83JA654	K. A. Parker and R. O'Fee, *J. Am. Chem. Soc.* **105**, 654 (1983).
83JA7754	L. Guerrier, J. Royer, D. S. Grierson, and H. P. Husson, *J. Am. Chem. Soc.* **105**, 7754 (1983).

83JCS(P1)2857	J. D. Coyle and L. R. B. Bryant, *J.C.S. Perkin 1*, 2857 (1983).
83MI1	P. Deslongchamps, "Stereoelectronic Effects in Organic Chemistry." Pergamon, Oxford, 1983.
83OMR203	Y. Takeuchi and T. A. Crabb, *Org. Magn. Reson.* **21**, 203 (1983).
83PHA512	H. Mohrle and C. Kamper, *Pharmazie* **38**, 512 (1983).
83TL1559	S. J. Croker, R. S. T. Loeffler, T. A. Smith, and R. B. Sessions, *Tetrahedron Lett.* **24**, 1559 (1983).
83TL4577	T. Shono, Y. Matsumura, and T. Kanazawa, *Tetrahedron Lett.* **24**, 4577 (1983).
83USP4400511	M. E. Freed, U.S. Pat. 4,400,511 (1983) [*CA* **100**, 34562 (1983)].
83USP4414389	M. E. Freed, U.S. Pat. 4,414,389 (1983) [*CA* **100**, 121108 (1983)].
84ACS(B)526	S. Brandaenge, L. H. Eriksson, and B. Rodriguez, *Acta Chem. Scand., Ser. B* **B38**, 526 (1984).
84AHC(36)1	T. A. Crabb and A. R. Katritzky, *Adv. Heterocycl. Chem.* **36**, 1 (1984).
84CPB3892	S. Takano and K. Shishido, *Chem. Pharm. Bull.* **32**, 3892 (1984).
84EUP104532	S. Shimano, S. Kobayashi, M. Yanagi, O. Yamada, M. Saito, and F. Futatsuya, Eur. Pat. Appl. 104532 (1984) [*CA* **101**, 110916 (1984)].
84FES718	E. Occelli, L. Mariani, L. Fontanella, and N. Corsico, *Farmaco, Ed. Sci.* **39**, 718 (1984).
84JCS(P1)1	B. Ringdahl, A. R. Pinder, W. E. Pereira, N. J. Oppenheimer, and J. Cymerman-Craig, *J.C.S. Perkin 1*, 1 (1984).
84JCS(P1)155	C. H. Hassall, A. Krohn, C. J. Moody, and W. A. Thomas, *J.C.S. Perkin 1*, 155 (1984).
84JOC1891	S. F. Nelsen, *J. Org. Chem.* **49**, 1891 (1984).
84JOC1909	H. Iida, M. Tanaka, and C. Kibayashi, *J. Org. Chem.* **49**, 1909 (1984).
84JOC2412	H. Iida, Y. Watanabe, M. Takano, and C. Kibayashi, *J. Org. Chem.* **49**, 2412 (1984).
84JOC3812	Z-K. Liao and H. Kohn, *J. Org. Chem.* **49**, 3812 (1984).
84KFZ1445	L. S. Nazarova, Y. B. Rozonov, A. M. Likhosherstov, T. V. Morozova, A. P. Skoldnikov, N. V. Kaverina, and V. A. Markin, *Khim. Farm. Zh.* **18**, 1445 (1984).
84MI1	J. J. Tufariello, in "1,3 Dipolar Cycloaddition Chemistry" (A. Padwa, ed.), Vol 2, p. 83. Wiley, New York, 1984.
84MI2	U. Nagai and K. Sato, *Pept. Chem.*, 207 (1984).
84MI3	S. Cerrini, W. Fedeli, G. Lucente, F. Mazza, F. Pinnen, and G. Zanotti, *Int. J. Pept. Protein Res.* **23**, 223 (1984).
84PHA812	A. P. Skoldnikov, K. S. Raevski, A. M. Likhosherstov, A. V. Stavrovskaya, C. Ruerger, A. Rostock, and H. Roehnert, *Pharmazie* **39**, 812 (1984).
85BSB187	J. J. M. Sleeckx and M. J. O. Anteunis, *Bull. Soc. Chim. Belg.* **94**, 187 (1985).
85EUP135837	P. Traxler, Eur. Pat. Appl. 135837 (1985) [*CA* **104**, 109354 (1985)].
85HCA724	R. O. Day, V. W. Day, D. M. Wheeler, P. A. Stadler, and H. R. Loosli, *Helv. Chim. Acta* **68**, 724 (1985).
85JA7776	A. I. Meyers, R. Hanreich, and K. T. Wanner, *J. Am. Chem. Soc.* **107**, 7776 (1985).
85JAP(K)97980	Mitsubishi Chemical Industries Co. Ltd., Jpn. Kokai 85/97980 [*CA* **103**, 141961 (1985)].
85JOC670	J. Royer and H. P. Husson, *J. Org. Chem.* **50**, 670 (1985).

85JOC3243	T. Shono, Y. Matsumura, K. Uchida, and H. Kobayashi, *J. Org. Chem.* **50**, 3243 (1985).
85MI1	A. Calgani, F. Mazza, G. Pochetti, D. Rossi, and G. Lucente, *Int. J. Pept. Protein Res.* **26**, 166 (1985).
85MI2	B. K. Sathyanarayana and J. Applequist, *Int. J. Pept. Protein Res.* **26**, 518 (1985).
85T2007	P. N. W. Van der Vliet, J. A. M. Hamersma, and W. N. Speckamp, *Tetrahedron* **41**, 2007 (1985).
85T2861	J. A. M. Hamersma and W. N. Speckamp, *Tetrahedron* **41**, 2861 (1985).
85TL647	U. Nagai and K. Sato, *Tetrahedron Lett.* **26**, 647 (1985).
85TL1515	J. Royer and H. P. Husson, *Tetrahedron Lett.* **26**, 1515 (1985).
85TL3167	C. J. Moody and C. J. Pearson, *Tetrahedron Lett.* **26**, 3167 (1985).
85TL3803	V. Ratovelomanana, J. Royer, and H. P. Husson, *Tetrahedron Lett.* **26**, 3803 (1985).
85TL5481	G. Zanotti, F. Pinnen, and G. Lucente, *Tetrahedron Lett.* **26**, 5481 (1985).
85TL6249	D. Lathbury and T. Gallagher, *Tetrahedron Lett.* **26**, 6249 (1985).
85USP4551450	P. Traxler, U.S. Pat. 4,551,450 (1985) [*CA* **105**, 24120 (1985).
86CC741	M. Lemaire, A. Guy, D. Imbert, and J. P. Guette, *J.C.S. Chem. Commun.*, 741 (1986).
86CC1287	W. Carruthers and M. J. Williams, *J.C.S. Chem. Commun.*, 1287 (1986).
86CC1717	S. Kano, Y. Yuasa, T. Yokomatsu, K. Asami, and S. Shibuya, *J.C.S. Chem. Commun.*, 1717 (1986).
86CPB105	Y. Terao, Y. Yasumoto, K. Ikeda, and M. Sekiya, *Chem. Pharm. Bull.* **34**, 105 (1986).
86CPB3142	H. Takechi, S. Tateuchi, M. Machida, Y. Nishibata, K. Aoe, Y. Sato, and Y. Kanaoka, *Chem. Pharm. Bull.* **34**, 3142 (1986).
86EUP190105	W. Foery, A. Nyffeler, H. R. Gerber, and H. Martin, Eur. Pat. Appl. 190105 (1986) [*CA* **105**, 166904 (1986)].
86EUP198348	P. F. Corey, Eur. Pat. Appl. 198348 (1986) [*CA* **106**, 32692 (1986)].
86EUP198352	P. F. Corey, Eur. Pat. Appl. 198352 (1986) [*CA* **106**, 32693 (1986)].
86H621	S. Kano, T. Yokomatsu, Y. Yuasa, and S. Shibuya, *Heterocycles* **24**, 621 (1986).
86H2835	K. Takahashi, A. Tachiki, K. Ogura, and H. Iida, *Heterocycles* **24**, 2835 (1986).
86JAP(K)27985	S. Matsumoto, T. Yotsuaya, H. Hanabe, and S. Suzuki, Jpn. Kokai 86/27985 [*CA* **104**, 296285 (1986)].
86JCS(P1)1733	T. Okawara, K. Nakayama, Y. Honda, T. Yamasaki, and M. Furukawa, *J.C.S. Perkin 1*, 1733 (1986).
86JMC784	J. W. Skiles, J. T. Suh, B. E. Williams, P. R. Mennard, J. N. Barton, B. Loev, H. Jones, E. S. Neiss, A. Schwab, W. S. Maun, A. Khandwala, P. S. Wolf, and I. Weinryla, *J. Med. Chem.* **29**, 784 (1986).
86JMC1814	C. Botre, F. Botre, G. Jommi, and R. Signorini, *J. Med. Chem.* **29**, 1814 (1986).
86JOC1936	A. I. Meyers, B. A. Lefker, K. T. Wanner, and R. A. Aitken, *J. Org. Chem.* **51**, 1936 (1986).
86JOC3290	K. Saoi and M. Ishizaki, *J. Org. Chem.* **51**, 3290 (1986).
86JOC4475	D. S. Grierson, J. Royer, L. Guerrier, and H. P. Husson, *J. Org. Chem.* **51**, 4475 (1986).

86LA859	H. Takechi, M. Machida, and Y. Kanaoka, *Liebigs Ann. Chem.* 859 (1986).
86MI1	T. G. Bird, K. Moschner, M. H. Robert, J. Collard-Motte, Z. Janousek, R. Merenyi, and H. G. Viehe, *Croat. Chem. Acta* **59**, 51 (1986).
86MI2	J. M. M. Smits, P. T. Beurskens, B. Zeegers and H. C. J. Ottenheijm, *J. Crystallogr. Spectrosc. Res.* **16**, 747 (1986).
86RTC360	J. J. G. S. Van Es, J. H. Koek, C. Erkelens, and J. Lugtenberg, *Recl. Trav. Chim. Pays-Bas* **105**, 360 (1986).
86TL1489	J. J. Tufariello and J. M. Puglis, *Tetrahedron Lett.* **27**, 1489 (1986).
86TL1727	A. Brandi, A. Guarna, A. Goti, and F. De Sarlo, *Tetrahedron Lett.* **27**, 1727 (1986).
86TL2683	A. Padwa, S. P. Carter, U. Chiacchio, and D. N. Kline, *Tetrahedron Lett.* **27**, 2683 (1986).
86TL5085	M. A. Ciufolini and C. Y. Wood, *Tetrahedron Lett.* **27**, 5085 (1986).
86USP4596819	E. D. Nicolaides, F. J. Tinney, J. S. Kaltenbronn, D. E. DeJohn, E. A. Lunney, W. H. Roark, and J. T. Repine, U.S. Pat. 4,596,819 (1986) [*CA* **105**, 153560 (1986)].
87BSB57	C. Hootele, W. Ibebeke-Bomangwa, F. Driessens, and S. Sabil, *Bull. Soc. Chim. Belg.* **96**, 57 (1987).
87CC125	S. Murahashi, Y. Mitsue, and K. Ike, *J.C.S. Chem. Commun.*, 125 (1987).
87GEP3600903	H. Scholz, A. Zeidler, B. Wuerzer, and N. Meyer, Ger. Pat. 3,600,903 (1987) [*CA* **107**, 154348 (1987)].
87H85	H. Matsuyama, M. Kobayashi, and H. H. Wasserman, *Heterocycles* **26**, 85 (1987).
87H755	M. Yamamoto, T. Suenaga, K. Suzuki, K. Naruchi, and K. Yamada, *Heterocycles* **26**, 755 (1987).
87H917	K. T. Wanner and A. Kaertner, *Heterocycles* **26**, 917 (1987).
87JAP(K)158280	T. Haga, H. Nagano, K. Morita, and M. Sato, Jpn. Kokai 87/158280 [*CA* **107**, 217620 (1987)].
87JCS(P1)885	G. Lawton, C. J. Moody, C. J. Pearson, and D. J. Williams, *J.C.S. Perkin 1*, 885 (1987).
87JMC498	D. E. Butler, J. D. Leonard, B. W. Caprathe, Y. J. L'Italien, M. R. Pavia, F. M. Hershenson, P. H. Poschel, and J. G. Marriott, *J. Med. Chem.* **30**, 498 (1987).
87JOC418	R. E. TenBrink, *J. Org. Chem.* **52**, 418 (1987).
87MI1	L. Banting, T. A. Crabb, and A. N. Trethewey, *Magn. Reson. Chem.* **25**, 352 (1987).
87MI2	L. Banting and T. A. Crabb, *Magn. Reson. Chem.* **25**, 696 (1987).
87TL755	A. Padwa, Y. Tomioka, and M. K. Venkatramanan, *Tetrahedron Lett.* **28**, 755 (1987).
87TL1573	T. L. Gilchrist, D. Hughes, and R. Wasson, *Tetrahedron Lett.* **28**, 1573 (1987).
87TL2801	P. F. Corey, *Tetrahedron Lett.* **28**, 2801 (1987).
88JCS(P1)1173	T. A. Crabb, A. N. Trethewey, and Y. Takeuchi, *J.C.S. Perkin 1*, 1173 (1988).
88MI1	T. A. Crabb, A. N. Trethewey, and Y. Takeuchi, *Magn. Reson. Chem.* **26**, 345 (1988).
88MI2	T. A. Crabb and A. N. Trethewey, *Magn. Reson. Chem.* **26**, 748 (1988).

Synthesis of Condensed 1,2,4-Triazolo[3,4-z]Heterocycles

MOHAMMED A. E. SHABAN AND ADEL Z. NASR

Department of Chemistry
Faculty of Science,
Alexandria University, Alexandria, Egypt

I. Introduction 279
II. Condensed 1,2,4-Triazolo-azoles 280
   A. Pyrrolo[2,1-c]1,2,4-triazoles 281
   B. 1,2,4-Triazolo[4,3-a]indoles 282
   C. 1,2,4-Triazolo[3,4-a]isoindoles 283
III. Condensed 1,2,4-Triazolo-diazoles 283
   A. 1,2,4-Triazolo-1,2-diazoles 283
     1. Pyrazolo[5,1-c]1,2,4-triazoles 283
     2. 1,2,4-Triazolo[4,3-b]indazoles 287
   B. 1,2,4-Triazolo-1,3-diazoles 287
     1. Imidazolo[2,1-c]1,2,4-triazoles 288
     2. Imidazolo[5,1-c]1,2,4-triazoles 288
     3. 1,2,4-Triazolo[4,3-a]benzimidazoles 289
IV. Condensed 1,2,4-Triazolo-oxazoles 290
   1,2,4-Triazolo-1,3-oxazoles 290
   1,2,4-Triazolo[3,4-b]1,3-benzoxazoles 291
V. Condensed 1,2,4-Triazolo-thiazoles 292
   A. 1,2,4-Triazolo-1,2-thiazoles 292
     1,2,4-Triazolo[4,3-b]1,2-benzothiazoles 292
   B. 1,2,4-Triazolo-1,3-thiazoles 293
     1. 1,3-Thiazolo[2,3-c]1,2,4-triazoles 293
     2. 1,2,4-Triazolo[3,4-a]1,3-benzothiazoles 297
VI. Condensed 1,2,4-Triazolo-selenazoles 299
   1,2,4-Triazolo-1,3-selenazoles 299
   1,2,4-Triazolo[3,4-b]1,3-benzoselenazoles 300
VII. Condensed 1,2,4-Triazolo-triazoles 300
   1,2,4-Triazolo-1,2,4-triazoles 300
     1. 1,2,4-Triazolo[4,3-b]1,2,4-triazoles 301
     2. 1,2,4-Triazolo[3,4-c]1,2,4-triazoles 304
VIII. Condensed 1,2,4-Triazolo-oxadiazoles 305
   1,2,4-Triazolo-1,3,4-oxadiazoles 305
   1,2,4-Triazolo[3,4-b]-1,3,4-oxadiazoles 305
IX. Condensed 1,2,4-Triazolo-thiadiazoles 307
   1,2,4-Triazolo-1,3,4-thiadiazoles 307
   1,2,4-Triazolo[3,4-b]1,3,4-thiadiazoles 307

X. Condensed 1,2,4-Triazolo-tetrazoles	309
1,2,4-Triazolo[4,3-d]tetrazoles	309
XI. Condensed 1,2,4-Triazolo-azines	310
A. 1,2,4-Triazolo[4,3-a]pyridines	311
B. 1,2,4-Triazolo[4,3-a]quinolines	314
C. 1,2,4-Triazolo[3,4-a]isoquinolines	316
D. 1,2,4-Triazolo[4,3-b]isoquinolines	317
XII. Condensed 1,2,4-Triazolo-diazines	318
A. 1,2,4-Triazolo-1,2-diazines	318
1. 1,2,4-Triazolo[4,3-b]pyridazines	318
2. 1,2,4-Triazolo[3,4-a]phthalazines	322
3. 1,2,4-Triazolo[4,3-b]cinnolines	325
B. 1,2,4-Triazolo-1,3-diazines	326
1. 1,2,4-Triazolo[4,3-a]pyrimidines	327
2. 1,2,4-Triazolo[4,3-c]pyrimidines	332
3. 1,2,4-Triazolo[4,3-a]quinazolines	332
4. 1,2,4-Triazolo[3,4-b]quinazolines	334
5. 1,2,4-Triazolo[4,3-c]quinazolines	335
C. 1,2,4-Triazolo-1,4-diazines	337
1. 1,2,4-Triazolo[4,3-a]pyrazines	337
2. 1,2,4-Triazolo[4,3-a]quinoxalines	339
XIII. Condensed 1,2,4-Triazolo-oxazines	341
A. 1,2,4-Triazolo-1,3-oxazines	341
1. 1,2,4-Triazolo[3,4-b]1,3-oxazines	342
2. 1,2,4-Triazolo[3,4-b]1,3-benzoxazines	342
B. 1,2,4-Triazolo-1,4-oxazines	343
1. 1,2,4-Triazolo[3,4-c]1,4-oxazines	343
2. 1,2,4-Triazolo[3,4-c]1,4-benzoxazines	344
XIV. Condensed 1,2,4-Triazolo-thiazines	344
A. 1,2,4-Triazolo-1,3-thiazines	344
1. 1,2,4-Triazolo[3,4-b]1,3-thiazines	345
2. 1,2,4-Triazolo[4,3-c]1,3-benzothiazines	346
B. 1,2,4-Triazolo-1,4-thiazines	346
1,2,4-Triazolo[3,4-c]1,4-benzothiazines	346
XV. Condensed 1,2,4-Triazolo-triazines	347
A. 1,2,4-Triazolo-1,2,3-triazines	347
1,2,4-Triazolo[4,3-c]1,2,3-benzotriazines	347
B. 1,2,4-Triazolo-1,2,4-triazines	348
1. 1,2,4-Triazolo[4,3-b]1,2,4-triazines	348
2. 1,2,4-Triazolo[3,4-c]1,2,4-triazines	350
3. 1,2,4-Triazolo[3,4-c]1,2,4-benzotriazines	352
4. 1,2,4-Triazolo[4,3-d]1,2,4-triazines	352
5. 1,2,4-Triazolo[3,4-f]1,2,4-triazines	354
C. 1,2,4-Triazolo-1,3,5-triazines	355
1,2,4-Triazolo[4,3-a]1,3,5-triazines	356
XVI. Condensed 1,2,4-Triazolo-thiadiazines	358
1,2,4-Triazolo-1,3,4-thiadiazines	358
1. 1,2,4-Triazolo[3,4-b]1,3,4-thiadiazines	359
2. 1,2,4-Triazolo[3,4-b]1,3,4-benzothiadiazines	361

XVII. Condensed 1,2,4-Triazolo-tetrazines 361
    1,2,4-Triazolo-1,2,4,5-tetrazines 361
        1,2,4-Triazolo[4,3-b]1,2,4,5-tetrazines 362
XVIII. Condensed 1,2,4-Triazolo-azepines 362
    A. 1,2,4-Triazolo[4,3-a]azepines 363
    B. 1,2,4-Triazolo[4,3-a]benzazepines 363
XIX. Condensed 1,2,4-Triazolo-diazepines 364
    1,2,4-Triazolo-benzodiazepines 364
        1. 1,2,4-Triazolo[4,3-a]1,4-benzodiazepines 365
        2. 1,2,4-Triazolo[4,3-d]1,4-benzodiazepines 366
        3. 1,2,4-Triazolo[4,3-a]1,5-benzodiazepines 366
XX. Condensed 1,2,4-Triazolo-triazepines 367
    1,2,4-Triazolo-1,2,4-triazepines 367
        1. 1,2,4-Triazolo[4,3-b]1,2,4-triazepines 368
        2. 1,2,4-Triazolo[4,3-d]1,2,4-triazepines 368
    References 369

I. Introduction

The structural characteristic of 1,2,4-triazoles (**1**) that distinguishes them from 1,2,3-triazoles (**2**) is the presence of only two adjacent nitrogen atoms in the former instead of the three that occur in the latter. When at least one of the three nitrogens of a 1,2,4-triazole ring is common with another ring, the system is called a fused or condensed 1,2,4-triazole system. Fusion of a 1,2,4-triazole ring with another heterocycle results in three possible isomeric systems having nitrogen bridgeheads:

(1) condensed 1,2,4-triazolo[1,2-z]heterocycles(**3**),
(2) condensed 1,2,4-triazolo[3,4-z]heterocycles(**4**),
(3) condensed 1,2,4-triazolo[1,5-z]heterocycles(**5**).

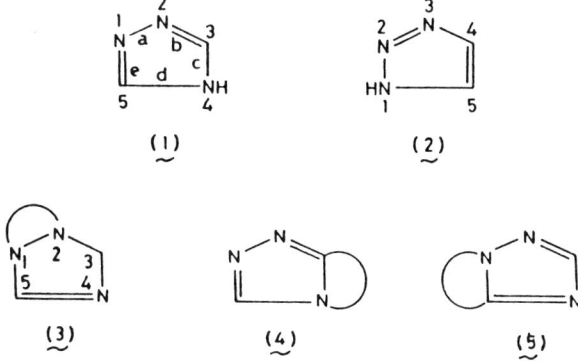

(The letter z designates the place of fusion of the heterocycle to the 1,2,4-triazole ring.) This chapter reviews the methods of synthesis of type (2) of the previously mentioned compounds, i.e., condensed 1,2,4-triazolo[3,4-z]heterocyclic systems (4).

The synthesis of condensed 1,2,4-triazoles has been of interest because of the multifarious biological activities they possess. Thus, for example, 1,2,4-triazolo[4,3-a]quinoxalines are active fungicides (77USP4008322) against *Piricularia oryazae*; 1,2,4-triazolo[4,3-b]pyridazines showed mild central nervous system activity (79PHA801); 8,9-dimethoxy-1,2,4-triazolo[4,3-c]quinazolines (77USP4053600) and 7,8,9,10-tetrahydro-1,2,4-triazolo[3,4-a]phthalazines [78JAP(K)7821197] exhibited antiinflammatory activity. Ishii *et al.* (79YZ533) reported that 2H-1,2,4-triazolo-[3,4-a]phthalazin-3-one, 3-methyl-1,2,4-triazolo[3,4-a]phthalazine, and 3-ethyl-1,2,4-triazolo[3,4-a]phthalazine, which are metabolites of the active hypotensive drugs ecarazine (1-ethoxycarbonylhydrazinophthalazine) and hydralazine (1-hydrazinophthalazine), were potent inhibitors of cyclic adenosine monophosphate phosphodiesterase, equal to theophylline in potency, and possess smooth muscle relaxant activity. Some pyrrolidino[2,1-c]1,2,4-triazoles, 1,2,4-triazolo[4,3-a]pyridines, and 1,2,4-triazolo[4,3-a]azepines are useful as analeptics and as central nervous system and respiratory system stimulants (59BP825514). 3-Amino-1,2,4-triazolo[4,3-c]pyrimidines (62BP898408) were found to be bronchodilators and respiratory stimulants. They also inhibited formation of granulomata and were useful in treating rheumatoid arthritis (62BP898408).

In this review, the synthesis of condensed 1,2,4-triazolo[3,4-z]-heterocycles (4) is arranged according to the complexity of the ring directly fused to the 1,2,4-triazole nucleus (irrespective of other rings that might be fused to it) beginning with those rings having one nitrogen atom in a five-membered ring and going to those that are more complex. The heterocycles have been arranged in the following order according to their types of heteroatoms: nitrogen, oxygen, sulfur, and other elements. Consideration has been given to the possible alternative nomenclature of some 1,2,4-triazoles[3,4-z]heterocycles such as heterocyclo[n,m-c]1,2,4-triazoles (n,m are numbers indicating linkage positions of the heterocycle to the traizole ring) in order to comply with the nomenclature rules of the IUPAC. The literature has been searched up to the end of April, 1988.

II. Condensed 1,2,4-Triazolo-azoles

The synthesis of the three following types of 1,2,4-triazolo-azoles (6–8) relevant to the topic of this review will be discussed.

Sec. II.A] CONDENSED 1,2,4-TRIAZOLO[3,4-z]HETEROCYCLES 281

(6) Pyrrolo[2,1-c]1,2,4-triazoles

(7) 1,2,4-Triazolo[4,3-a]indoles

(8) 1,2,4-Triazolo[3,4-a]isoindoles

A. PYRROLO[2,1-c]1,2,4-TRIAZOLES

Condensation of imidate ester 9 with acid hydrazides gave intermediate hydrazides 10, which underwent dehydrative cyclization to give the 3-substituted pyrrolidino[2,1-c]1,2,4-triazoles (11) (57CB909; 59BP825514). Dehydrative cyclization of 2-aroylhydrazonopyrrolidines (12) with hexamethyldisilazane (HMDS) and trifluoromethanesulfonic acid (H$^+$) afforded the 3-arylpyrrolidino[2,1-c]1,2,4-triazoles (14) (87MI1). Arylhydrazones (15) of pyrrolidin-2-one undergo cyclization with one-carbon cyclizing agents (e.g., methyl chloroformate, phosgene, and thiophosgene) to give the pyrrolidino[2,1-c]1,2,4-triazoles (16) and (17) (80USP-4213773). Reaction of pyruvaldehyde phenylhydrazonyl chloride (18) with pyrrole (19) at room temperature gave, among other products, pyrazolo[3,4:4,5]pyrrol[2,1-c]1,2,4-triazole (20) and its isomer 21 (78-JHC1485).

B. 1,2,4-TRIAZOLO[4,3-*a*]INDOLES

The 1,2,4-triazolo[4,3-*a*]indoles (**25**) have been synthesized (75JHC717) by the cyclization of 2-amidobenzophenones (**22**) with hydrazine through the intermediates 3-amino-4-hydroxyquinazolines (**23**) and triazolylbenzophenones (**24**).

C. 1,2,4-TRIAZOLO[3,4-*a*]ISOINDOLES

Direct cyclocondensation of 1-hydrazinoisoindole (26) with orthoesters gave (76UKZ1159) 3-substituted 5*H*-1,2,4-triazolo[3,4-*a*]isoindoles (27). Some pyrrolidino[2,1-*c*]1,2,4-triazoles are useful (59BP825514) as herbicidals, analeptics, and central nervous system and respiratory system stimulants.

III. Condensed 1,2,4-Triazolo-diazoles

A. 1,2,4-TRIAZOLO-1,2-DIAZOLES

Of the various possible 1,2,4-triazolo-1,2-diazoles, only the two fused-ring system, (28 and 29) are related to this review.

1. *Pyrazolo[5,1-c]1,2,4-triazoles*

Pyrazolo[5,1-*c*]1,2,4-triazoles 33 were prepared [71GEP(0)1810462; 77JCS(P1)2047] by the dehydrative cyclization of pyrazol-3-ylhydrazides

(28) Pyrazolo [5,1-c] 1,2,4-triazolos

(29) 1,2,4-Triazolo [4,3-b] indazoles

31 as well as by the oxidative cyclization of aldehyde pyrazole-3-ylhydrazones **32**. Coupling of the 3-pyrazolodiazonium salts (**34**) with 2-chloro-1,3-diketones (**35**) gave hydrazonyl chlorides **36**, which were cyclized to the pyrazolo[5,1-c]1,2,4-triazoles (77JHC227) (**37**) by treatment with triethylamine. Adopting the approach of constructing the triazole ring onto a pyrazole, German investigators (70CB3284; 79TL1567; 87MI2, 87MI3) synthesized pyrazolo[5,1-c]1,2,4-triazoles (**39–41**) through the addition of diazopyrazoles **38** to nitrogen, phosphorus, or sulfur ylides, respectively.

Reaction of N-phenylbenzoylhydrazonyl chloride (**43**) with suitably substituted pyrazoles (**42**) in the presence of triethylamine gave [82JCS(P1)2663] the corresponding pyrazolo[5,1-c]1,2,4-triazoles (**45**). Heating the triazolotriazepines (**47**) with acetic anhydride resulted in ring contraction to give (74JHC751; 85MI1) the pyrazolo[5,1-c]1,2,4-triazoles (**48**). Ring transformation of 7-diazo-6-methyl-1,2,4-triazolo[4,3-b]-

Sec. III.A]   CONDENSED 1,2,4-TRIAZOLO[3,4-z]HETEROCYCLES   285

pyridazin-8-one (**49**) by ultraviolet irradiation in acidic medium gave (72JPR55) 7-carboxy-6-methyl-1*H*-pyrazolo[5,1-*c*]1,2,4-triazole (**50**).

The pyrazolo[5,1-*c*]1,2,4-triazoles **56** have recently been prepared (86JHC43) from 8-thioxo-1,3-benzothiazines (**52**) and thiocarbohydrazide (**51**) through the mesoionic intermediates (**53**) and triazolothiadiazines (**54**).

2. 1,2,4-Triazolo[4,3-b]indazoles

A reported example of these compounds, namely 1,2,4-triazolo[4,3-b]indazole (**59**), was synthesized (76H1655) by the cycloaddition of C-acetyl-N-phenylnitrilimine (**58**) to 1-methylindazole (**57**).

B. 1,2,4-TRIAZOLO-1,3-DIAZOLES

Compounds of this class pertaining to this review include the three fused-ring systems (**60–62**), the synthesis of which will be discussed.

Imidazolo [2,1-c] 1,2,4-triazoles (**60**)

Imidazolo [5,1-c] 1,2,4-triazoles (**61**)

1,2,4-Triazolo [4,3-a] benzimidazoles (**62**)

1. Imidazolo[2,1-c]1,2,4-triazoles

Scott et al. [70TL4083; 72JHC(P1)2224] reported the condensation of hydrazonyl chlorides of type **63** with 1,2-diaminoethane in aqueous dioxane to produce 3-aryl-5,6,7-trihydroimidazolo[2,1-c]1,2,4-triazoles (**65**). Fusion of an imidazole ring onto a triazole nucleus was the approach used (78UKZ725) in the reaction of 3-amino-2-methyl-5-substituted-1,2,4-triazole (**66**) with α-haloketones in the presence of perchloric acid to give 1-methylimidazolo-[2,1-c]1,2,4-triazolium ions (**67**).

2. Imidazolo[5,1-c]1,2,4-triazoles

The substituted imidazolo[5,1-c]1,2,4-triazole (**70**) has been obtained (68JOC1097) by the reaction of diethyl azodicarboxylate (**69**) with the strained bicyclic imidazolidine **68**.

3. *1,2,4-Triazolo[4,3-a]benzimidazoles*

In one of the approaches [82JCS(P1)2663] for the synthesis of this ring system (e.g., **72**), pyruvaldehyde phenylhydrazonyl chloride (**18**) was used to furnish the two nitrogens and one-carbon fragment necessary for its formation from 2-aminobenzimidazole (**71**). Coupling of benzimidazol-2-yl diazonium chloride (**73**) with α-chloroacetylacetone (**74**) gave intermediate hydrazonyl chloride **75** which, upon base-catalyzed cyclization, furnished (83MI1) 3-acetyl-1*H*-1,2,4-triazolo[4,3-*a*]benzimidazole (**76**). Cyclization of 2-hydrazinobenzimidazoles (**77**) with carboxylic acids (84MI1) or orthoesters (59JOC1478; 84MI1) gave 3-substituted 1,2,4-triazolo[4,3-*a*] benzimidazoles (**78**), while cyclization with carbon disulfide (65ZOR136; 65ZOR139; 84MI3) or phenyl isothiocyanate (59JOC1478) gave corresponding 3-thiones **79**.

IV. Condensed 1,2,4-Triazolo-oxazoles

The synthesis of 1,2,4-triazolo-1,2-oxazoles related to this review, has never been reported.

1,2,4-TRIAZOLO-1,3-OXAZOLES

Of the three possible 1,2,4-triazolo-1,3-oxazoles (**80–82**) which may be found in this review, only the synthesis of 1,2,4-triazolo[3,4-*b*]1,3-benzoxazoles (**82**) has been recorded.

1,3-Oxazolo [2,3-*c*] 1,2,4-triazoles (**80**)

1,3-Oxazolo [4,3-*c*] 1,2,4-triazoles (**81**)

1,2,4-Triazolo [3,4-*b*] 1,3-benzoxazoles (**82**)

1,2,4-Triazolo[3,4-b]1,3-benzoxazoles

1,2,4-Triazolo[3,4-*b*]1,3-benzoxazole-3-thione (**85**) has been prepared by cyclizing 2-hydrazino-1,3-benzoxazole (**83**) with carbon disulfide (58USP2861076) or phenyl isothiocyanate (59JOC1478). The alternative approach (62TL1193) of fusing the 1,3-benzoxazole onto a 1,2,4-triazole was also used in the synthesis of 3-phenyl-1,2,4-triazolo[3,4-*b*]-benzoxazole (**88**) from 3-hydroxy-5-phenyl-4-(2-methoxyphenyl)-1,2,4-triazole (**86**).

V. Condensed 1,2,4-Triazolo-thiazoles

A. 1,2,4-TRIAZOLO-1,2-THIAZOLES

This class of compounds includes the two possible ring systems **89** and **90**. Only the synthesis of **90** has been recorded in the literature.

1,2,4-Triazolo [4,3-b] 1,2-thiazoles (**89**)

1,2,4-Triazolo [4,3-b] 1,2-benzothiazoles (**90**)

1,2,4-Triazolo[4,3-b]1,2-benzothiazoles

An example of this ring system was prepared (78USP4108860) through cyclization of 3-hydrazido-1,2-benzothiazole-1,1-dioxide (**91**) with phosphoryl chloride in dimethylformamide (DMF). 3-Methyl-1,2,4-triazolo-[4,3-*b*]1,2-benzothiazole-5,5-dioxide (**95**) was obtained (79USP4140693) when 3-chloro-1,2-benzothiazole-1,1-dioxide (**93**) was treated with 3-methyltetrazole (**94**) in the presence of pyridine.

B. 1,2,4-TRIAZOLO-1,3-THIAZOLES

Relevant to this review are the three possible 1,2,4-triazolo-1,3-thiazoles (**96–98**). Only the synthesis of the two systems **96** and **98** will be reviewed; the synthesis of **97** has not been reported.

1,3-Thiazolo [2,3-c] 1,2,4-triazoles (**96**)

1,3-Thiazolo [4,3-c] 1,2,4-triazoles (**97**)

1,2,4-Triazolo [3,4-b] 1,3-benzothiazoles (**98**)

1. *1,3-Thiazolo[2,3-c]1,2,4-triazoles*

Kendall *et al.* (50BP634951) reported as early as 1950 that synthesis of this ring system occurred by the reaction of 3-mercapto-5-methyl-1,2,4-triazole (**99**) with chloroacetic acid to give the S(triazolyl)thioglycolic acid derivative **100**, which afforded 5-hydroxy-3-methyl-1,3-thiazolo[2,3-c]-1,2,4-triazol-5-one (**101**) on hydrative cyclization with acetic anhydride in the presence of pyridine. No rigorous proof was given to exclude the possible alternative structure 1,3-thiazolo[2,3-b]1,2,4-triazole (**102**). A similar synthetic pathway has been reported by Potts and Husain (71JOC10) in which 3-mercapto-5-methyl-1,2,4-triazole (**99**) reacted with α-haloketones to give **103**, which was cyclized to the 1,3-thiazolo[2,3-c]1,2,4-triazoles (**104**). Surprisingly, 3-mercapto-5-phenyl-1,2,4-triazole (**105**), under the same reaction conditions, gave 1,3-thiazolo[3,2-b]1,2,4-triazole (**106**). The direction of cyclization of this reaction, i.e., the triazole-ring nitrogen at which cyclization took place, was found to be affected by the acidity of the medium (71JOC10; 85JHC1185), the polarity of the solvent (71JOC10; 85JHC1185), and the nature of the substitutent [67MI1; 71JCS(C)1667, 71JOC10; 72JMC332; 74IJC485; 77IJC(B)1143; 77JAP(K)7753883; 78JHC401; 79IJC(B)364; 80JHC1321; 82IJC(B)243; 82IJC(B)732, 82MI1; 85IJC(B)1221]. 1,3-Thiazolo[2,3-c]1,2,4-triazoles

(108) and 1,3-thiazolo[3,2-b]1,2,4-triazoles (109) were also prepared (79S52) by the acid- or base-catalyzed cyclization of the S-propynyl-1,2,4-triazoles (107).

A variety of one-carbon cyclizing agents have been used (71JOC10) to prepare 3-substituted 1,3-thiazolo[2,3-c]1,2,4-triazoles (111) from 2-hydrazino-1,3-thiazoles (110). The acid hydrazides (113) of these hydrazines, prepared either directly from 110 and carboxylic acid derivatives or indirectly from 1-acylthiosemicarbazides (112) and α-haloketones, also undergo facile dehydrative cyclization to 111. [71JOC10; 78JHC401;

Sec. V.B] CONDENSED 1,2,4-TRIAZOLO[3,4-z]HETEROCYCLES

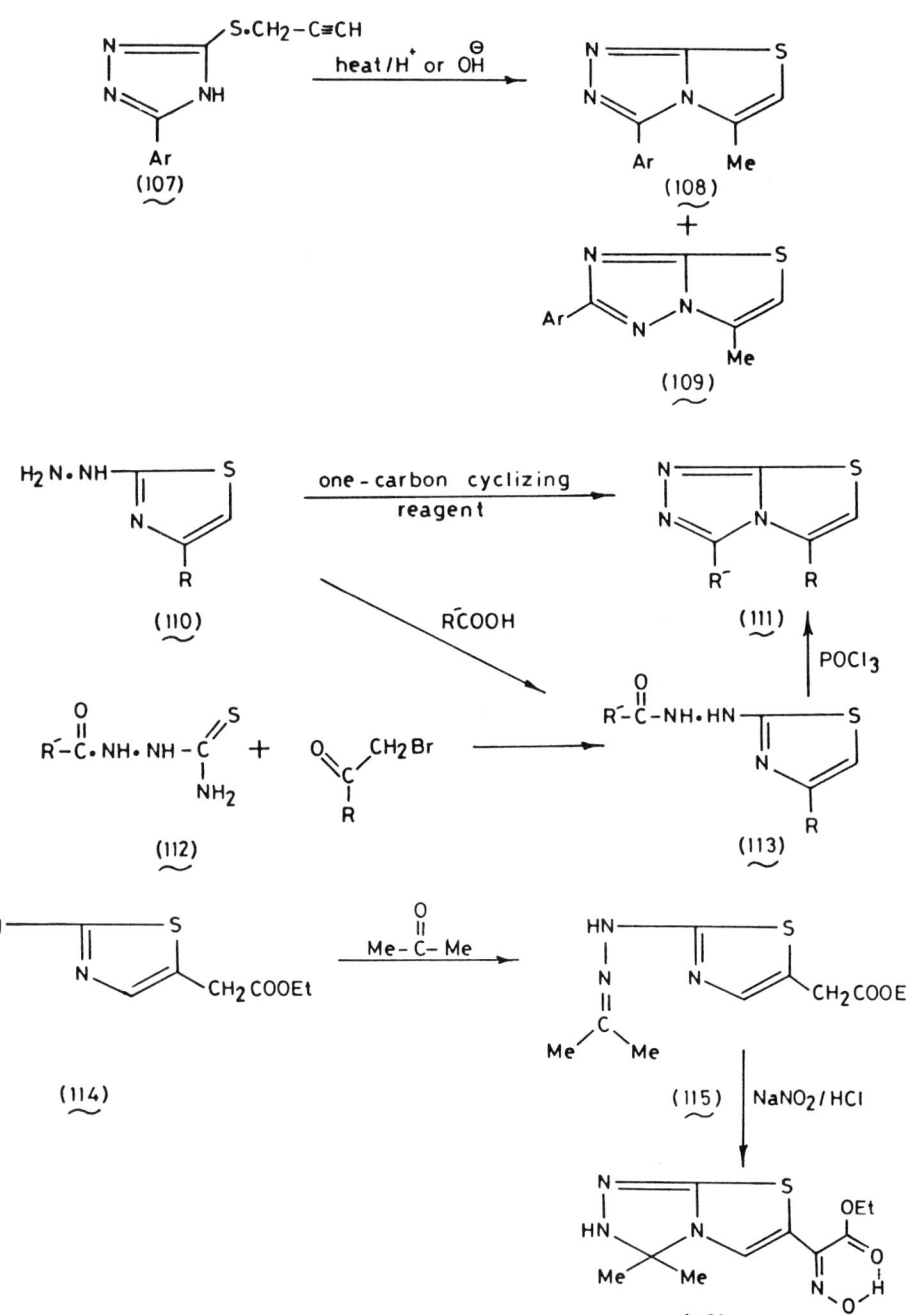

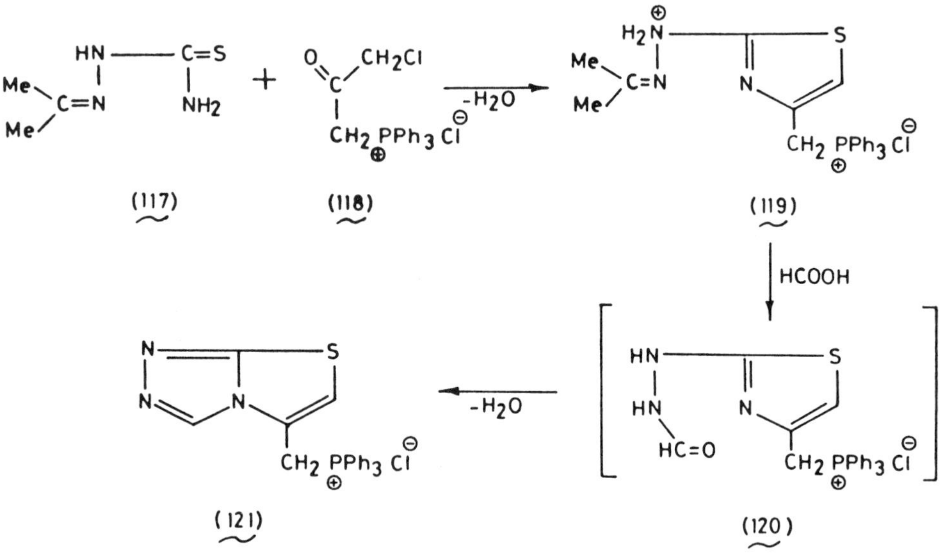

81IJC(B)132; 86JHC1439]. Attempted nitrosation of the methylene group of 1,3-thiazol-2-ylhydrazone (**115**) proceeded with the simultaneous formation (80JHC1321) of the title fused-ring system to give **116**.

Henichart et al. (85JHC1185) prepared **121** by the cyclocondensation of acetone thiosemicarbazone (**117**) and phosphonium salt **118** to form **119** which, upon treatment with formic acid, afforded 1,3-thiazolo[2,3-c]1,2,4-triazole salt **121**. Sasaki et al. (82JOC2757; 84CPB5040) synthesized the 3,5-disubstituted 1,3-thiazolo[2,3-c]1,2,4-triazoles (**111**) by ring transformation of 2-mercapto-5-substituted-1,2,4-oxadiazoles (**122**) (82JOC2757 and 2-methanesulfonyl-5-phenyl-1,3,4-oxadiazole (84CPB5040). The following scheme illustrates the transformation of **122** using α-chloroketones and ammonia.

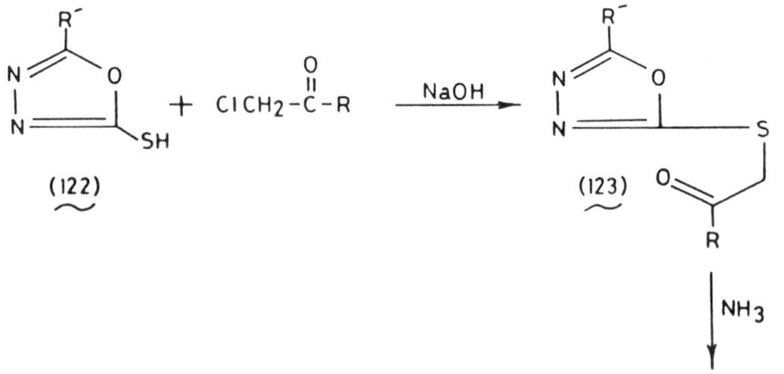

2. 1,2,4-Triazolo[3,4-a]1,3-benzothiazoles

Suitably functionalized phenyl isothiocyanates (e.g. **127**) have been used as starting materials for the synthesis of this fused heterocyclic system; the rest of the skeleton is provided by acid hydrazides [74JOC3506 75GEP(O)2509843; 79H1171]. The intermediate 1-acyl-4-arylthiosemicarbazides (**128**) and 3-mercapto-1,2,4-triazoles (**129**) underwent base-catalyzed cyclization to 1,2,4-triazolo[3,4-a]1,3-benzothiazoles **130**.

The reaction of 2-hydrazino-1,3-benzothiazoles (**131**) with formic acid (80MI1; 84KGS40), acetic anhydride (57USP2786054), aromatic esters (59JOC1478), orthoesters [59JOC1478; 73GEP(O)2250077], carbon disulfide (57JCS727; 58USP2861076), phenylisocyanate (59JOC1478), and phenyl isothiocyanate (59JOC1478) offers an unequivocal pathway for synthesizing 3-substituted 1,2,4-triazolo[3,4-*b*]1,3-benzothiazoles (**132**).

Bower and Doyle (57JCS727) reported that dehydrogenative cyclization of hydrazones **133** derived from aromatic aldehydes and 2-hydrazino-1,3-benzothiazole with lead tetraacetate gave 3-aryl-1,2,4-triazolo[3,4-*b*]1,3-benzothiazoles (**134**). Reinvestigation [71JCS(C)2265], however, showed these products were actually the *N*-acetylated hydrazides (**135**), which could be cyclized to **134** through the removal of an acetic acid by boiling with phenol. In addition to lead tetraacetate [71JCS(C)2265; 79JIC742], a wide variety of oxidizing agents have been reported to affect this cyclization, such as ferric chloride (64E200, 70MI1) in ethanol, bromine in acetic acid (70MI1), and bromine in the presence of sodium carbonate (72JCS(P1)1519].

1,2,4-Trizaolo[3,4-b]1,3-benzothiazoles (**134**) were also obtained (59JOC1478) by cyclizing hydrazides **136**, derived from 2-hydrazinobenzothiazole by boiling in phenol. Hydrazides **136** were claimed as intermediates in the one-step synthesis (59JOC1478) of **134** from 2-chloro-1,3-benzothiazole (**137**) and acid hydrazides. Some 1,3-thiazolo[2,3-c]-1,2,4-triazoles and 1,2,4-triazolo[3,4-b]1,3-benzothiazoles were found to exhibit insecticidal [77JAP(K)7753881, 77JAP(K)7753882], miticidal [77JAP(K)7753881, 77JAP(K)7753882], fungicidal [73GEP(O)2250077; 75GEP(O)2509843; 77JAP(K)7753881, 77JAP(K)7753882], and bactericidal [75GEP(O)2509843] activities.

VI. Condensed 1,2,4-Triazolo-selenazoles

Members of 1,2,4-triazolo-1,2-selenazoles that may be included in this chapter await synthesis.

1,2,4-Triazolo-1,3-selenazoles

There are three possible fused systems (**138–140**) that are related to the title of this chapter. Examples of compounds belonging only to system **140** have been synthesized.

1,3-Selenazolo[2,3-c] 1,2,4-triazoles (**138**)

1,3-Selenazolo [4,3-c] 1,2,4-triazoles (139)

1,2,4-Triazolo [3,4-b] 1,3-benzoselenazoles (142)

1,2,4-Triazolo[3,4-b]1,3-benzoselenazoles

The only reported (59JOC1478) examples (**142**) of these compounds were synthesized by condensative cyclization of 2-hydrazino-1,3-benzoselenazole (**141**) with aliphatic orthoesters or esters of aromatic acids.

(141) → RC(OR)$_3$ → (142)

VII. Condensed 1,2,4-Triazolo-triazoles

1,2,4-Triazolo-1,2,3-triazoles relevant to the topic of this chapter have not been synthesized.

1,2,4-TRIAZOLO-1,2,4-TRIAZOLES

Two 1,2,4-triazole rings may be fused together to form triazolo-triazoles having either one or two nitrogen bridgeheads as in **143** and **144**, respectively; the latter type is out of the scope of this chapter. Fusion of the 1,2,4-triazole rings through a carbon and a nitrogen (i.e., with one nitrogen bridgehead as in **143**) may take place according to the three following patterns (**145–147**). Syntheses of only the first two systems (**145** and **146**), which are pertinent to the subject of this chapter, will be reviewed.

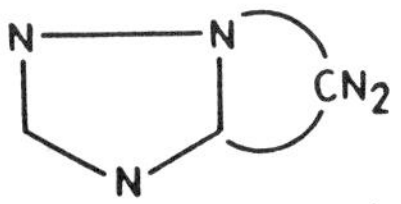

(143)

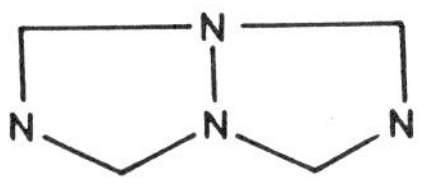

(144)

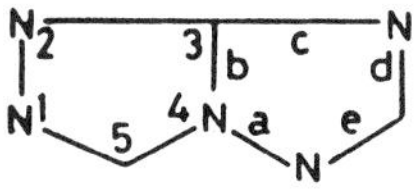

1,2,4-Triazolo [4,3-b] 1,2,4-triazoles (145)

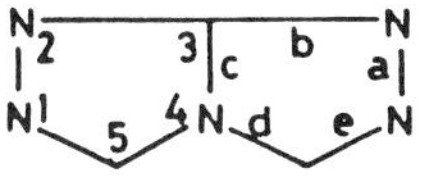

1,2,4-Triazolo [3,4-c] 1,2,4-triazoles (146)

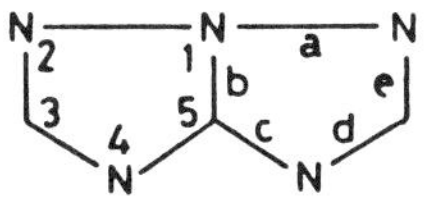

1,2,4-Triazolo [1,5-b] 1,2,4-triazoles (147)

1. *1,2,4-Triazolo[4,3-b]1,2,4-triazoles*

Condensation of the two amino functions of 3,4-diamino-1,2,4-triazole derivatives (**46**) with one-carbon cyclizing agents, such as acetic anhydride (62LA148; 65JHC302) or benzoyl chloride [50JCS614; 72JCS(P1)1319; 74JCS(P2)997], gave the 1,2,4-triazolo[4,3-*b*]1,2,4-triazoles (**148** and **149**).

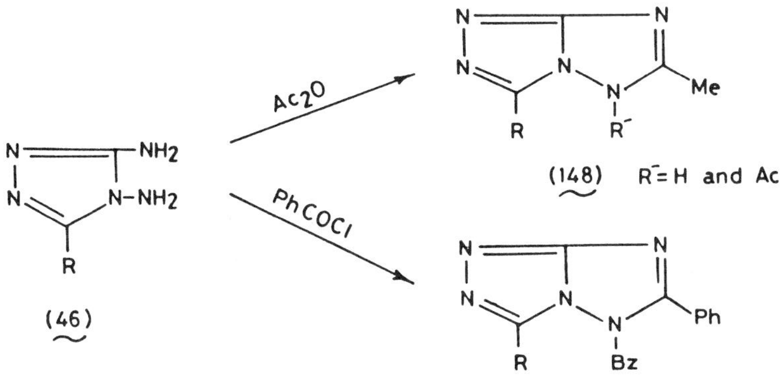

(148) R⁻= H and Ac

Cyclization of 1,2,4-triazol-5-ylhydrazonyl bromides (**150**) by heating with acetic acid and sodium acetate was reported (65TL841) to give 3-aryl-6-phenyl-1,2,4-triazolo[4,3-*b*]1,2,4-triazoles (**151**). On the other hand, cyclization of **150** in organic solvents was found [74JCS(P2)997] to depend on the pH of the medium; in slightly acidic media, 3-aryl-5-phenyl-1,2,4-triazolo[3,4-*c*]1,2,4-triazoles (**152**) were exclusively formed, while in highly basic media, 3-aryl-6-phenyl-1,2,4-triazolo[4,3-*b*]1,2,4-triazoles (**151**) predominated. Oxidative cyclization of aromatic aldehyde 1,2,4-triazolo-3-ylhydrazones (**153**) with lead tetraacetate gave [57JCS727; 70TL1841; 72JCS(P1)1319] 3-aryl-6-phenyl-1,2,4-triazolo[3,4-*b*]1,2,4-triazoles (**151**) in addition to the N-acetylated hydrazides **154**.

Pyrolysis of 3-azido-4-benzylideneamino-5-methyl-1,2,4-triazole (**157**) afforded (65JOC711; 66JHC119) 3-methyl-6-phenyl-5H-1,2,4-triazolo[4,3-*b*]1,2,4-triazole (**158**). 3,6,7-Triamino-1,2,4-triazolo[4,3-*b*]1,2,4-triazoles (**161**) were prepared by double ring closure of triaminoguanidine (**159**) with two equivalents of cyanogen bromide. The reaction of 4-amino-3-hydrazino-1,2,4-triazoles (**162**) with cyanogen bromide or carbon disulfide gave (68JOC143) 3-amino- or 3-mercapto-1,2,4-triazolo[4,3-*b*]1,2,4-triazoles (**163**), respectively. Ring closure of 3-hydrazino-5-phenyl-1,2,4-

triazole (**164**) with cyanogen bromide in methanol at room temperature gave (68JOC143) 3-amino-5-phenyl-7H-1,2,4-triazolo[3,4-c]1,2,4-triazole (**165**), while in boiling methanol, 3-amino-5-phenyl-5H-1,2,4-triazolo[4,3-b]1,2,4-triazole (**166**) was obtained (68JOC143).

Cyclizing 4-amino-3-methylthio-1,2,4-triazoles (**167**) (83S415) or 4-amino-2-methyl-1,2,4-triazolo-3-thiones (**169**) (85H2613) with aromatic (83S415; 85H2613) or heterocyclic nitriles (85H2613) in the presence of potassium *t*-butoxide afforded the corresponding 1,2,4-triazolo[4,3-b]-

1,2,4-triazoles (**168**). These reactions did not take place with aliphatic nitriles.

2. 1,2,4-Triazolo[3,4-c]1,2,4-triazoles

1,2,4-Triazolo[3,4-c]1,2,4-triazoles (**173**) were obtained [76ACS(B)463] by the dehydrative cyclization of 2-substituted 3-aroylhydrazino-1,2,4-triazoles (**172**). Blocking of N-2 in **171** forces the cyclization to take place at the less nucleophilic N-4 to give the title compounds.

VIII. Condensed 1,2,4-Triazolo-oxadiazoles

None of the 1,2,4-triazolo-1,2,3-oxadiazoles or 1,2,4-triazolo-1,2,4-oxadiazoles that should be included in this review have been synthesized.

1,2,4-Triazolo-1,3,4-oxadiazoles

The only possible fused system of this type that is pertinent to this review is 1,2,4-triazolo[3,4-b]1,3,4-oxadiazoles (**174**).

1,2,4-Triazolo [3,4-b] 1,3,4-Oxadiazoles (174)

1,2,4-Triazolo[3,4-b]1,3,4-oxadiazoles

1,2,4-Triazolo[3,4-*b*]1,3,4-oxadiazoles (**178**) have been obtained [71TL1729; 72JCS(P1)269], as in the following outline, through three steps involving the formation of the 1,3,4-oxadiazolyl hydrazidic bromides (**177**), which were then cyclized with triethylamine. Merging these reactions in one step was achieved by heating the tribromo-diazobutadiences (**179**) with benzoylhydrazine and three equivalents of triethylamine (71TL1729). Thermal dehydrogenative cyclization of the 1,3,4-oxadiazol-

2-ylhydrazones (**176**) by heating in nitrobenzene also gave (71IJC901) the title compounds **178**.

The reaction of N-(α-chlorobenzylidene)carbamyl chloride (**180**) with sodium azide gave (74JOC1226), among other products, 3,6-diphenyl-1,2,4-triazolo[3,4-b]1,3,4-oxadiazole (**182**) and not the other possible product **181**. The former product was unequivocally prepared (74JOC1226) by cyclization of 4-benzoylamino-5-chloro-3-phenyl-1,2,4-triazole (**184**) with aqueous sodium carbonate and found to be identical with that obtained by the previous method. 2-Hydrazino-1,3,4-oxadiazoles (**185**) with carboxylic acid [79IJC(B)499] or carbon disulfide [79IJC(B)499; 84MI4] afforded the disubstituted 1,2,4-triazolo[3,4-b]1,3,4-oxadiazoles **186** and **187**.

IX. Condensed 1,2,4-Triazolo-thiadiazoles

Like their oxygen counterparts, no members of 1,2,4-triazolo-1,2,3-thiadiazoles or 1,2,4-triazolo-1,2,4-thiadiazoles, which are related to this review, have been synthesized.

1,2,4-TRIAZOLO-1,3,4-THIADIAZOLES

1,2,4-Triazolo[3,4-b]1,3,4-thiadiazoles (**188**) is the only system of this type which conforms with the topic of this chapter.

1,2,4-Triazolo[3,4-b]1,3,4-thiadiazoles

The 1,2,4-triazolo[3,4-b]1,3,4-thiadiazole ring system was first described by Kanaoka (56JPJ1133), who synthesized various 3-alkyl and 3-aryl derivatives (**191**) by the dehydrative cyclization of 4-acylamine-1,2,4-triazole-5-thiols (**190**). Several similar syntheses were performed by starting with 4-amino-1,2,4-triazole-5-thiols (**189**) and constructing the thiadiazole ring using carboxylic acids [73JHC387; 77JHC567; 80JIC1112; 82JIC769; 86IJC(B)566], acetic anhydride (65JHC302), acid chlorides (56JPJ1113), aldehydes (69IJC959; 82JIC900), aryl isothiocyanates (83S411), cyanogen bromide (66JOC3528; 81JHC1353; 86JHC1439), and carbon disulfide (66JOC3528; 81IJC(B)369]. Cyclocondensation of thiocarbohydrazide (**51**) with two equivalents of carboxylic acids (73JHC387) gave 3,6-disubstituted 1,2,4-triazolo[3,4-b]1,3,4-thiadiazoles (**193**), while with carbon disulfide and pyridine, it gave (61ACS1295), in addition to 4-amino-2,5-dimercapto-1,2,4-triazole (**194**), 3,6-dimercapto-1,2,4-triazolo[3,4-*b*]1,3,4-thiadiazole (**195**).

6-Substituted-3-substituted-amino-1,2,4-triazolo[3,4-*b*]1,3,4-thiadiazoles (**198**) were prepared (80MI2) by cyclocondensation of 2-mercapto-5-substituted-1,3,4-thiadiazoles (**197**) with 4-substituted thiosemicarbazides (**196**). 2-Hydrazino-1,3,4-thiadiazoles (**199**) have also been used as thiadiazole precursors on which the triazole ring was formed by cyclocondensation with orthoesters (57MI2; 66JOC3528), cyangogen bromide (66JOC3528), carbon disulfide (66JOC3528; 86JHC1339), and aryl isothiocyanates (86JHC1339). Many of the prepared 1,2,4-triazolo[3,4-*b*]1,3,4-thiadiazoles possess various biological activities such as analgesic [86IJC(B)566], anti-inflammatory [86IJC(B)566], fungicidal (80MI2;

86JHC1339), antimicrobial (86JHC1339), central nervous system depressant (82JIC769), mild hypocholesterolemic (82JIC769), and hypotensive activity (82JIC769).

X. Condensed 1,2,4-Triazolo-tetrazoles

This fused ring system comprises only one type of compound related to this review, namely, 1,2,4-triazolo[4,3-*d*]tetrazoles (**201**).

1,2,4-Triazolo[4,3-*d*]tetrazoles

Examples (**204**) were prepared by cyclization of the hydrazidoyl bromides **203** [65AG963; 67JCS(C)239; 68JCS(C)1711]. Oxidative cyclization of the 1-methyl-tetrazol-5-ylhydrazones (**202**) with lead tetraacetate also gave [66JCS(C)1202] **204** in addition to the N-acetylated hydrazides **205**.

XI. Condensed 1,2,4-Triazolo-azines

Belonging to this class of compounds are the four systems (**206–208**). The first member of this class, 1,2,4-triazolo[4,3-*a*] quinolines (**207**), was prepared as early as 1900 by Marckwald and Meyer (1900CB1885). The 1,2,4-triazolo[4,3-*a*]pyridines (**206**) were also reported by Marckwald and Rudzik (03CB1111) in 1903. The last two systems, **208** and **209**, were only obtained after 1960. Synthesis of these compounds may be achieved either by assembling the 1,2,4-triazole ring onto the azine nucleus or by constructing the azine onto the 1,2,4-triazole ring. Judging from the amount of work published on the synthesis of this ring system, the former approach is much more widely used than the latter.

1,2,4-Triazolo [4,3-*a*] pyridines (**206**)

1,2,4-Triazolo [4,3-*a*] quinolines (**207**)

Sec. XI.A] CONDENSED 1,2,4-TRIAZOLO[3,4-z]HETEROCYCLES 311

1,2,4-Triazolo [3,4-a] isoquinolines (208)

1,2,4-Triazolo [4,3-b] isoquinolines (209)

A. 1,2,4-TRIAZOLO[4,3-a]PYRIDINES

Synthesis of this system according to the first approach usually utilizes an azine precursor having a hydrazino function on the carbon adjacent to the ring nitrogen. Thus, condensative cyclization of 2-hydrazinopyridines (210) using a variety of one-carbon cyclizing agents such as carboxylic acids (03CB1111; 15JCS688; 59USP2917511; 62USP3050525; 66CB2593, 66JOC251, 66MI1; 76FES126), carboxylic acid anhydrides (57JCS4510; 59JOC1478; 85CPB4769), carboxylic acid esters (59JOC1478; 76FES126), orthoesters (59JOC1478; 66JOC251; 80EUP17438; 81USP4244953; 86JHC1071), acid chlorides (57JCS727; 66JOC251), cyanogen bromide (66CB2593, 66JOC251), carbon disulfide (48JA1381; 66JOC251), phosgene (70CB1934; 85CPB4769), urea (59USP2917511; 66CB2593, 66JOC251), isothiocyanates (59JOC1478), isocyanates (59JOC1478; 85CPB4769), ethyl chloroformate (66JOC251), thioimidates (76JOC3124), trithiocarbonic acid (23JCS312), and N-dichloromethylene benzamide (70S433) affords the 3-substituted 1,2,4-triazolo[4,3-a]pyridines **211**.

(210) → one-carbon cyclizing reagent → (211)

For example, 1-benzoyl-2-(pyrid-2-yl)hydrazine (213), obtainable from the reaction of 2-hydrazinopyridines (212) with benzoyl chloride (57JCS727; 66JOC251) or benzoic anhydride (59JOC1478), undergoes dehydrative cyclization in boiling phenol (59JOC1478) or phosphoryl

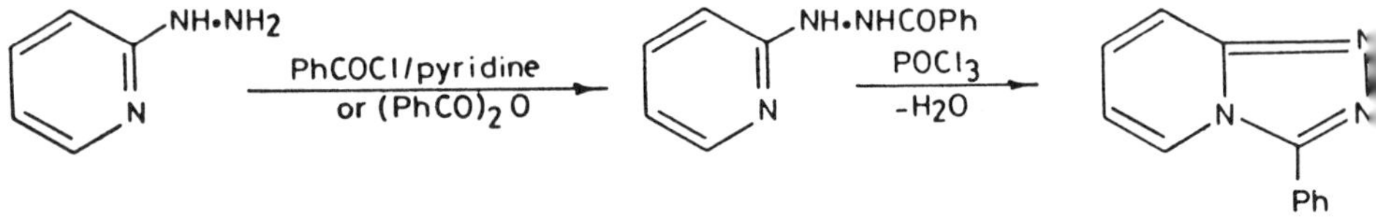

(212) → (213) → (214)

chloride (57JCS727; 66JOC251) or on treatment with thionyl chloride (70CB1918) to give 3-phenyl-1,2,4-triazolo[4,3-*a*]pyridine (**214**). Hydrazones **215** derived from aromatic aldehydes and 2-hydrazinopyridines have also been cyclized to the title compounds (**211**) by chemical and electrochemical oxidation as well as by thermal dehydrogenation. Thus, in addition to anodic oxidation (81H699) and thermal dehydrogenation (65IJC162), chemical oxidations with lead tetraacetate (57JCS727;

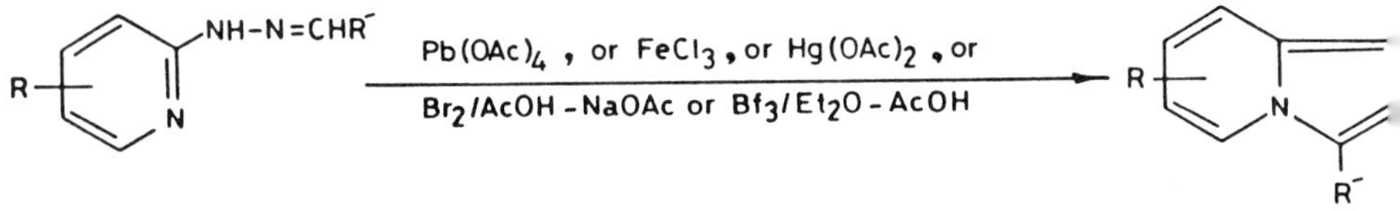

(215) → (211)

68BP1131590; 75JHC337), ferric chloride (65IJC162), mercuric acetate (81CC376), bromine in acetic acid and sodium acetate (63T1587), sulfur (75JHC337), and boron trifluoride etherate in acetic acid (68DOK127) were also used to affect this cyclization.

Reaction of arylhydrazones **216** of piperidin-2-ones with methyl chloroformate, phosgene, or thiophosgene yielded (80USP4213773) the fused-ring system **217**. Reaction of 2,2′-azopyridine (**218**) with diazoalkanes afforded [66JCS(C)78; 73LA2088] the 2-(pyrid-2-yl)-1,2,4-triazolo[4,3-*a*]-

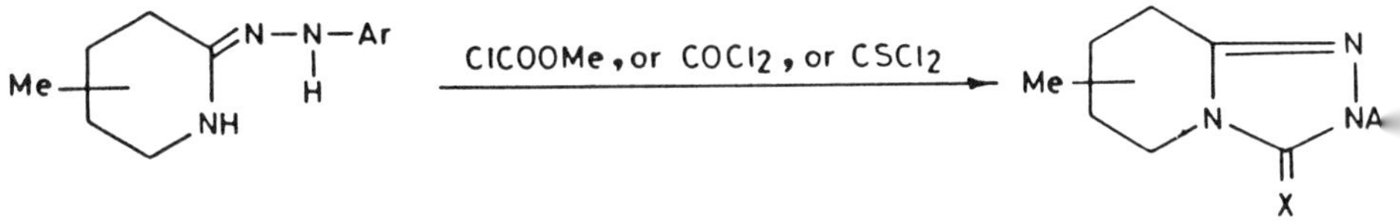

(216) → (217) X = O or S

pyridines (**219**). Pyrido[2,1-*c*]1,2,4-triazines (**220**) were reported to undergo transformation (68DOK127) into 1,2,4-triazolo[4,3-*a*]pyridines (**221**) by the action of boron trifluoride etherate in acetic acid.

When 5-substituted tetrazoles (**223**) were treated with 2-chloropyridine (**222**) in the presence of pyridine or triethylamine, they gave (61CB1555; 78ZC175) the 3-substituted 1,2,4-triazolo[4,3-*a*]pyridines (**226**) through the 3-(pyrid-2-yl)tetrazole (**224**) and nitrilimine (**225**) intermediates. 1,2,4-Triazolo[4,3-*a*]pyridine C-nucleoside (**228**) was obtained by cyclizing 2-hydrozinopyridine (**212**) with 2,5-anhydroaldonic acid (78MI1) or with the C-glycosyl thiomidate (76JOC3124) **227**.

B. 1,2,4-TRIAZOLO[4,3-*a*]QUINOLINES

3-Substituted derivatives **230** were prepared by cyclizing 2-hydrazinoquinolines (**229**) with different cyclizing agents, namely, carboxyclic acids (1900CB1885, 1900CB1895; 57USP2786054; 73FRP2149467), phenyl salicylate (59JOC1478), orthoesters [59JOC1478; 72GEP(O)2203782,

72USP3681343], isothiocyanates (1900CB1885, 1900CB1895; 59JOC1478), carbon disulfide (58USP2861076), and *N*-dichloromethylene benzamide (70S433). Thermal Dehydrogenative cyclization or oxidative cyclization of aromatic aldehyde quinol-2-ylhydrazones **231** with ferric chloride gave (65IJC162) the 2-aryl derivatives. **232**.

Using the approach of constructing the quinoline nucleus onto a triazole skeleton, 3-hydroxymethyl-1,2,4-triazolo-4-ylbenzophenone (**233**) was oxidized with manganese dioxide to the corresponding aldehyde **234** which, on treatment with hydrazine at room temperature, gave (75JHC717) the 1,2,4-triazolo[4,3-*a*]quinoline (**237**) and its 9,10-dehydro derivative **238**. Reaction of *C*-ethoxycarbonyl-*N*-phenylhydrazidoyl chloride (**240**) with quinoline (**239**) in the presence of triethylamine gave (67TL3071) 3-ethoxycarbonyl-1-phenyl-1,2,4-triazolo[4,3-*a*]quinoline (**241**).

C. 1,2,4-Triazolo[3,4-a]isoquinolines

Similar to the preparation of quinoline isomers **230**, condensative cyclization of 1-hydrazinoisoquinolines (**242**) with carboxylic acids (66-JHC158; 67USP3354164; 76USP426639), carboxylic acid anhydrides (67USP3354164; 70CB1960; 76USP426639), carboxylic acid esters (70CB1960), acid chlorides (67USP3354164; 70CB1960), phosgene (70CB1960), thiophosgene (70CB1960), and ammonium isothiocyanate (70CB1960) gave the title compounds **243**.

Dehydrative cyclization of isoquinolin-1-ylhydrazides (**244**) with thionyl chloride gave (70CB1918) 3-substituted 1,2,4-triazolo[3,4-a]isoquinolines (**246**) via the 1,3,4-oxadiazole-S-oxides **245**. Cyclization of aldehyde isoquinol-1-ylhydrazones (**247**), thermally (65IJC162; 81KFZ44) or by oxidation with ferric chloride (65IJC162), led to **248**. Reaction of 1-chloroisoquinoline (**249**) with 5-substituted tetrazoles **223** took place (70CB1918) according to the mechanism given for preparing 3-substituted 1,2,4-triazolo[4,3-a]pyridines **226** to give **250**.

D. 1,2,4-Triazolo[4,3-b]isoquinolines

In contrast to 1-hydrazinoisoquinolines (**229**), boiling 3-hydrazinoisoquinoline (**251**) with acetic anhydride gave (78JHC463) the triacetyl derivative **252** and not the triazolo[4,3-b]isoquinoline (**254**). Nevertheless, **254** was obtained (78JHC463) by the dehydrative cyclization of monoacetyl derivative **253** with polyphosphoric acid. In contrast to the stable 1,2,4-triazolo[3,4-a]isoquinolines (**208**), 1,2,4-triazolo[4,3-b]isoquinolines (**209**) are unstable because of their "quinonoid" structure.

Various 1,2,4-triazoloazines possess bactericidal (73FRP2149467), fungicidal (73FRP2149467), anxiolytic (81USP4244953), anti-inflammatory (62USP3050525; 76USP426639), antibacterial (68BP1131590), insecticidal [76GEP(O)2438789], anticonvulsant (59USP2917511), tranquilizing (59USP2917511), analgesic (76FES126), and coronary dilating (67USP3354164) activities.

XII. Condensed 1,2,4-Triazolo-diazines

A. 1,2,4-TRIAZOLO-1,2-DIAZINES

Under this heading, the synthesis of 1,2,4-triazolo[4,3-*b*]pyridazines (**255**) as well as the closely related and medicinally very important 1,2,4-triazolo[3,4-*a*]phthalazines (**256**), in addition to 1,2,4-triazolo[4,3-*b*]-cinnolines (**257**), will be discussed.

1,2,4-Triazolo [4,3-b] pyridazines (**255**)

1,2,4-Triazolo [3,4-a] phthalazines (**256**)

1,2,4-Triazolo [4,3-b] cinnolines (**257**)

1. *1,2,4-Triazolo[4,3-b]pyridazines*

1,2,4-Triazolo[4,3-*b*]pyridazines (**255**) were first obtained by Bulow (09CB2209) in 1909, yet most of the work has been published after 1950. Two general routes have been used: (a) starting with 4-amino-1,2,4-triazole (**258**) and forming the fused pyridazine ring and (b) starting with 3-hydrazinopyridazine and forming the fused 1,2,4-triazole ring. According to the first route, 4-amino-1,2,4-triazole (**258**) was condensed with various 1,3-dicarbonyl compounds (09CB2209, 09CB2595; 10CB1975; 58BEP566543; 58ZOB2773; 59JA6289; 61USP2933388; 62BSF355; 63USP3096329, 68T2687; 79PHA801), azo-β-esters (45USP2390707; 47USP2432419), β-keto acetals [79JCS(P1)3085], cyclic β-keto aldehydes [79JCS(P1)3085], 4,4-dimethyl-2-butanone (58USP2837521), and tetraethyl monoorthomalonate (71JPR780) to yield the corresponding 1,2,4-triazolo[4,3-*b*]pyridazines (e.g., **259** and **260**).

The products of this reaction, obtained using unsymmetrical 1,3-dicarbonyl compounds such as 4,4-dimethoxybutanone (**261**), may be assigned [79JCS(P1)3085] structure **262** or **263**. Rigorous proof that the product, indeed, has structure **263** was given by Liberman and Jacquier

Sec. XII.A] CONDENSED 1,2,4-TRIAZOLO[3,4-z]HETEROCYCLES 319

(62BSF355) who unequivocally prepared **262** by condensative cyclization of 3-hydrazino-6-methylpyridazine (**264**) with formic acid.

The first synthesis of triazolo-pyridazines, using the route of forming the fused triazole ring onto a pyridazine function, was reported by Takahayashi (55JPP1242) in 1955 when 3-hydrazino-6-chloropyridazine was cyclized by reaction with ethyl formate. Since then, this route has been more widely used than that utilizing the 4-amino-1,2,4-triazole, simply because of the ready availability of 3-hydrazinopyridazines. 3-Hydrazinopyridazines (**265**) may be cyclized to the 1,2,4-triazolo[4,3-*b*]-pyridazines (**266**) using various one-carbon cyclizing agents such as carboxylic acids (55JPJ1242; 56JPJ1296; 57MI1; 62ACS2389; 62BSF355;

(265) → (266)

64JHC42; 65CPB586; 68T2687; 69M671; 70CR1990; 77MI1; 81EUP29130, 81JHC1523; 84JHC1389; 87MI4), carboxylic acid anhydrides (57MI1; 60BP839020; 65CPB586; 78FES565; 85JHC1045), acid chlorides [58MI1; 61MI1; 82IJC(B)317; 83FES842; 84MI1], carboxylic acid esters [55JPJ1242; 82IJC(B)317], orthoesters (55JPJ1242; 67CB875; 81JHC1523; 85EUP156734, 85JHC1045; 86M867; 87MI4), ethyl chloroformate (76USP3994898; 86M867), carbon disulfide (63JCS5660; 67T387; 78KGS1571), cyanogen chloride (63JCS5660), isothiocyanates (67JOC-1139; 67T387; 70CR1990; 77MI1), 1-(cyanoacetyl)-3,5-dimethylpyrazole [78IJC(B)1000], benzylthioacetaimidates (81JHC893), methyl diethoxyacetate (67JOC1139), and 1-carbomethoxy-2-methoxypseudo-urea [83GEP(O)3222342]. Some C-13 and C-14 labeled triazolo-pyridazines were prepared (84MI2) according to this route using suitable C-labeled chlorides.

In many cases, the intermediate hydrazides 269, prepared either directly (66T2073) from 3-hydrazinopyridanzines (267) and acid derivatives or indirectly [78GEP(O)2741763, 78USP4112095] from 3-halopyridazines (268) and acid hydrazides, were isolated and subsequently cyclized. Belonging to the route of forming the fused 1,2,4-triazole onto a pyridazine, and bearing a close resemblance to the utilization of hydrazinopyridazines, is the reaction of methyl phenylhydrazonochloroacetate (**272**) with pyridazine (**271**) to give (75JHC1133) the 1,3-disubstituted 1,2,4-triazolo-[4,3-b]pyridazine (**273**).

Pyridazin-3-ylhydrazones of aromatic aldehydes (**274**), on oxidative cyclization with bromine (66T2073; 67T387; 80ACH405; 83H749) or lead tetraacetate (66T2073; 68SAP6706255), gave the corresponding 3-substituted 1,2,4-triazolo[4,3-b]pyridazines (**275**). Subjecting ketone pyridazin-3-ylhydrazones (**276**) to thermal cyclization may yield either or both of the two possible products **227** and **278**. Almost always, however, only one product is obtained [76JCS(P1)1363] that retains the smaller of the two groups of the original ketone; the larger group is eliminated as the corresponding hydrocarbon.

In carrying out such a dehydrogenative cyclization on alkyl hydrazones **280** derived from 6-hydrazinotetrazolo[1,5-b]pyridazine (**279**) using lead tetraacetate, it was interesting to find (70JOC1138) that the fused 1,2,4-triazole ring was formed with concomitant opening of the tetrazole ring to an azido function to yield 3-alkyl-6-azido-1,2,4-triazolo[4,3-b]pyridazines (**281**, R = alkyl). Cyclization of **279** with cyanogen bromide also led (68JHC351) to opening of the tetrazole ring, giving 3-amino-6-azido-1,2,4-triazolo[4,3-b]pyridazine (**281**, R = NH_2).

2. *1,2,4-Triazolo[3,4-a]phthalazines*

1,2,4-Triazolo[3,4-a]phthalazines (**256**) were first obtained by Hartmann and Druey (49USP2484029) in 1949. The most widely used method usually depends on 1-hydrazinophthalazines (**282**) as the starting material and formation of the fused 1,2,4-triazole ring. Thus, condensative cyclization of **282** with various one-carbon cyclizing agents such as carboxylic acids [49USP484029; 51HCA195; 69JOC3221; 78JAP(K)7821197; 87MI5], carboxylic acid anhydrides (75JOC2901), acid chlorides (49USP2484029; 51HCA195; 54BP711756; 58BAP227; 75JOC2901; 87JHC667), carboxylic acid esters (75JOC2901), orthoesters (69JOC3221; 75JOC2901), cyanogen bromide (69JOC3221), carbon disulfide (69JOC3221), urea (69JOC3221), ethyl chloroformate (51HCA195), phthalaldehydic acid (81JHC1625), 1,3-dicarbonyl compounds (82PHA451; 83JHC1231; 87H1177), aroyl (acyl) pyruvate (83JHC1231; 87H1177), isothiocyanates (81JHC499), substituted nitriles [75JOC2901; 78JAP(K)7821197; 87JHC667], and 2-methyl-2-thiopseudourea (78JHC311) gave a wide variety of 3-substituted 1,2,4-triazolo[3,4-a]phthalazine (283).

Double ring closure occurred when 1,4-dihydrazinophthalazine (284) was cyclized with cyanogen bromide or triethyl orthoformate to give (69JOC3221) 1,2,4-triazolo[3,4-a]1,2,4-triazolo[4,3-c]phthalazines (285). Double ring closure failed, however, when cyclization of 284 was made with formic acid (59JOC1205) or ethyl acetoacetate (56GEP951992); only one ring closure took place to give 286.

Dehydrogenative cyclization of aldehyde phthalazin-1-ylhydrazones (287) afforded [51HCA195; 82MI2; 83JAP(K)58124786] the corresponding 1,2,4-triazolo[3,4-a]phthalazines (288). A one-pot synthesis of 3-aryl-1,2,4-triazolo[3,4-a]phthalazines (290) was made (84S602) by the oxidation of o-phthalaldehyde bis(aroylhydrazones) (289) with lead tetraacetate. 6-Hydrazino-1,2,4-triazolo[3,4-a]phthalazine (292) was obtained (56GEP-951993) upon heating a mixture of 1-amino-3-iminoisoindolenine (291), hydrazine hydrate, and formic acid in DMF.

When 1-hydrazinophthalazine (**294**) was allowed to react with 3,4,6-tri-*O*-benzoyl-2,5-anhydroaldonic acid (**293**), it gave (78MI1) the 1,2,4-triazolo[3,4-*a*]phthalazine C-nucleoside **295**. 3-(Alditol-1-yl)-1,2,4-

triazolo[3,4-*a*]phthalazines (**301**) were synthesized through two different routes: (a) catalytic dehydrogenative cyclization of *aldehydo*-sugar phthalazinyl hydrazones (81MI1; 83MI2, 83MI3) (**297**) and (b) autodehydrative cyclization of aldonic acid phthalazinyl hydrazides (**300**) (89UP1).

3. 1,2,4-Triazolo[4,3-b]cinnolines

Cyclization of 3-hydrazino-5,6,7,8-tetrahydrocinnoline (**302**) with carboxylic acid, urea, cyanogen bromide, and carbon disulfide gave (71BSF3043) the corresponding 3-substituted 6,7,8,9-tetrahydro-1,2,4-triazolo[4,3-*b*]cinnolines (**303**). The same products (**303**) were also obtained [79JCS(P1)3085], together with the 1,2,4-triazolo-phthalazines (**306**), by the reaction of 2-hydroxymethylenecyclohexanone (**304**) with 4-amino-1,2,4-triazoles (**305**). Compounds of this class were found to possess multifarious activities such as antihistaminic (81EUP29130, 81USP4293554), hypotensive (76USP3994898; 77MI1; 78USP4112095),

[Reaction scheme: (302) + one-carbon cyclizing reagent → (303)]

[Reaction scheme: (304) + (305) → (303) + (306)]

anti-inflammatory [76USP3994898; 78JAP(K)7821197], antimicrobial [73GEP(O)2161586; 75GEP(O)2418435], inotropic (83H749; 87PHA304), anthelmintic [83GEP(O)3222342], anxiolytic [78GEP(O)2741763; 79MI1; 81USP4260756], benzodiazepine tranquilizing antagonist (85EUP156734), and mild central nervous system stimulants (79PHA801).

B. 1,2,4-TRIAZOLO-1,3-DIAZINES

The title compounds include five fused-ring systems (**307–311**) within the scope of this chapter.

1,2,4-Triazolo[4,3-a]pyrimidines (**307**)

1,2,4-Triazolo[4,3-c]pyrimidines (**308**)

1,2,4-Triazolo[4,3-a]quinazolines (**309**)

Sec. XII.B] CONDENSED 1,2,4-TRIAZOLO[3,4-z]HETEROCYCLES 327

1,2,4-Triazolo[3,4-b] quinazolines (310)

1,2,4-Triazolo[4,3-c] quinazolines (311)

1. *1,2,4-Triazolo[4,3-a]pyrimidines*

Construction of the pyrimidine ring of 1,2,4-triazolo[4,3-*a*]pyrimidines has been achieved by condensing 3-amino-1,2,4-triazoles with 1,3-dicarbonyl compounds (57JCS727; 60JOC361; 62BSF355), propiolate esters (70CB3266; 71CB2702), acrylate esters (53CB1401; 71CB3961), and alkoxyacrolines (83S44). The reaction of 3-amino-1,2,4-triazoles (**312**) with β-ketoesters may yield the following possible isomeric products **313–316**. Yet, only one product of these isomers will be obtained depending on the substitutent on the 3-amino-1,2,4-triazole (**312**). Thus, changing the substitutent to phenyl from hydrogen, methyl, or methylmercapto, changes (09CB4638; 52MI1; 57JCS727; 58BEP566543; 59JOC787; 60JCS1829, 60JOC361, 60YZ(79)899; 61USP2933388; 87T2497] the product from 1,2,4-triazolo[1,5-*a*]pyrimidin-5-ones (**314**) to 1,2,4-triazolo[4,3-*a*]pyrimidin-7-ones (**315**). Ultraviolet absorption spectrometry has been used (59JOC779) to differentiate between the four possible isomers **313–316**, but ^{13}C-NMR spectra gave (87T2497) the unequivocal proof of these structures.

Reiter and his group (88JHC173) have used Schiff bases (e.g. **317**) derived from 1-substituted 5-amino-3-methylthio-1,2,4-triazoles as precursors for the synthesis of the title compounds through their reaction with phenoxyacetyl chloride in the presence of triethylamine or with a mixture of dichloroacetic acid and phosphoryl chloride in DMF to give **318** and **319**, respectively. 1,2,4-Triazolo[4,3-*a*]pyrimidines (**321**) were also synthesized from 2-hydrazinopyrimidines (**320**) by condensation with carboxylic acids [57JCS727; 59JAP3326, 59JOC787, 59JOC793, 59YZ(79)899, 59YZ(79)1487; 60YZ(80)956; 62BSF355, 62FRPP1308696; 70CB3266; 71CB2702, 71ZC422; 86JPR237], orthoesters [57JCS727; 59JOC779, 59JOC787, 59JOC793, 59YZ903, 59YZ(79)1482, 59YZ(79)1487; 60JCS1829, 60JOC361, 60YZ(80)952; 68T2839; 81AJC2635; 85CPB3113; 86JPR237; 87JOC2220], acetic anhydride (68ZC421), carbon disulfide (57JCS727; 58JAP8072; 60JCS1829, 60JOC361; 68T2839), phenyl isothiocyanate (59JOC787, 60JOC361), and ethyl imidates [67JCS(C)498].

The direction of cyclization of the unsymmetrical 2-hydrazino-4-hydroxy-6-methylpyrimidine (**322**) seems to be affected by the acidity as well as the nature of the cyclizing agent (59JOC787). Thus, with nonacidic formylating agents such as triethyl orthoformate, **322** gave (59JOC779, 59JOC787, 59JOC793; 86JPR237) the two isomeric 1,2,4-triazolo[4,3-*a*]pyrimidines **323** and **324**. Cyclization with formic acid, however, gave (59JOC787, 59JOC793) **323** only.

Hydrazones of symmetrically disubstituted pyrimidin-2-ylhydrazines (57JCS727; 77CPB3137) (e.g. **325**) or 1-N-substituted pyrimidin-2-ylhydrazones [67JCS(C)498] undergo dehydrogentative cyclization photochemically (77CPB3137) or chemically with lead tetraacetate [57JCS727; 67JCS(C)498; 77CPB3137] to give 1,2,4-triazolo[4,3-*a*]pyrimidines (**326**). On the other hand, hydrazones (e.g. **327**) of the unsymmetrically disubstituted pyrimidine-2-ylhydrazines may theoretically yield either or both of the two possible products **328** and **329**. Bower and Doyle (57JCS727) reported that the dehydrogenative cyclization of benzaldehyde (4-hydroxy-6-methylpyrimidin-2-yl)hydrazone (**327**) with lead tetraacetate gave 3-phenyl-7-methyl-1,2,4-triazolo[4,3-*a*]pyrimidin-5-one (**328**), whereas Allen *et al.* (60JOC361) claimed that this product is 3-phenyl-5-methyl-1,2,4-triazolo[4,3-*a*]pyrimidin-7-one (**329**) and rationalized this conclusion on the basis that the same product was obtained from the reaction of 5-amino-3-phenyl-1,2,4-triazole (**330**) with ethyl acetoacetate (60JOC361). Obviously, this rationale is irrelevant since the last reaction may also yield either of the two structures.

2-Ethylmercaptopyrimidine derviative **331** was used as a starting material to obtain the 2-acylhydrazinopyrimidines (**332**) which underwent dehydrative cyclization in boiling phenol (59JOC793; 60JOC361) to 1,2,4-triazolo[4,3-*a*]pyrimidines **313** or **315**, depending on the acyl group of the acid hydrazide. Differentiation between the two structures has been achieved through comprehensive study of their ultraviolet absorption spectra (60JOC361).

Krenzel (86H93, 86KGS1350; 87JPR331) prepared the 3-oxo-2,3,5,6,7,-8-hexahydro-1,2,4-triazolo[4,3-a]pyrimidines (**337**) by reacting 1-ethoxy-

Sec. XII.B   CONDENSED 1,2,4-TRIAZOLO[3,4-z]HETEROCYCLES   331

carbonyl-2-methylthio-1,4,5,6-tetrahydropyrimidine (**333**) with hydrazine or aryl hydrazines according to the following scheme. Bitha *et al.* (87JOC2220) reported that pyrimidino[1,2-*a*]1,2,4,5-tetrazin-6-ones (**339**) undergo acid-catalyzed ring contraction to yield the 1,2,4-triazolo[4,3-*a*]pyrimidin-7-ones (**342**). The latter were also obtained directly by cyclizing the corresponding 2-hydrazinopyrimidin-4-ones (**338**) with orthoesters.

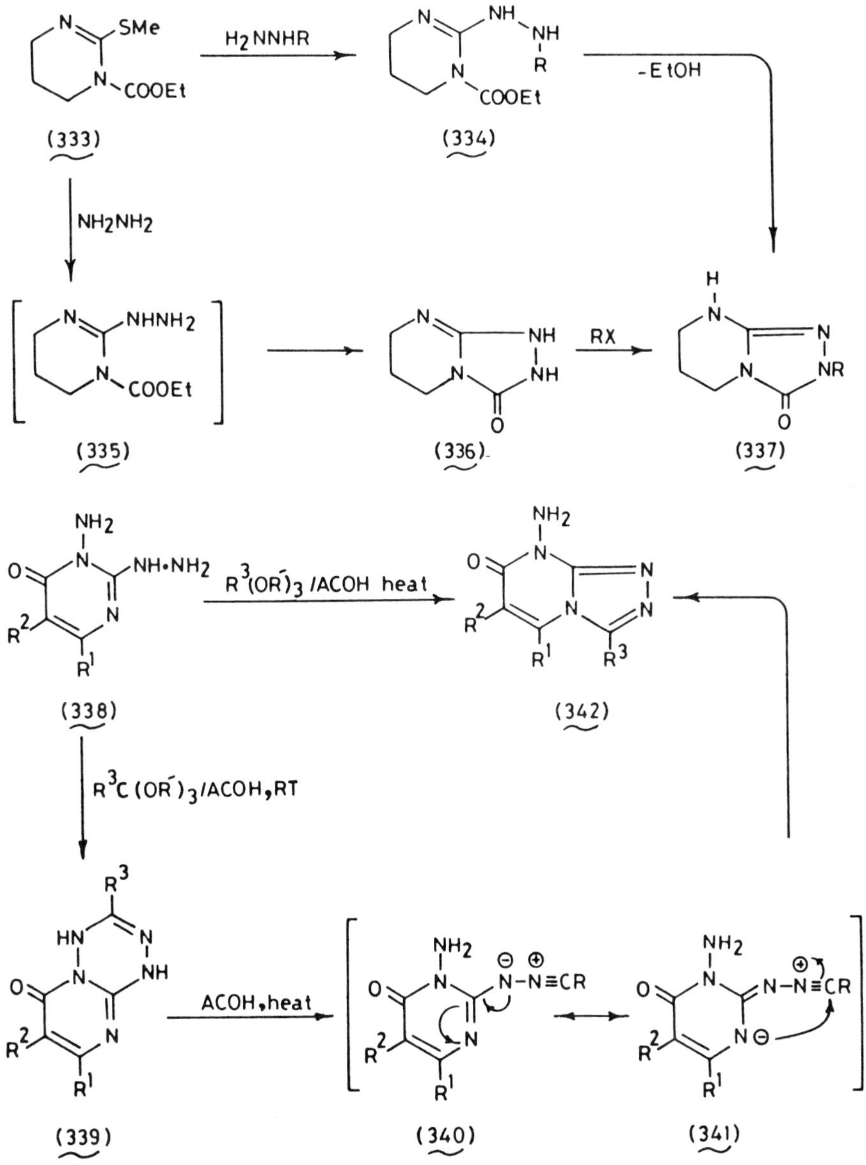

2. 1,2,4-Triazolo[4,3-c]pyrimidines

The procedure of forming the pyrimidine ring of this system was used in the reaction of 1,3-diamino-4,6-bis(5-phenyl-1,2,4-triazol-3-yl)benzene (**343**) with benzoyl chloride to give (76MI1) benzo[bis(1,2,4-triazolopyrimidine)] **344** through cyclization of the intermediate di-*N*-benzoyl derivative. A variety of substituted 1,2,4-triazolo[4,3-*c*]pyrimidines (**346**) were synthesized by cyclocondensation of 4-hydrazinopyrimidines (**345**) with carboxylic acids (56PJ804; 60G1821), acetic anhydride (87MI6), 1,1-diethoxyethyl acetate (81USP4269980), orthoesters (83EUP121341; 85USP4532242; 87MI7), acid chlorides (70CB3278), cyanogen chloride (62BP898408; 63JCS5642), ethyl chloroformate (65JCS3357), carbon disulfide (65JCS3369; 82AJC1263), and diethyl oxalate (87MI6).

3. 1,2,4-Triazolo[4,3-a]quinazolines

The quinazoline ring of 1,2,4-triazolo[4,3-*a*]quinazoline **348** was formed (62TL1193) by the dehydrative cyclization of 3-benzoylamino-4,5-diphenyl-1,2,4-triazole (**347**). Alternatively, 1,2,4-triazolo[4,3-*a*]-quinazolines were prepared by assembling the fused 1,2,4-triazole ring onto the quinazoline nucleus when 2-hydrazinoquinazolines (e.g. **349**) were cyclized with formic acid or acetic anhydride to give (86JHC833) angular 1,2,4-triazolo[4,3-*a*]quinazolinones **350** in addition to linear 1,2,4-triazolo[3,4-*b*]quinazolinones **351**.

(347) →POCl₃, −H₂O→ (348)

(349) →HCOOH or Ac₂O→ (350) + (351)   R = H or Me

An unambiguous synthesis which leads to 1,2,4-triazolo[4,3-*a*]-quinazolines only (e.g. 353) involves the cyclization of 3-substituted-2-hydrazinoquinazolines (e.g. 352) with one-carbon cyclizing reagents such as carboxylic acids [80PHA800; 83GEP(D)158549; 83PHA25], acid chlorides (83EUP76199, 83PHA25), orthoesters [80GEP(D)139715;

(352) →one-carbon cyclizing reagent→ (353)

86JCR(S)232, 86JHC833], cyanogen bromide (83PHA25), carbon disulfide (78PHA124), ethyl chloroformate (78PHA125), dicarboxylic esters (83PHA25), urea (80PHA800; 83PHA25), phosgene (83PHA25), thiophosgene (83PHA25), isocyanates (80PHA800), isothiocyanates (80PHA800), and 1,1-dicarbonyldiimidazole (86JHC833). Oxidative cyclization of quinazolin-4-on-2-ylhydrazones 354 with ferric chloride gave (86JHC833) angular 1,2,4-triazolo[4,3-*a*]quinazolinones 356 and not linear 1,2,4-

triazolo[3,4-b]quinazolinones 355. The evidence concerning (86JHC833) the structure 356 was provided by preparing the compounds' ethyl derivatives 358, which were also obtained directly from 3-ethyl-2-hydrazinoquinazolin-4-one (357).

4. *1,2,4-Triazolo[3,4-b]quinazolines*

An attempt to prepare substituted 1,2,4-triazoloquinazolin-5-one 359 and/or 360 by cyclizing 2-hydrazinoquinazolin-4-one (349) with benzoyl chloride in DMF obtained, interestingly enough, (86JHC833) unsubstituted 1,2,4-triazolo[3,4-b]quinazolin-5-one 362 as a result of the formylating effect of the DMF in the presence of acid chlorides (Vilsmeier reagent). Assignment of linear structure 362 and not the alternative angular structure, 361, to the product was based (86JHC833) on the fact that its ethylation product, 363, was different from angular ethylated product 364, obtained by direct cyclization of 3-ethyl-2-hydrazinoquinazoline-4-one (357).

5. 1,2,4-Triazolo[4,3-c]quinazolines

Condensative cyclization of 3-(pyrid-2-yl)-5-(2-aminophenyl)-1,2,4-triazole (**365**) with benzoic acid, in the presence of polyphosphoric acid, or with benzoyl chloride, followed by thermal cyclization of intermediate **366** gave (72JHC131) 5-phenyl-3-(pyrid-2-yl)-1,2,4-triazolo[4,3-c]quinazoline (**367**). Cyclization of 4-hydrazinoquinazolines (**368**) with formic acid (81PHA62), acetic anhydride (81PHA62), carbon disulfide (63N732), and orthoesters (77USP4053600) also gave the 1,2,4-triazolo[4,3-c]-quinazolines (**369**). Aldehyde quinazolin-4-ylhydrazones (**370**) and quinazolin-4-ylhydrazides (**371**) have been used as starting materials to obtain (63N732; 89UP2, 89UP3) 3-substituted 1,2,4-triazolo[4,3-c]quinazolines **372**. 4-Chloroquinazoline (**373**) reacted with 5-substituted tetrazoles (**223**) in pyridine to afford (61CB1555) 3-substituted 1,2,4-triazolo[4,3-c]quinazolines (**372**).

(365) → PhCOOH, PPA → (367)

(365) → PhCOCl → (366) → Heat (-H₂O) → (367)

(368) → one-carbon cyclizing reagent → (369)

(370) → FeCl₃ → (372)

(371) → POCl₃, -H₂O → (372)

Sec. XII.C]   CONDENSED 1,2,4-TRIAZOLO[3,4-z]HETEROCYCLES   337

(373) + (223) —pyridine, −HCl, −N₂→ (372)

Many 1,2,4-triazolopyrimidines and 1,2,4-triazoloquinazolines are useful in various applications such as antihistaminics (83EUP76199), acaricidals (77GEP(O)2533120], insecticidals [77GEP(O)2533120], nematocidals [77GEP(O)2533120], analgesics [74GEP(O)2508333; 84IJC(B)1293], tranquilizers [77USP4053600), anti-inflammatory agents [73GEP(O) 2261095; 74GEP(O)2508333; 77USP4053600], antihypertensives [73GEP(O)2261095], antagonists for nucleic acid metabolism (58JPA8072; 59JPA3326), inhibitors of passive cutaneous anaphylaxis [80GEP(D)139715; 83GEP(D)158549), granulomata (62BP898408), and in the treatment of rheumatoid arthritis (62BP898408).

C. 1,2,4-TRIAZOLO-1,4-DIAZINES

Under this title, the synthesis of the following two ring systems **374** and **375**, will be discussed.

1,2,4-Triazolo[4,3-*a*] pyrazines (374)

1,2,4-Triazolo[4,3-*a*] quinoxalines (375)

1. *1,2,4-Triazolo[4,3-a]pyrazines*

Cyclization of 2-hydrazinopyrazines (**376**) with one-carbon cyclizing reagents such as formic acid (62JOC3243), benzoyl chloride (62JOC-3243), orthoesters [62JOC3243; 69JCS(C)1593; 79JOC1028; 81AJC2635], ethyl chloroformate [66JCS(C)2038; 69BP1146770], carbon disulfide

(69JCS(C)1593; 81AJC2635], urea [66JCS(C)2038], phenylisothiocyanate (62JOC3243), phosgene [66JCS(C)2038; 69JCS(C)1593; 79JOC1028], halonitriles [66JCS(C)2038; 69BP1146770, 69JCS(C)1593], diethoxymethyl acetate (77JOC4197; 83USP4402958), and thioacetamidates (79JOC1028; 84JMC924; 86MI1) gave the title compounds (**377**). 3-Aryl-1,2,4-triazolo-[4,3-*a*]pyrazine (**380**) was obtained (62JOC3243) by cyclizing 1-aroyl-2-(pyrazin-2-yl) hydrazines (**378**) with phosphoryl chloride. Hydrazides **378** were also intermediates in the one-step synthesis (62JOC3243), using 2-chloropyrazine (**379**) and aroylhydrazines.

Aromatic aldehyde pyrazin-2-ylhydrazones (e.g. **381**) undergo (83USP4402958) dehydrogenative cyclization with lead tetra-acetate to yield the corresponding 1,2,4-triazolo[4,3-*a*]pyrazines (e.g. **382**). Fusion of a triazole ring to a pyrazine has also been accomplished by the reaction (75JHC1133) of methyl phenylhydrazonochloroacetate (**272**) with

Sec. XII.C] CONDENSED 1,2,4-TRIAZOLO[3,4-z]HETEROCYCLES 339

pyrazine (**383**) in the presence of triethylamine to give 1-phenyl-3-methoxycarbonyl-1,2,4-triazolo[4,3-*a*]pyrazine (**384**).

2. *1,2,4-Triazolo[4,3-a]quinoxalines*

Cyclization of 3-substituted 2-hydrazinoquinoxalines **385** with carboxylic acids (60AJ4044), acid chlorides (60JA4044), carboxylic anhydrides (60JA4044; 76GEP(O)2515494], orthoesters (78ZC92; 85H2025), dicarboxylic esters (77PHA687), ethyl acetoacetate (60JA4044), pyruvic acid (60JA4044), ethyl chloroformate (60JA4044; 77PHA687), carbon disulfide (77PHA687), phenyl cyanate (77PHA687), and 2-methyl-2-thiopseudourea (78JHC311) gave 3,10-disubstituted 1,2,4-triazolo[4,3-*a*]quinoxalines **386**. 1,2,4-Triazolo[4,3-*a*]1,2,4-triazolo[3,4-*c*]quinoxalines **389** were obtained either by double ring closure of 2,3-dihydrazinoquinoxaline (**387**) with orthoesters (68ZC302) or by single ring closure of 10-

hydrazino-1,2,4-triazolo[4,3-a]quinoxalines (**388**) using triethyl orthoformate (78ZC92), ethyl chloroformate (85ZC366), and carbon dilsulfide (85ZC366).

Thermal cyclization of aldehyde quinoxalin-2-yl-hydrazones (**390**) gave (60JA4044) corresponding 1,2,4-triazolo[4,3-a]quinoxalines **391**. Ketone quinoxalin-2-ylhydrazones (**392**), however, yield (60JA4044) either **391** or **393**, whichever carries the smaller of the two groups of original ketone; the larger group will be eliminated as the corresponding hydrocarbon.

The reaction of 2-chloroquinoxalines (**394**) with acid hydrazides led directly (60JA4044) to 1,2,4-triazolo[4,3-*a*]quinoxalines **386** without isolating the corresponding intermediate hydrazides. 4-Chloro-1,2,3,4-tetrazolo[1,5-*a*]quinoxaline (**395**) was reported (78ZC175) to react with 5-substituted tetrazoles (**223**) in the presence of triethylamine to give 1,2,4-triazolo[4,3-*a*]tetrazolo[5,1-*c*]quinoxalines **396**. Members of this class of compounds were found to possess fungicidal (85H2025), virustatic (77PHA687), antiallergic (85JMC363), antiviral (84JMC924; 86MI1), anxiolytic (83USP4402958), coronary vasodilating (84JMC924), and bronchodilating properties (73MI2) along with inhibition of passive cutaneous anaphylaxis (81EUP39920) and selective adenosine antagonist (88JMC1011) activities.

XIII. Condensed 1,2,4-Triazolo-oxazines

Syntheses of relevant 1,2,4-triazolo-1,2-oxazines have not appeared.

A. 1,2,4-Triazolo-1,3-oxazines

Of the four theoretically possible modes of fusing the two heterocyclic rings of this system (**397–400**), only two were synthesized (**397** and **399**) and will be reviewed.

1,2,4-Triazolo[3,4-b]1,3-oxazines (397)

1,2,4-Triazolo[4,3-c]1,3-oxazines (398)

1,2,4-Triazolo[3,4-b]1,3-benzoxazines (399)

1,2,4-Triazolo[4,3-c]1,3-benzoxazines (400)

1. *1,2,4-Triazolo[4,3-b]1,3-oxazines*

Sasaki *et al.* (84CPB5040) synthesized an example of this group by the thermal cyclization of 1-benzoyl-2-(4,5,6-trihydro-1,3-oxazin-2-yl)hydrazine (**401**) to give 3-phenyl-5,6,7-trihydro-1,2,4-triazolo[3,4-*b*]1,3-oxazine (**402**).

(401) →[Heat, -H₂O] (402)

2. *1,2,4-Triazolo[3,4-b]1,3-benzoxazines*

1,3-Dipolar cycloaddition of the *N*-aryl-*C*-phenylnitrilimines (**404**) with 1,3-benzoxazines **403** gave [86JCR(S)200] 1,2,4-triazolo[3,4-*b*]1,3-benzoxazines **405**.

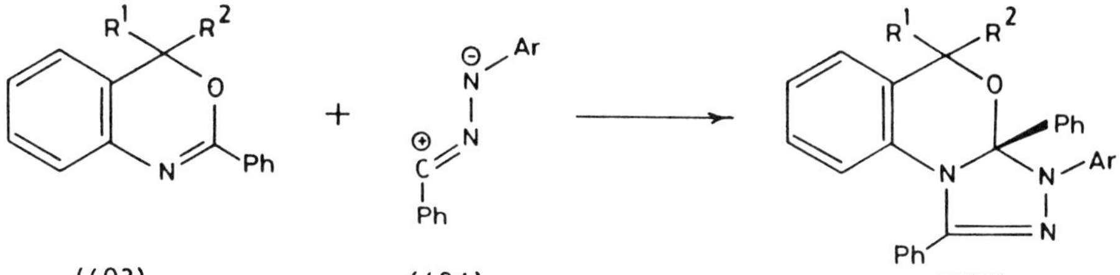

Sec. XIII.B] CONDENSED 1,2,4-TRIAZOLO[3,4-z]HETEROCYCLES 343

(403) + (404) → (405)

B. 1,2,4-TRIAZOLO-1,4-OXAZINES

In this class of fused compounds, the two possible ring systems, **406** and **407**, are within the scope of this chapter.

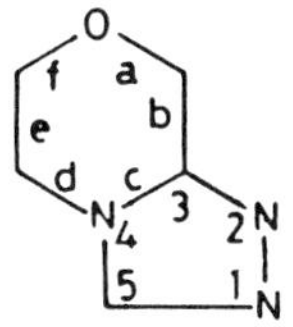

1,2,4-Triazolo [3,4-c] 1,4-oxazines (406)

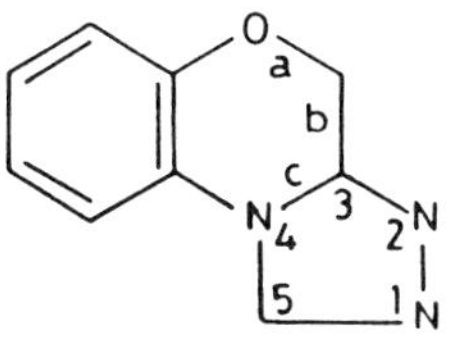

1,2,4-Triazolo [3,4-c] 1,4-benzoxazines (407)

1. *1,2,4-Triazolo[3,4-c]1,4-oxazines*

Dehydrative cyclization of 1-acyl-3-(2,5,6-trihydro-1,4-oxazin-3-yl)hydrazines (**408**) led [81IJC(B)132] to the formation of the 3-substituted 5,6,8-trihydro-1,2,4-triazolo[3,4-c]1,4-oxazines (**409**).

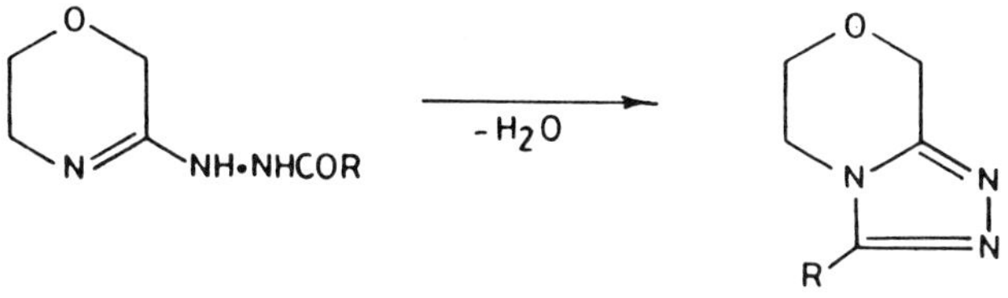

(408) —$-H_2O$→ (409)

2. 1,2,4-Triazolo[3,4-c]1,4-benzoxazines

Reactive chlorines of 3-chloro-1,4-benzoaxazin-2-ones (81UPS4276292; 85H871) and 3-chloro-1,4-benzoxazin-2-thiones (81USP4276292), or the thiol group of 1,4-benzoxazin-3-thiones (**410**), (75USP3929783) condense with acid hydrazides to give corresponding 1,2,4-triazolo[3,4-c]1,4-benzoxazines **411**. Thermal cyclization of 1-(1,4-benzoxazin-3-yl)-2-ethoxycarbohydrazine (**412**) led (85USP4547499) to the elimination of an ethanol molecule along with the formation of 1,2,4-triazolo[3,4-c]-1,4-benzoxazin-3-one (**413**). Few reports documented antiallergic (81USP4276292; 85H871; 85USP454799), antianxiety (75USP3929783), or anti-inflammatory [81IJC(B)132] activities.

XIV. Condensed 1,2,4-Triazolo-thiazines

1,2,4-Triazolo-1,2-thiazines have not been synthesized.

A. 1,2,4-TRIAZOLO-1,3-THIAZINES

Of the four possible ring systems, **414–417**, pertaining to this chapter, the synthesis of only **414** and **417** has been reported.

1,2,4-Triazolo[3,4-b]1,3-thiazines (414)

1,2,4-Triazolo[4,3-c]1,3-thiazines (415)

1,2,4-Triazolo[3,4-b]1,3-benzothiazines (416)

1,2,4-Triazolo[4,3-c]1,3-benzothiazines (417)

1. *1,2,4-Triazolo[3,4-b]1,3-thiazines*

Reaction of 3-chloropropyl isothiocyanate (**418**) with acid hydrazides in the presence of triethylamine gave (79H1171) 3-substituted 5,6,7-trihydro-1,2,4-triazolo[3,4-*b*]1,3-thiazines **419**. Cyclization of cinnamyltriazolethiones **420** with HBr afforded (83MI4) the 3,7-disubstituted 5,6-dihydro-1,2,4-triazolo[3,4-*b*]1,3-thiazines (**421**) in addition to rearranged isomeric products **422**.

2. 1,2,4-Triazolo[4,3-c]1,3-benzothiazines

Cycloaddition of 1,3-benzothiazines (**423**) to hydrazonyl halides **424** led (84H537) to the formation of examples (**425**) of the title compounds.

B. 1,2,4-TRIAZOLO-1,4-THIAZINES

Of the possible condensed systems, **426** and **427**, only system **427** has been synthesized.

1,2,4-Triazolo[3,4-c]1,4-thiazines (**426**)

1,2,4-Triazolo[3,4-c]1,4-benzothiazines (**427**)

1,2,4-Triazolo[3,4-c]1,4-benzothiazines

Reactions of 1,4-benzothiazin-3-thione (75USP3929783) or 3-chloro-1,4-benzothiazin-2-ones (**428**) (85H871) with acid hydrazide gave **429**. Some of

Sec. XV.A] CONDENSED 1,2,4-TRIAZOLO[3,4-z]HETEROCYCLES 347

these compounds showed anxiolytic (75USP3929783) and antiallergic (85H871) activities.

XV. Condensed 1,2,4-Triazolo-triazines

A. 1,2,4-TRIAZOLO-1,2,3-TRIAZINES

Belonging to this class are ring systems **430** and **431**. Synthesis of system **431** only has been reported.

1,2,4-Triazolo [4,3-c] 1,2,3-triazines (**430**)

1,2,4-Triazolo [4,3-c] 1,2,3-benzotriazines (**431**)

1,2,4-Triazolo[4,3-c]1,2,3-benzotriazines

Cyclization of 4-hydrazino-1,2,3-benzotriazine (**432**) with triethyl orthoformate gave (70JOC3448) 1,2,4-triazolo[4,3-*c*]1,2,3-benzotriazine **433**. Compound **433** has also been synthesized (70JOC3448) by cyclizing 3-(2-aminophenyl)-1,2,4-triazole (**434**) with nitrous acid.

B. 1,2,4-TRIAZOLO-1,2,4-TRIAZINES

Under this heading, the synthesis of the five isomeric 1,2,4-triazolo-1,2,4-triazines, **435–439**, having nitrogen bridgeheads will be reviewed.

1,2,4-Triazolo [4,3-b] 1,2,4-triazines (**435**)

1,2,4-Triazolo [3,4-c] 1,2,4-triazines (**436**)

1,2,4-Triazolo [3,4-c] 1,2,4-benzotriazines (**437**)

1,2,4-Triazolo [4,3-d] 1,2,4-triazines (**438**)

1,2,4-Triazolo [3,4-f] 1,2,4-triazines (**439**)

The first (**435**) was prepared as early as 1934 by Stolle and Dietrich (34JPR139), while the last four (**436–439**) were only obtained after 1960.

1. *1,2,4-Triazolo[4,3-b]1,2,4-triazines*

The first example 3-aminonaphtho[1,2-*e*]1,2,4-triazolo[4,3-*b*]1,2,4-triazine (**442**) was synthesized by Stolle and Dietrich (34JPR193) in the reaction of 2-aminonaphthalene (**440**) with 3,5-dichloroimino-3,5-dihydro-1,2,4-triazole (**441**). Extensively studied was the cyclization of 3,4-diamino-

Sec. XV.B] CONDENSED 1,2,4-TRIAZOLO[3,4-z]HETEROCYCLES 349

(440)   (441)   (442)

1,2,4-triazoles (46) with a wide variety of two-carbon cyclizing reagents such as 1,2-dicarbonyl compounds [50JCS614, 50JCS1579; 52JCS4817; 54JA619; 55MI1; 73MI1; 76JCS(P1)1492; 77JOC1018; 79JHC1393], α-ketoacids (70JOC3448; 83MI4; 84H537), α-ketocarboxylate (64BEP642615, 64FRP1379480) dicarboxylic esters (70AP650), and ethyl diethoxyacetate (64BEP642615) to give 1,2,4-triazolo[4,3-b]1,2,4-triazines 443.

(46)   (443)

3-Hydrazino-1,2,4-triazines (444) were used as starting materials for synthesizing this ring system by condensation with the common one-carbon cyclizing reagents, namely, carboxylic acids [55MI1; 64CB-2647, 64FRP1379480; 68CB3969; 72JCS(P1)1221; 79JHC1393; 83MIP1; 86JHC721; 89UP3), benzoyl chloride (55MI1; 60MI1), acetic anhydride (55MI1; 60MI1), orthoesters [68CCC2513; 69BSF2492; 72JCS(P1)1221; 76JCS(P1)1492; 79JHC1393], carbon disulfide (64BEP642615; 83MI4), cyanogen bromide [76JCS(P1)1492; 79JHC1393; 81JHC1353], urea [64CB2179; 76JCS(P1)1492], and formamide (60MI1). That the products of this reaction possessed structure 443 and not the alternative isomeric structure 445 was proved by direct comparison with the corresponding structure 443, unequivocally prepared [64BEP642615, 64FRP1379480, 69BSF2492; 76JCS(P1)1492; 79JHC1393] from 3,4-diamino-1,2,4-triazoles (46). In some cases, however, the structures of the cyclization products were not decisively assigned. Thus, the products obtained from cyclization of 3-hydrazino-1,2,4-triazino[5,6-b]indoles (446) were claimed by some investigators (71ZOR173; 87JHC1435; 89UP4) to be the linear 4',3':2,3]-1,2,4-triazino[5,6-b]indoles 447 or the angular [3',4':3,4]-1,2,4-triazino[5,6-b]indoles 448 (87AP1191, 87AP1196; 88MI1).

3-Acylhydrazino-1,2,4-triazines **450**, obtained from 3-methylmercapto-1,2,4-triazines (**449**) and acid hydrazides, were also reported (64BEP-642615; 69BSF2492) to undergo dehydrative cyclization to 1,2,4-triazolo-[4,3-*b*]1,2,4-triazines **443**.

2. *1,2,4-Triazolo[3,4-c]1,2,4-triazines*

Cyclocondensation of 3-hydrazino-1,2,4-triazoles (**451**) with 1,2-dicarbonyl compounds may involve either N-2 or N-4 of the triazole ring, according to the reaction conditions, to give (77JOC1018) either the

1,2,4-triazolo[3,4-c]1,2,4-triazines (**452**) or the 1,2,4-triazolo[5,1-c]1,2,4-triazines (**455**). Assignment of structure **452** to the product obtained under controlled conditions was possible (77JOC1018) when it was isomerized to **455**. The latter was also obtained (77JOC1018) by cyclization of 1-acetonyl-3-chloro-1,2,4-triazoles (**453**) with hydrazine followed by lead tetraacetate. Daunis *et al.* (69BSF2492) reported that ring closure of 5-hydroxy-3-hydrazino-1,2,4-triazines (**456**) with ethyl formate took place with the formation of the two possible isomeric structures 1,2,4-triazolo[4,3-c]1,2,4-triazines (**457**) and 1,2,4-triazolo[3,4-b]1,2,4-triazines (**458**). Sasaki and Ito (81JHC1353), however, noted that ring closure of **456** with cyanogen bromide gave only **459**.

As expected, cyclocondensation of 3-hydrazino-1,2,4-triazines, having their N-2 blocked (e.g. **460**), gave (64CB2185) the only possible product, namely, 1,2,4-triazolo[3,4-c]1,2,4-triazines (**461**).

3. *1,2,4-Triazolo[3,4-c]1,2,4-benzotriazines*

1,2,4-Triazolo[3,4-c]1,2,4-benzotriazines (**463**) were prepared by cyclization of 3-hydrazino-1,2,4-benzotriazines (**462**) with carboxylic acids (69CB3818), ortho esters (80MI3), carbon disulfide (69CB3818), and phenyl isothiocyanate (80MI3).

4. *1,2,4-Triazolo[4,3-d]1,2,4-triazines*

The title compounds have been prepared by cyclization of 5-hydrazino-1,2,4-triazines (67CB3467; 68CB2747; 70BSF1606; 71RRC311; 74MIP1; 81JHC1353; 82JHC1345) (e.g. **464**) with carboxylic acids (67CB3467; 68CB2747; 70BSF1606; 71RRC311; 74MIP1), orthoesters (82JHC1345), carbon disulfide (67CB3467), and cyanogen bromide (81JHC1353) and by the reaction (68CB2747) of 3-mercapto-1,2,4-triazine (e.g. **465**) with acid hydrazides followed by dehydrative cyclization of intermediate hydrazides **466** with polyphosphoric acid.

The two nitrogen bridgeheads of the bis(1,2,4-triazolo)-1,2,4-triazines **472** were concomitantly formed by double ring closure of 3,5-dihydrazino-1,2,4-triazines (69BSF3670) (**468**) with carboxylic acids or by a single ring closure of either 7-hydrazino-1,2,4-triazolo[4,3-b]1,2,4-triazines **469**

Sec. XV.B] CONDENSED 1,2,4-TRIAZOLO[3,4-z]HETEROCYCLES 353

[67AC(R)366; 68CB3969; 69BSF3670] or 5-hydrazino-1,2,4-triazolo[4,3-d]1,2,4-triazines **470** (69BSF3670). Dehydrogenative cyclization of aldehyde 1,2,4-triazolo[4,3-b]1,2,4-triazin-7-ylhydrazones (**471**) also gave **472** (69BSF3670).

5. 1,2,4-Triazolo[3,4-f]1,2,4-triazines

Reaction of 1,2,4-triazol-4-ylamidines (**473**) with diethyl carbonate in the presence of sodium ethoxyethoxide led (70JPR669) to intermediate **474** which, upon cyclization, gave the 1,2,4-triazolo[3,4-f]1,2,4-triazino-8-ones (**475**). Kurasawa and his group (85JHC1715) reported that the reaction of 3-(α-hydroxyimino-4-amino-5-methyl-4H-1,2,4-triazol-3-ylmethyl)2-oxo-1,2-dihydroquinoxaline (**476**) with orthoesters and iron powder in acetic acid gave 1,2,4-triazolo[3,4-f]1,2,4-triazines **480** and its 7,8-dihydro isomers **479**.

Subjecting 6-hydrazino-1,2,4-triazino-5-ones (**481**) to cyclization with carboxylic acids (79JHC555), acetic anhydride (79JHC555), benzoyl chloride (79JHC555), orthoesters (79JHC555), carbon disulfide (79JHC555), cyanogen bromide (79JHC555), or 2,5-anhydroallonic acid (86JMC2231) gave examples of the title compounds (**482**). Reaction of

(**483**) + RCONHNH$_2$ $\xrightarrow{-\text{HBr} -\text{H}_2\text{O}}$ (**482**)

6-bromo-1,2,4-triazin-5-ones (**483**) with acid hydrazides directly afforded (74JPR667) the 1,2,4-triazolo[3,4-*f*]1,2,4-triazin-8-ones (**482**) without isolation of the corresponding intermediate hydrazides.

(**481**) $\xrightarrow{\text{one-carbon cyclizing reagent}}$ (**482**)

Many 1,2,4-triazolo-1,2,4-triazines showed biological activities such as antibacterial (74MIP1), antiviral (74MIP1), antimetabolic (74MIP1), antitumour (86JMC2231), and anti-inflammatory (83MIP1) activities as well as inhibition of lymphoid leukemia (71RRC311).

C. 1,2,4-Triazolo-1,3,5-triazines

There is only one possible combination of this fused ring system (**484**) related to this chapter.

1,2,4-Triazolo [4,3-*a*] 1,3,5-triazines (**484**)

1,2,4-Triazolo[4,3-a]1,3,5-triazines

One route to these compounds is the reaction of 3-amino-1,2,4-triazoles (**312**) with dicyanodiamide (**485**) to yield (49USP2473797; 53JOC1610) 5,7-diamino-1,2,4-triazolo[4,3-*a*]1,3,5-triazines (**486**). The 1,3,5-triazine ring of 7-methoxy-3-phenyl-1,2,4-triazolo[4,3-*a*]1,3,5-triazine (**489**) was formed (70T3357) by the direct cyclocondensation of *O*-methyl-*N*-(5-phenyl-1,2,4-triazol-3-yl)urea (**488**) with methyl diethoxyacetate (**487**) or triethyl orthoformate. 3,5,7-Triamino-1,2,4-triazolo[4,3-*a*]1,3,5-triazine (**490**) was prepared (53JOC1610; 63ZOB1355) in one step by the reaction of two equivalents of dicyanodiamide (**485**) with one equivalent of hydrazine dihydrochloride.

Starting with 2-hydrazino-1,3,5-triazines (e.g. **491**), workers have formed the fused 1,2,4-triazole (**492**) (70T3357) through ring closure with methyl diethoxyacetate (**487**). Dehydrogenative cyclization of bis(4,6-substituted)-1,3,5-triazin-2-ylhydrazones **493** gave (66M1713; 70T3357; 77CPB3137) only one product (**494**) because of the symmetry of the starting hydrazone. On the other hand, the product of dehydrogenative cyclization of the unsymmetrically substituted hydrazones, **495, 496,** or **497,** was found to be influenced (70T3357) by the electronic and steric factors of the substitutents. Thus, subjecting aldehyde 6-methoxy-4-methyl-1,3,5-triazin-2-ylhydrazones (**498**) to dehydrogenative cyclization with lead tetraacetate afforded mainly (70T3357) the 5-methoxy-7-methyl-1,2,4-triazolo[4,3-a]1,3,5-triazines (**499**) as a result of cyclization with the more nucleophilic nitrogen adjacent to the more electron-releasing methoxy group. Alternative isomer **501** was obtained (70T3357) by the reaction of **500** with triethyl orthoacetate.

(500) → (501)

Reaction of 2,4,6-trichloro-1,3,5-triazine (502) with three equivalents of 5-substituted tetrazoles 223 led (61CB1555) to the formation of 1,2,4-triazolo[4,3-a]1,2,4-triazolo[4,3-c]1,2,4-triazolo[4,3-e]1,3,5-triazines (503).

XVI. Condensed 1,2,4-Triazolo-thiadiazines

1,2,4-Triazole rings may, theoretically, be fused to 1,2,3-, 1,2,4-, 1,2,5-, 1,2,6-, 1,3,4-, or 1,3,5-thiadiazine rings. However, only 1,2,4-triazolo-1,3,4-thiadiazines have been prepared.

1,2,4-TRIAZOLO-1,3,4-THIADIAZINES

Two condensed ring systems, 504 and 506, have been synthesized and will be reviewed; system 505 has not been synthesized.

1,2,4-Triazolo [3,4-b] 1,3,4-thiadiazines (504)

Sec. XVI] CONDENSED 1,2,4-TRIAZOLO[3,4-z]HETEROCYCLES 359

1,2,4-Triazolo [4,3-d] 1,3,4-thiadiazines (505)

1,2,4-Triazolo [3,4-b] 1,3,4-benzothiadiazines (506)

1. *1,2,4-triazolo[3,4-b]1,3,4-thiadiazines*

Cyclization of 4-amino-3-mercapto-1,2,4-triazoles **189** with α-haloketones (52JCS4811; 69IJC959; 74IJC287; 78GEP(O)2818395; 78IJC(B)-481; 80JIC1112; 81IJC(B)369; 83JPS45; 86IJC(B)283, 86JCR(S)70, 86JHC1439], β-haloacetals [52JCS4811; 69IJC959; 74IJC287; 78IJC-(B)481], chloroacetic acid (74IJC287; 80JIC1112), chloracetyl chloride ([81IJC(B)369], 2,3-dichloroquinoxaline [74IJC287; 78IJC(B) 481], dimethyl acetylenedicarboxylate [78IJC(B)737; 81OPP123], and benzoin [81IJC(B)369] led to the formation of the title compounds.

Dehydrative cyclization of 3-(*S*-carboxy)-5-phenyl-4-(pyrrol-1-yl)1,2,4-triazole (**508**) with polyphosphoric acid gave (82G345) the 1,2,4-triazolo[3,4-*b*]1,3,4-thiadiazine (**509**). Ring transformation of 1,3,4-oxadiazoles to 7*H*-1,2,4-triazolo[3,4-*b*]1,3,4-thiadiazines (**507**) was reported by Sasaki *et al.* (82JOC2757) through the reaction of [(1,3,4-oxadiazol-2-yl)thio]ketones (**123**) with hydrazine hydrate in the presence of acetic acid.

Phenacyl bromides react (86JHC43) with anhydro 1-amino-5-aryl-2-mercapto-1,2,4-triazolo[3,2-c]quinazolin-4-ium hydroxides (**53**) through pyrimidine ring-opening and simultaneous formation of the 1,3,4-thiadiazine nucleus to give **54**. Compound **514** was also obtained [86JAP(K)61260085] when 2-hydrazino-5-methyl-6*H*-1,3,4-thiadiazine (**513**) was cyclized with aryloxyacetyl chlorides.

2. *1,2,4-triazolo[3,4-b]1,3,4-benzothiadiazines*

Examples of those compounds (**517**) having saturated benzene rings were prepared by reacting 4-amino-3-mercapto-1,2,4-triazoles (**516**) with 2-halocyclohexanones (**515**) (78JHC209), 2-halocyclohexan-1,3-diones (78JHC209), dimedone [81IJC(B)369], and 2-bromo-3,6-dihydroxy-5-undecyl-1,4-benzophenone (85MI2). Some 1,2,4-triazolo[3,4-*b*]1,3,4-thiadiazines exhibited antibacterial (86JHC1439), antiparasitic (83JPS45), and fungicidal [78IJC(B)481] activities.

XVII. Condensed 1,2,4-Triazolo-tetrazines

1,2,4-Triazole rings may form condensed ring systems with 1,2,3,4-, 1,2,3,5-, or 1,2,4,5-tetrazines, but only 1,2,4-triazolo-1,2,4,5-tetrazines have been synthesized.

1,2,4-Triazolo-1,2,4,5-Tetrazines

1,2,4-Triazolo[4,3-*b*]1,2,4,5-tetrazines (**518**) pertain to this review.

1,2,4-triazolo[4,3-b]1,2,4,5-tetrazines

Dickinson and Jacobsen [74AC298; 75JCS(P1)975] prepared 1,2,4-triazolo[4,3-*b*]1,2,4,5-tetrazine **520** by reacting 6-phenyl-3-hydrazino-1,2,4,5-tetrazine (**519**) with carbon disulfide or by reacting 4-amino-5-hydrazino-1,2,4-triazol-3-thione (**522**) with benzaldehyde in alkaline medium. The reaction involved the air oxidation of tetrahydrotriazolo-tetrazine intermediate **522**.

XVIII. Condensed 1,2,4-Triazolo-azepines

The synthesis of the two categories of compounds **523** and **524** will be reviewed.

1,2,4-Triazolo[4,3-*a*] azepines (**523**)

1,2,4-Triazolo[4,3-*a*] benzazepines (**524**)

A. 1,2,4-TRIAZOLO[4,3-a]AZEPINES

These compounds were obtained (70S433) by cyclization of 2-hydrazinoazepines such as 2-hydrazino-3,4,5,6,7-pentahydroazepine (**525**) with N-dichloromethylene benzamide to give 3-benzoylamino-1,2,4-triazolo[4,3-a]5,6,7,8,9-pentahydroazepine (**526**). Reaction of methyl chloroformate, phosgene, or thiophosgene with the 3,4,5,6,7-pentahydroazepin-2-one arylhydrazones (**527**) afforded (80USP4213773) the 3-oxo- or 3-thioxo-5,6,7,8,9-pentahydro-1,2,4-triazolo[4,3-a]azepines (**528**).

3-Substituted 5,6,7,8,9-pentahydro-1,2,4-triazolo[4,3-a]azepines (**531**) were prepared (57CB909; 59BP811765; 59BP825514) from the imidate ester **529** by reaction with acid hydrazides to give disubstituted hydrazines **530**, which undergo direct thermal dehydrative cyclization to **531**. Cyclization (70CB1934) of **530** with phosgene afforded 2-acyl-3-oxo-derivatives **532**.

B. 1,2,4-TRIAZOLO[4,3-a]BENZAZEPINES

Condensation of the thiol group of 5-aryl-1,3,4,5-tetrahydrobenzazepin-2-thiones (**533**) with acid hydrazides afforded (81CCC148) fused 1,2,4-triazolobenzazepines **534**. Compounds of this system exhibited analeptic (59BP811765, 59BP825514), herbicidic (80USP4213773), anticonvulsant (81CCC148), and central nervous system (59BP825514) and respiratory system stimulating activities (59BP825514).

XIX. Condensed 1,2,4-Triazolo-diazepines

1,2,4-Triazolo-1,2-, 1,3-, or 1,4-diazepines have not been synthesized.

1,2,4-TRIAZOLO-BENZODIAZEPINES

Of the various possible 1,2,4-triazolo-benzodiazepines relevant to this chapter, only the three types, **535–537**, have been synthesized.

1,2,4-Triazolo[4,3-a]1,4-benzodiazepines (**535**)

Sec. XIX]   CONDENSED 1,2,4-TRIAZOLO[3,4-z]HETEROCYCLES   365

1,2,4-Triazolo $\left[4,3-\underline{d}\right]$ 1,4-benzodiazepines (**536**)

1,2,4-Triazolo $\left[4,3-\underline{a}\right]$ 1,5-benzodiazepines (**537**)

1. *1,2,4-triazolo[4,3-a]1,4-benzodiazepines*

Numerous reports have described [72GEP(O)2203782, 72USP3681343; 73CPB1619, 73JAP(K)7334199; 75JHC717; 76GEP(O)2601400; 77USP-3994940; 80JHC575, 80JMC392, 80JMC643] the synthesis of compounds such as **539** by the condensation of 2-(3-halomethyl-1,2,4-triazol-4-yl) benzophenones **538** with ammonia, as well as by condensation of 2-hydrazino-1,4-benzodiazepines **540** with carboxylic acids (77BP1469197;

79JMC1390), acid chlorides (79JMC1390, 79USP4141902; 80JMC392), orthoesters (70TL4039), cyanogen bromide (79JMC1390), or thiophosgene (79JMC1390). The reactive thiol group of 1,4-benzodiazepin-2-thiols **541** condenses with acid hydrazides to give [70GEP(O)2012190; 71TL1609; 72SAP7201487, 72SAP7201488, 72USP3681343; 75GEP(O)2426305; 80-JMC392] the title compounds.

2. *1,2,4-Triazolo[4,3-d]1,4-benzodiazepines*

Compounds such as **543** and **545** were prepared either from 5-hydrazino-1,4-benzodiazepines (e.g. **542**), by cyclization (78JHC1127) with carboxylic acid derivatives, or from 1,4-benzodiazepine-5-imidates (e.g. **544**, Z = OMe) (71TL1609), 5-thioimidates (e.g. **544,** Z = SMe) (78JHC1127), or 5-imidoyl chlorides (e.g. **544** Z = Cl) (76TL1931; 77TL1699) with carboxylic acid hydrazides.

3. *1,2,4-Triazolo[4,3-a]1,5-benzodiazepines*

Reaction of 5-phenyl-2-thio-1,4-dihydro-1,5-benzodiazepin-4-ones (**546**) with acid hydrazides afforded [75GEP(O)2426305] 11-*H*-1,2,4-triazolo[4,3-*a*]1,5-benzodiazepin-10-ones (**547**). Members of this ring system have been

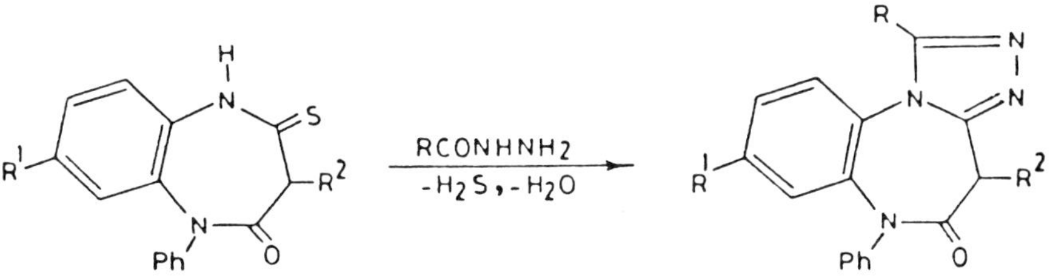

Sec. XX]   CONDENSED 1,2,4-TRIAZOLO[3,4-z]HETEROCYCLES   367

(546) → (547)

RCONHNH$_2$, $-H_2S$, $-H_2O$

found to be useful as sedatives [70GEP(O)2012190; 71GEP(O)2036324; 72SAP7201489, 72USP3681343; 73GEP(O)2251673; 77BP1469197, 77-USP4032535; 78JMC1290; 79USP4141902; 80JHC575], tranquilizers [72USP3681343; 73GEP(O)2251673; 77BP1469197; 79USP4141902; 86-JHC43], nictonine antagonists (72USP3681343), antianxiolytics (72MI1; 79JMC1390, 79USP4180668; 80JMC392, 80JMC643, 80JMC873), antidepressants (71JMC1078; 77BP1469197; 79JMC1390; 80JMC392, 80JMC-402), antispasmodics [70GEP(O)2012190], muscle relaxants [73GEP(O)-2251673; 77BP1469197, 77USP4032535; 79USP4141902], hypnotics [73GEP(O)2251673], psychotropics (79JMC1390), and anticonvulsants (77USP4032535; 78JMC1290).

XX. Condensed 1,2,4-Triazolo-triazepines

Of the various possible 1,2,4-triazolo-triazepines, only 1,2,4-triazolo-1,2,4-triazepines have been synthesized.

1,2,4-Triazolo-1,2,4-triazepines

Two ring systems, **548** and **549**, have been synthesized.

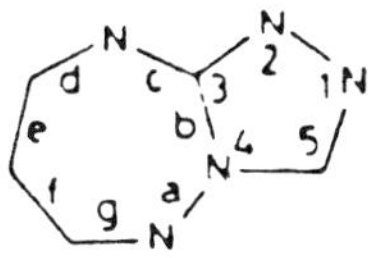

1,2,4-Triazolo[4,3-b]1,2,4-triazepines (548)

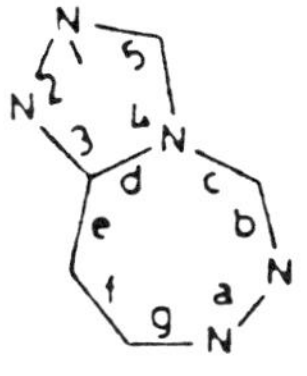

1,2,4-Triazolo[4,3-d]1,2,4-triazepines (549)

1. *1,2,4-Triazolo[4,3-b]1,2,4-triazepines*

These compounds were prepared (70AP709; 72JHC153; 74JHC751; 75JHC661; 85MI1) by cyclizing 3,4-diamino-1,2,4-triazoles (46) with β-ketoesters. The reaction of 46 with ethyl acetoacetate was reported (74JHC751) to yield products that were assigned 6-methyl-7,9-dihydro-1,2,4-triazolo[4,3-*b*]1,2,4-triazepin-8-one structures (550) by some authors (74JHC751) or 8-methyl-5,9-dihydro-1,2,4-triazolo[4,3-*b*]1,2,4-triazepin-6-ones (551) by others (70AP709). Kochlar (72JHC153), however, was able to isolate both products.

2. *1,2,4-Triazolo[4,3-d]1,2,4-triazepines*

Cyclization of 2,7-dimethyl-5-hydrazino-6*H*-1,2,4-triazepin-3-thione (552) with orthoesters gave [78JCR(S)190] 3-substituted 6,8-dimethyl-9*H*-1,2,4-triazepin-5-thiones 553.

Acknowledgment

This chapter is dedicated to Professors G. Soliman and H. S. El Khadem.

References

1900CB1885	W. Marckwald and E. Meyer, *Chem. Ber.* **33**, 1885 (1900).
1900CB1895	W. Marckwald and M. Chain, *Chem. Ber.* **33**, 1895 (1900).
03CB1111	W. Marckwald and K. Rudzik, *Chem. Ber.* **36**, 1111 (1903).
09CB2209	C. Bulow, *Chem. Ber.* **42**, 2209 (1909).
09CB2595	C. Bulow, *Chem. Ber.* **42**, 2595 (1909).
09CB4638	C. Bulow and K. Haas, *Chem. Ber.* **42**, 4638 (1909).
10CB1975	C. Bulow and K. Haas, *Chem. Ber.* **43**, 1975 (1910).
15JCS688	R. G. Fargher and R. Furness, *J. Chem. Soc.*, 688 (1915).
23JCS312	W. H. Mills and H. Schindler, *J. Chem. Soc.*, 312 (1923).
34JPR193	R. Stolle and W. Dietrich, *J. Prakt. Chem.* **139**, 193 (1934).
45USP2390707	N. Heimbach, U.S. Pat. 2,390,707 (1945) [*CA* **40**, 20791 (1946)].
47USP2432419	N. Heimbach, U.S. Pat. 2,432,419 (1947) [*CA* **42**, 2193 (1948)].
48JA1381	D. S. Tarbell, C. W. Todd, M. C. Paulson, E. G. Lindstrom, and V. P. Wystrach, *J. Am. Chem. Soc.*, **70**, 1381 (1948).
49USP2473797	D. W. Kaiser, U.S. Pat. 2,473,797 (1949) [*CA* **43**, 7975 (1949)].
49USP2484029	M. Hartmann and J. Druey, U.S. Pat. 2,484,029 (1949) [*CA* **44**, 4046 (1950)].
50BP634951	J. D. Kendall, G. F. Duffin, and Ilford Ltd. Br. Pat. 634,951 (1950) [*CA* **44**, 9287 (1950)].
50JCS614	E. Hoggarth, *J. Chem. Soc.*, 614 (1950).
50JCS1579	E. Hoggarth, *J. Chem. Soc.*, 1579 (1950).
51HCA195	J. Druey and B. H. Ringier, *Helv. Chim. Acta* **34**, 195 (1951) [*CA* **45**, 10248 (1951)].
52JCS4811	E. Hoggarth, *J. Chem. Soc.*, 4811 (1952).
52JCS4817	E. Hoggarth, *J. Chem. Soc.*, 4817 (1952).
52MI1	E. J. Birr, *Z. Wiss. Photogr., Photophys. Photochem.* **47**, 2 (1952) [*CA* **47**, 2617 (1953)].
53CB1401	E. J. Birr and W. Walther, *Chem. Ber.* **86**, 1401 (1953).
53JOC1610	D. W. Kaiser, G. A. Peters, and V. P. Wystrach, *J. Org. Chem.* **18**, 1610 (1953).
54BP711756	Roche Products Ltd., Br. Pat. 711,756 (1954) [*CA* **49**, 11723 (1955)].
54JA619	E. C. Taylor, Jr., W. H. Gumprecht, and R. F. Vance, *J. Am. Chem. Soc.* **76**, 619 (1954).
55JPJ1242	N. Takahayashi, *J. Pharm Soc. Jpn.* **75**, 1242 [*CA* **50**, 8655 (1956)].
55MI1	R. Fusco and S. Rossi, *Rend. Ist. Lomb. Sci. Lett., Cl Sci. Mat. Nat.* **88**, 173 (1955) [*CA* **50**, 10742 (1956)].
56GEP951992	C. F. Mainkur, Ger. Pat. 951,992 (1956) [*CA* **53**, 4314 (1959)].
56GEP951993	C. F. Mainkur, Ger. Pat. 951,993 [*CA* **53**, 5298 (1959)].
56JPJ804	D. Shiho, S. Tagami, N. Takahayashi, and R. Honda, *J. Pharm. Soc. Jpn.* **76**, 804 (1956) [*CA* **51**, 1196 (1957)].
56JPJ1133	M. Kanaoka, *J. Pharm. Soc. Jpn.* **76**, 1133 (1956) [*CA* **51**, 5390 (1957)].

56JPJ1296	N. Takahayashi, *J. Pharm. Soc. Jpn.* **76**, 1296 (1956) [*CA* **51**, 6645 (1957)].
57CB909	S. Peterson and E. Tietze, *Chem. Ber.* **90**, 909 (1957).
57JCS727	J. D. Bower and F. P. Doyle, *J. Chem. Soc.*, 727 (1957).
57JCS4510	J. D. Bower, *J. Chem. Soc.*, 4510 (1957).
57MI1	N. Takahayashi, *Pharm. Bull.* **5**, 229 (1957) [*CA* **52**, 6359 (1958)].
57MI2	M. Kanaoka, *Pharm. Bull.* **5**, 385 (1957) [*CA* **52**, 5390 (1958)].
57USP2786054	L. G. S. Brooker and E. J. V. Lare, U.S. Pat. 2,786,054 (1957) [*CA* **51**, 9385 (1957)].
58BEP566543	Anonymous, Kodak Soc., Belg. Pat. 566,543 (1958) [*CA* **53**, 14796 (1959)].
58JAP8072	K. Shirakawa, Jpn. Pat. 8072 (1958) [*CA* **54**, 4035 (1960)].
58MI1	S. Biniecki, A. Haase, J. Izdebski, E. Kesler, and L. Rylski, *Bull. Acad. Pol. Sci., Ser. Sci., Chim., Geol. Geogr.* **6**, 227 (1958) 227 [*CA* **52**, 18424 (1958)].
58USP2837521	D. M. Burness, U.S. Pat. 2,837,521 (1958) [*CA* **53**, 2262 (1959)].
58USP2861076	E. B. Knott and L. A. Williams, U.S. Pat. 2,861,076 (1958) [*CA* **53**, 9251 (1959)].
58ZOB2773	A. N. Kost and F. Gents, *Zhr. Obshch. Khim.* **28**, 2773 (1958) [*CA* **53**, 9197 (1959)].
59BP811765	Farbenfabriken Bayer Akt.-Ges, Br. Pat. 811,765 (1959) [*CA* **54**, 577 (1960)].
59BP825514	Farbenfabriken Bayer Akt.-Ges., Br. Pat. 825,514 (1959) [*CA* **55**, 7450 (1961)].
59JA6289	E. A. Steck and P. Brundage, *J. Am. Chem. Soc.* **81**, 6289 (1959).
59JAP3326	K. Shirakawa, Jpn. Pat. 3326 (1959) [*CA* **54**, 14277 (1960)].
59JOC779	C. F. Allen, H. R. Beilfuss, D. M. Burness, G. A. Reynolds, J. F. Tinker, and J. A. VanAllan, *J. Org. Chem.* **24**, 779 (1959).
59JOC787	C. F. H. Allen, H. R. Beilfuss, D. M. Burness, G. A. Reynolds, J. F. Tinker, and J. A. VanAllan, *J. Org. Chem.* **24**, 787 (1959).
59JOC793	C. F. Allen, H. R. Beilfuss, D. M. Burness, G. A. Reynolds, J. F. Tinker, and J. A. VanAllan, *J. Org. Chem.* **24**, 793 (1959).
59JOC1205	G. A. Reynolds, J. A. VanAllan, and J. F. Tinker, *J. Org. Chem.* **24**, 1205 (1959).
59JOC1478	G. A. Reynolds and J. A. VanAllan, *J. Org. Chem.* **24**, 1478 (1959).
59USP2917511	J. B. Bicking, U.S. Pat. 2,917,511 (1959) [*CA* **54**, 8854 (1960)].
59YZ(79)899	K. Shirakawa, *Yakugaku Zasshi* **79**, 899 (1959) [*CA* **54**, 556 (1960)].
59YZ(79)903	K. Shirakawa, *Yakugaku Zasshi* **79**, 903 (1959) [*CA* **54**, 556 (1960)].
59YZ(79)1482	K. Shirakawa, *Yakugaku Zasshi* **79**, 1482 (1959) [*CA* **54**, 11039 (1960)].
59YZ(79)1487	K. Shirakawa, Yakugaku Zasshi **79**, 1487 (1959) [*CA* **54**, 11039 (1960)].
60BP839020	Ilford Ltd., Br. Pat. 839,020 (1960) [*CA* **55**, 2322 (1961)].
60GI821	B. Camerino, G. Palamidessi, and R. Sciaky, *Gazz. Chim. Ital.* **90**, 1821 (1960) [*CA* **57**, 7261 (1962)].
60JA4044	D. I. Shiho and S. Tagami, *J. Am. Chem. Soc.* **82**, 4044 (1960).
60JCS1829	L. A. Williams, *J. Chem. Soc.*, 1829 (1960).
60JOC361	C. F. H. Allen, G. A. Reynolds, J. F. Tinker, and L. A. Williams, *J. Org. Chem.* **25**, 361 (1960).
60MI1	J. Hadacek, *Spisy Prirodoved. Fak. Univ. J. E. Purkyne Brne* **22**, 29 (1960–1961) [*CA* **55**, 25977 (1961)].

60YZ(80)952	K. Shirakawa, *Yakugaku Zasshi* **80,** 952 (1960) [*CA* **54,** 24761 (1960)].
60YZ(80)956	K. Shirakawa, *Yakugaku Zasshi* **80,** 956 (1960) [*CA* **54,** 24761 (1960)].
61ACS1295	J. Sandstrom, *Acta Chem. Scand.* **15,** 1295 (1961) [*CA* **57,** 12471 (1962)].
61CB1555	R. Huisgen, H. J. Sturm, and M. Seidel, *Chem. Ber.* **94,** 1555 (1961).
61MI1	A. Haase and S. Biniecki, *Acta Pol. Pharm.* **18,** 461 (1961) [*CA* **61,** 3103 (1964)].
61USP2933388	E. B. Knott, U.S. Pat. 2,933,388 (1960) [*CA* **55,** 1254 (1961)].
62ACS2389	S. Linholter and R. Rosenoern, *Acta Chem. Scand.* **16,** 2389 (1962) [*CA* **59,** 1632 (1963)].
62BP898408	Imperial Chemical Industries Ltd., Br. Pat. 898,408 (1962) [*CA* **57,** 11209 (1962)].
62BSF355	D. Libermann and R. Jacquier, *Bull. Soc. Chim. Fr.,* 355 (1962).
62FRP1308696	D. Libermann, Fr. Pat. 1,308,696 (1962) [*CA* **58,** 12579 (1963)].
62JOC3243	P. J. Nelson and K. T. Potts, *J. Org. Chem.* **27,** 3243 (1962).
62LA148	H. Genlen and G. Roebisch, *Justus Liebigs Ann. Chem.* **660,** 148 (1962) [*CA* **58,** 9051 (1963)].
62TL1193	S. Naqui and V. R. Srinivasan, *Tetrahedron Lett.,* 1193 (1962).
62USP3050525	J. B. Bicking, U.S. Pat. 3,050,525 (1962) [*CA* **58,** 1480 (1963)].
63JCS5642	G. W. Miller and F. L. Rose, *J. Chem. Soc.,* 5642 (1963).
63JCS5660	N. K. Basu and F. L. Rose, *J. Chem. Soc.,* 5660 (1963).
63N732	G. S. Sidhu, G. Thyarajan, and N. Rao, *Naturwissenschaften* **50,** 732 (1963) [*CA* **60,** 6841 (1964).
63T1587	M. S. Gibson, *Tetrahedron* **19,** 1587 (1963).
63USP3096329	E. A. Steck, U.S. Pat. 3,096,329 (1963) [*CA* **59,** 14004 (1963)].
63ZOB1355	V. A. Titkov, and I. D. Plentnev, *Zh. Obshch. Khim.* **33,** 1355 (1963) [*CA* **59,** 12952 (1963).
63BEP642615	Ilford Ltd., Belg. Pat. 642,615 (1964) [*CA* **63,** 18127 (1965)].
64CB2179	A. Dornow, W. Abele, and H. Menzel, *Chem. Ber.* **97,** 2179 (1964).
64CB2185	A. Dornow, H. Menzel, and P. Marx, *Chem. Ber.* **97,** 2185 (1964).
64CB2647	A. Dornow, H. Pietsch, and P. Marx, *Chem. Ber.* **97,** 2647 (1964).
64E200	V. R. Rao and V. R. Srinivasan, *Experientia* **20,** 200 (1964) [*CA* **60,** 14496 (1964)].
64FRP1379480	Ilford Ltd., Fr. Pat. 1,379,480 (1964) [*CA* **62,** 11838 (1965)].
64JHC42	T. Kuraishi and R. N. Castle, *J. Heterocycl. Chem.* **1,** 42 (1964).
65AG963	F. L. Scott, R. N. Butler, and D. A. Cronin, *Angew. Chem.* **77,** 963 (1965) [*CA* **64,** 2081 (1966)].
65CPB586	Y. Nitta, I. Matsuura, and F. Yoneda, *Chem. Pharm. Bull.* **13,** 586 (1965) [*CA* **63,** 9939 (1965)].
65IJC162	S. Naqui and V. R. Srinivasan, *Indian J. Chem.* **3,** 162 (1965).
65JCS3357	G. W. Miller and F. L. Rose, *J. Chem. Soc.,* 3357 (1965).
65JCS3369	W. Broadbent, G. W. Miller, and F. L. Rose, *J. Chem. Soc.,* 3369 (1965).
65JHC302	R. G. Child and A. S. Tomcufcik, *J. Heterocycl. Chem.* **2,** 302 (1965).
65JOC711	H. H. Takimoto, G. C. Denault, and S. Hotta, *J. Org. Chem.* **30,** 711 (1965).
65TL841	F. L. Scott and J. B. Aylward, *Tetrahedron Lett.,* 841 (1965).

65ZOR136	G. N. Tyurenkova and N. P. Bednyagina, *Zh. Org. Khim.* **1**, 136 (1965) [*CA* **62**, 16234 (1965)].
65ZOR139	N. P. Bednyagina and I. N. Gestsova, *Zh. Org. Khim.* **1**, 139 (1965) [*CA* **62**, 16234 (1965)].
66CB2593	T. Kauffmann, K. Voget, S. Barck, and J. Schulz, *Chem. Ber.* **99**, 2593 (1966).
66JCS(C)78	A. R. Katritzky and S. Musierowicz, *J. Chem. Soc. C*, 78 (1966).
66JCS(C)1202	F. L. Scott and R. N. Butler, *J. Chem. Soc. C*, 1202 (1966).
66JCS(C)2038	S. E. Mallett and F. L. Rose, *J. Chem. Soc. C*, 2038 (1966).
66JHC119	H. H. Takimoto, G. C. Denault, and S. Hotta, *J. Heterocycl. Chem.* **3**, 119 (1966).
66JHC158	G. S. Sidhu, S. Naqui, and D. S. Iyengar, *J. Heterocycl. Chem.* **3**, 158 (1966).
66JOC251	K. T. Potts and H. R. Burton. *J. Org. Chem.* **31**, 251 (1966).
66JOC3528	K. T. Potts and R. M. Huseby, *J. Org. Chem.* **31**, 3528 (1966).
66M1713	M. Jelenc, J. Kobe, B. Stanovnik, and M. Tisler, *Monatsh. Chem.* **97**, 1713 (1966) [*CA* **66**, 104972 (1967)].
66MI1	S. Takase and T. Demura, *Kogyo Kagaku Zasshi* **69**, 1417 (1966) [*CA* **66**, 37838 (1967)].
66T2073	A. Pollak and M. Tisler, *Tetrahedron* **22**, 2073 (1966).
67AC(R)366	F. D'Alo and A. Masserini, *Ann. Chim. Rome* **57**, 366 (1967) [*CA* **67**, 43784 (1967)].
67CB875	C. E. Voelcker, J. Marth, and H. Beyer, *Chem. Ber.* **100**, 875 (1967).
67CB3467	T. Sasaki and K. Minamoto, *Chem. Ber.* **100**, 3467 (1967).
67JCS(C)239	R. N. Butler and F. L. Scott, *J. Chem. Soc. C*, 239 (1967).
67JCS(C)498	R. G. W. Spickett and S.H.B. Wright, *J. Chem. Soc. C*, 498 (1967).
67JOC1139	S. Stanovink, A. Krbavcic, and M. Tisler, *J. Org. Chem.* **32**, 1139 (1967).
67MI1	V. W. Dymek, B. Janik, A. Cygankiewiez, and H. Gawron, *Acta Pol. Chim.* **24**, 97 (1967) [*CA* **68**, 28208 (1968)].
67T387	B. Stanovnik and M. Tisler, *Tetrahedron* **23**, 387 (1967).
67TL3071	R. Fusco, P. D. Croce, and A. Salvi, *Tetrahedron Lett.*, 3071 (1967).
67USP3354164	J. E. Francis, U.S. Pat. 3,354,164 (1967) [*CA* **69**, 36142 (1968)].
68BP1131590	C. F. Boehringer and G.M.B.H. Soehne, Br. Pat. 1,131,590 (1968) [*CA* **70**, 28926 (1969)].
68CB2747	T. Sasaki, K. Minamato, and S. Fukuda, *Chem. Ber.* **101**, 2747 (1968).
68CB3969	T. Sasaki, K. Minamoto, and M. Murata, *Chem. Ber.* **101**, 3969 (1968).
68CCC2513	K. Kalfus, *Collect. Czech. Chem. Commun.* **33**, 2513 (1968) [*CA* **69**, 59160 (1968)].
68DOK127	L. N. Yakhontov, E. V. Pronina, B. V. Rozynov, and M. V. Rubtsov, *Dokl. Akad. Nauk SSSR* **178**, 127 (1968) [*CA* **69**, 27343 (1968)].
68JCS(C)1711	R. N. Butler and F. L. Scott, *J. Chem. Soc. C*, 1711 (1968).
68JHC351	A. Kovacic, B. Stanovnik, and M. Tisler, *J. Heterocycl. Chem.* **5**, 351 (1968).
68JOC143	K. T. Potts and C. Hirsch, *J. Org. Chem.* **33**, 143 (1968).
68JOC1097	H. W. Heine, A. B. Smith, III, and J. D. Bower, *J. Org. Chem.* **33**, 1097 (1968).

68SAP6706255	H. Berger, K. Stach, and W. Voemel, S. Afr. Pat. 67/06,255 (1968) [*CA* **70**, 57869 (1969)].
68T2687	H. G. O. Becker and H. Boettcher, *Tetrahedron* **24**, 2687 (1968).
68T2839	A. H. Beckett, *Tetrahedron* **24**, 2839 (1968).
68ZC302	H. Paul, G. Richmann, and E. Mantey, *Z. Chem.* **8**, 302 (1968) [*CA* **69**, 77236 (1968)].
68ZC421	A. Spasov, E. Golovinski, and G. Rusev, *Z. Chem.* **8**, 421 (1968([*CA* **70**, 28885 (1969)].
69BP1146770	J. A. Maguire and F. L. Rose, Br. Pat. 1,146,770 (1969) [*CA* **71**, 39005 (1969)].
69BSF2492	J. Daunis, R. Jacquier, and P. Viallefont, *Bull. Soc. Chim. Fr.*, 2492 (1969).
69BSF3670	J. Daunis, R. Jacquier, and P. Viallefont, *Bull. Soc. Chim. Fr.*, 3670 (1969).
69CB3818	T. Sasaki and M. Murata, *Chem. Ber.* **102**, 3818 (1969).
69IJC959	T. George, R. Tahilramani, and D. A. Dabholkar, *Indian J. Chem.* **7**, 959 (1969) [*CA* **72**, 3470 (1970)].
69JCS(C)1593	J. Maguire, D. Paton, and F. L. Rose, *J. Chem. Soc. C*, 1593 (1969).
69JOC3221	K. T. Potts and C. Lovelette, *J. Org. Chem.* **34**, 3221 (1969).
69M671	M. Japelj, B. Stanovnik, and M. Tisler, *Monatsh. Chem.* **100**, 671 (1969) [*CA* **71**, 22094 (1969)].
70AP650	H. Gehlen, R. Drohla, *Arch. Pharm. (Weinheim Ger.)* **303**, 650 (1970) [*CA* **73**, 87862 (1970)].
70AP709	H. Genlen and R. Drohla, *Arch. Pharm. (Weinheim Ger.)* **303**, 709 (1970) [*CA* **73**, 130946 (1970)].
70BSF1606	J. Daunis, K. Diebel, R. Jacquier, and P. Viallefont, *Bull. Soc. Chim. Fr.*, 1606 (1970).
70CB1918	H. Reimlinger, J. J. M. Vandewalle, G.S.D. King, W.R.F. Lingier, and R. Merenyi, *Chem. Ber.* **103**, 1918 (1970).
70CB1934	H. Reimlinger, J. J. M. Vandewalle, and W.R.F. Lingier, *Chem. Ber.* **103**, 1934 (1970).
70CB1960	H. Reimlinger, J. M. Vandewalle, and W.R.F. Lingier, *Chem. Ber.* **103**, 1960 (1970).
70CB3266	H. Reimlinger and M. A. Peiren, *Chem. Ber.* **103**, 3266 (1970).
70CB3278	H. Reimlinger, *Chem. Ber.* **103**, 3278 (1970).
70CB3284	H. Reimlinger and R. Merenyi, *Chem. Ber.* **103**, 3284 (1970).
70CR1990	M. Robba, G. Dore, and M. Bonhomme, *C.R. Acad. Sci.*, Ser. C **271**, 1990 (1970) [*CA* **74**, 112021 (1971)].
70GEP(O)2012190	J. B. Hester, Jr., Ger. Pat. Offen. 2,012,190 (1970) [*CA* **35**, 109801 (1970)].
70JOC1138	B. Stanovnik, M. Tisler, M. Ceglar, and V. Bah, *J. Org. Chem.* **35**, 1138 (1970).
70JOC3448	K. T. Potts and E. G. Brugel, *J. Org. Chem.* **35**, 3448 (1970).
70JPR669	H. G. O. Becker, D. Beyer, G. Israel, R. Mueller, W. Riediger, and H. J. Timpe, *J. Prakt. Chem.* **312**, 669 (1970).
70MI1	V. R. Rao and V. R. Srinivasan, *Symp. Synth. Heterocycl. Comp. Physiol. Interest [Proc.]*, *1968*, 137 (1970) [*CA* **69**, 27312 (1968)].
70S433	H. Reimlinger, W. R. F. Lingier, J. J. M. Vandewalle, and E. Goes, *Synthesis*. 433 (1970) [*CA* **95**, 98876 (1971)].
70T3357	J. Kobe, B. Stanovnik, and M. Tisler, *Tetrahedron* **26**, 3357 (1970).

70TL1841	F. L. Scott and T. A. F. O'Mahony, *Tetrahedron Lett.*, 1841 (1970).
70TL4039	K. Meguro and Y. Kuwada, *Tetrahedron Lett.*, 4039 (1970).
70TL4083	F. L. Scott and J. K. O'Halloran, *Tetrahedron Lett.*, 4083 (1970).
71BSF3043	J. Daunis, M. Guerret-Rigail, and R. Jacquier, *Bull. Soc. Chim. Fr.*, 3043 (1971).
71CB2702	H. Reimlinger, R. Jacquier, and J. Daunis, *Chem. Ber.* **104**, 2702 (1971).
71CB3961	H. Reimlinger, E. De Ruiter, and A. P. Maurits, *Chem. Ber.* **104**, 3961 (1971).
71GEP(O)1810462	J. Bailey, E. B. Knott, and P. A. Marr, Ger. Pat. Offen. 1,810,462 (1971) [*CA* **76**, 47395 (1972)].
71GEP(O)2036324	J. B. Hester, Jr., Ger. Pat. Offen. 2,036,324 (1971) [*CA* **74**, 100113 (1971)].
71IJC901	T. R. Vakula and V. R. Srinivasan, *Indian J. Chem.* **9**, 901 (1971) [*CA* **75**, 151733 (1971)].
71JCS(C)1667	R. S. Shadbolt, *J. Chem. Soc. C*, 1667 (1971).
71JCS(C)2265	R. N. Butler, P. O'Sullivan, and F. L. Scott, *J. Chem. Soc. C*, 2265 (1971).
71MC1078	J. B. Hester, Jr., A. D. Rudzik, and B. V. Kamdar, *J. Med. Chem.* **14**, 1078 (1971).
71JOC10	K. T. Potts and S. Husain, *J. Org. Chem.* **36**, 10 (1971).
71JPR780	H. G. O. Becker, H. Boettcher, R. Ebisch, and G. Schmoz, *J. Prakt. Chem.* **312**, 780 (1971).
71RRC311	C. Cristescu, *Rev. Roum. Chim.* **16**, 311 (1971) [*CA* **75**, 20358 (1971)].
71TL1609	J. B. Hester, Jr., D. J. Duchamp, and C. G. Chidester, *Tetrahedron Lett.*, 1609 (1971).
71TL1729	F. L. Scott, T. M. Lambe, and R. N. Butler, *Tetrahedron Lett.*, 1729 (1971).
71ZC422	A. V. Spasov and Z. Raikov, *Z. Chem.* **11**, 422 (1971) [*CA* **76**, 85785 (1971)].
71ZOR173	I. S. Ioffe, A. B. Tomchin, and E. N. Zhukova, *Zh. Org. Chim.* **7**, 173 (1971) [*CA* **74**, 112020 (1971)].
72GEP(O)2203782	J. B. Hester, Jr., Ger. Pat. Offen. 2,203,782 (1972) [*CA* **77**, 126708 (1972)].
72JCS(P1)269	R. N. Butler, T. M. Lambe, and F. L. Scott, *J.C.S. Perkin 1*, 269 (1972).
72JCS(P1)1221	M. F. G. Stevens, *J.C.S. Perkin 1*, 1221 (1972).
72JCS(P1)1319	T. A. F. O'Mahony, R. N. Butler, and F. L. Scott, *J.C.S. Perkin 1*, 1319 (1972).
72JCS(P1)1519	R. N. Butler, P. O'Sullivan, and F. L. Scott, *J.C.S. Perkin 1*, 1519 (1972).
72JCS(P1)2224	F. L. Scott, J. K. O'Halloran, J. O'Driscoll, and A. F. Hegarty, *J.C.S. Perkin 1*, 2224 (1972).
72JHC131	P. M. Hergenrother, *J. Heterocycl. Chem.* **9**, 131 (1972).
72JHC153	M. M. Kochlar, *J. Heterocycl. Chem.* **9**, 153 (1972).
72JMC332	M. M. Kochar and B. B. Williams, *J. Med. Chem.* **15**, 332 (1972).
72JPR55	H. G. O. Becker and H. Boettcher, *J. Prakt. Chem.* **314**, 55 (1972) [*CA* **77**, 139949 (1972)].
72MI1	T. M. Itil, B. Saletu, J. Marasa, and A. N. Mucciardi, *Pharmakopsychiatr./Neuro-Psychopharmakol.* **5**, 225 (1972) [*CA* **78**, 67157 (1973)].

72SAP7201487	J. B. Hester, Jr., *S. Afr. Pat.* 72/01,487 (1972) [*CA* **79**, 42577 (1973)].
72SAP7201488	J. B. Hester, Jr. S. Afr. Pat. 72/01,488 (1972) [*CA* **79**, 42578 (1973)].
72SAP7201489	J. B. Hester, Jr., *S. Afr. Pat.*, 72/01,489 (1972) [*CA* **79**, 3515 (1973)].
72USP3681343	J. B. Hester, Jr., U.S. Pat. 3,681,343 (1972) [*CA* **77**, 126707 (1972)].
73CPB1619	K. Meguro, H. Tawada, and Y. Kuwada, *Chem. Pharm. Bull.* **21**, 1619 (1973).
73FRP2149467	B. A. Dreikorn and K. E. Kramer, Fr. Demande 2,149,467 (1973) [*CA* **79**, 78814 (1973)].
73GEP(O)2161586	H. Berger, R. Gall, H. Merdes, W. Voemel, and W. Sauer, Ger. Pat. Offen. 2,161,586 (1973) [*CA* **79**, 66381 (1973)].
73GEP(O)2250077	C. J. Paget, Jr. and E. Lilly, Ger. Pat. Offen., 2,250,077 (1973) [*CA* **79**, 18721 (1973)].
73GEP(O)2251673	R. B. Moffett, Ger. Pat.Offen. 2,251,673 (1973([*CA* **79**, 32114 (1973)].
73GEP(O)2261095	G. E. Hardtmann and F. G. Kathawala, Ger. Pat. Offen. 2,261,095 (1973) [*CA* **79**, 66385 (1973)].
73JAP(K)7334199	K. Meguro, H. Tawada, and Y. Kuwada, *Jpn. Kokai* 73/34,199 (1973) [*CA* **79**, 32116 (1973)].
73JHC387	H. Golgolab, I. Lalezari, and L. Hosseini-Gohari, *J. Heterocycl. Chem.* **10**, 387 (1973).
73LA2088	E. Fahr, P. Juergen, N. Pelz, and T. Erlenmaier, *Justus Liebigs Ann. Chem.* **12**, 2088 (1973) [*CA* **80**, 120841 (1974)].
73MI1	A. D. Sinegibskaya, G. A. Vshivkova, and I. Ya. Postovskii, *Khim. Khim. Tekhnol., Obl. Nauchno-Tekh. Konf.* [Mater.], *4th, 1973*, Vol. *2*, p. 41 (1973) [*CA* **82**, 125368 (1975)].
73MI2	S. Yurugi, A. Miyake, and M. Tomimoto, *Takeda Kenkyusho ho* **32**, 111 (1973) [*CA* **80**, 37073 (1974)].
74AC298	N. W. Jacobsen and R. G. Dickinson, *Anal. Chem.* **46**, 298 (1974) [*CA* **80**, 78195 (1974)].
74GEP(O)2508333	M. Yamamoto, S. Morooka, M. Koshiba, S. Inaba, and H. Yamamoto, Ger. Pat. Offen. 2,508,333 (1974) [*CA* **83**, 206322 (1975)].
74IJC287	K. S. Dhaka, J. Mohan, V. K. Chadha, and H. K. Pujari, *Indian J. Chem.* **12**, 287 (1974) [*CA* **81**, 136118 (1974)].
74IJC485	K. S. Dhaka, J. Mohan, V. K. Chadha, and H. K. Pujari, *Indian J. Chem.* **12**, 485 (1974) [*CA* **81**, 136112 (1974)].
74JCS(P2)997	A. F. Hegarty, P. Quain, T.A.F. O'Mahony, and F. L. Scott, *J.C.S. Perkin 2*, 997 (1974).
74JHC751	R. M. Claramunt, J. M. Fabrega, and J. Elguero, *J. Heterocycl. Chem.* **11**, 751 (1974).
74JOC1226	O. Tsuge, M. Yoshida, and S. Kanemasa, *J. Org. Chem.* **39**, 1226 (1974).
74JOC3506	J. H. Wikel and C. J. Paget, Jr., *J. Org. Chem.* **39**, 3506 (1974).
74JPR667	L. Heinisch, *J. Prakt. Chem.* **316**, 667 (1974).
74MIP1	C. Cristescu, Rom. Pat., 56,269 (1974) [*CA* **82**, 16868 (1975)].
75GEP(O)2418435	H. Berger, R. Gall, K. Stach, W. Voemel, and R. Hoffmann, Ger. Pat. Offen 2,418,435 (1975) [*CA* **84**, 44109 (1976)].
75GEP(O)2426305	B. R. Vogt, Ger. Pat. Offen 2,426,305 (1975) [*CA* **84**, 44182 (1976)].
75GEP(O)2509843	C. J. Paget, Jr. and J. H. Wikel, Ger. Pat. Offen., 2,509,843 (1975) [*CA* **84**, 44070 (1976)].
75JCS(P1)975	R. G. Dickinson and N. W. Jacobsen, *J.C.S. Perkin 1*, 975 (1975).

75JHC337	L. Kramberger, P. Lorencak, S. Polance, B. Vercek, B. Stanovnik, and M. Tisler, *J. Heterocycl. Chem.* **12**, 337 (1975).
75JHC661	E. M. Essassi, J. P. Lavergne, and P. Viallefont, *J. Heterocycl. Chem.* **12**, 661 (1975)).
75JHC717	A. Walser, T. Flynn, and R. I. Fryer, *J. Heterocycl. Chem.* **12**, 717 (1975).
75JHC1133	P. O. Croce, *J. Heterocycl. Chem.* **12**, 1133 (1975).
75JOC2901	H. Zimmer, J. M. Kokosa, and K. J. Shah, *J. Org. Chem.* **40**, 2901 (1975).
75USP3929783	J. Krapcho and C. F. Turk, U.S. Pat. 3,929,783 (1975) [*CA* **84**, 105628 (1976)].
76ACS(B)463	P. Wolkoff, *Acta Chem. Scand., Ser. B* **B30**, 463 (1976) [*CA* **85**, 78060 (1976)].
76FES126	J. Moragues, A. Vega, J. Prieto, M. Marquez, and D. J. Roberts, *Farmaco, Ed. Sci.* **31**, 126 (1976) [*CA* **84**, 180138 (1976)].
76GEP(O)2438789	W. Lorenz, I. Hammann, and B. Homeyer, Ger. Pat. Offen. 2,438,789 (1976) [*CA* **85**, 33022 (1976)].
76GEP(O)2515494	W. Lorenz, I. Hammann, B. Homeyer, and W. Stendel, Ger. Pat. Offen. 2,515,494 (1976) [*CA* **86**, 72700 (1977)].
76GEP(O)2601400	J. B. Hester, Jr., Ger. Pat. Offen. 2,601,400 (1976) [*CA* **86**, 29899 (1977)].
76H1655	M. Ruccia, N. Vivona, and G. Cusmano, *Heterocycles* **4**, 1655 (1976).
76JCS(P1)1363	P. Yui-How and J. Parrick, *J.C.S. Perkin 1*, 1363 (1976).
76JCS(P1)1492	E. J. Gray and M. F. G. Stevens, *J.C.S. Perkin 1*, 1492 (1976).
76JOC3124	T. Huynh-Dinh and J. Igolen, *J. Org. Chem.* **41**, 3124 (1976).
76MI1	V. V. Korshak, L. A. Rosanov, M. K. Kereselidze, and T. K. Dzhashiashvili, *Izv. Akad. Nauk Gruz. SSR, Ser. Khim.* **2**, 39 (1976) [*CA* **85**, 123862 (1976)].
76TL1931	B. R. Vogt, P. C. Wade, and M. S. Puar, *Tetrahedron Lett.*, 1931 (1976).
76UKZ1159	F. S. Babichev, N. N. Romanov, and V. P. Shmailova, *Ukr. Khim. Zh.* **42**, 1159 (1976) [*CA* **86**, 89707 (1977)].
76USP426639	F. G. F. Eloy and R. W. Shanahan, U.S. Pat. 426,639 (19760 [*CA* **84**, 164787 (1976)].
76USP3994898	H. Hoehn and E. Schuelze, U.S. Pat. 3,994,898 (1976) [*CA* **86**, 106629 (1977)].
77BP1469197	J. B. Hester, Jr., Br. Pat. 1,469,197 (1977) [*CA* **87**, 102387 (1977)].
77CPB3137	T. Tsujikawa and M. Tatsuta, *Chem. Pharm. Bull.* **25**, 3137 (1977).
77GEP(O)2533120	W. Lorenz, I. Hammann, and B. Homeyer, Ger. Pat. Offen. 2,533,120 (1977) [*CA* **87**, 53352 (1977)].
77IJC(B)1143	R. P. Gupta, M. L. Sachdeva, and H. K. Pujari, *Indian J. Chem., Sect. B*, **15B**, 1143 (1977) [*CA* **88**, 190698 (1978)].
77JAP(K)7753881	K. Ozawa, M. Hatano, M. Ando, and M. Mizuno, *Jpn. Kokai* 77/53,881 (1977) [*CA* **87**, 117873 (1977)].
77JAP(K)7753882	M. Hatano, M. Ando, and M. Mizuno, *Jpn. Kokai* 77/53,882 (1977) [*CA* **87**, 135346 (1977)].
77JAP(K)7753883	Y. Morishima and M. Mizuno, *Jpn. Kokai* 77/53,883 (1977) [*CA* **87**, 117874 (1977)].
77JCS(P1)2047	J. Bailey, *J.C.S. Perkin 1*, 2047 (1977).

77JHC227	M. H. Elnagdi, M. R. H. Elmoghayar, E. M. Kandeel, and M. K. A. Ibrahim, *J. Heterocycl. Chem.* **14,** 227 (1977).
77JHC567	A. Shafiee, I. Lalezari, M. Mirrashed, and D. Nercesian, *J. Heterocyl. Chem.* **14,** 567 (1977).
77JOC1018	J. Daunis and H. Lopez, *J. Org. Chem.* **42,** 1018 (1977).
77JOC4197	J. Bradac, Z. Furek, D. Janezic, S. Molan, I. Smerkolj, B. Stanovnik, M. Tisler, and B. Vercek, *J. Org. Chem.* **42,** 4197 (1977).
77MI1	E. Gafitanu, M. Caprosu, R. Danet, and M. Petrovanu, *Rev. Med. Chir.* **81,** 469 (1977) [*CA* **88,** 136551 (1978)].
77PHA687	G. Westphal, H. Wasicki, A. Esser, U. Zielinski, F. G. Werber, M. Tonew, and E. Tonew, *Pharmazie* **32,** 687 (1977).
77TL1699	R. C. Wade, B. R. Vogt, and M. S. Puar, *Tetrahedron Lett.*, 1699 (1977).
77USP3994940	J. B. Hester, Jr., U.S. Pat. 3,994,940 (1976) [*CA* **86,** 106673 (1977)].
77USP400322	B. A. Dreikorn and T. D. Thibault, U.S. Pat. 4,008,322 (1977) [*CA* **86,** 166387 (1977).
77USP4032535	R. I. Fryer and A. Walser, U.S. Pat. 4,032,535 (1977) [*CA* **87,** 102388 (1977)].
77USP4053600	G. E. Hardtmann and F. G. Kathawala, U.S. Pat. 4,053,600 (1977) [*CA* **88,** 22970 (1978)].
78FES565	E. Bellasio and A. Campi, *Farmaco, Ed. Sci.* **33,** 565 (1978) [*CA* **89,** 163525 (1978).
78GEP(O)2741763	G. R. Allen, Jr., J. W. Hanifin, Jr., D. B. Moran, and J. D. Albright, Ger. Pat. Offen. 2,741,763 (1978) [*CA* **88,** 190876 (1978)].
78GEP(O)2818395	J. C. Pascal and H. Pinhas, Ger. Pat. Offen. 2,818,395 (1978) [*CA* **90,** 152246 (1979)].
78IJC(B)481	S. Bala, R. P. Gupta, M. L. Sachdeva, A. Singh, and H. K. Pujari, *Indian J. Chem., Sect. B* **16B,** 481 (1978) [*CA* **90,** 38886 (1979)].
78IJC(B)737	V. P. Upadhyaha and V. R. Srinivasan, *Indian J. Chem., Sect. B* **16B,** 737 (1978) [*CA* **90,** 3888 (1979)].
78IJC(B)1000	H. Jahine, H. A. Zaher, Y. Akhnookh, and Z. El-Gendy, *Indian J. Chem., Sect. B* **16B,** 1000 (1978) [*CA* **91,** 56951 (1979)].
78JAP(K)7821197	M. Kadota and K. Honda, *Jpn. Kokai* 78/21,197 (1978) [*CA* **89,** 43471 (1978)].
78JCR(S)190	A. Hasnaoui, J. P. Lavergne, and P. Viallefont, *J. Chem. Res., Synop.* **6,** 190 (1978) [*CA* **89,** 215380 (1978)].
78JHC209	W. L. Albrecht, W. D. Jones, Jr., and F. W. Sweet, *J. Heterocycl. Chem.* **15,** 209 (1978).
78JHC311	Y. Lin, T. L. Fields, and S. A. Lang, Jr., *J. Heterocycl. Chem.* **15,** 311 (1978).
78JHC401	E. Campaigne and T. P. Selby, *J. Heterocycl. Chem.* **15,** 401 (1978).
78JHC463	G. Hajos and A. Messmer, *J. Heterocycl. Chem.* **15,** 463 (1978).
78JHC1127	R. Madronero and S. Vega, *J. Heterocycl. Chem.* **15,** 1127 (1978).
78JHC1485	M. Ruccia, N. Vivona, G. Cusmano, and G. Macaluso, *J. Heterocycl. Chem.* **15,** 1485 (1978).
78JMC1290	M. Gall, B. V. Kamdar, and R. J. Collins, *J. Med. Chem.* **21,** 1290 (1978).
78KGS1571	I. B. Lundina, N. N. Vereshchagina, G. S. Melkozerova, and D. V.

	Kiryaeva, *Khim. Geterotsikl. Soedin.*, 1571 (1978) [*CA* **90**, 87379 (1979)].
78MI1	A. Rosenthal and H. H. Lee, *J. Carbohydr., Nucleosides, Nucleotides* **5**, 559 (1978) [*CA* **91**, 57390 (1979)].
78PHA124	K. Kottke and H. Kuehmstedt, *Pharmazie* **33**, 124 (1978).
78PHA125	K. Kottke and H. Kuehmstedt, *Pharmazie* **33**, 125 (1978).
78UKZ725	F. S. Babichev, L. Al Yusofi, and V. N. Bubnovskaya, *Ukr Khim. Zh.* **44**, 725 (1978) [*CA* **89**, 163506 (1978)].
78USP4108860	P. C. Wade, B. R. Vogt, and T. P. Kissick, U.S. Pat. 4,108,860 (1978) [*CA* **90**, 87473 (1979)].
78USP4112095	G. R. Allen, Jr., J. W. Hanifin, Jr., D. B. Moran, and G. D. Albright, U.S. Pat. 4,112,095 (1978) [*CA* **90**, 152221 (1979)].
78ZC92	A. Koennecke and E. Lippmann, *Z. Chem.* **18**, 92 (1978) [*CA* **89**, 24252 (1978)].
78ZC175	A. Koennecke and E. Lippmann, *Z. Chem.* **18**, 175 (1978) [*CA* **89**, 146869 (1978)].
79H1171	L. G. Payne, M. T. Wu, and A. A. Patchett, *Heterocycles* **12**, 1171 (1979).
79IJC(B)364	Sangita and H. K. Pujari, *Indian J. Chem., Sect. B* **17B**, 364 (1979) [*CA* **92**, 215351 (1980)].
79IJC(B)499	H. Singh, L.D.S. Yadav, and B. K. Battacharya, *Indian J. Chem., Sect. B* **17B**, 499 (1979) [*CA* **94**, 30659 (1981)].
79JCS(P1)3085	J. S. Bajwa and P. J. Sykes, *J.C.S. Perkin 1*, 3085 (1979).
79JHC555	C. A. Lovelette, *J. Heterocycl. Chem.* **16**, 555 (1979).
79JHC1393	R. I. Trust and J. D. Albright, *J. Heterocycl. Chem.* **16**, 1393 (1979) 1393.
79JIC742	D. S. Desphande, *J. Indian Chem. Soc.* **56**, 742 (1979).
79JMC1390	J. B Hester, Jr. and P. von Voigtlander, *J. Med. Chem.* **22**, 1390 (1979).
79JOC88	P. C. Wade, B. R. Vogt, B. Toeplitz, and M. S. Puar, *J. Org. Chem.* **44**, 88 (1979).
79JOC1028	T. Huynh-Dinh, R. S. Sarfati, C. Gouyette, and J. Igolen, *J. Org. Chem.* **44**, 1028 (1979).
79MI1	A. s. Lippa, J. Goupet, E. N. Greenblatt, C. A. Klepner, and B. Beer, *Pharmacol. Biochem. Behav.* **11**, 99 (1979) [*CA* **91**, 204433 (1979)].
79PHA801	K. C. Joshi and K. Dubey, *Pharmazie* **34**, 801 (1979).
79S52	A. Mignot, H. Moskowitz, and M. Miocque, *Synthesis*, 52 (1979) [*CA* **90**, 103904 (1979)].
79TL1567	G. Ege and K. Gilbert, *Tetrahedron Lett.*, 1567 (1979).
79USP4140693	P. C. Wade, B. R. Vogt, and T. P. Kissick, U.S. Pat. 4,140,693 (1979) [*CA* **90**, 204108 (1979)].
79USP4141902	J. B. Hester, Jr., U.S. Pat. 4,141,902 (1979) [*CA* **91**, 5253 (1979)].
79USP4180668	J. B. Hester, Jr., U.S. Pat. 4,180,668 (1979) [*CA* **92**, 146817 (1980)].
79YZ533	A. Ishii, K. Kubo, T. Deguchi, and M. Tanaka, *Yakugaku Zasshi* **99**, 533 (1979) [*CA* **91**, 133826 (1979)].
80ACH405	J. Kosary and P. Sohar, *Acta Chim. Acad. Sci. Hung.* **103**, 405 (1980) [*CA* **94**, 47248 (1981)].
80EUP17438	R. I. Trust and J. D. Albright, Eur. Pat., 17,438 (1980) [*CA* **94**, 139815 (1981)].

80GEP(D)139715	K. Kottke, H. Kuehmstedt, V. Hagen, R. Helga, and S. Schnitzler, Ger. Pat. DD 139,715 (1980) [*CA* **96,** 69016 (1982)].
80JHC575	J. B. Hester, Jr., *J. Heterocycl. Chem.* **17,** 575 (1980).
80JHC1321	E. Moskowitz, A. Mignot, and M. Miocque, *J. Heterocycl. Chem.* **17,** 1321 (1980).
80JIC1112	V. K. Chadha and G. R. Sharma, *J. Indian Chem. Soc.* **57,** 1112 (1980) [*CA* **94,** 208821 (1981)].
80JMC392	J. B. Hester, Jr., A. D. Rudzik, and P. F. von Voigtlander, *J. Med. Chem.* **23,** 392 (1980).
80JMC402	J. B. Hester, Jr., A. D. Rudzik, and P. von Voigtlander, *J. Med. Chem.* **23,** 402 (1980).
80JMC643	J. B. Hester, Jr., A. D. Rudzik, and P. F. von Voigtlander, *J. Med. Chem.* **23,** 643 (1980).
80JMC873	J. B. Hester, Jr., P. von Voigtlander, and G. N. Evenson, *J. Med. Chem.* **23,** 873 (1980).
80MI1	S. Singh, L. D. S. Yadav, and H. Singh, *Bokin Bobai* **8,** 385 (1980) [*CA* **94,** 103250 (1981)].
80MI2	D. S. Desphande, *Acta Cienc. Indica [Ser.] Chem.* **6,** 80 (1980) [*CA* **94,** 65568 (1981)].
80MI3	A. Messmer, G. Hajos, P. Benko, and L. Pallos, *Magy. Kem. Foly.* **86,** 471 (1980) [*CA* **94,** 175058 (1981)].
80PHA800	K. Kottke and H. Kuehmstedt, *Pharmazie* **35,** 800 (1980).
80USP4213773	A. D. Wolf, U.S. Pat. 4,213,773 (1980) [*CA* **94,** 15744 (1981)].
81AJC2635	D. J. Brown and K. Shinozuka, *Aust. J. Chem.* **34,** 2635 (1981).
81CC376	R. N. Butler and S. M. Johnston, *J.C.S. Chem. Commun.*, 376 (1981).
81CCC148	Z. Vejdelek, E. Svatek, J. Holubek, J. Metys, M. Bartosova, and M. Protiva, *Collect. Czech. Chem. Commun.* **46,** 148 (1981) [*CA* **95,** 7163 (1981)].
81EUP29130	N. P. Peet and S. Sunder, Eur. Pat., 29,130 (1981) [*CA* **95,** 150692 (1981)].
81EUP39920	R. E. Brown, V. S. Georgiev, P. Kropp, and B. Loev, Eur. Pat., 39,920 (1981) [*CA* **96,** 69038 (1982)].
81H699	I. Tabakovic and S. Crljenak, *Heterocycles* **16,** 699 (1981) [*CA* **95,** 115396 (1981)].
81IJC(B)132	D. R. Shridhar, M. Jogibhukta, P. P. Joshi, and P. G. Reddy, *Indian J. Chem., Sect. B* **20B,** 132 (1981) [*CA* **95,** 80849 (1981)].
81IJC(B)369	B. Dash, E. K. Dora, and C. S. Panda, *Indian J. Chem., Sect. B* **20B,** 369 (1981) [*CA* **95,** 132765 (1981)].
81JHC499	A. M. M. E. Omar, M. G. Kasem, and I. M. Laabota, *J. Heterocycl. Chem.* **18,** 499 (1981).
81JHC893	M. Legraverend, E. Bisagni, and J. M. Lhoste, *J. Heterocycl. Chem.* **18,** 893 (1981).
81JHC1353	T. Sasaki and E. Ito, *J. Heterocycl. Chem.* **18,** 1353 (1981).
81JHC1523	R. D. Thompson and R. N. Castle, *J. Heterocycl. Chem.* **18,** 1523 (1981).
81JHC1625	A. Amer and H. Zimmer, *J. Heterocycl. Chem.* **18,** 1625 (1981).
81KFZ44	V. F. Knyazeva, V. G. Granik, R. G. Glushkov, and G. S. Arutyunyan, *Khim.-Farm. Zh.* **15,** 44 (1981) [*CA* **95,** 115238 (1981)].

81MI1	M. A. E. Shaban, R. S. Ali, and S. M. El-Badry, *Carbohydr. Res.* **95**, 51 (1981).
81OPP123	N. D. Heindel and J. R. Reid, *Org. Prep. Proced. Int.* **13**, 123 (1981) [*CA* **95**, 43055 (1981)].
81PHA62	M. El Enany and S. Botros, *Pharmazie* **36**, 62 (1981).
81USP4244953	R. I. Trust and J. D. Albright, U.S. Pat. 4,244,953 (1981) [*CA* **94**, 156942 (1981)].
81USP4260756	D. B. Moran, J. P. Dusza, and J. D. Albright, U.S. Pat. 4,260,756 (1981) [*CA* **95**, 97829 (1981)].
81USP4269980	R. A. Hardy, Jr., J. S. Baker, and N. Q. Quinones, U.S. Pat. 4,269,980 (1981) [*CA* **95**, 62258 (1981)].
81USP4276292	V. S. Georgiev, B. Loev, R. Mack, and J. H. Musser, U.S. Pat. 4,276,292 (1981) [*CA* **95**, 132905 (1981)].
81USP4293554	G. R. Allen, Jr., J. W. Han, Jr., D. B. Moran, and J. D. Albright, U.S. Pat. 4,293,554 (1981) [*CA* **96**, 866 (1982)].
82AJC1263	D. J. Brown and K. Shinozuka, *Aust. J. Chem.* **35**, 1263 (1982).
82G345	A. K. El-Shafei, A. B. Ghattas, A. Sultan, H. S. El-Kashef, and G. Vernin, *Gazz. Chim. Ital.* **112**, 345 (1982) [*CA* **98**, 89267 (1983)].
82IJC(B)243	J. Mohan, *Indian J. Chem., Sect. B* **21B**, 243 (1982) [*CA* **97**, 55739 (1982)].
82IJC(B)317	N. A. Shams, A. M. Kaddah, and H. Moustafa, *Indian J. Chem., Sect. B* **21B**, 317 (1982) [*CA* **97**, 183330 (1982)].
82IJC(B)732	K. Jain and R. N. Handa, *Indian J. Chem., Sect. B* **21B**, 732 (1982) [*CA* **98**, 89259 (1983)].
82JCS(P1)2663	H. A. Elfahham, K. U. Sadek, G. E. H. Elgemeie, and M. H. Elnagdi, *J.C.S. Perkin 1*, 2663 (1982).
82JHC1345	C. A. Lovelette and K. Geagan, *J. Heterocycl. Chem.* **19**, 1345 (1982).
82JIC769	M. K. Mody, A. R. Prasad, T. Ramalingam, and P. B. Sattur, *J. Indian Chem. Soc.* **59**, 769 (1982) [*CA* **97**, 216131 (1982)].
82JIC900	H. K. Gakhar, A. Jain, and S. B. Gupta, *J. Indian Chem. Soc.* **59**, 900 (1982) [*CA* **98**, 16624 (1983)].
82JOC2757	T. Sasaki, E. Ito, and I. Shimizu, *J. Org. Chem.* **47**, 2757 (1982).
82MI1	G. D. Gupta and H. K. Pujari, *Ann. Soc. Sci. Bruxelles, Ser. 1* **96**, 155 (1982) [*CA* **99**, 22386 (1983)].
82MI2	M. Z. A. Badr, H. A. H. El-Sherief, G. M., El-Naggar, and S. A. Mahgoub, *Bull. Fac. Sci., Assuit Univ.* **11**, 25 (1982) [*CA* **98**, 53807 (1983)].
82PHA451	H. Zimmer, A. Amer, and L. Baldwin, *Pharmazie* **37**, 451 (1982).
83EUP76199	W. R. Tully, R. Westwood, D. A. Rowlands, and S. Clements-Jewery, Eur. Pat., 76,199 (1983) [*CA* **99**, 194989 (1983)].
83EUP121341	J. J. Wade, Eur. Pat. 121341 (1983) [*CA* **102**, 95663 (1985)].
83FES842	G. Auzzi, F. Bruni, L. Cecchi, A. Costanzo, L. Pecori Vettori, and F. De Sio, *Farmaco, Ed. Sci* **38**, 842 (1983) [*CA* **100**, 103277 (1984)].
83GEP(D)158549	K. Kottke, H. Kuehmstedt, H. Volker, H. Renner, and S. Schnitzler, Ger. Pat. DD 158,549 (1983) [*CA* **99**, 70757 (1983)].
83GEP(O)3222342	M. Roesner, D. Duewel, and R. Kirsch, Ger. Pat. Offen. 3,222,342 (1983) [*CA* **100**, 156626 (1984)].
83H749	J. Kosary, E. Kasztreiner, and M. Soti, *Heterocycles* **20**, 749 (1983) [*CA* **99**, 70663 (1983)].

83JAP(K)58124786	Gruppo Lepetit S. P. A., *Jpn. Kokai* 58/124,786 (1983) [*CA* **99**, 194985 (1983)].
83JHC1231	A. Amer and H. Zimmer, *J. Heterocycl. Chem.* **20**, 1231 (1983).
83JPS45	M. A. El-Dawy, A. M. M. E. Omar, A. M. Ismail, and A. A. B. Hazzaa, *J. Pharm. Sci.* **72**, 45 (1983) [*CA* **98**, 143384 (1983)].
83MI1	E. M. Knadeel, H. H. S. Alnima, and M. H. Elnagdi, *Pol. J. Chem.* **57**, 327 (1983) [*CA* **101**, 7093 (1984)].
83MI2	M. A. E. Shaban and M. A. M. Nassr, *Abstr. Pap., 186th Natl. Meet., Am. Chem. Soc.*, Washington, D.C., CARB-23 (1983).
83MI3	M. A. E. Shaban, M. A. M. Nassr, and M. A. M. Taha, *Carbohydr. Res.* **113**, C16 (1983).
83MI4	L. Strzemecka, *Pol. J. Chem.* **57**, 881 (1983) [*CA* **102**, 95593 (1985)].
83MIP1	W. P. Heilman and J. M. Gullo, *PCT Int. Appl.* WO 83/00,864 (1983) [*CA* **99**, 105285 (1983)].
83PHA25	K. Kottke, H. Kuehmstedt, and D. Knoke, *Pharmazie* **38**, 25 (1983).
83S44	M. Kuenstlinger and E. Breitmaier, *Synthesis*, 44 (1983) [*CA* **98**, 179290 (1983)].
83S411	P. Molina and A. Tarraga, *Synthesis*, 411 (1983) [*CA* **99**, 70638 (1983)].
83S415	P. Molina, M. Alajarin, and M. J. Vilaplana, *Synthesis*, 415 (1983) [*CA* **99**, 70640 (1983)].
83USP4402958	R. A. Hardy, Jr. and J.R.A. Hardy, U.S. Pat. 4,402,958 (1983) [*CA* **100**, 6546 (1984)].
84CPB5040	T. Sasaki, M. Ohno, and E. Ito, *Chem. Pharm. Bull.* **32**, 5040 (1984).
84H537	L. Fodor, M. S. El-Gharib, J. Szabo, G. Bernath, and P. Sohar, *Heterocycles* **22**, 537 (1984).
84IJC(B)1293	A. Kamal and P. B. Sattur, *Indian J. Chem., Sect. B* **23B**, 1293 (1984) [*CA* **102**, 16690 (1985)].
84JHC1389	N. P. Peet, *J. Heterocycl. Chem.* **21**, 1389 (1984).
84JMC924	S. W. Schneller, R. D. Thompson, J. G. Cory, R. A. Olsson, E. De Clercq, I. K. Kim, and P. K. Chiang, *J. Med. Chem.* **27**, 924 (1984).
84KGS40	A. Rutavicius and S. Iokubaitite, *Khim. Geterotsikl. Soedin.*, 40 (1984) [*CA* **100**, 191774 (1984)].
84MI1	K. C. Liu, J. L. Chang, and C. F. Chen, *T'ai-wan Yao Hsueh Tsa Chih* **36**, 33 (1984) [*CA* **102**, 6310 (1985)].
84MI2	M. K. May and A. E. Lanzilotti, *J. Labelled Compd. Radiopharm.* **21**, 489 (1984) [*CA* **102**, 6359 (1985)].
84MI3	K. C. Liu and J. L. Chang, *T'ai-wan Yao Hsueh Tsa Chih* **36**, 57 (1984) [*CA* **102**, 6320 (1985)].
84MI4	S. A. Zayed, I. E. Fakhr, F. A. Gad, and A. Abdulla, *J. Chin. Chem. Soc.* **31**, 315 (1984) [*CA* **102**, 78786 (1985)].
84S602	S. Pekkas, N. Rodios, and N. E. Alexanrou, *Synthesis*, 602 (1984) [*CA* **102**, 62171 (1985)].
85CPB3113	T. Nagamatsu, M. Ukai, F. Yoneda, and D. J. Brown, *Chem. Pharm. Bull.* **33**, 3113 (1985).
85CPB4769	J. C. Lancelot, D. Laduree, S. Rault, and M. Robba, *Chem. Pharm. Bull.* **33**, 4769 (1985).
85EUP156734	J. Bourguignon, J. P. Chambon, and C. G. Wermuth, Eur. Pat. 156,734 (1985) [*CA* **104**, 68878 (1986)].

85H871	J. H. Musser, V. S. Georgiev, R. Mack, B. Loev, R. E. Brown, and F. C. Huang, *Heterocycles* **23**, 871 (1985).
85H2025	K. Makino, G. Sakata, K. Morimoto, and Y. Ochiai, *Heterocycles* **23**, 2025 (1985).
85H2613	P. Molina, M. Alajarin, and M. J. Perez de Vega, *Heterocycles* **23**, 2613 (1985).
85IJC(B)1221	S. Narayan, R. N. Handa, and H. K. Pujari, *Indian J. Chem., Sect. B* **24B**, 1221 (1985) [*CA* **106**, 32931 (1987)].
85JHC1045	I. Sircar, *J. Heterocycl. Chem.* **22**, 1045 (1985).
85JHC1185	R. Houssin, J. L. Bernier, and J. P. Henichart, *J. Heterocycl. Chem.* **22**, 1185 (1985).
85JHC1715	Y. Kurasawa, Y. Okamoto, and A. Takada, *J. Heterocycl. Chem.* **22**, 1715 (1985).
85JMC363	B. Loev, J. H. Musser, R. E. Brown, H. Jones, R. Kahen, F. C. Huang, A. Khandwala, P. Sonnino-Goldman, and M. J. Leibowitz, *J. Med. Chem.* **28**, 363 (1985).
85MI1	J. M. Fabrega and R. M. Claramunt, *Afinidad* **42**, 485 (1985) [*CA* **105**, 114981 (1986)].
85MI2	M. S. Rao, V. R. Rao, and T.V.P. Rao, *Sulfur Lett.* **4**, 19 (1985) [*CA* **106**, 4983 (1987)].
85USP4532242	J. J. Wade, U.S. Pat. 4,532,242 (1985) [*CA* **104**, 5891 (1986)].
85USP4547499	J. B. Hester, Jr., U.S. Pat. 4,547,499 (1985) [*CA* **104**, 75033 (1986)].
85ZC366	E. Lippmann and E. M. Tober, *Z. Chem.* **25**, 366 (1985) [*CA* **105**, 172403 (1986)].
86H93	I. Krezel, *Heterocycles* **24**, 93 (1986).
86IJC(B)283	G. S. Dhindsa and R. K. Vaid, *Indian J. Chem., Sect. B* **25B**, 283 (1986) [*CA* **106**, 49491 (1987)].
86IJC(B)566	A. R. Prasad, T. Ramalingam, A. B. Rao, P. V. Diwan, and P. B. Sattur, *Indian J. Chem., Sect. B* **25B**, 566 (1986) [*CA* **106**, 84500 (1987)].
86JAP(K)61260085	Y. Kawashima, H. Ishikawa, S. Kida, T. Tanaka, and K. Masuda, *Jpn. Kokai* 61/260,085 (1986) [*CA* **106**, 138475 (1987)].
86JCR(S)70	P. Molina and M. J. Vilaplana, *J. Chem. Res., Synop.*, 70 (1986) [*CA* **105**, 6495 (1986)].
86JCR(S)232	P. D. Kennewell, R. M. Scrowston, I. G. Schenouda, W. R. Tully, and R. Westwood, *J. Chem. Res., Synop.*, 232 (1986) [*CA* **106**, 32978 (1987)].
86JCR(S)200	G. Capozzi, G. Ottana, G. Romeo, and G. Valle, *J. Chem. Res., Synop.*, 200 (1986) [*CA* **106**, 3295 (1987)].
86JHC43	P. Molina, A. Arques, I. Cartagena, and M. V. Valcarcel, *J. Heterocycl. Chem.* **23**, 43 (1986).
86JHC721	N. Vinot and P. Maitte, *J. Heterocycl. Chem.* **23**, 721 (1986).
86JHC833	R. Murdoch, W. R. Tully, and R. Westwood, *J. Heterocycl. Chem.* **23**, 833 (1986).
86JHC1071	D. B. Moran, G. O. Morton, and J. D. Albright, *J. Heterocycl. Chem.* **23**, 1071 (1986).
86JHC1339	A. M. M. E. Omar and O. M. AboulWafa, *J. Heterocycl. Chem.* **23**, 1339 (1986).
86JHC1439	B. N. Goswami, J.C.S. Kataky, and J. N. Baruah, *J. Heterocycl. Chem.* **23**, 1439 (1986).

86JMC2231	K. Ramasamy, B. G. Ugarkar, P. A. McKernan, R. K. Robins, and G. R. Revankar, *J. Med. Chem.* **29**, 2231 (1986).
86JPR237	U. Burkhardt and S. Johne, *J. Prakt. Chem.* **328**, 237 (1986).
86KGS1350	I. Krezel, *Khim. Geterotsikl. Soedin.*, 1350 (1986) [*CA* **107**, 39742 (1987)].
86M867	J. Kosary, G. Jerkovich, K. Polos, and E. Kasztreiner, *Monatsh. Chem.* **117**, 867 (1986) [*CA* **106**, 213902 (1987)].
86MI1	S. W. Schneller, J. L. May, and E. De Clercq, *Croat. Chem. Acta* **59**, 307 (1986) [*CA* **107**, 7490 (1987)].
87AP1191	M. I. Younes, H. H. Abbas, and S. A. Metwally, *Arch. Pharm. (Weinheim Ger.)* **320**, 1191 (1987) [*CA* **108**, 75358 (1988)].
87AP1196	M. I. Younes, A.A.M. Abdel-Alim, and S. A. Metwally, *Arch. Pharm. (Weinheim, Ger.)* **320**, 1196 (1987) [*CA* **108**, 75359 (1988)].
87H1177	H. Zimmer and A. Amer, *Heterocycles* **26**, 1177 (1987).
87JHC667	H. M. Faid Allah and R. Soliman, *J. Heterocycl. Chem.* **24**, 667 (1987).
87JHC1435	V. J. Ram, V. Dube, and A. J. Vlietinck, *J. Heterocycl. Chem.* **24**, 1435 (1987).
87JOC2220	P. Bitha, J. J. Hlavka, and Y. Lin, *J. Org. Chem.* **52**, 2220 (1987).
87JPR331	I. Krenzel, *J. Prakt. Chem.* **328**, 331 (1986) [*CA* **107**, 7159 (1987)].
87MI1	B. Rigo, P. Cauliez, S. Taisne, D. Valigny, and D. Couturier, *Int. Congr. Heterocycl. Chem.*, *11th*, Heidelberg, Germany, p. 501 (1987).
87MI2	G. Ege, J. Fischer, K. Gilbert, and K. Maurer, *Int. Congr. Heterocycl. Chem.*, *11th*, Heidelberg, Germany, p. 68 (1987).
87MI3	K. Gilbert, K. Maurer, and G. Ege, *Int. Congr. Heterocycl. Chem.*, *11th*, Heidelberg, Germany, p. 60 (1987).
87MI4	M. A. Hassan and A. F. Fahmy, *Int. Congr. Heterocycl. Chem.*, *11th*, Heidelberg, Germany, p. 283 (1987).
87MI5	A. F. M. Fahmy, E. S. H. Eltamany, and M. O. Orabi, *Int. Congr. Heterocycl. Chem.*, *11th*, Heidelberg, Germany, p. 284 (1987).
87MI6	A. Dlugosz, *Int. Congr. Heterocycl. Chem.*, *11th*, Heidelberg, Germany, p. 481 (1987).
87MI7	J. J. Wade and D. B. Olson, *Int. Congr. Heterocycl. Chem.*, *11th*, Heidelberg, Germany, p. 126 (1987).
87PHA304	J. Kosary, K. Polos, and E. Kasztreiner, *Pharmazie* **42**, 304 (1987).
87T2497	J. Reiter, L. Pongo, and P. Dvortsak, *Tetrahedron* **43**, 2497 (1987).
88JHC173	J. Reiter, L. Ponge, P. Sohar, and P. Dortsak, *J. Heterocycl. Chem.* **25**, 173 (1988).
88JMC1011	B. K. Trivedi and R. F. Bruns, *J. Med. Chem.* **31**, 1011 (1988).
88MI1	F. F. Abdel-Latif, S. A. Mahgoub, R. M. Shaker, and M. Z. A. Badr, *Abstr. Pap., Chem. Conf., 2nd*, Alexandria, Egypt, p. 36 (1988).
89UP1	M. A. E. Shaban, and M. A. M. Taha, unpublished results.
89UP2	M. A. E. Shaban, M. A. M. Taha, and E. M. Sharchira, unpublished results.
89UP3	M. A. E. Shaban, M. A. M. Taha, and H. A. M. Hamouda, unpublished results.
89UP4	M. A. E. Shaban, M. A. M. Taha, and A. E. Morgan, unpublished results.

Advances in Pyridazine Chemistry

MIHA TIŠLER AND BRANKO STANOVNIK

*Department of Chemistry,
E. Kardelj University,
61000 Ljubljana, Yugoslavia*

I. Introduction	385
III. Synthetic Methods	386
A. From Carbonyl Compounds	386
B. Cycloaddition Reactions	391
C. From Other Heterocycles	398
D. Miscellaneous Syntheses	402
E. Syntheses of Labelled Pyridazines	403
III. Transformations of Pyridazines	404
A. Reactions on the Ring Carbon Atoms	404
B. Reactions at the Ring Nitrogen Atoms	407
C. Reactions of Functional Groups	409
D. Radical Reactions	419
E. Oxidations and Reductions: Reduced Pyridazines	420
F. Ring Opening and Rearrangements	421
G. Photochemical Transformations	423
IV. Theoretical Aspects and Physical Properties	424
A. Calculations	424
B. Dielectric Properties, Basicity, and Tautomerism	425
C. Spectra	427
D. X-Ray Analysis	429
E. Complexes	431
V. Polymers	432
VI. Biological Activity	433
A. Natural Products	433
B. Synthetic Compounds	435
C. Plant Growth and Protection	438
VII. Other Applications	439
VIII. Analysis	439
References	440

I. Introduction

This review covers the last ten years of progress in pyridazine chemistry. The last report appeared in this series in 1979 (79AHC363). Numerous syntheses of pyridazines follow the well known approaches, but there are some significant developments and new methods, such as syntheses of pyridazines by cycloaddition, which have become a powerful synthetic tool in the hands of organic chemists. Pyridazines have been investigted intensely, from a theoretical standpoint, by spectroscopic methods. They have also been investigated for their applications in agriculture and in particular for their biological activity for use as potential drugs.

Several aspects of pyridazine chemistry and activity were included in review articles of general interest. Such reviews deal with aromaticity and antiaromaticity of azines (80MI29) and substituent effects in azines (80MI18). Pyridazines were mentioned in a review covering unusual organic compounds including azoalkanes (80AG815), cycloadditions of azadienes (83T2869; 87MI28), diazoquinones [78H(9)1771; 79H(13)389], thermal and photochemical decomposition of azoalkanes (80CRV99), N-methyl inversion barriers of six-membered heterocycles (81AG567), steric effects in heteroaromatics (88AHC173), reactions of annular nitrogens of azines with electrophiles (88AHC127), direct amination (86MI26), homolytic reactions (87H481), one-electron oxidations (81ACR131), and π-complexes of transition metals (78JHC1057). Reviews also appeared on pyridazine nucleosides (86MI28), some biologically active pyridazines (84MI23), antihypertensive 3-hydrazinopyridazines (80MI25), and pyridazines with antimicrobial activity (77PHA555). References that have appeared by mid-1988 are included here.

II. Synthetic Methods

A. From Carbonyl Compounds

Simple ketones were used as starting material for synthesis in a few cases. Condensation of 1-chloroketones (80PHA140; 81UKZ657), 1,2-unsaturated ketones (85AKZ720), enaminoketones, or vinamidinium salts (79S385; 84JOC4769) with hydrazines afforded pyridazines. From saturated or unsaturated 1,4-diketones, 3-mono or 3,6-disubstituted pyridazines were synthesized by using hydrazine, substituted hydrazines, or hydrazides (81H1705, 81JOC5156, 81MI23; 84JHC1297; 86MI13, 86MI19; 87JHC1745). Similarly, a 1,4-dicarbonyl derivative of furanophane was converted into a furano(3,6)pyridazinophane (**67**) (80JOC4584).

A major source for pyridazines is still 1,4-keto acids or esters. Except for one case when a 1-formyl-4-ester was used as starting material (83MI25), 6-alkyl- or 6-aryl-2,3,4,5-tetrahydro-3(2H)-pyridazinones were synthesized from 1,4-keto acids or esters and hydrazine or substituted hydrazines [76MI4; 77IJC(B)436, 77MI5; 78JHC881, 78MI3, 78MI5,

78MI9; 79IJC(B)136, 79KGS943, 79MI8, 79RRC1381; 80RRC1375; 81IJC(B)424, 81IJC(B)845, 81IJC(B)1097, 81JMC592; 83JHC1473, 83MI7; 84JMC1099, 84MI15; 85AF784; 85MI6, 85MI23, 85MI29, 85MI30; 86MI24, 86RRC387, 86ZC21; 87IJC(B)348, 87JHC63; 88JHC799, 88JMC461]. From β-acyl lactic acids and hydrazine, depending on reaction conditions, either 4-hydroxy-4,5-dihydro-3($2H$)-pyridazinones or 3($2H$)-pyridazinones can be prepared (85BSF865). From benzoyl acrylic acids and hydrazines in the presence of sodium acetate, pyridazinones were obtained, whereas in alkaline solution, pyrazoles are formed (78MI9). For aromatization of the obtained 4,5-dihydro-3($2H$)-pyridazinones, bromine is used preferentially, but if bromination of the side chain has to be avoided, sodium m-nitrobenzenesulfonate can be used (81JMC592).

From heterocyclic 1,4-keto acids or esters, a variety of heteroaryl substituted 3($2H$)-pyridazinones can be prepared (80AKZ862; 81JHC425, 81MI14; 83MI19; 84AKZ572, 84MI22; 85MI15, 85MI18, 85MI27, 85MI31; 87MI21).

The reaction between a cyclic 1,4-keto ester and hydrazine was used to synthesize [7]-, [8]-, and [9]-(3,5)pyridazinophanes (**1**) (84G). From esters of 1,4-dioxo-2-carboxylic acids, which are at the same time 1,4-diketones and 1,4-keto esters, esters of 1,4-dihydropyridazine-5-carboxylic acid were obtained if ethanol was used as solvent. In glacial acetic acid, however, 1-aminopyrroles were formed (81CB564).

$n = 3, 4, 5$

(1)

Saturated or unsaturated 1,4-dicarboxylic acids also serve as starting material for 3-hydroxy-6($1H$)-pyridazinones. A detailed investigation of the reaction between succinic anhydride and phenylhydrazine revealed that, in glacial acetic acid at room temperature, five products are formed; two of them are pyridazines **2** and **3** [84IJC(B)439]. Other hydrazines and

(2)   (3)

this acid or its anhydride were used to prepare various pyridazines (83MI17; 86MI18).

Various 3-hydroxy-6(1H)-pyridazinones were obtained from maleic anhydride and various hydrazines (78MI2; 81PS323; 82MI3; 82MI4; 85AKZ743; 86ZOR711). From maleic anhydride and hydrazine, 1,2-bis(3-carboxyacryloyl)hydrazine can be obtained; this undergoes acid-catalyzed cyclization into maleic hydrazide (84MI11, 84MI12). In a similar manner, pyridazines were prepared from fluoromaleic anhydride (84MI28; 86MI1) or chloro- and dichloromaleic anhydride (79MI32). From mucochloric acid and heterocyclic hydrazines, 1-heteroaryl-4,5-dichloro-6(1H)-pyridazinones were prepared [81JCR(S)103; 85H2603].

Pyridazines were also prepared from hydrazones or semicarbazones of carbonyl compounds. Semicarbazones of 1,4-keto esters are transformed into tetrahydropyridazines, giving further either imidazo[1,5-b]pyridazines or 1-ureidopyroles (83H551). ω-Bromoacetophenone semicarbazone reacts with enamines of various methylcyclohexanones to give, depending on reaction conditions, tetrahydro-, dihydro-, and pyridazines (84G521). Semicarbazones of phenacyl bromides are reduced electrochemically into 4 to give, upon heating in dimethylformamide (DMF), 3,6-diarylpyridazines 5 (85SC939; 87MI1).

$$2\ Ar-C\overset{N-NHCONH_2}{\underset{CH_2Br}{\diagdown}} \longrightarrow Ar-C\overset{N-NHCONH_2}{\underset{CH_2-CH_2C\overset{Ar}{\underset{N-NHCONH_2}{\diagdown}}}{\diagdown}} \overset{\Delta}{\longrightarrow} \underset{Ar}{\diagdown}\overset{Ar}{\diagdown}_{N-N}$$

(4)   (5)

Pyridazinones were obtained from an attempted Fischer indolization of a cyclopentanone phenylhydrazone [85CI(L)697], from hydrazones of unsaturated 1,4-keto esters (87H2101) or from 3-arylhydrazones of 4-(4-methoxyphenyl)glutaconic anhydride in the presence of seconday amines [83IJC(B)512]. Pyridazines 6 can be synthesized from 2-phenylhydrazones of 1,2,3-tricarbonyl compounds and phosphacumulenylides (80TL2939; 85CB1709).

$$R-\overset{O}{\overset{\|}{C}}-\underset{\underset{NHPh}{N}}{C}-\overset{O}{\overset{\|}{C}}-R^1 + Ph_3\overset{+}{P}\overset{-}{C}=C=X \longrightarrow \underset{\underset{Ph}{N}}{\overset{R}{\diagdown}}\overset{O}{\underset{N}{\diagdown}}R^1$$

$$X = O, NPh, etc.$$

(6)

Isobutyraldehyde phenylhydrazone, when treated with an excess of methyl methacrylate at 120°C, gave a mixture of an azoalkane, pyridazinone **7**, and pyrazoline **8** in a ratio of 2:2:1, as a result of different cyclization paths (79JOC218).

In a similar manner, pyridazines were obtained from various arylhydrazones and acrylonitrile (84ZOR416; 87PHA695). Also, hydrazones of α-haloacetophenones, when transformed into transient azoalkenes and intercepted with enol ethers, are transformed into 1,4,5,6-tetrahydropyridazines. These, in turn, were transformed with hot ethanolic sodium ethoxide into 3-arylpyridazines [85JCR(S)310]. A dihydropyridazine was formed by thermal decomposition obtained by silver oxide oxidation of bishydrazones of 1,4-dicarbonyl compounds [87JCR(S)170]. Pyridazines can also be prepared from 1-nitro-2,2-bismethyl-mercaptoethene, after its conversion into the corresponding hydrazone, and subsequent treatment with 1,2-keto aldehydes (77TL3619).

Azines were also used as starting material. Acetophenone azine, when treated with lithium diisopropylamide (LDA), is transformed in its dianion which is cyclized at room temperature into 3,6-diphenyl-1,4,5,6-tetrahydropyridazine as the sole product. The *p*-anisyl analog, however, gives either the corresponding pyridazine, pyrrole, or a mixture of both compounds depending on reaction conditions (78JOC3370). Thermal cyclization of ketazines also yields pyridazines (87ZOR1063). An azine phosphorane, upon benzoylation, gave the expected benzoylated product and a pyridazine and pyrazole derivative. Pyridazine ylide **9** is transformed, upon heating at ~230°C, into acetylenic derivative **10** (82JOC2768).

Many synthetic approaches are based on reactions with reactive methylene compounds and hydrazones, a reaction that was first introduced by Schmidt and Druey in 1954. Benzil or monohydrazones of benzil or related compounds were transformed into various substituted pyridazines after treatment with esters of substituted acetic acids (79MI2, 79TL2921; 80JMC1398), ethyl acetoacetate (81JPS419; 87JHC23), diethyl malonate (79MI22), 2,2-dimethyl-1,3-dioxan-4,6-dione (Meldrum's acid) [82JCS(P1)1845; 83JCS(P1)1203], activated nitriles (78MI1; 79MI41; 83UKZ1095, 83UKZ1197; 88H1579), or malonodinitrile (79JPR71).

Aryldiazonium salts couple to various compounds with a reactive methylene group, and subsequent cyclization leads to pyridazine derivatives. In this manner, pyridazines were prepared from a dimer of ω-cyanoacetophenone (85LA1492), a dimer of ethyl cyanoacetate (diethyl 3-amino-2-cyano-2-pentenedioate) [82JCS(P1)989, 82S490; 85ZN(B)664], 3-amino-2-cyano-4-ethoxycarbonylcrotononitrile (87H899, 87LA889), and other activated crotononitriles [84S62; 85H1999, 85LA1492, 85MI1, 85OPP107, 85S1135; 86H101, 86ZN(B)105]. However, dicyano compound **11,** when treated with an aryldiazonium salt in ethanol and in the presence of sodium acetate, forms hydrazone **12** which, upon heating in acetic acid, is transformed into **13**. The same reaction proceeded differently in acetic acid and in the presence of sodium acetate to give **15** via **14** (86H1219; 87BCJ4486).

Other examples of coupling diazonium salts to β-arylglutaconic anhydrides with subsequent ring opening and cyclization to pyridazines, [80IJC(B)648] and coupling to 1,3-dicarbonyl or 1-cyano-3-carbonyl compounds (80S623); 81JHC333, 86JHC93) or diesters of 2-pentenoic acid (82MI8) are reported.

B. Cycloaddition Reactions

A pyridazine ring can be formed in a cycloaddition reaction of diazo compounds or diazonium salts, with alipahtic or cyclic azo compounds. It can also be formed by cycloaddition of various 1,2,4,5-tetrazines and dienophiles in inverse electron demand Diels–Alder reactions. As shown later, this synthetic approach has been used extensively and with success in the 1980s.

Diamino cyclopropenylium salts and phosphorylated or carbonylated diazomethanes react in the presence of ethyldiisopropylamine to give 4,5-diaminopyridazines in moderate yield (86JHC385). Tetrachlorocyclopropene first gives addition products **16**. Their stability varies, and they rearrange into 3,4,5-trichloro pyridazines (**17**) [78JCR(S)40, 78MI10]. Other cyclopropenes react similarly [80LA590; 85ZOR1026; 87ZN(B)210].

2-Diazo-4,5-dicyanoimidazole undergoes cycloaddition with 2,3-dimethylbuta-1,3-diene at room temperature to give pyridazinylimidazole **18**. It has been proposed that the reaction proceeds by initial attack of the terminal nitrogen atom to give a transient aziridine, followed by ring opening and hydrogen transfer. The reaction may occur in a stepwise manner (84CC295).

In 1975 Carlson, Sheppard and Webster showed that the products of an old, known reaction between dienes and aromatic diazonium salts are not coupling products, but rather, they are dihydropyridazines. A detailed investigation of this reaction, in particular with electron-rich dienes, revealed the cycloaddition is concerted, and 3,6-dihydropyridazines **19** are formed first. They are transformed further into either 1,6-dihydropyridazines **20** or pyridazinium ions (84TL57).

$ArN_2^+ X^-$ + → (19) $\xrightarrow{-HX}$ (20)

Upon heating a mixture of benzaldehyde and hydrazine salt in the presence of styrene (2 : 1 : 6 molar mixture), in addition to 1,5-diazabicyclo [3.3.0] octane, 3,5,6-triphenyl-1,4,5,6-tetrahydropyridazine was obtained as byproduct. The latter was the only product when a 2 : 1 : 2 ratio was used. The pyridazine formation is explained on the basis of cycloaddition of benzaldehyde azine to styrene, followed by tautomerization (87JOC2277). Similarly, other diazadienes add ethyl vinyl ether to give also 1,4,5,6-tetrahydropyridazines [83JCS(P1)1803].

Several azadienophiles were employed in cycloadditions. Pyridazines were thus obtained from dienes and diethyl azodicarboxylate (82S958; 84TL1769) or cyclic azo compounds (80LA1307), arenediazocyanides, and related compounds, preferentially in a crown ether-catalyzed synthesis (79CC1019; 82JA548). With unsymmetrical dienes, a variable degree of regio-selectivity, depending on the particular aryl substituent of the azo compound, was observed (82TL3875). Thus, cycloaddition of methyl vinyl ketone and 1-isopropyl-3,4-dimethyl-1,2-diaza-1,3-butadiene afforded a mixture of four stereoisomeric 1,4,5,6-tetrahydropyridazines (21–24) (78KGS1684). 1,3-Diketones, when converted into bis-(silyloxy)-1,3-butadienes, react with various aza dienophiles. The cycloadducts are relatively unstable and are immediately desilylated. With unsymmetrical aza dienophiles, two regioisomers are theoretically possible. It was found that only one is formed (87CB1597).

Since the report by Carboni and Lindsey in 1959 on the cycloaddition reaction of tetrazines to multiple bonded molecules as a route to pyridazines, such reactions have been extensively studied. In addition to acetylenes and ethylenes, enol ethers, ketene acetals, enol esters and enamines, and even aldehydes and ketones have been used as starting materials for pyridazines. A detailed investigation of various 1,2,4,5-tetrazines in these syntheses revealed the following facts. In [4 + 2] cycloaddition reactions of 3,6-bis(methylthio)-1,2,4,5-tetrazine with dienophiles, which lead to pyridazines, the following order of reactivity was observed (in parenthesis the reaction temperature is given): ynamines (25°C) > enamines (25–60°C) > ketene acetals (45–100°C) > enamides (80–100°C) > trimethylsilyl or alkyl enol ethers (100–140°C) > enol

(25%) (17%) (50%) (8%)
(21) (22) (23) (24)

R = i−Pr
R¹ = MeCO

acetates (130–140°C). Raney-nickel reductive desulfurization provided an approach to the 3,6-unsubstituted pyridazines, whereas oxidation to the bis-sulfone opened a way of preparing monosulfones. The second sulfone group can be easily displaced to give an amino or methoxy derivative. Dimethyl 1,2,4,5-tetrazinedicarboxylate was found to be more reactive than the previously mentioned 3,6-bis(methylthio) analog (88JOC1415).

In this manner, starting from an appropriate tetrazine with two electron-withdrawing groups, these can be introduced at positions 3 and 6 in the formed pyridazine. It is well known that such groups can be introduced directly into the pyridazine only with great difficulty. This method allows the introduction, for example, of trifluoromethyl groups [78JCS(P1)378; 87CZ81] or heterocyclic rings such as 2-pyridyl (80DOK1392; 81AJC1223, 81JOC881, 81SC655). On the other hand, with properly substituted dienophiles, such groups can be introduced in the pyridazine ring at positions 4 and/or 5. 4-Nitro- or 4-cyano-pyridazines, usually not readily accessible, can be prepared in this manner (85H683).

In a study of the reaction between dimethyl 1,2,4,5-tetrazine-3,6-dicarboxylate and electron-rich ethylenes, it was found that, upon cycloaddition, the loss of nitrogen is fast. The slow step is the final aromatization of the resulting dihydropyridazine by loss of either morpholine, pyrrolidine, alcohol, or silylol, these groups being attached to the starting dienophile. Hydrolysis of the ester groups and subsequent decarboxylation give, at positions 3 and 6, unsubstituted pyridazines (84JOC4405).

It has also been established that dihydropyridazines, obtained by this reaction and having alkyl or aryl groups at positions 3 and 6, exist in solution in an equilibrium between the 1,4- and 4,5-dihydro tautomers. For example, 3,4-diphenyl-1,4-dihydropyridazine exists in chloroform solution in an 8:1 ratio of the 1,4- and 4,5-tautomers (83JHC855).

As dienophiles, ethyl vinyl ether (87JHC1285), *cis,cis*-cycloocta-1,5-diene (79T277), cyclopropenes (79BSB905), fulvenes (78T2509; 79LA675), acetamidines in their very activated ketene-*N,N*-acetal form (81SC655), and aldehydes or ketones were used. In this particular case, aldehydes were found to be more reactive than ketones, and cyclic ketones were more reactive than the open-chain analogs. Moreover, a stepwise mechanism is preferred, although cycloaddition is not excluded (79JOC629).

Ketene acetals react with various 3-substituted aryl 1,2,4,5-tetrazines to give almost exclusively the ortho isomers (**25**). An exception is 1,1-bis-dimethylaminoethene, which afforded both regio isomers **25** and **26**, the ratio being solvent dependent (84TL2541).

R^1 = SMe , OMe , OEt , NMe$_2$

Diketene reacted to give spiro-bipyridazine **27** accompanied by a small amount of substituted pyridazine **28** (83CZ172).

Transformations with *N,N*-dimethylhydrazones of crotonaldehyde or cinnamaldehyde to give 1,4-dihydropyridazines are known (81AP376), and an interesting application of enamines forms compound **29** (79LA675). Dienamines were also used (87H337).

(29)

The double bond of some unsaturated sugars has been involved in this reaction to give the corresponding pyridazines after prolonged heating. A mixture of two products, **30** and **31,** was obtained, the bicyclic compound being formed only in 2–3% yield (85LA628).

R^1	R^2	
AcO	H	R = COOMe
H	AcO	

There are several examples of alkynes as dienophiles. Silylalkynes give silylated pyridazines, and when a double bond is present in addition to the triple bond, only the triple bond reacts (81CB3154). Using 3-phenyl-1,2,4,5-tetrazines and silylalkynes, cycloaddition proceeds highly stereospecifically (82CB2574). Bis(trimethylstannyl)acetylene was transformed with 3,6-bis(trifluoromethyl)-1,2,4,5-tetrazine into 4,5-bis-trimethylstannyl-3,6-bis (trifluoromethyl)pyridazine. With iodine or chlorine, both stannyl groups can be replaced to give the corresponding 4,5-dihalo derivatives [78JCS(P1)378].

An acetylenic side chain in 5-ethynyl-2′-deoxyuridine or its O-acetyl derivative is converted by this method into a new pyridazine ring (86JOC950). Intramolecular cycloaddition of 1,2,4,5-tetrazines with acetylenic side chain has been used to prepare various condensed heterocyclic rings (**32**) (85TL4355; 86TL2747; 87CZ16; 88JOC1415).

These cycloadditions can also be extended to pyridazines (87TL6027) and were used to prepare the indoline ring (**33**) as a part of PDE I and PDE

(32)

X = O, NH
R = SMe, SOMe, SO$_2$Me, Ph

II, two 3′,5′-cAMP phosphodiesterase inhibitors (87JA2717). If these transformations are employed twice, the inverse electron-demand cycloaddition reactions represent a valuable synthetic strategy for the following conversions: 1,2,4,5-tetrazine → pyridazine → benzene (or indoline).

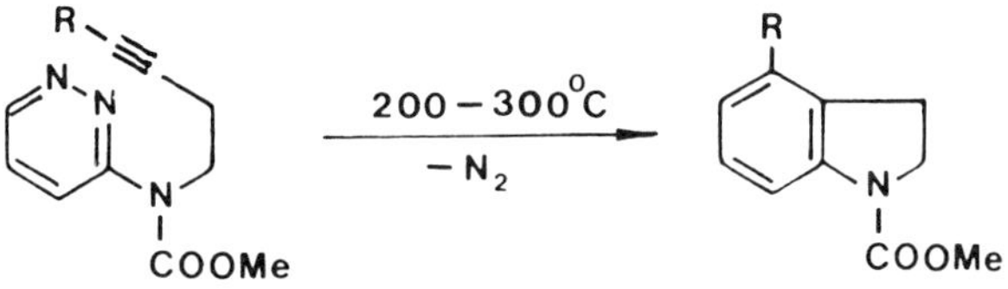

(33)

On the other hand, trimethyl 3,4,6-pyridazinetricarboxylate and tetramethyl 3,4,5,6-pyridazinetetracarboxylate undergo cycloaddition with norbornene and related compounds to give esters of benzenepolycarboxylic acids (85LA853). In a similar manner, xanthenes were obtained from allyloxypyridazines (80CPB198).

Finally, various heterocycles have been used as dienophiles in these cycloadditions. From vinylindoles, a pyridazine ring was formed as side chain (87C125), and 3,5-bis(trifluoromethyl)-1,2,4,5-tetrazine reacts with some benzoazoles. The cycloadducts formed undergo [4 + 2]-cycloreversion with nitrogen elimination to give condensed pyridazines as shown in the case of indole and benzothiophene. However, the cycloadducts from N-methylindole and benzofuran may undergo ring opening to give substituted pyridazines 34 (87CZ81).

2,6-Dimethyl-1,4-dihydropyridine-3,5-dicarboxylates react with 3,6-dipyridyl-1,2,4,5-tetrazines in two ways. If a 1-unsubstituted starting pyridine was used, instead of cycloaddition, a hydrogen transfer reaction took place to give substituted pyridines; tetrazine acted as hydrogen acceptor. In the case of a 1-methyl analog, 1,2-dihydropyridine derivative 35 formed, which in turn reacted further with tetrazine leading to spiro intermediate

36. This undergoes aromatization to pyridazines with unsaturated side chain **37**. The latter are easily converted in aqueous ethanol into pyridonylpyridazines **38**. Structures have been confirmed by X-ray analysis (87JOC2026).

An interesting example of this cycloaddition is its application to identifying 2- and 3-acetyl arsabenzenes, which are formed together with 3,6-dipyridylpyridazine **39** (81JOC881). From a cycloaddition reaction of 2,7-

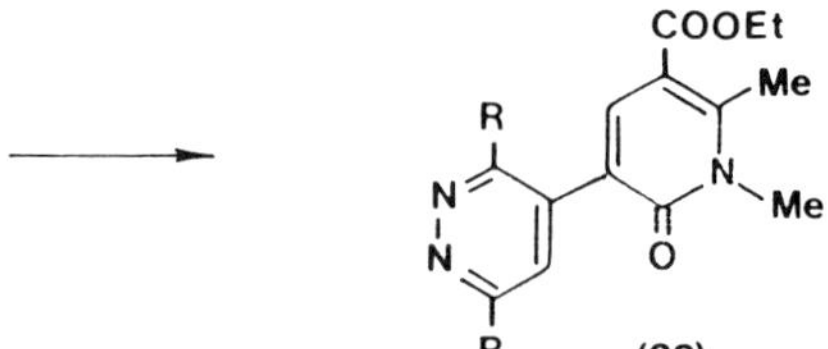

[Structure (38) with COOEt, Me, N-Me, R groups on pyridazine-pyridinone system]

dihydrothiepin-1,1-dioxide, an approximately equal mixture of the bis-pyridazine adduct and bis-pyridazinylethene **40** were isolated, the latter being a product of sulfur dioxide extrusion (81AP892).

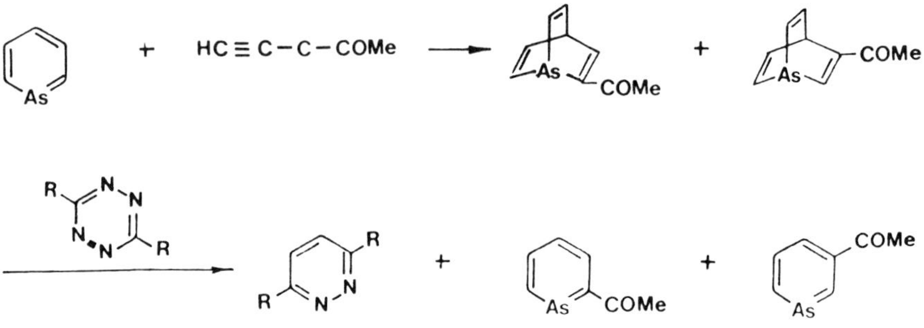

[Reaction schemes showing arsabenzene + HC≡C-C-COMe giving bicyclic adducts with As and COMe groups]

[Reaction of tetrazine (R = 2-pyridyl) giving pyridazine (39) + arsabenzene-COMe products]

(39)

R = 2-pyridyl

[Structure (40): bis-pyridazinylethene]

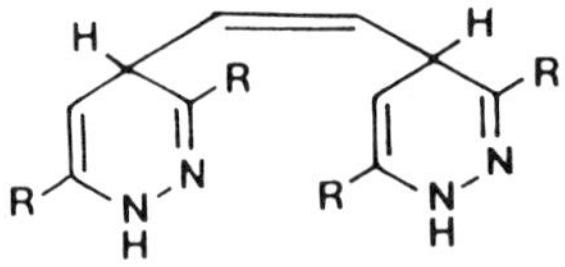

(40)

N-Methoxycarbonyl-Δ-^{2}-azetine reacted with a tetrazine at room temperature to give predominantly a substituted pyridazine. The reaction involves ring scission of the azetidine ring (77CC806).

C. From Other Heterocycles

Furans and 2-furanones (butyrolactones and butenolides) are main sources of heterocycles for pyridazine syntheses. A furan ring attached to a ribofuranosyl ring is readily transformed with hydrazine to give pyridazine-3-*C*-nucleoside (83JOC2998; 87JOC4521). Similarly, photooxidation of polyhydroxyalkylfuran derivatives and subsequent treatment with hydrazine afforded a pyridazine with a 3-polyhydroxyalkyl side chain (84MI16, 84MI17).

Various saturated or unsaturated 2-furanones were transformed with hydrazines into pyridazines [80IJC(B)1038, 80MI8; 81JPR164, 81MI9; 82ACH(111)387, 82IJC(B)763; 85FES200, 85JCS(P1)1627, 85MI7, 85MI21; 86M231; 87MI15]. From 4-acetyl-4,5-dihydrofuran-2(3H)ones and hydrazine, pyridazines **41** are formed. The synthesis proceeds by ring opening and rearrangement (84JHC305). Dehydration with TsOH or sulfuric acid transform the pyridazines into 5-arylidene **42** or 5-methyl derivatives **43** (85JHC1615; 87FES585).

2-Acetoxy-3(2H)-furanones react with hydrazine to give 1,4-dihydro-4-pyridazinones, whereas with monosubstituted hydrazines, anhydro-5-hydroxypyridazinium hydroxide derivatives **44** are obtained. Sodium borohydride reduction of these betaines provides an easy entry to 5-hydroxy-1,6-dihydropyridazines **45,** which can be alkylated to 1,2-disubstituted derivatives (79JOC3053).

A new synthetic approach uses 2-arylazo-2,5-dimethyl-3-oxo-2,3-dihydrofurans. After hydrogenolysis and cyclization, 1,4-dihydropyridazines **46** are obtained (79S790). Pulvinic lactone, a lichen metabolite,

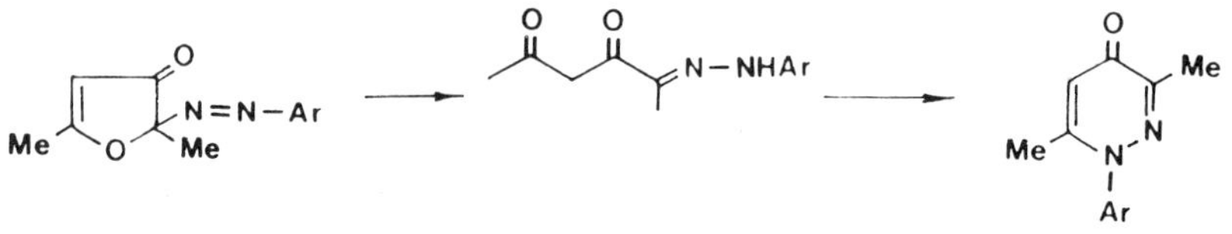

(46)

gives pyridazinone **47** with hydrazine [85IJC(B)785]. Pyridazines were also prepared from some 2-aminothiophene precursors. The reaction with hydrazine proceeds evidently by ring opening, elimination of hydrogen sulfide, and ring closure (88CB573). Pyridazines can also be obtained from 1-aminopyrrolidines, which undergo facile oxidative enlargement (even by column chromatography on silica or after stirring with silica gel) (79TL5025). Arylaminomaleimides, under the influence of bromine in acetic acid, are isomerized into pyridazinones (80MI20), and pyrrolidinediones or pyrrolopyridazines are also converted with hydrazine into pyridazines (81HCA1930).

(47)

Other five-membered heterocycles such as pyrazolines, oxaxolin-5-ones (83H2385; 84H2483), isoxazoles or isoxazolines [82ACS(B)1; 83TL1285], and 1,2,3-triazoline [78H(9)243] could be converted into pyridazine derivatives. There are few examples of six-membered heterocycles being transformed into pyridazines. 3,4-Dicyano-2,4,4,6-tetramethyl-1,4-dihydropyridine reacts with nitric/acetic acid below −10°C to give, in addition to a pyridine and a pyrazole, pyridazine **48**. It was postulated that pyridazine formation requires the presence of acetic acid; this was confirmed (84CCC2620).

(48)

3-Phenacylpyridinium methiodide, when treated with hydrazine under Wolff–Kishner reduction conditions, is transformed into 6-phenyl-4-propylpyridazine. Some related pyridazines were prepared in a similar manner, but 3-benzoylpyridinium methiodide failed to react [80JCS(P1)72]. 2-Dimethylamino-5-phenyl-1,3,4-thiadiazin-6-one reacts with the electron-rich 1-diethylaminopropyne to give two pyridazines, **49** and **50**, in 72% and 2% yield, respectively. The formation of both compounds is explained by the initial addition of ynamine, ring opening, and cyclization in two ways (82CC1003).

3,6-Diaryl-1,2,4,5-tetrazines, when treated with bulky amides such as lithium di-isopropylamide, undergo two competing reactions. In the first, tetrazine is reduced with concomitant formation of an imine from the amide. The imine is then attacked further by amide to give a pyridazine. For example, 3,6-diphenyl-1,2,4,5-tetrazine is converted with lithium diethylamide into 3,6-diphenylpyridazine in low yield. With lithium di-isopropylamide, 4-methyl-3,6-diphenylpyridazine is obtained in moderate yield [83JCS(P1)1601].

4H-Pyran-thiones, when treated with hydrazine, can give pyridazines, 1-amino-4(1H)-pyridinethiones, or pyrazoles. Pyridazines are the exclusive or main product when the starting pyranthiones are substituted at position 3 with a hydroxy group. Perhaps hydrazine attacks position 6, followed by elimination of hydrogen sulfide to give **51** (78BCJ179).

1,2-Diazepines were also transformed into pyridazines either with hydrochloric acid, NBS, or hydrazine (77BCJ2153; 86H3059). A 1,2,5-triazepine could also be transformed into a pyridazine derivative (85H1675).

Many condensed pyridazines undergo ring opening of a fused five- or six-membered ring to give substituted pyridazines. Such transformations are known with furo[2,3-d]pyridazines and pyrano [2,3-d]pyridazines (79JHC245, 79JHC249), oxazolo[3,2-b]pyridazinium salts (82CPB1557), imidazo [1,2-b]pyridazines or their 1-oxides [78H(10)269], an oxadiazolo-pyridazine (80CPB3570), and pyridazino[3,4-d]-1,3-oxazine-4,5-dione after reduction with sodium borohydride (79JHC1213).

Pyridazines are also formed from some particular bicyclic azoalkanes such as 1,2-diazabicyclo[3,1,0]hexene, 2,3-diazabicyclo[2,2,1]heptene, and related systems (80TL1009, 80TL4615; 84JOC1261; 87JOC5498). The synthesis of 5-hydroxy-piperazic acid from a tricyclic diazaanthracene precursor by hydrazinolysis is described (77H119). The lactam ring in 4-formyl-6,7-dihydroxanthyletin is transformed with phenylhydrazine into a pyridazinone ring as substituent (**52**) [82IJC(B)273].

(52)

D. Miscellaneous Syntheses

Several synthesis start from cyclopropanes. Silyloxy or benzoyl alkyl cyclopropanecarboxylates react with hydrazines to give substituted 3(2H)-pyridazinones (80JHC541; 84S786). Also, cyclopropenylium salts, when treated with phosphorylated or carbalkoxylated diazomethanes, give 4-amino- or 4,5-diamino-pyridazines (86JHC385, 86MI7). On the other hand, a diazomethyl group may be present as substituent at the cyclopropane ring, and isomerization also leads to pyridazines (82CB2965). By this method, sterically hindered pyridazines **53** were obtained. Upon irradiation these are rearranged into 1,2-Dewar pyridazines **54** in high yields. These bicyclic pyridazines are thermally very stable compounds and no rearomatization takes place thermally (84CB445).

Pyridazines were obtained also by photolysis of 1-phenyl-1-vinyl azide in the presence of iron pentacarbonyl (3,6-di-phenylpyridazine was obtained in 1.1% yield) (78HCA589) or by thermal decomposition of an allenic hydrazonate (81JA7011). Acetylenic hydrazides can be transformed into pyridazines [84BSF(2)129], and thermal cyclization of dialkali metal salts of ω-hydroxyketone tosylhydrazones afforded pyridazines in moderate yield (85TL655). Propionyl phenylhydrazine, after reaction with 4-bromobutyronitrile, converts into a pyridazine (87SC1253).

Diimine has been trapped on a polymeric material to give polymer bound pyridazine which, after reduction and hydrolysis, afforded 3-hydroxyhexahydropyridazine (79TL1333). Electro-chemical reduction of azobenzenes has been reported to give 1,2-diarylhexahydropyridazines (78TL4955; 86ZC438).

Pyridazines were also formed from dioximes by oxidative cyclization [78H(9)1367; 79JOC3524; 82MI18]. As oxidants, lead tetraacetate or phenyliodoso bis(trifluoroacetate) were used, and substituted pyridazine 1,2-dioxides 55 are obtained accompanied by isoxazoles, isoxazoloisoxazoles, and open chain compounds. It was also established that a compound to which the structure of 3,6-diphenylpyridazine 1,2-dioxide was previously assigned is, in fact, a dihydroisoxazoloisoxazole derivative (79JOC3524).

E. Syntheses of Labelled Pyridazines

Mainly for the purposes of biological studies, many labelled pyridazines were synthesized either by conventional methods or by some modification of these methods. A review on synthesis and radioactive labelling of amezinium, a pyridazin drug (4-amino-6-methoxy-1-phenylpyridazinium methyl sulfate) has appeared (81AF1529). At positions 4 and 6, deuterated 3(2H)-pyridazinones were prepared by gas-phase decarboxylation of the corresponding carboxylic acids or by cyclization [83JCS(P1)1203], and a stereospecific synthesis of cis-3,4,5,6-tetrahydropyridazine-3,4,d_2 is described (79JA3663). By reductive tritiation, another tritium-labelled pyridazine drug (SR95531) was synthesized (87MI13).

^{13}C- or ^{14}C-labelled pyridazine drugs or active compounds were prepared. Syntheses of 4-acetyl-5,6-bis(4-chlorophenyl)-2-(2-hydroxyethyl)-3(2H)-pyridazinone (81MI4), N-(2,5-dimethyl-1H-pyrrol-1-yl)-6-(4-morpholinyl)-3-pyridazinamine hydrochloride (85MI4), pirizidilol dihydrochloride (85MI26), a triazolopyridazine (86MI17), LY195115 (a pyridazinyl indolone) (86MI22), a pyridazinyl 1,4-benzodioxan derivative (87MI3), and, of CI-930, an aryl 3(2H)-pyridazinone (87MI9) were reported.

III. Transformations of Pyridazines

A. Reactions on the Ring Carbon Atoms

A great majority of reactions involving pyridazine carbon atoms are alkylations and arylations. Cross-coupling reaction between chloropyridazines and Grignard reagents in the presence of nickel-phosphine complexes as catalysts afforded alkyl- and arylpyridazines in various yields. Methylmagnesium iodide showed higher reactivity than ethylmagnesium iodide. Otherwise unaccessible naphthyl- and thienyl-pyridazines have been prepared in this manner (78CPB2550). Methylation was successful also via reaction with dimethyl- and pentamethylcuprates (85JPR536). Alkylation and arylation with Grignard reagents were reported (77MI2; 85JHC927). 6-Phenyl-2-(p-toluenesulfonyl)-3(2H)-pyridazinone reacted with 10 equimolar amounts of methylmagnesium iodide at room temperature to give 5-substituted 4,5-dihydro derivatives **56** together with 1,2-dihydro- (**57**) and 2,5-dihydro derivatives **58,** the ratio of these products being dependent on the reactions. With other Grignard reagents, only 5-substituted 4,5-dihydro-3(2*H*)-pyridazinones were isolated (85JHC927).

Aromatic aldehydes react with 4,5-dihydro-6-methyl-3(2*H*)-pyridazinone in basic medium to give 4-arylmethyl derivatives [80IJC(B)203]. Also, the benzylidene derivative of 3-hydrazinopyridazine, when heated at 250°C in the presence of sodium ethoxide, is transformed into a mixture of six compounds, two of them being 3-ethylamino- (25%) and 3-amino-pyridazine (4%) (78KGS1120). 6-Aryloxy-4,5-dihydro-3(2*H*)-pyridazinones were condensed with aldehydes to give the corresponding 4-alkyl or 4-arylalkyl derivatives [78IJC(B)631]. Nitromethane or nitroethane were also found to be alkylating agents, and various 3(2*H*)-pyridazinones were alkylated in the presence of a basic catalyst to give the corresponding 5- or 4-methyl (ethyl) derivatives in variable yields (77CPB1856).

Esters of pyridazinyl-3-acetic acid **60** can be prepared when, for example, 3-methoxypyridazine 1-oxide reacts with methyl β-aminocrotonate in the presence of benzoyl chloride to give **59**. Since β-aminocrotonic esters are tautomeric with the imines of acetoacetic esters, mild acid hydrolysis converts them into the corresponding acetoacetic ester and then to ester **60** (78JHC1425).

4,5-Dihydropyridazine-3,6-diones have reactive methylene groups, and they react with aromatic aldehydes to give 4,5-bis-arylidene derivatives (86JPR932; 87JPR525). Pyridazines have been shown to undergo 1,3-dipolar cycloaddition with diazomethane [82H(18)175; 88H1431].

There are many examples of arylation or heteroarylation of pyridazines. Pyridazine or its 3-phenyl analog, when treated with phenyllithium, yields adducts almost exclusively at positions 3 or 6. However, in the presence of a complexing agent such as *N,N,N′,N′*-tetramethylethylenediamine (TMEDA), a mixture of adducts at positions 3 and 4 in the ratio of 1:4 is formed. Upon hydrolysis, dihydropyridazines **61** are obtained (78RTC116). Mono-, di-, and tri-substituted 3(2*H*)-pyridazinones react similarly. The 6-aryl analogs gave 6,6-diaryl derivatives **62,** and the 4-phenyl derivative reacted similarly. From 6-aryl-2-phenyl derivatives, the 2,6-diphenyl compounds **63** were obtained, but with Grignard reagents,

(61) (62) (63) (64)

(65) (66)

position 4 was attacked to give **64**. From 6-aryl-2,4-diphenyl analogs with both reagents, the 2,3,3,4,6-pentaaryl derivatives (e.g. **65**) were obtained (80S457) as main products. 6-Aryl-4-phenyl-4,5-dihydro derivatives reacted with phenyllithium by 1,2-addition, the substituent enters at position 6 to give **66** (80JPR617). Substituted 3(2H)-pyridazinones are transformed with Grignard reagents into 3-alkylated or 3-arylated products (85MI28).

Thienylpyridazines were prepared from 2- or 3-thienyllithium and pyridazine. In diethyl ether, the thienyl ring is attached at position 3 to give first the corresponding 2,3-dihydro derivatives accompanied by a small amount of 1,4-dihydro isomers, whereas in tetrahydrofuran and at low temperature, position 4 of the pyridazine ring is attacked. Aromatization can be achieved with various oxidants, preferentially with chloranil [81JCR(S)104; 82CJC2668].

Phenylmagnesium bromide has been used to introduce a phenyl group in various substituted pyridazinones [80MI12; 81IJC(B)502, 81MI12].

An interesting example of carbon-carbon bond formation represents the reaction between 3(2H)-pyridazinone and pyridine N-oxide in the presence of platinized palladium-carbon catalyst at 150°C to give, in 2% yield, 6-(pyridyl-2')-3(2H)-pyridazinone. Pyridazine itself did not react (78YZ67). 4,5-Diacylpyridazines were prepared from pyridazine and the corresponding aldehydes according to the Minisci reaction (78M63).

There are also some reports on amination. It was found that an effective reagent for direct amination of pyridazines is $KNH_2/NH_3/KMnO_4$. 4-Aminopyridazine is thus obtained in an excellent yield (82JHC1285). 3-Phenylpyridazine, with the same reagent, gave a mixture of 4-amino (49%), 5-amino (18%) and 6-amino (5%) derivatives (82JHC1285). 4-Nitropyridazines were aminated by liquid ammonia and potassium permanganate to give 5-amino derivatives. 4-Cyanopyridazines are aminated only in the presence of sodium amide, liquid ammonia, and permanganate to also give the 5-amino derivatives (88JHC831), and 4-nitropyridazine 1-oxides react in the same way. 3-Methoxy and 3,6-dimethoxypyridazines give the corresponding 4-aminopyridazines, and in addition, dimerization took place to give 3,3'-dimethoxy-4,4'-bipyridazine in 23% yield (86JHC621). It was found that 6-aryl-3(2H)-pyridazinones react on prolonged heating with hydrazine hydrate to give 4-amino derivatives. Maleic hydrazide, however, is transformed into 4-hydrazino derivative (84H1801). Benzylidene derivative of 6-chloro-3-hydrazinopyridazine, when heated with sodium ethoxide at 300°C for 15 min, afforded 3-amino-6-chloropyridazine in 18% yield (78KGS1120).

In addition to the traditional introduction of a halogen atom via a lactam group, maleic hydrazide was transformed into 3,6-dichloropyridazine with thionyl chloride or thionyl chloride and methyl sufonyl chloride [87CI(L)694]. Bromine in glacial acetic acid has been found to add at the 4,5-double bond of 1,2-disubstituted-3,6-pyridazinediones (82MI5).

An interesting reactivity is observed with 2H-cyclopenta(d)pyridazine and its 2-methyl and 2-phenyl analogs. Monochlorination of these compounds revealed the relative reactivities are 7,1 : 1,7 : 1 [78H(11)155]. If the cyclopentyl part is substituted with chlorine atoms, the 2-methyl compounds are transformed with N-chlorosuccinimide (NCS) into the 2-chloromethyl derivative (80JOC1695). In contrast to other perfluoroazines, perfluoroalkylpyridazines, when fluorinated with cobalt trifluoride, lose nitrogen to give perfluorinated olefins [81JCS(P1)2059].

A direct introduction of a cyano group in the pyridazine ring was also studied. Pyridazine or its 3- and 4-methyl analogs react with trimethylsilyl cyanide and tosyl chloride to give 2,3-dihydro derivatives which are transformed with 1,8-diazabicyclo[5.4.0]undec-7-ene (DBU) into the corresponding 3- (or 6-) pyridazinecarbonitriles (86H793). Alternatively, 6-aryl-3(2H)-pyridazinone or its 4-chloro analog, after treatment with potassium cyanide in dimethyl sulfoxide, yield the same 4,5-dicyano derivative (86JHC1515).

B. Reactions at the Ring Nitrogen Atoms

Protonation and basicity studies are reported in Section IV,B. Alkylations, like methylation, which proceed in the case of pyridazinones as N- and/or O-methylation, are treated in Section III,C. Only N-alkylations are mentioned in this section.

An interesting case of N-oxidation is given by [2.2](2.5)furano(3,6)pyridazinophane (**67**). Oxidation with m-CPBA gave the chiral N-oxide **68** together with a smaller amount of diketone **69**, resulting from furan ring

(67)   (68)   (69)

cleavage (80JOC4584). N-Amination of maleic hydrazide with hydroxylamine-O-sulfonic acid was reported (83MI24).

3-Alkynyl pyridazines, with m-CPBA exclusively, gave the 1-oxides. It is suggested that the alkynyl group at position 3 of a pyridazine acts as an electron-withdrawing group and reduces the nucelophilicity of the N-2 atom [78H(19)1397; 80CPB3488].

3(2H)-Pyridazinones are N-alkylated under phase-transfer conditions to give 2-substituted products (81S631). Reactants include dimethyl sulfate, ethyl bromoacetate [80IJC(B)203], ethyl chloroformate (84JPR599), phenacyl bromide (77MI6), or with chloroacetonitrile (87MI5). They are vinylated (80KGS394) and also react with vinyl cyanide (79RRC899). At the same position, hydroxymethylation (81MI13) or a Mannich reaction [77IJC(B)1025; 85MI14] takes place. 3-Phenylpyridazine was treated with p-bromophenacyl bromide and, upon dehydrobromination, the corresponding phenacylide was obtained (79MI7). 4,5-Diamino-3(2H)-pyridazinethione, when treated with an excess of methyl iodide gave mixed products of S-methylation and quaternization at position 1 (81H9).

Pyridazine analogs of naturally occurring nucleosides were synthesized, the sugar moiety being attached at position 2 of the 3(2H)-pyridazinone ring (82JMC813; 83JHC369; 86UKZ1087; 87UKZ1099). Alkylation was achieved either with acetobromo sugars or with 1-O-acetyl derivatives by a stannic chloride catalyzed procedure (82JMC813).

The first example of a mesoionic pyridazine was found during a metabolite study of 4-cyano-3(2H)-pyridazinone. The isolated mesoionic compound **70** was synthesized at low temperature as indicated. When the reaction mixture was heated, the normal N2-riboside (**71**) was formed. A series of analogous compounds were prepared. Both types of nucleosides are highly sensitive to base, and deblocking of acetyl groups could be achieved with methanolic $NaHCO_3$ (84CC422, 84JMC1613).

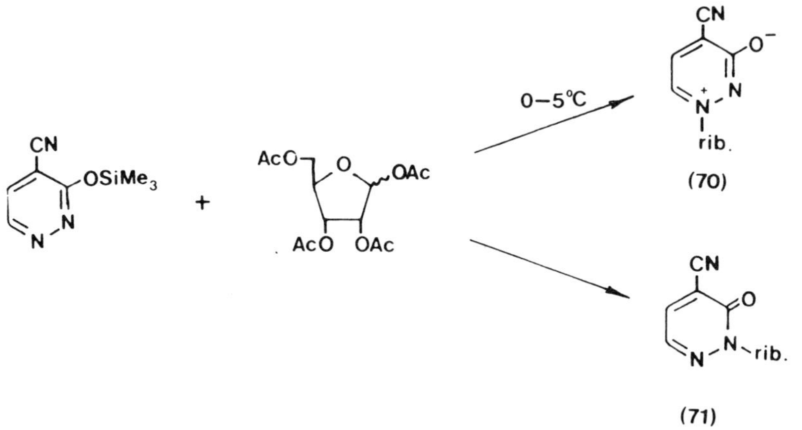

(70)

(71)

C. Reactions of Functional Groups

In general, halogen atom replacement reactions still constitute the most useful way of introducing various functionalities in the pyridazine ring. These transformations will be reviewed first, although those reactions that give the corresponding amino, hydrazino, thio, and related functional groups will be mentioned later when these groups are discussed.

As already known, a halogen atom at position 4 in the pyridazine ring is more reactive towards nucleophiles then one at position 3. This has been confirmed also with 3,4-dichloro-5,6-diphenylpyridazine and various nucleophiles (81MI2). The high reactivity of 6-halo-1,3-dimethylpyridazinium halides or phosphorodichloridates has been used to prepare alkyl halides from alcohols (80S746).

Chlorination of pyridazinones into halopyridazines by means of phosphorus oxychloride remains the method most used (79MI5; 80NKK33; 81MI21; 82MI6), although chlorosulfonyl isocyante or chlorocarbonyl isocyanate were also successful (86SC543; 87OPP80). 4,5-Dihydro-6-phenyl-2-(phenylmethyl)-3(2H)-pyridazinone, when treated with phosphorus pentachloride, afforded 3,4-dichloro-6-phenylpyridazine and ~10% of the 3-chloro analog. Evidently the reaction proceeded not only by introduction of the chloro atom, but also with dehydrogenation and debenzylation (83JHC1473). Bromination of a 3-hydroxypyridazine 1-oxide with bromine in water yielded the 4-bromo derivative (78CPB3884). Iodinated pyridazines were prepared from the chloro analogs with either sodium iodide (77AJC2319) or iodine and ammonium persulfate, but in this case, the benzene ring was iodinated preferentially (87SC1907). 3-Fluoro-6-

methylpyridazine was obtained from the 3-amino precursor with fluoroboric acid and sodium nitrite, but is quite unstable (77AJC2319). Dehalogenations are performed catalytically as in the case of 3,6-dichloro-4-pyrrolidinopyridazine (78T2069).

In a cross-coupling reaction of 3- or 4-chloropyridazines and trimethylsilylacetylene in the presence of $Pd(PPh_3)_2Cl_2$-CuI as catalyst, the trimethylsilyl derivatives were obtained (82S312). The same transformation with monosubstituted acetylenes afforded 3-alkynylpyridazines or their 1-oxides (80CPB3488). In a modified approach (Sonogashira method) 3-chloro- (or iodo) pyridazines or their 1-oxides can be alkynylated by monosubstituted acetylenes using Pd-phosphine complex and diethylamine [78H(19)1397]. 3,6-Dichloropyridazine can be monocyanomethylated by sodium derivatives of cyanomethyl compounds (82ZVK591) and with a Grignard reagent prepared from 2-bromothiophene. The 3,6-dithienyl-2'-pyridazine was prepared in good yield (85JHC719).

3,4,6-Trichloropyridazine is converted, upon dimethylation, into a 6-chloro-3,4-dimethoxy derivative, a molecular complex of the latter compound, and 3-chloro-4,6-dimethoxypyridazine in a ratio of 1:1. The complex had been described previously as pure 3-chloro-4,6-dimethoxypyridazine (68MI5) and could not be separated into components (87CPB350). On the other hand, methoxylation of 3,4,5-trichloropyridazine with an equivalent amount of sodium methoxide afforded three monomethoxy derivatives, i.e., 3-, 4-, and 5-methoxy-dichloropyridazines in the ratio of 1:3:6 (87CPB421).

From 4- or 5-bromo- (or -chloro)- 3(2H)-pyridazinones and alcohols, several alkoxypyridazinones were prepared by phase-transfer catalysis. 4-Halo derivatives reacted slower than 5-halo analogs (81MI8). In the case of 4,5-dihalo compounds, substitution is regioselective, and 5- and 4-alkoxy derivatives are formed in ratio of 9:1 (85LA1465).

3,6-Dichloropyridazine was transformed with sodium 2,2,2-trifluoroethoxide into mono- or di-alkoxy derivatives, but with other sodium fluoroalkoxides, polysubstitution occurred (85CJC3037). 4,5-Dichloro-2-methyl-3(2H)-pyridazinone was used to prepare herbicidal pyridazino-1,2,4-triazines (87MI16).

3,4,5-Trichloropyridazine, with the sodium salt of o-(p-tolylaminomethyl)phenoxide in addition to the expected ether (reaction at position 4), gave the N-pyridazinyl derivative **72** in a ratio of about 4:1 (79KFZ34). 2-Aryl-4,5,6-trichloro-3(2H)-pyridazinones with hydroxide gave the 5-hydroxy derivative (79MI3), and 3-chloropyridazines reacted with phenols (80MI21).

Methylation of 3(2H)-pyridazinones afforded mainly a mixture of N- and O-methyl derivatives. This is the case with diazomethane (79MI2), but

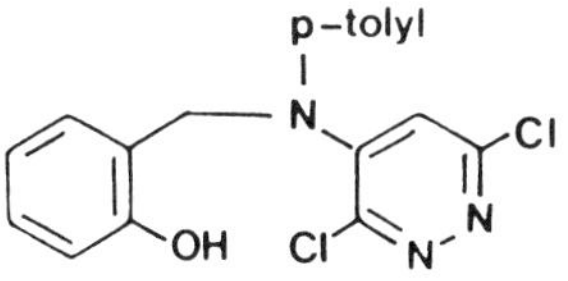

(72)

methylation of 4-nitro-5,6-diphenyl-3(2H)-pyridazinone with methyl iodide gives a mixture of the N-methyl derivative and the 2-methyl-4-methoxy derivatives, indicating a facile displacement of the nitro group (80MI2, 80MI3). Methylation of 4-hydroxy-5,6-diphenyl-3(2H)-pyridazinone afforded four products (80MI3). Methylation of monochloromaleic hydrazide afforded mainly the O-methyl derivative (87MI23). An interesting case is methylation of 3(2H)-pyridazinone 1-imides. The unsubstituted compound was transformed with diazomethane into N-methyl derivative **73,** whereas the 6-methyl analog yielded O-methyl derivative **74** (85CPB3540).

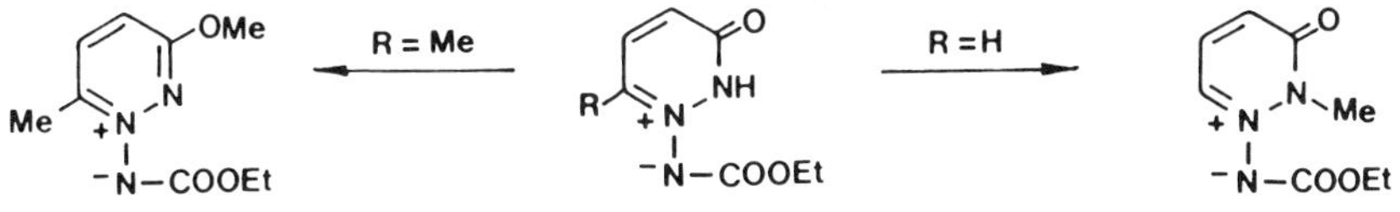

(74) (73)

Protected β-D-ribofuranose and 3(2H)-pyridazinones gave the N-glucosides, whereas 4-amino-3-chloro-6(1H)-pyridazinone afforded the O-glycoside (87MI27). In the formation of pyridazine cyclonucleosides, either the 2'- (**75**) or 5'-hydroxy group (**76**) was involved in ring formation (83JOC3765).

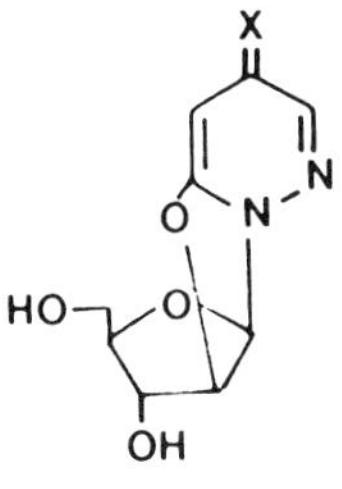

X = O, NH

(75)   (76)

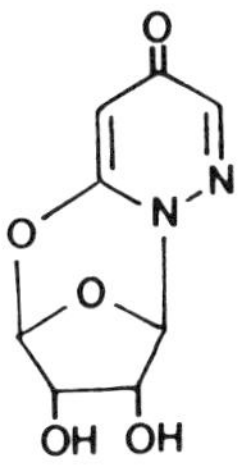

Alkylations with triethyloxonium tetrafluoroborate (82ZOR1557; 84ZOR1760), dialkyl sulfates, ethyl chloroacetate (81RRC1285; 84MI9; 86MI23), or allyl bromide (82MI17) were reported. Several papers describe alkylation or arylation of the N-side-chain of pyridazinium ylides (77MI9, 77MI10; 78MI11; 79MI25). Thermal decomposition of 3-ethoxypyridazines was studied with regard to decomposition rates of ethene and pyridazinones [86JCS(P2)1255].

There are many examples of modifying the carbon side chain. Alkylpyridazines can be lithiated at the side chain with LDA and subsequently alkylated or acylated (78CPB2428, 78CPB3633; 80YZ774). With ketones, the otherwise unaccessible tertiary alcohols are formed. These reverted to methylpyridazines and ketones under the influence of a base, but upon heating in neutral solvent, the corresponding pyridazinylethenes were formed (79CPB916). Substituted pyridazinylethenes were also prepared from 3- or 4-methylpyridazines and aromatic or heterocyclic aldehydes (78JHC749; 80AP53; 82MI11), sometimes in low yield. 3-Methylpyridazine, when hydroxymethylated and then thermally dehydrated, afforded 3-vinylpyridazine (85CC1632).

A new approach for introducing a carbon side-chain at position 3 in the pyridazine ring is displacement of a methylthio group of quaternized pyridazine with reactive methylene compounds to give **77** (79TL4837). Alternatively, a 3-chloropyridazine reacts with dimethylsulfoxonium methylide

(77)

to give pyridazine methylide **78**. When this is acetylated and desulfurized with Raney nickel, the corresponding alcohol (**79**) is obtained. This contrasts with other azinyl methylides which are transformed into the acetonyl derivatives under these conditions (81CPB2837).

(78)

(79)

Resonance-stabilized anions are also obtained if methyl groups at positions 1 or 4 in the pyridazine part of 2H-cyclopenta[d]pyridazine are deprotonated [78H(11)387]. Pyridazinium ylides readily undergo 1,3-dipolar cycloaddition reactions [77CPB192; 79H(12)661; 81JCS(P1)73; 87BCJ3645].

Pyridazine 3-carbaldehyde can be prepared from 3-hydroxymethylpyridazine by oxidation with selenium dioxide (82BSB153). Its 2-oxide was obtained similarly (78JMC1333). Contrary to aromatic aldehydes, pyridazine-4-carbaldehydes behave differently under the reaction conditions typical for benzoin condensation or cyanhydrin formation. In the first case, a crossed Cannizzaro reaction takes place and, with an equivalent amount of hydrocyanic acid, compound **80** is obtained. This is transformed with acetic anhydride into a mixture of E- and Z-isomers of **81**. On attempted chromatographic separation, the mixture is converted into methyl 4-pyridazinecarboxylate (78JHC637).

(80)   (81)

Both 3- and 4-pyridazinecarbaldehyde undergo Wittig reactions, and the corresponding *trans*-β-pyridazinylpropenals were prepared (82AP175). An aldehyde group reacted normally with a reactive methylene group as shown in the reaction with 1-acetyl-3-indolinone (83JHC101).

The trimethylsilyl ester of 3-pyridazinecarboxylic acid reacts with aldehydes and ketones through *ipso* substitution of the ester group to give **82**. The silyl group can be removed in hot ethanol or with pyridinium trifluoroacetate to give **83** (88T3281).

o-Aroylpyridazinecarboxylic acids react with thionyl chloride to give corresponding lactones **84,** which are transformed by ring opening to the corresponding amides or azides (**85**) (85LA167).

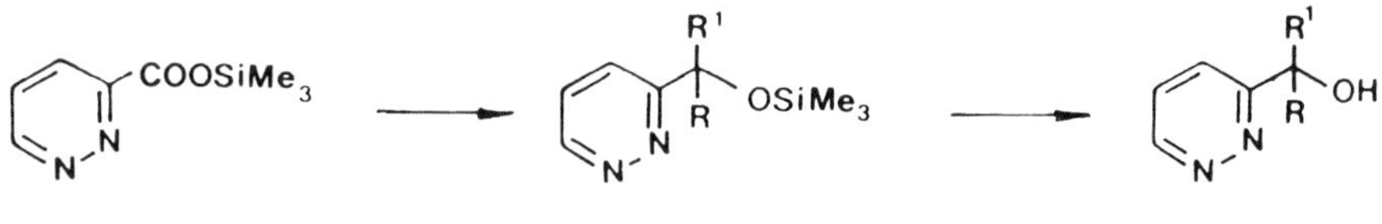

(82) (83)

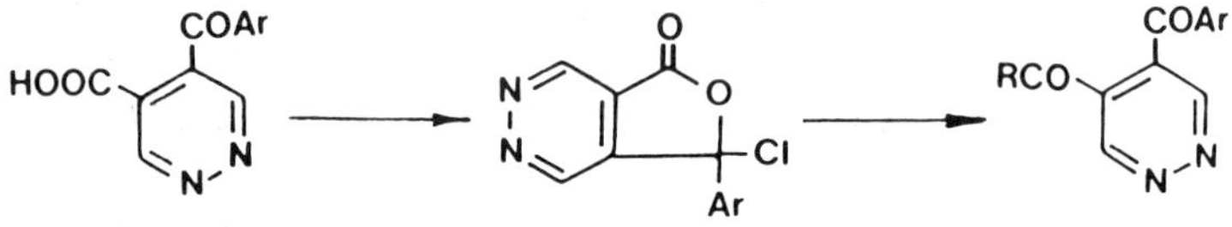

(84) (85) R=NHR¹, N₃

4,5-Pyridazinedicarboxylic acid anhydride reacts smoothly with various 1,3- or 1,4-difunctional nucleophiles to give spiro compounds **86** (reaction with thioureas) [84JCS(P1)2491].

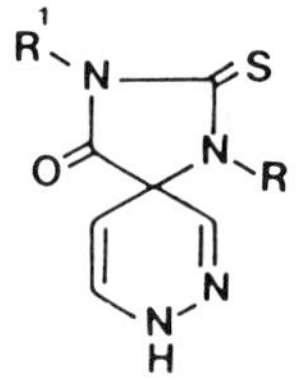

(86)

The general method of introducing an amino functionality in the pyridazine ring, i.e., by replacement of a halogen atom, was used in many cases. Transformations involved primary or secondary amines or ammonia [77CPB1708; 81IJC(B)78, 81SC835; 83JCR(S)37; 84MI19; 85ACH177, 85MI13; 87AP1222; 88JHC119] or heterocyclic amines [81IJC(B)1084, 81JHC803; 83H765; 85FES517]. 3,4,5-Trichloropyridazines react with cyclic secondary amines to give either 5-mono or 3,5-disubstituted products, depending on reaction conditions (84JHC1389). Although 5-bromo-1-(*p*-bromophenyl)-2-methyl-3,6-pyridazinedione was known to react with some amines to give products of cine substitution, with morpholine, the normal substitution product was obtained (81JHC1109). Dichloromethoxypyridazines react with primary or secondary amines, for example with morpholine, to give substitution and demethylation products **87** and **88** (87H1).

Sec. III.C] ADVANCES IN PYRIDAZINE CHEMISTRY 415

[Structures: 3,4-dichloro-6-methoxypyridazine → 3-chloro-6-morpholino-pyridazinone (77%) (87) + 3-chloro-4,5-dimorpholino-pyridazine (7%) (88); Mo = morpholino]

A peculiar case was observed when 3-chloro-4-methyl-6-phenylpyridazine reacted with primary aliphatic amines. The autocatalyzed reaction proceeds some hundred time faster than the noncatalyzed reaction. A mechanistic interpretation is proposed (83NJC667). Also, 4-cyano-3,6-dichloropyridazine, when treated with primary amines at 0°C, did not afford the 3- or 6-amino derivatives, but rather, afforded the 5-amino derivatives (77CPB1708). Perfluoro-3,5-diisopropylpyridazine reacted with dimethylamine to give substitution product **89** which, on standing, transformed into bicyclic product **90**. This transformation is accelerated by water (82CC1412).

[Structures: R = CF(CF$_3$)$_2$; difluoro-pyridazine → bis(dimethylamino) pyridazine (89) → bicyclic product (90)]

There are also examples of formation of an amine group by Hofmann degradation (81JHC1465) or by amination of pyridazinones with benzylamine and octamethylcyclotetrasilazane (via a silylated intermediate) (84CB1523) with an azirine (78HCA2116) or with pyrrolylmagnesium bromide (79RRC453).

Aminopyridazines or their N-oxides can be further functionalized on the amino group to give useful synthons, mainly for formation of bicyclic products (for reviews, see 83H1591; 85MI32). In this manner, an amino group can be transformed with amide acetals into amidines **91**, amidoximes **92**, (which can be dehydrated into cyanoamino derivatives **93**), or simultaneously methylated to **94**. Hydrolysis of the latter affords the substituted urea derivative **95**, and either hydrolysis of amidine **91** or formylation of the aminopyridazine give formylamino compounds **96** (81H2173; 82JHC577; 84H1545, 84MI10; 85M1447; 86M221, 86S807). The formylamino group can be reduced into the methylamino group with borane-methyl sulfide (85H2651).

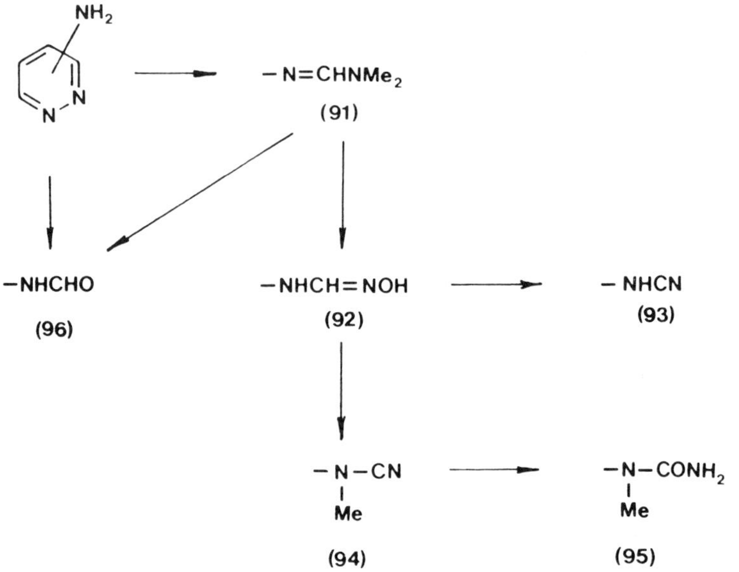

```
    NH₂
     |
  [pyridazine]  ⇌  —N=CHNMe₂
                      (91)
```

—NHCHO   —NHCH=NOH   ⟶   —NHCN
 (96)   (92)   (93)

                 ↓

              —N—CN   ⟶   —N—CONH₂
                | |
                Me   Me
               (94)   (95)

The amino group of 5-aminopyridazin-4-yl-*o*-fluorophenyl ketone was converted into the methylamino or ethylamino group by treatment with an ortho ester, reduction with $NaBH_4$, and subsequent oxidation with permanganate [88JCS(P1)401]. 3-Aminopyridazine and its 6-chloro analog react with chloroketene diethyl acetal in ethyl acetate to give a chloroimidate intermediate which reacted with another molecule of the amine to give **97** (88T3149). Aminopyridazines have been acylated by protected allonic acid [84JCS(P1)229] to give further C-nucleosides or a 1,2-benzothiazin-1,1-dioxide-3-carboxylate (78LA635). 1-Pyridazinylpyrazoles were acylated in the pyrazole ring (80JHC781). With isothiocyanates and isocyanates, the corresponding (thio)ureas were prepared (77MI11; 87H689).

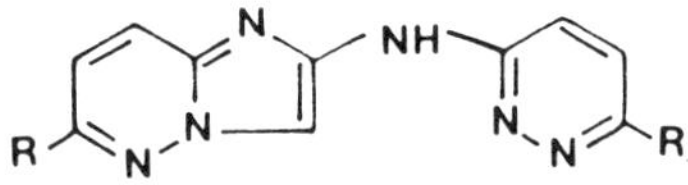

[Structure **97**: imidazo[1,2-b]pyridazine–NH–pyridazine with R substituents]

(97)

Stable diazonium salts were obtained upon nitrosation of 5-amino-2-phenyl-4-chloro-6(1*H*)-pyridazinone **98**, whereas the diazonium salt prepared from 4-amino-5-chloro analog **99** is unstable and highly reactive (83MI15). Diazotization of 5-aminopyridazin-4-yl-*p*-chlorophenyl ketone gave the corresponding 4-one [86JCS(P1)169].

(98)   (99)

Reaction of halopyridazines with hydroxylamine to give hydroxylaminopyridazine has been achieved only with activated compounds [79H(12)1157, 79JHC861]. This has been ascribed to the low nucleophilicity of hydroxylamine when compared to that of amines or hydrazines. A 3-hydroxylamino group reacted with bromoacetaldehyde at room temperature to give the first example of an imidazo[1,2-b]pyridazine 1-oxide (81T1787).

Most hydrazinopyridazines were obtained by substitution of a halogen atom with hydrazine [78IJC(B)1000; 81MI22; 85ACH221] or with ethyl hydrazinecarboxylate (81S608). By the ^{15}N-labelling technique, it could be demonstrated that displacement of the halogen atom in 3-halopyridazines with hydrazine occurs both by the normal S_N reaction (**100**) and also by the ANRORC mechanism (30%) to give **101** (83JHC1259). The hydrazino group was used to construct heterocyclic rings such as pyrazoles [80-JHC1527; 81MI18; 82IJC(B)317; 83JHC193; 84H513, 84JMC1077; 86JHC-193] or pyridazines [78IJC(B)1000; 81MI18; 82IJC(B)317; 84JMC1077].

3-Hydrazinopyridazine was transformed into pyridazine by the aza-transfer reaction with tosyl azide (78TL3059). Hydrazinopyridazines react with ethoxycarbonylisothiocyanate to give the corresponding thiosemicarbazides (88M83). They react with 2-chlorosulfonyl acetate to also give, in addition to the expected sulfonyl derivative, the pyridazinylhydrazone of glyoxylic acid (80ACH405). Hydrazones were prepared (80MI16), and reaction with diethyl pyrocarbonate gave the more stable ring ethoxycarbonylated derivatives **102**, although in one case, the side-chain substituted derivative **103** was obtained [81ACH205; 82ACH(109)237]. Oxidation of ethyl pyridazine-3-hydrazinecarboxylates by air in the presence of palladized carbon gave the corresponding ethyl pyridazinediazenecarboxylates (81S608).

An azido group was introduced into the pyridazine ring using an aza-transfer reaction with some diazonium or diazo compounds (81MI24). The Curtius rearrangement was used to convert the azido into the amino group (85LA167). 4-Azidopyridazine 1-oxides, when treated with potassium cyanide, were transformed into the corresponding 4-(3-cyano-1-triazeno)pyridazine 1-oxides. In hot ethanol, the cyanotriazeno group is split off, and in hot hydrochloric acid, it is replaced with a chlorine atom; 4-aminopyridazine 1-oxides were also formed (80CPB529).

The first synthesis of pyridazine-3,4,6-trithiol was achieved from 3,4,6-trichloropyridazine and thiourea (78AJC389). This method was used also in another case (80S410) and the thiophenyl group was introduced with thiophenol alone (85MI25) or under phase-transfer catalysis (PTC) in the presence of 18-crown-6 (87AP1222). Depending on the reaction conditions, 3,4,5,6-tetrafluoropyridazine reacted with thiocarbonyldifluoride as a gas, or as its liquid trimer, in the presence of cesium fluoride to yield selectively a 5-trifluoromethylthio- or 4,5-bis-trifluoromethylthio derivative [88JCS(P1)1179]. Thiation of pyridazinones was successful with Lawesson reagent [80ACS(B)597; 82H(19)2283].

3-Chloro-4,5-diaminopyridazine, when treated with sodium methylmercaptide, did not afford the anticipated 3-methylthio derivative, but rather, a mixture of imidazo[4,5-d]pyridazine derivatives. The extra carbon atom needed for the formation of the imidazole ring originated from DMF, which was used as solvent (81JHC303). In an alternative approach, 4,5-diamino-3(2H)-pyridazinethione was treated with methyl iodide, but in this case, 4,5-diamino-1-methyl-3-(methylthio)pyridazinium iodide was formed. The required 3-methylthio compound was obtained only when an exact amount of methyl iodide was used (81JOC2467).

6-Chloro-3(2H)-pyridazinethione with N,N-dimethylformamide–dimethyl acetyl (DMF-DMA) gives the S-methyl derivative (81AJC1729), whereas pyridazinethiones react with 3,6-dichloropyridazine to give the double substituted product (78JOC1190). A 3-acylmethylthio group in pyridazines can be replaced with an ethoxy group, but the phenacylthio group was transformed with DMF and potassium carbonate, with extrusion of sulfur, into an acylalkylidene (**104**) (83T2295).

(**104**)

D. RADICAL REACTIONS

Pyridazine and its derivatives were alkylated homolytically, and the methyl or alkyl radicals were generated by oxidative decarboxylation of acetic and other carboxylic acids. Substitution occurred mainly at positions β to the nitrogen atoms, and regioselectivity is lower when compared to homolytic benzylation or acylation of pyridazines (84H1395; 88OPP117). Pyridazines also react with 1,3,5-trioxanyl radicals generated from 1,3,5-trioxan and in the presence of ferrous sulfate and hydrogen peroxide in acid solution. Pyridazine gave a mixture of the 4-mono and 4,5-disubstituted products, the former being hydrolyzed into 4-pyridazinecarbaldehyde (80JHC1501). The orientation of radical attack at pyridazines differs significantly from that of other π-deficient nitrogen heteroaromatics (87H481). Several examples of alkoxycarbonylation of pyridazines were elaborated [85AG(E)692, 85T1199; 87H731; 88T2449]. Homolytic acylation of ethyl 4-pyridazinecarboxylate yielded the corresponding 5-acyl derivatives. After hydrolysis and decarboxylation 4-acylpyridazines are obtained (79M365; 83AP508).

Pyridazine and its derivatives were substituted with nucleophilic radicals. They react either with 1-formylpyrrolidine or with N-acetylproline in the presence of radical generators to give 5-substituted pyridazines (78TL619; 86MI6). Also, reactions of 3-chloro-6-methoxypyridazine with ketone enolates in liquid ammonia show typical characteristics of a radical chain ($S_{RN}1$) mechanism, and ketones **105** are obtained (81JOC294).

1,2,3,6-Tetrahydropyridazines were shown to undergo radical addition of thioacetic acid to the 4,5-double-bond to yield the corresponding 4-thioacetoxy derivatives (80JHC1465). One electron oxidation of maleic

(105)

hydrazide generated a cyclic α-carbonylhydrazyl as determined by electron spin resonance (ESR) spectra. Because of this ready oxidation to a radical, it is possible that maleic hydrazide plant growth inhibition activity is connected with such an intermediate (79TL2821).

E. OXIDATIONS AND REDUCTIONS: REDUCED PYRIDAZINES

For the synthesis of aromatic pyridazines, reduced pyridazines were oxidized preferentially with bromine, but the use of *m*-nitrobenzenesulfonic acid (78JHC881) or selenium dioxide [82IJC(B)371] were also reported. Iodine (87MI11) or electrochemical oxidation (84CC1627) gave polypyridazine. When 3,6-disubstituted pyridazines are oxidized by molecular oxygen under basic conditions in dimethyl sulfoxide and at room temperature, the corresponding maleic acids were isolated. This transformation is accompanied with chemilluminescent light emission [82H(19)1415; 84MI25]. Maleic hydrazide, when oxidized to 3,6-pyridazinedione, undergoes cycloaddition with ergostatrienes (81ZOR-1909).

6-Aryl-3(2*H*)-pyridazinones are selectively reduced with zinc dust in acetic acid into the 4,5-dihydro compounds (84CC1373), and these can be reduced to the hexahydro derivatives by means of LiAlH$_4$ (80KGS1287). Whereas the vinyl group of 1-vinyl-3-methyl-6(1*H*)-pyridazinones is reduced by Et$_3$SiH/CF$_3$COOH or catalytically over Pt catalyst, the corresponding 1-allyl derivatives are reduced only in the presence of Pt catalyst (79IZV803). 3(2*H*)pyridazinone-6-carboxylic acid, when catalytically reduced in the presence of D$_2$ was transformed into deuterated glutamine-2,3,4-^{2}H$_3$ in moderate yield (81MI25).

Some pyridazines were investigated by cyclic voltametry and/or preparative scale electrolysis to give mixtures of reduced compounds [77NKK990; 81ACS(B)185]. The half-wave potentials of some 3(2*H*)-pyridazinones were determined at different pH and correlated with Hammett σ-constants (77KGS668). Maleic hydrazide is stable, but reduced pyridazines undergo reductive N—N bond cleavage upon electroreduction (78-LA1505).

With molybdenum(III) species, pyridazine *N*-oxides are deoxygenated

(80S129), and 3-hydrazinopyridazines are reduced to the corresponding amines (80S830). With nickel-aluminum alloy in potassium hydroxide solution, reduction proceeds with N—N bond cleavage to the 1,4-diaminobutanes (87JOC1043). Similarly, 1,2,3,6-tetrahydropyridazines were cleaved hydrogenolytically and transformed into a derivative of 2,5-diamino-2,5-didesoxyribose (80LA1307).

Pyridazine or 3-methylpyridazine with trimethylsilyl cyanide and benzoyl cyanide gave Reissert compound **106,** which is easily rearranged into **107.** In addition, compound **108** is formed, resulting from the reaction of one mole of cyanide and three moles of benzoyl chloride (81JHC443;

(106)   (107)   (108)

85JHC1543). Also, the adduct of methyllithium and pyridazine, when treated with electrophiles such as methyl iodide, methyl chloroformate, or tosyl chloride is transformed into 2-substituted 3-methyl-2,3-dihydropyridazines (80RTC234).

When 1-phenylthiocarbamoyl-2-phenyl-6-methylhexahydropyridazine was treated with hot hydrochloric acid, a tetrahydropyridazine (most probably 1,4,5,6-tetrahydro) was obtained, among other products, as result of elimination (80KGS228).

F. Ring Opening and Rearrangements

Thermal decomposition of 3,4,5,6-tetrahydropyridazines gives various carbon fragments. The process implies 1,4-biradical intermediates (79JA2069; 80JA3863), and a linear relationship was observed between the solvent polarity and the rates of thermal decomposition (78CB596). Similarly, perfluoroalkylpyridazines with CoF_3–CaF_2, give perfluoro-alkanes and -alkenes (78CC304). 3(2H)-Pyridazinone is degraded by a Mycobacterium that uses it as a sole carbon source (80MI19).

Pyridazine N-oxides undergo ring opening with Grignard reagents [79JCS(P1)2136; 83TL489] or with tosymethyl isocyanide (TOSMIC) and butyllithium (80TL3723) to give polyunsaturated aliphatic compounds. Under conditions of flash vacuum pyrolysis, pyridazine N-oxides are transformed at 750°C into aliphatic nitriles, pyrroles, or oxazoles (87H2677). Ring contractions to pyrroles were observed when 4,5- or 1,6-dihydro compounds were reduced with zinc dust in acetic acid (84CC1373; 86H101) or when 1,2,3,6-tetrahydropyridazines were treated with LDA at low temperature (84TL1769).

4,5-Diethoxy-3(2H)-pyridazinones are rearranged under the influence of sodium ethoxide into 4,5-diethoxypyrazoles (78YZ413), and these are also formed from aryl(4-pyridazinyl) methanols after treatment with p-toluenesulfonic acid at elevated temperatures (84JHC1727). Diazopyridazine **109** is converted in the solid state under strictly anhydrous conditions into lactone **110**, which is extremely reactive and, with water or ethanol, gives pyrazole **111** [79H(12)457].

1,3-Diphenyl-1,4,5,6-tetrahydropyridazine, when heated in PPA, gives, as the main product, 4-benzoyl-1,2,3,4-tetrahydroquinoline. This transformation is assumed to proceed via [3,3]-sigmatropic rearrangement typical of the Fischer indole synthesis (84TL3101). When ^{15}N-labelling was used, it was shown that fluorinated pyridazines are thermally rearranged into fluorinated pyrimidines and, to lesser extent, fluorinated pyrazines via diazabenzvalenes as the most probable intermediates [79CC445, 79CC446; 81JCS(P1)1071].

Chloro compound **112**, when treated with sodium methoxide, is methoxylated at the ring (**113**); if an ortho methyl group is present, this is methoxylated to give **114**. This transformation is considered to be an allylic rearrangement with participation by π-electrons of the heteroaromatic system (84M1171).

G. PHOTOCHEMICAL TRANSFORMATIONS

Although 3,6-diphenylpyridazine is little affected by photooxidation [78IJC(B)980], pyridazine or its 3-methyl derivative gives radicals upon photolysis (84JMR334, 84MI3; 85MP1). Irradiation of sterically hindered pyridazine **53** gives a very stable 1,2-Dewar pyridazine **54**, but flash photolysis causes fragmentation, giving a mixture of small molecules (84CB445). Investigation of the photokinetic trans–cis isomerization and cis-cyclization of 3- and 4-styrylpyridazines revealed that isomerization runs with participation by a triplet state but cyclization runs with participation by a singlet state (78T1971).

Photochemical oxygen transfer from pyridazine N-oxides to oxygen acceptors such as cyclohexane, cyclohexene, anisole (81TL2277, 81TL3637), or phosphine sulfides [81CI(L)365; 83JCS(P2)1113] has been studied. It was suggested that the reaction involves an oxygen atom elimination to give "oxene," atomic oxygen, followed by its reaction with acceptors. Also, 3(2H)-pyridazinethiones are transformed in a photosensitized reaction into the corresponding pyridazinones (85MI19). Various products were obtained upon irradiation of pyridazine 1,2-dioxides, such as isoxazolo[5,4-d]isoxazoles **115**; from 3-phenyl analogs oxazoles **116** and diphenylfuroxan **117** (79T1267) were also formed.

3,4,5,6-Tetrahydropyridazines resist denitrogenation on radiation at 350 nm, but at 185 nm from 3,3,6,6-tetramethyl derivative, a mixture of isobutene (22%) and 1,1,2,2-tetramethylcyclobutane (12%) was obtained (80JA7131). Irradiation of 3-oxidopyridazinium betaines gave fused stable diaziridines **118**, whereas 4-oxido analogs are isomerized into pyrimidinones **119** [77CL1005; 79JCS(P1)1199].

(119)

Pyridazine N-ethoxycarbonylylides, prepared from pyridazines by N-amination and subsequent treatment with ethyl chloroformate, are transformed photochemically into 1-ethoxycarbonyl-pyrroles (80CPB2676) or 3-pyrrolin-2-one derivatives (85CPB3540). On the other hand, 1,2-bis(carbethoxy)pyridazines are transformed photochemically into 1-ethoxycarbonylpyrrolin-3-ones (87CB1597).

Fluorinated or chlorinated pyridazines are photochemically rearranged into halogenated pyrazines [78JCS(P1)378; 82JOC398]. Photoaddition of olefins to 3(2H)-pyridazinethiones gives the disulfides and also thieno[2,3-c]pyridazines, which can be desulfurized to give 4-alkylated pyridazines **120,** thus providing a new C—C bond at position 4 (79MI15; 86CPB3061).

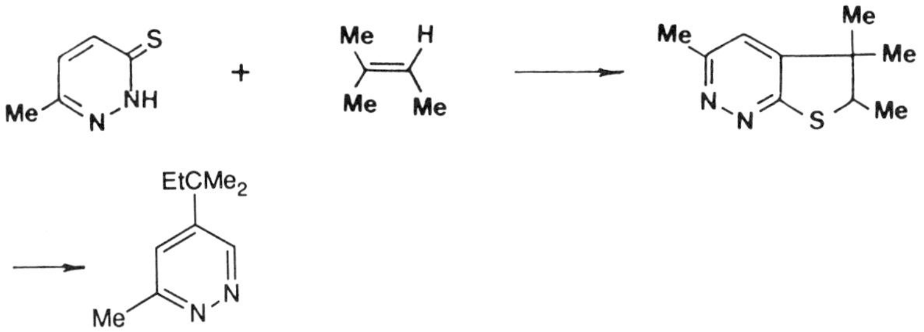

(120)

IV. Theoretical Aspects and Physical Properties

A. CALCULATIONS

Since the crystal structure of pyridazine is unknown, theoretical studies were performed and a structure proposed (85JCP5892). Application of a new parametric quantum chemical model, AM1, has been applied to pyridazine (85JA3902; 88JA6297) for which complete lack of differential overlap (CNDO/2) and intermediate neglect of differential overlap (INDO) calculations were also performed (78BCJ3443; 79T1595). Perturbation theory and the graph-theoretical definition of resonance energy were applied to pyridazine and compared to valence bond calculations. Results are remarkably similar (23 vs. 22 kcal/mol) (78T3419). Simple symmetry-dependent rules in conjugated systems were applied to pyridazine (79MI33), and excited state and pairing relations in pyridazine were discussed (82MI16). The SCF-ASMO-LCI method and the Pariser–Parr–Pople (PPP) approximation were used to calculate the energy of $\Pi \to \Pi^*$ and $n \to \Pi^*$ transitions and other excited states of pyridazine (79MI23). Using *ab initio* orbital theory at the STO-3G level, the relationship between the influence of substituents on energy and on proton-transfer equilibria for pyridazine were reported (84JA6552).

Other calculations include charge densities of 4-substituted pyridazines (79CPB2105), 3,6-dihydropyridazine and its 3-oxo and 3,6-dioxo analogs [84JCS(P2)1465], geometries and energies of the excited states of pyridazine (86MI25), direction of alkylation of 3(2*H*)-pyridazinones (85ZOR2445), adsorption of pyridazine on graphitized carbon black (86MI8), and the heats of formation of the ground state of perfluoropyridazines (84MI6).

B. Dielectric Properties, Basicity, and Tautomerism

The dipole moment of pyridazine (4.41 D), dielectric constant, refractive indices, density, and partial molar volume were determined (82MI1). A correlation of the dipole moment with the effect of the pyridazinyl group as substituent was given (87KGS672). The relative gas phase and solution basicity of pyridazine (86JA3237) and 1,2-dimethylhexahydropyridazine (88JA6303) have been determined. Ionization constants for various pyridazines were determined and calculated [78MI13; 85JCS(P2)417; 85JCS(P2)359], and a correlation of substituent effects on the basicity was presented (77MI12). Carbon acidity of 3- and 4-methylpyridazine was reported (83ZOR465).

There are several studies related to tautomerism of pyridazines. It was found that 3,6-diphenyl-1,4-dihydropyridazine exists in CHCl$_3$ solution in a 8:1 ratio of the 1,4- and 4,5-dihydro forms (83JHC855). Derivatives of 3-pyridazinylmethanes were investigated for tautomeric equilibria (78CPB3633; 83IZV1687, 83MI20; 84KGS832). For **121,** it was found that in CDCl$_3$ 30% of the NH form is present (83IZV1687).

(121)

Investigations on 3- and 4-pyridazinyl substituted β-keto esters showed that a nitrogen atom in a γ-position enhanced the enol content (81KGS530).

Spiropyridazines **122** and **123** exist in equilibrium with the open-chain structures. The sulfur analog of **122** exists in dimethyl sulfoxide 5% in the open-chain form (82G249), whereas the nitrogen analog exists almost exclusively in the spiro form [80JCS(P2)1339]. On the other hand, compound **123** exists in chloroform almost exclusively as the open-chain amide (79G117). Spectral and dipole measurements show that the hydrazone form of 3-chloro-6-hydrazinopyridazine is not present (82KGS1536).

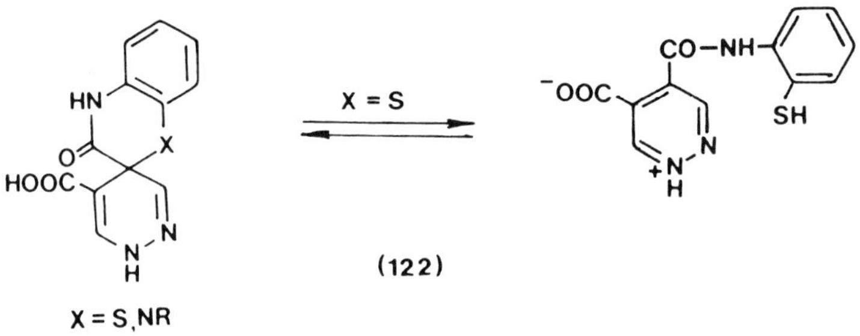

(122)

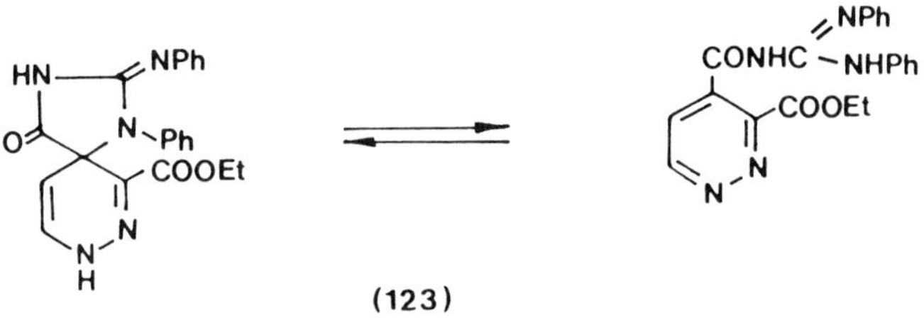

(123)

Spectroscopic studies of 3-aryl-5-hydroxy-1-pyridazinyl-pyrazoles have shown they exist as hydroxy tautomers (80MI27), while studies of protonation of 3(2H)-pyridazinone and maleic hydrazide indicate that by changing acidity, several neutral and/or cationic species can be present (79AJC2297).

For thiones, the thione form is generally present (86JPR522), although it was claimed without proof that 4-arylmethyl-6-methyl-3-mercaptopyridazine exists in the thiol form (80S410). For pyridazine-3,4,6-trithiol, no firm conclusion could be reached about its preponderant tautomeric form, but it was concluded that it does not exist in the trithiol form nor as the 6(1H)- or 3(2H)-thione (78AJC389).

The free energies of the rotational barriers of pyridazinylamidines with the —C(Me)NMe$_2$ side chain were determined; they are in the range of 10–14 kcal/mol, which is lower than similar amidines with a =CHNMe$_2$ side chain (78JHC1105).

C. Spectra

Many spectra were recorded and correlated in connection with structural investigations of pyridazines. Pyridazine itself was investigated by UV, IR and Raman spectroscopy. Studies involve the first three Π → Π* singlet transition energies (80MP519), two nearby electronic states (81CL873), gas-phase time-resolved spectroscopy (86JCP1996), an electric field-induced spectrum (77CPL290), triplet lifetimes in the solid phase at low temperature (88CPL286), the contour analyses of type-C bands (78JSP394), the vibrational structure of S$_1$ electronic states of gaseous pyridazine (79MI40), emission spectroscopy of gas-phase pyridazine (83JCP4083), and solvent effects on the electronic absorption spectrum of pyridazine N-oxide (82MI9).

There are also infrared studies of hydrogen-bonded complexes of pyridazine and phenols (84JCP2132), complexes of pyridazine with metal(II) ions (81JST191), Raman spectra of pyridazine (80MI1; 86JST33), pyridazine adsorbed on silica (78BCJ3063) or on silver electrode (87JCP11), or as silver sols (88JCP954). The surface-enhanced Raman spectroscopic response of maleic hydrazide was also examined (87MI8).

Ultraviolet, infrared and ^{1}H-NMR spectra for a number of 1-methyl (or phenyl) 4-(or 5-)substituted 4-(or5-)hydroxy-6(1H)-pyridazinones were reported (80CCC127) along with spectral data for 3-methyl-pyridazine (83MI1) and 3,6-dichloropyridazine [79PIA(A)279]. The pH dependence of the electronic spectra of some pyridazines was measured [86JCS(P2)359], and the effect of substitution on the resolved multiple peaks in ultraviolet spectra of 6-substituted and 4,6-disubstituted-3-chloropyridazines was evaluated (83MI18). Ultraviolet spectra of several pyridazinium ylides were recorded (76MI1, 76MI2, 76MI3). The infrared and Raman spectra of several pyridazines were measured over a wide acidity range (86JST59). Infrared spectra of 3,6-dichloropyridazine (79MI11) and other deuterated or substituted pyridazines were also recorded [85MI12, 85SA(A)703; 86BCJ2997]. The infrared and Raman spectra of 3(2H)-pyridazinethione show that only the thione form is observed in apolar solvents [83SA(A)367].

NMR spectra were applied to structural investigations of monamycins D$_1$ and H$_1$; the data support a single conformation for each

[77JCS(P1)2369]. From dynamic NMR data, the rotational barriers of the dimethylamino group in reduced pyridazines were obtained (~11 kcal/mol) (81JA3300). ^{1}H- and ^{13}C-NMR chemical shifts were measured, and electronic effects in 3- and 4-substituted pyridazines were evaluated (79CPB1169; 86KGS951). ^{1}H-NMR and ^{57}Fe-Moessbauer spectra of a cyclopentadienyliron carbonyl complex with pyridazine were investigated (86JOM207).

A quantitative measure of electron distribution has been developed using ^{13}C-NMR chemical shifts. These correlations also allow the prediction of ^{13}C chemical shifts in substituted pyridazines (82OMR192). ^{13}C-NMR spectra of various pyridazines were recorded and analyzed. They were used to determine the most stable conformations of reduced pyridazines (87T2443) or the ring-chain isomerization of heterospiro compounds of type **122** (83OMR42). Substituent effects of several 3(2H)-pyridazinones were correlated [83JCS(P1)1203], and nuclear relaxation rates of ^{13}C and quadrupole relaxation of ^{14}N have been measured for pyridazine (78MP997).

^{15}N Chemical shifts for pyridazine (80HCA504; 84OMR201; 85MI22), its N-oxide (80HCA504; 81JMR387), and 3-methylpyridazine (84OMR201) were reported. A relationship exists between the changes in ^{15}N chemical shifts of azine N-oxides and changes in the ^{13}C chemical shifts of the carbons α to the N-oxide function. On comparison of the ^{15}N chemical shift changes of the N-oxides and their oxygen protonated analogs, a measure of the extent of back donation, related to the π-deficiency of the N-oxides, is established. For pyridazine N-oxide, the relative back bonding amounts to 70% [82H(19)93]. ^{15}NMR chemical shifts are reported for 1,2-dimethylhexahydropyridazine (80JOC3609), and cyclic voltametry together with NMR was used for studies of conformational changes of 1-ethyl-2-methyl- and 1,2-dimethyl-hexahydropyridazine (78JA4012). ^{13}C–^{15}N nuclear-spin coupling constants and ^{15}N–^{15}N spin coupling constants of 3-methyl- and 3,6-dimethyl-pyridazines were observed [85JCS(P2)1533]. Also, the ^{17}O chemical shift data for pyridazine N-oxide at natural abundance were reported (85JHC981).

For pyridazine, a time-resolved electron paramagnetic resonance (EPR) study (85CPL321), hyperfine coupling in a radical anion [83ZN(A)415], and spin-lattice relaxation rate (84JA557) were reported. ESR studies, found that the radical, obtained through one-electron oxidation, has a semiquinone-type structure [88JCS(P2)1259], and similar investigations on cation radicals of some tetra- and hexahydropyridazines revealed that these species exist in half-chair conformations (85JOC4749).

There are also several electron-impact mass spectral studies of derivatives of 3- and 4-styrylpyridazines (81JHC255; 84JHC435), some

4(1*H*)-pyridazinones (85OMS483), 2-benzo-azolyl-3(2*H*)-pyridazinones (87OMS477), and some 3,4-diazanorcaradienes (79MI14). 3,4- and 4,5-Didehydropyridazines are postulated as intermediates in the vapor phase during electron impact or pyrolytic fragmentation of pyridazine-3,4- and -4,5-dicarboxylic acid anhydride [82H(19)1427]. Fragmentation of pyridazine by electron-impact luminiscence was studied, and cleavage of the N—N bond was found to be important (81CL1631). Van der Waals clusters of pyridazine with methane, ammonia, water, and methanol were generated in a supersonic, molecular jet expansion and investigated by mass spectrometry. No spectral evidence for water methanol clusters could be observed (87JCP6707).

Pyridazine was investigated by photoelectron spectroscopy (79MI42, 79MP1381), and photoelectron spectra of 1-(4-pyridazinyl)- and 1-(3-pyridazinyl)-2-(3-pridyl)ethene, in their trans configuration, were recorded [80ZN(A)844]. Conformation of 1,2-dimethylhexahydropyridazine was examined by variable-temperature photoelectron spectroscopy. The diequatorial conformation with diaxial ion pairs is more stable than the other conformer by ~1.2 kcal/mol, but in solution, the energy difference is only 0.4 kcal/mol (80JA7438).

The fluorescence lifetime of pyridazine was measured; it is so short that it is not possible to differentiate between the fluorescence emission and the flash lamp pulse (78AJC1889).

D. X-Ray Analysis

Structures of many pyridazines were determined by X-ray analysis, particularly those compounds that exhibit biological activity. X-Ray structural analysis has been applied to many compounds resulting from syntheses or transformations of pyridazines, such as structures **38** (87JOC2026), **41** (as erythro) (84JHC305), **49,** and **50** (82CC1003).

For 5-benzoyl-2,6-diphenyl-4-hydroxy-3(2*H*)-pyridazinone, the 4-hydroxy form was confirmed (86M231). 4-Amino-5-cyano-1,6-dihydro-6-imino-1-phenyl-3-pyridazinecarboxylic acid exists as an inner salt (87LA889). For pyridazine nucleoside 4-hydroxy-1-(β-D-ribofuranosyl)-6(1*H*)-pyridazinone, glycosyl torsional angle was found to be in the high anti region [78BBA229; 81AX(B)1576]. For the condensation product of acetonylacetone with hydrazine, several structures were earlier proposed, but an x-ray analysis confirmed structure **124** (78JOC3615).

There are crystallographic data for hexahydro-3,6-pyridazinedione at $-165°$C [78ACS(A)219]; its 1,2-dimethyl analog has a "twisted boat" conformation at the same temperature [77ACS(A)808]. A slightly twisted

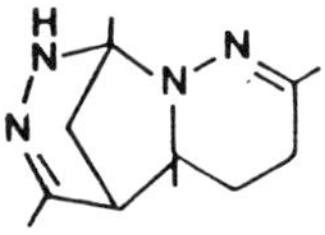

(124)

boat was also found in the case of 3,5-di-*tert*-butyl-4-diisopropylamino-1*H*-pyridazinium-6-phosphonate [85AX(C)1130]. 4,5-Dichloro-6(1*H*)-pyridazinone was found to exist as two polymorphs, each in the oxo form as a hydrogen-bonded dimer (85AX(C)1807].

X-Ray structures were also determined for pyridazine-3,6-dicarboxylic acid (87JHC1285), 4,5-diamino-1-methyl-3-methylthiopyridazinium iodide (81H9; 82AX(B)135], 1-(*p*-bromophenyl)-5-bromopyridazine-3,6-dione (79JHC855), 3,6-bis(hydroxymethyl)pyridazine [88AX(C)1267], 6-diethylamino-*N*,*N*,5-trimethyl-4-oxo-3-phenyl-1,4-dihydro-1-pyridazinecarbothioamide, 3-dimethylamino-*N*,*N*-diethyl-4-methyl-5-oxo-6-phenyl-4,5-dihydro-4-pyridazinecarbothioamide [85AX(C)1277], 4-bromo-1-(p-nitrophenyl)-3-phenylpyridazinium-5-olate [83AX(C)1415],5-amino-6-chloro-4-nitro-2-(β-D-ribofuranosyl)-3(2*H*)-pyridazinone (81-JHC1551), (6-methoxy-3-pyridazinyl)sulfanilamide [81BSF(1)153, 81MI17; 86IJC-(A)707], bis(hexahydro-1-pyridazinyl)thiophosphoric acid *O*-phenyl ester [84AX(C)441], and 1,3-dihydro-3,3-dimethyl-5-(1,4,5,6-tetrahydro-6-oxo-3-pyridazinyl)-2*H*-indol-2-one (87JMC623).

The structure for pyridazine derivative **126,** obtained from bicyclic ketone **125** with methanolic sodium methoxide, was determined by X-ray analysis (84JOC1261). Structures of **127** (81JA7011) and **128** (81AJC1223) were also determined.

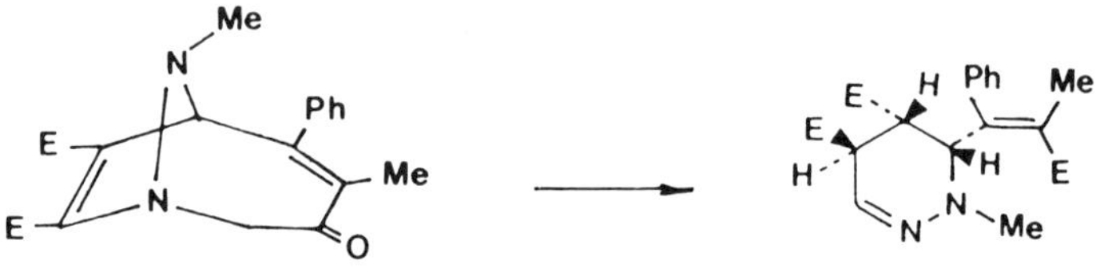

E = COOMe

(125)

(126)

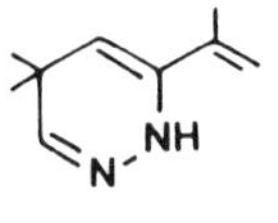

(127)

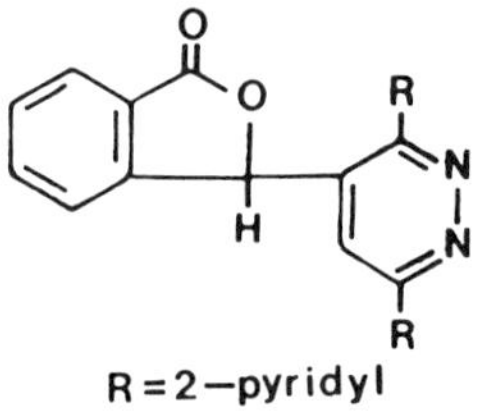

R = 2—pyridyl
(128)

Electron diffraction studies were used to determine the molecular structure of gaseous pyridazine and 3,6-dichloropyridazine [77ACS(A)63]. Structures of several complexes were examined, such as the complex between 6-chloro-3,4-dimethoxy- and 3-chloro-4,6-dimethoxypyridazine (87CPB350), the palladium(II) complex with 1,3-dimethyl-1,4,5,6-tetrahydropyridazine (80MI22), and bridged binuclear Cu(II)pyridazine complexes (85CC1709; 86MI15; 88CJC348).

The crystal structure of the fungicide, O-(2-chloroethyl)-O-isobutyl-O-(2-phenyl-4-methylthio-3-oxo-2H-pyridazin-5-yl)thiophosphate (87CCC-696), and two cardiotonic pyridazines, CI-914 (**129**, R = H) and CI-930

(**129**)

(**129**, R = Me) with essentially planar geometries (87JMC1963), were determined. Two GABA A receptor-site antagonists, 2-(3-carboxypropyl)pyridazine derivatives [85AX(C)1532; 87MI4], and various minaprine analogs [85BSB261; 86AX(C)1206], derivatives of 3-morpholinoethylamino-[85AX(C)1532; 86AX(C)214] or 3-morpholinoethylthiopyridazine, were examined for their molecular structure by X-ray analysis.

E. COMPLEXES

Pyridazine forms hydrogen-bonded complexes with phenol (77MI1) or chloromethylsilane as (pyridazine)$_n$ (Me$_3$SiCl)$_m$ (n,m = 1–2) [87ZN(A)341]. Vapor pressure measurements have been carried out on a pentacarbonyl complex containing pyridazine (80MI5). Other polycarbonyl compounds were used to give complexes with metals such as chromium [81JCS(D)1524, 81MI2], manganese(I) (87MI2), iron (86JOM207), ruthenium (77JA6588), rhodium, iridium (79JOM97), tungsten (78IC1093), osmium (77IC2820), chromium, molybdenum, tungsten (79MI26), 3,6-bis(2'-pyridyl)pyridazine, and group VI metal tetracarbonyls (88MI1).

Borane (BH$_3$) adducts of pyridazine (85MI17) were prepared. Pyridazine has been studied as a ligand with an ML$_5$ group, and barriers for the haptotropic shift of monocoordinated pyridazine were found to increase in the order Cr < Mn < Fe < Co (87JA5316). Hydrazones prepared from 3-hydrazino-4-benzyl-6-phenylpyridazine were examined as chelating agents for various metals (83MI21; 85M463). Complexes of transition

metals with sulfapyridazine (79MI9, 79MI39) and complexes with pyridazine and 4,4'-bipyridyl ligands (81MI5) were prepared.
The formation and properties of many other complexes are reported in the literature. These include a cyclopentadienyl-titanium(III) complex with maleic hydrazide (81IC2084), a nickel(II) and cobalt (II) complex of 3,6-bis(1'-pyrazolyl)pyridazine [86JCS(D)625], copper complexes with pyridazine or pyridazine derivatives [79BCJ3420; 83MI6; 84MI7; 86JCS(D)2381; 87IC2384, 87MI22], niobium(V) and tantalum(V) complexes with maleic hydrazide (87MI19, 87MI20), stability constants of manganese, cobalt, nickel, copper, and zinc complexes of Schiff bases from sulfamethoxypyridazine (79JIC749), a phosphorodithioato-molybdenum(V) complex with pyridazine [83JCS(D)649], ruthenium, rhodium, palladium, iridium, and gold complexes with pyridazine, pyrazolyl-1-pyridazine or 3,4,5-pyridazinetrithiol [79IJC(A)276; 84MI30; 86JCR(S) 214], potentiometric studies on silver(I) complexes with pyridazine and 3-methylpyridazine (86MI9), coordination compounds of sulfamethoxypyridazine with rare earth elements (87MI7), osmium(III) pentaamine complex with pyridazine [79ACS(A)125; 82AJ7658], iridium [82BSF(1)433] and platinum complexes with pyridazine (77IC2618), or 3,6-bis(2'-pyridyl)pyridazine (85MI9).

Complexes of pyridazine N-oxides with cupric chloride were prepared, and depending on the particular N-oxide 2 : 1 or 1 : 1 complexes were formed (78TL1979).

V. Polymers

Electrochemical oxidation of pyridazine gave a blue poly(pyridazine) that shows electrical conductivity (84CC1627, 84MI21). Also, iodine induced polymerization and oxidation of pyridazine (87MI11). Plasma polymer films of tetrafluoropyridazine were examined (84MI5), and poly(3-vinylpyridazine) was obtained by ionic polymerization and formed a semiconducting adduct with iodine (85CC1632). Solvent effects on radical polymerization of vinylpyridazinones were studied (86MI2) along with the kinetics of radiation-induced polymerization of allylpyridazinones (82MI10). A phthalocyanine-iron-dipyridazine polymer has been shown to have semiconductor properties (83AG804).

There are several studies on copolymerization of N-vinyl-pyridazinones (86MI3), 2,6-dimethyl-3(2H)-pyridazinone (79MI10), and various pyridazinones (80MI6, 80MI7; 81MI6, 81MI7). Polydihydropyridazinones were also prepared by cyclocondensation of bis-(3-benzoylpropionic acids) and aryldihydrazines (78MI6).

VI. Biological Activity

A. Natural Products

Contrary to other heterocycles found in many important natural compounds, such pyridazines were discovered only after 1970, the first representatives being monamycins. These are now a big group of over 15 related compounds, and monamycin X (**130**), which was isolated from *Streptomyces jamaicensis* [79JCS(P1)1451], is just one example of these cyclohexadepsipeptide antibiotics.

(130)

Antrimycins (**131**) from *Streptomyces xanthocidiens* (81MI26, 81MI27; 82MI20) and cirratiomycins (**132**) from *Streptomyces ciratus* (82ABC865, 82ABC1885, 82ABC1891) are tuberculostatic peptides that also contain nonribosomal amino acids. Other examples are luzopeptins (**133**) (*from Actinomycetes luzonensis*) (85MI35), which are antitumor antibiotics active against Gram-positive bacteria.

	R^1	R^2
(131)	Me, Et, n-Pr, i-Bu	Me, Et
(132)	Me, i-Bu	Et

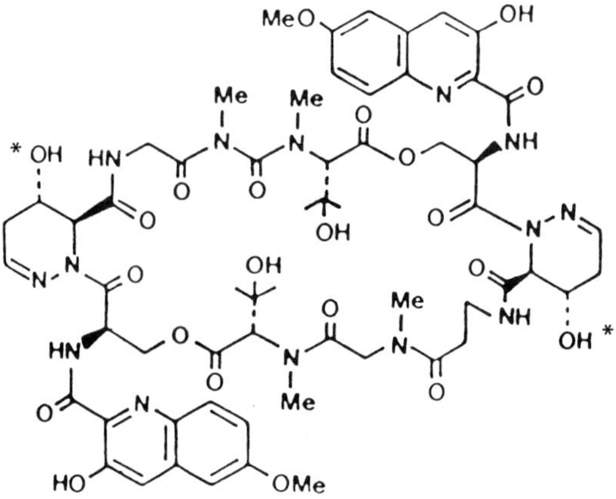

LUZOPEPTIN C

LUZOPEPTIN A* (di-Ac)

LUZOPEPTIN B* (mono-Ac)

(133)

Pyridazinomycin (**134**) is a new antifungal antibiotic produced by *Streptomyces violaceoniger*. The amino acid side chain can be viewed as L-ornithine, whose end nitrogen atom is part of the pyridazine ring (88MI2).

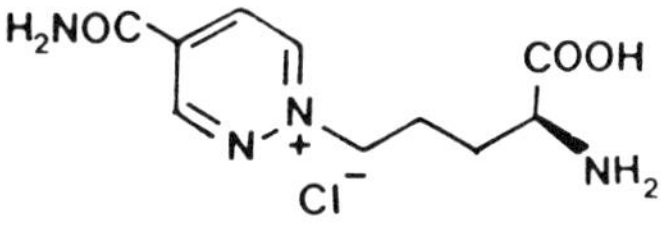

(**134**)

NMR studies of monamycins D_1 and H_1 revealed a single conformation for each, and the hydroxypiperazic acid residue is hydrogen bonded with the NH group of the valine part [77JCS(P1)2369]. Unnatural congener, monamycin X [79JCS(P1)1451] and deoxymonamycin-B_3 (77CC635) were synthesized.

B. SYNTHETIC COMPOUNDS

Many pyridazines have been tested for their biological activity and eventual application in medicine. A great number of experiments are connected with the cardiovascular system, followed by examination for analgetic and anti-inflammatory properties or antibacterial activity. Several review articles appeared. A summary of the therapeutic trials of 4-amino-6-methoxy-1-phenylpyridazinium methyl sulfate (Amezinium sulfate, Regulton) and results of multicenter study on 477 patients are presented (81AF1657). Moreover, several aspects of the pharmacology of this drug have been published (81AF1533, 81AF1544, 81AF1558, 81AF1566, 81AF1574, 81AF1580, 81AF1589, 81AF1594, 81AF1605, 81AF1616, 81AF1623, 81AF1638, 81AF1647, 81AF1653). Review articles on pharmacological effects of some aminopyridazines related to minaprine (**135**), a psychotropic pyridazine derivative, were published (85MI33; 86MI27).

(135)

Several studies relate the analgetically active and anti-inflammatory 4-ethoxy-2-methyl-5-morpholino-3(2H)-pyridazinone (M73101) (78MI14, 78MI15, 78MI16, 78MI17; 79MI29, 79MI30, 79MI34, 79YZ1091) and 3-amino-4-mercapto-6-methylpyridazine (82PHA208, 82PHA285, 82PHA441, 82PHA502). Several other 3(2H)-pyridazinones were investigated for analgesic, anti-inflammatory and antipyretic activities (78YZ1421, 78YZ1472; 79JMC53, 79MI28; 81PHA775; 82PHA136; 85FES921).

6-Substituted 2,3,4,5-tetrahydro-3-pyridazinones possess sedative effects (84MI27), and 3-(4-hydroxypiperidyl)-6-(2',4'-dichlorophenyl)pyridazine and its 2'-chlorophenyl analog were investigated for their sedative and anticonflict activities (86MI20). The last compound possesses anticonvulsant properties (85MI3). Anticonvulsant activity was found also with ureido derivatives of N-monosubstituted hexahydropyridazines (81JHC293) and various 6-aryl-3-(hydroxypolymethyleneamino)pyridazines (86JMC369).

As potential central nervous system depressants, various pyridazinyl ureas, semi- and thiosemicarbazide derivatives (83MI16), 4-ethoxy-2-methyl-5-morpholino-3(2H)-pyridazinone (79MI31) and various other pyridazines (81PHA698) were tested.

Several pyridazinyl analogs of γ-aminobutyric acid, GABA, i.e., SR 95103 (**136**), SR 42641 (**137**) (87MI26) or SR 95531 (86MI4) have been found to be potent, selective, reversible and competitive GABA A antagonists. Several related pyridazines were synthesized and investigated for their actvity (85MI24, 85PNA1832; 86MI5; 87JMC239, 87MI14, 87MI25) and their graphic models were studied (85MI34).

(136)   (137)

Structural studies on analogs of minaprine (**135**) were performed in order to establish structure–activity relationship (82BSB49, 82BSB123). Several pyridazinones with a side-chain urea (83MI26), thioamide (80MI24; 81CPB3433; 82JMC975; 83JMC373), or 2-cyanoguanidine group (83JMC1144) were tested for gastric antisecretory activity.

Pyridazinones (**129**) (84JMC1099) and their arylaminophenyl analogs (87JMC1157) show potent positive inotropic activity. Also, some substituted tetrahydro-6-oxo-3-pyridazinyl-2*H*-indol-2-ones are very potent (86JMC1832; 87JMC623, 87JMC824), and many related 3(2*H*)-pyridazinones were also tested for inotropic activity (84MI13, 84MI20; 85JMC1405; 86JMC261, 86JMC2142; 87JMC1955; 88JMC461). The pyridyl analog of **138** was found to be a potent cardiotonic compound (87AF398, 87MI10, 87MI12).

Several 3(2*H*)-pyridazinones (88JMC345) or their 2,3,4,5-tetrahydro derivatives were evaluated as combined vasodilator and β-adrenoacceptor antagonists (88JMC352, 88MI4, 88MI5). Modified steroids having a pyridazine or pyridazine N-oxide ring at position 17 were prepared, and some exhibited cardiotonic activity (83MI9, 83MI10).

Numerous derivatives of hydrazinopyridazine were prepared and tested for their antihypertensive activity. The hydrazine group in these compounds exists either as such (78FES99; 79FES299; 80MI14, 80MI15; 81JMC59; 82PHA51; 85MI8, 85MI20, 85MI23), in the form of hydrazones (80MI13; 84MI26), or as part of a 1-aminopyrrole (83MI8; 85AF508; 86AF84) or pyrazole ring (79MI36). Also, several 3(2*H*)-pyridazinones as their hydrazino analogs were tested for this activity (77MI4; 79MI4, 79MI12, 79MI35; 80JMC1445; 85MI11; 86AP60).

Some 3(2*H*)-pyridazinones inhibit blood platelets aggregation

(83JMC800; 84MI18; 87FES585), and pyridazine analogs of prostacyclin showed potent inhibition (79JA766). Some 3(2H)-pyridazinones and their 5-methylsulfonyl analogs were prepared for examination as haemostatics (78YZ1530); other were found to possess monoamine oxidase-inhibiting properties (77FES404; 88MI3). 2-Aryl-6-substituted-2,3,4,5-tetrahydro-3-pyridazinones were investigated and found to possess mild diuretic activity (79YZ211).

Bactericidal activity was found with some 2-(pyridazinyl-3′)-3-pyridazinones (84MI8) and S-(6-arylpyridazinyl-3)thioglycolic acids [78-IJC(B)936], whereas mesoionic pyridazine nucleosides have been shown to possess antibacterial activity *in vivo* (84JMC1613). Cephalosporin analogs with substituted pyridazines in the C-3 side chain showed good antimicrobial activity (77MI7, 77MI8), and this activity was investigated with several other pyridazines (81MI20; 84MI4, 84MI29; 86PHA460).

Amino derivatives of 2-phenyl-3(2H)-pyridazinone were examined for antifungal activity (87YZ819). Several 1-methyl-5-nitroimidazoles substituted with pyridazines at position 2 were investigated against various protozoa (78AF351). With 4-ethoxy-2-methyl-5-morpholino-3(2H)-pyridazinone, teratological or mutagenicity studies were performed (78MI12; 79MI37, 79MI38; 80MI26, 80MI28, 80MI30), and cytostatic activity was determined in the case of several 3(2H)-pyridazinones (78JMC1333; 80MI14; 81MI10, 81MI11; 82MI12; 83MI12; 84CCC2541; 84MI24, 84MI31, 84MI32).

Among other activities, investigations were concerned with radioprotective effects (83MI3), antitrypanosomal activity (80JMC578), and amplifiers of phleomycin (85AJC1685). Pyridazines were shown to be potent tyrosine hydroxylase dopamine β-hydroxylase inhibitors (83MI11).

There are several studies on the metabolism of pyridazines. The metabolism in rabbits of 4-ethoxy-2-methyl-5-morpholino-3(2H)-pyridazinone, an analgesic and anti-inflammatory agent, revealed 10 metabolic compounds that were isolated and identified. All were derivatives of pyridazines and the major metabolite route was oxidative cleavage of C—N and C—O bonds of the morpholino group (78CPB3124). As decomposition products of the antihypertensive 3-hydrazino-6-[bis-(2-hydroxyethyl)-amino]pyridazine, three pyridazines were isolated from urine and revealed that the hydrazino group was either displaced or cyclized into a triazolo ring (78FES565; 79MI1, 79MI6). Also, the related Propildiazine (**138**), an antihypertensive drug, is hydrolytically degraded into three compounds resulting from transformation of the hydrazino group to either an oxo group or a hydrogen atom and from formation of a diazene (81JPS334). Pharmacokinetics of an analog of **139** (88AF237) and toxicity of 4-ethoxy-2-methyl-5-morpholino-3(2H)-pyridazinone were evaluated (79MI24).

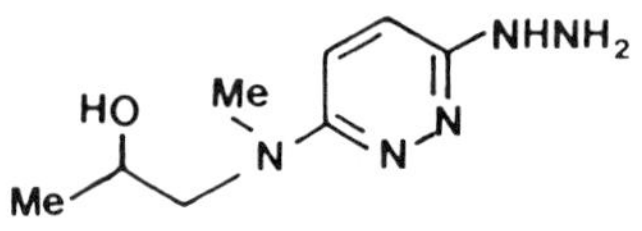

(138)

C. Plant Growth and Protection

There are at least four known actions of pyridazines on plants involving a one or more of the following effects: (a) inhibition of the Hill reaction (photolysis of water during the photosynthesis), (b) inhibition of pigment formation (chlorophylls, carotenoids), (c) change in the linolenic/linoleic acid ratio, and (d) influence on the chloroplast ribosomes. In addition, pyridazines may be used in plant protection as biocides.

Several 2-phenyl-3(2*H*)-pyridazinones were found to influence or inhibit electron transport in plant chloroplasts [81MI15; 83MI16; 85MI10; 86MI16; 87MI16, 87MI18, 87ZN(C)808]; they inhibit photosynthesis [79MI13, 79MI43; 82MI19; 87PIA415]. The influence of pyridazinones on the biosynthesis of carotenoids and the metabolism of lipids in plants has been surveyed [79ZN(C)1052]. Among various herbicides, SAN 6706 (**139**) has been extensively investigated; it blocks chlorophyll and carotenoid formation simultaneously [77ZN(C)236; 79MI27; 80MI17; 82MI13,

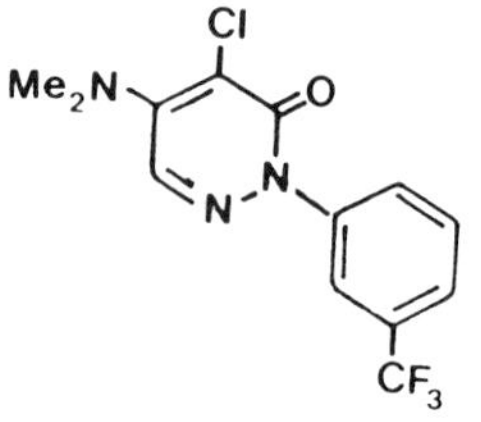

(139)

82MI15; 86MI10, 86ZN(C)585]. Some related pyridazinones exhibit similar activity [82ZN(C)1092; 85MI16; 88ZN(C)418], although the reverse activity was also observed, i.e., they stimulate the biosynthesis of fatty acids, chlorophylls, and carotenoids in Chlorella cultures (80MI10).

Substituted pyridazinones induce a specific decrease in the linolenic acid content accompanied by an increase in the linoleic acid content of plant membranes. The most distinct effect among many 5-substituted pyridazinones (78MI7; 83MI5; 86MI21; 87MI17) was shown by 4-chloro-5-

dimethylamino-2-phenyl-3(2H)-pyridazinone (SANDOZ 9785, BASF 13 338) (79MI16; 81MI1; 83MI22; 84MI2; 85P1923; 87MI24).

To study pesticidal activity, various 4,5-dichloro-, 4- or 5-alkylamino, or 1,4,5-trisubstituted 3(2H)-pyridazinones were prepared (84MI14; 85CCC492, 85MI15). Pyridazinyl thiophosphates were prepared and tested (78CCC2415, 78MI8; 79CCC1761, 79MI17, 79MI18, 79MI19, 79MI20, 79MI21; 80CCC2247, 80CCC2343; 83MI13, 83MI14). Some pyridazines display fungicidal activity (83MI12; 85MI5).

Labelled maleic hydrazide was studied for degradation in tobacco plants, and the major metabolite was found to be its β-D-glucoside (78MI4). 1,1-Dimethyl hexahydropyridazinium bromide has been studied as a growth regulator (77MI3) along with some pyridazinone herbicides for their influence on aflatoxin production (83MI2). A number of pyridazines and pyridazinones were prepared and investigated for their herbicidal activity (80MI9; 81MI3, 81MI15).

VII. Other Applications

A series of 3,5-disubstituted pyridazines and their 3,4,5,6-tetrahydro derivatives were prepared from aryl γ-oxo-butyric acids and were tested for photostability and eventual properties as liquid crystals (86ZC21). Liquid crystalline phases of 3-phenylpyridazines (77ZC333), derivatives of 3(2H)-pyridazinethione (79JPR629), pyridazin-3-yl benzoate (83ZC296), and other pyridazines (83MI23; 86MI11, 86MI12) were investigated. Investigations of new antiozonants showed that substituted pyridazines were weak antiozonants (85MI2).

VIII. Analysis

4-Acetyl-2-(2'-hydroxyethyl)-5,6-bis(4-chlorophenyl)-3(2H)-pyridazinone, an antihypertensive agent, was purified by a combination of low-pressure liquid chromatography and preparative thin-layer chromatography (TLC); byproducts were identified (81JPS419). High-pressure liquid chromatography (HPLC) has been used to determine maleic hydrazide and its β-D-glucoside in foods (80MI23). This method was also used to determine some 3(2H)-pyridazinones in waste water (82MI14) or for separation (80MI4). The retention behavior of eight herbicidal pyridazinones in a reversed-phase HPLC system has been examined (85ACH221).

Thin-layer chromatography was applied to study the relationship between the cytostatic activity of 3,6-pyridazinedione derivatives and their hydrophobic properties (87MI20), TLC photodensitometry was applied to the drug Cadralazine (83MI4). Direct and reverse isotopic dilution and cation-exchange methods were used to determine the drug Amezinium in body fluids (81AF1589), and for Azintamide, which is a potent choleretic pyridazine, a spectrophotometric method has been developed (84MI1). There are also analytical studies of some pyridazinones with iron(III) (82MI7) and nickel(II) salts (81MI19).

ACKNOWLEDGMENTS

We express our appreciation to M. Kastelic and T. Kozamernik for drawing formulas.

References

68M15	R. Schönbeck and H. Kloimstein, *Monatsh. Chem.* **99**, 15 (1968).
76MI1	D. Dorohoi, E. Lupu, and Mai Van Tri, *An. Stiint. Univ. "Al. I. Cuza" Iasi, Sect. 1b* **22**, 35 (1976) [*CA* **90**, 203314 (1979)].
76MI2	M. Petrovanu, D. Dorohoi, and Mai Van Tri, *An. Stiint, Univ. "Al. I. Cuza" Iasi, Sect. 1b* **33**, 41 (1976) [*CA* **91**, 73793 (1979)].
76MI3	D. Dorohoi, M. Rotariuc, and Mai Van Tri, *An. Stiint, Univ. "Al. I. Cuza" Iasi, Sect. 1b* **22**, 45 (1976) [*CA* **91**, 4844 (1979)].
76MI4	A. A. Nada, *Egypt. J. Chem.* **19**, 621 (1976) [*CA* **91**, 175283 (1979)].
77ACS(A)63	A. Almenningen, G. Bjørnsen, T. Ottersen, R. Seip, and T. G. Strand, *Acta Chem. Scand., Ser. A* **A31**, 63 (1977).
77ACS(A)808	T. Ottersen and U. Sørensen, *Acta Chem. Scand., Ser. A* **A31**, 808 (1977).
77AJC2319	G. B. Barlin and C. Y. Yap, *Aust. J. Chem.* **30**, 2319 (1977).
77BCJ2153	O. Tsuge, K. Kamata, and S. Yogi, *Bull. Chem. Soc. Jpn.* **50**, 2153 (1977).
77CC635	C. H. Hassall, W. H. Johnson, and C. J. Theobald, *J. C. S. Chem. Commun.*, 635 (1977).
77CC806	R. N. Warrener, G. Kretschmer, and M. N. Padon-Row, *J. C. S. Chem. Commun.*, 806 (1977).
77CL1005	Y. Maki, M. Kawamura, H. Okamoto, M. Suzuki, and K. Kaji, *Chem. Lett.*, 1005 (1977).
77CPB192	H. Hasegawa, H. Arai, and H. Igeta, *Chem. Pharm. Bull.* **25**, 192 (1977).
77CPB1708	M. Yanai, S. Takeda, and T. Mitsuoka, *Chem. Pharm. Bull.* **25**, 1708 (1977).
77CPB1856	M. Yanai, S. Takeda, and M. Nishikawa, *Chem. Pharm. Bull.* **25**, 1856 (1977).
77CPL290	J. R. Lombardi, *Chem. Phys. Lett.* **52**, 290 (1977).
77FES404	F. Buffoni, C. Ignesti, and R. Pirisino, *Farmaco, Ed. Sci.* **32**, 404 (1977).
77H119	C. H. Hassall and K. L. Ramachandran, *Heterocycles* **7**, 119 (1977).
77IC2618	K. R. Dixon, *Inorg. Chem.* **16**, 2618 (1977).
77IC2820	F. A. Cotton and B. E. Hanson, *Inorg. Chem.* **16**, 2820 (1977).

77IJC(B)436	M. A. Elkasaby, M. Y. Elkady, and N. A. Nour Eldin, *Indian J. Chem., Sect. B* **15B**, 436 (1977).
77IJC(B)1025	A. M. Kaddah and A. M. Khalil, *Indian J. Chem., Sect. B* **15B**, 1025 (1977).
77JA6588	F. A. Cotton, B. E. Hanson, and J. D. Jamerson, *J. Am. Chem. Soc.* **99**, 6588 (1977).
77JCS(P1)2369	C. H. Hassall, W. A. Thomas, and M. C. Moschidis, *J. C. S. Perkin 1*, 2369 (1977).
77KGS668	V. T. Glezer, J. Stradins, and L. Avota, *Khim. Geterotsikl. Soedin.*, 668 (1977).
77MI1	M. I. Baraton, S. Besnainou, and J. Gerbier, *Adv. Mol. Relaxation Interact. Processes* **11**, 309 (1977).
77MI2	M. A. F. El-Kaschef, F. M. E. Abdel-Hegeid, and M. A. Michael, *Egypt. J. Chem.* **20**, 117 (1977).
77MI3	J. Jung and J. Dressel, *Z. Pflanzenernaehr. Bodenkd.* **140**, 375 (1977).
77MI4	V. M. Kulkarni, *Curr. Sci.* **46**, 801 (1977) [*CA* **88**, 98934 (1978)].
77MI5	N. Latif, A. A. Nada, and N. B. Hanna, *Egypt. J. Chem.* **20**, 217 (1977) [*CA* **93**, 8115 (1980)].
77MI6	Mai Van Tri, M. Petrovanu, and I. Druta, *Bul. Inst. Politeh. Iasi, Sect. 2: Chim. Ing. Chim.* **23**, 45 (1977) [*CA* **88**, 10528 (1978)].
77MI7	T. Naito, J. Okumura, K. I. Kasai, K. Masuko, H. Hoshi, H. Kamachi, and H. Kawaguchi, *J. Antibiot.* **30**, 691 (1977).
77MI8	T. Naito, J. Okumura, K. I. Kasai, K. Masuko, H. Hoshi, H. Kamachi, and H. Kawaguchi, *J. Antibiot.* **30**, 698 (1977).
77MI9	E. Stefanescu and M. Petrovanu, *Rev. Med.-Chir.* **81**, 291 (1977) [*CA* **88**, 62354 (1978)].
77MI10	E. Stefanescu and M. Petrovanu, *Rev. Med.-Chir.* **81**, 475 (1977) [*CA* **88**, 152534 (1978)].
77MI11	M. Petrovanu, E. Gafitanu, M. Caprosu, and R. Danet, *Rev. Med.-Chir.* **81**, 659 (1977) [*CA* **89**, 163524 (1978)].
77MI12	P. Tomasik and R. Zalewski, *Chem. Zvesti* **31**, 246 (1977).
77NKK990	M. Murayama and K. Murakami, *Nippon Kagaku Kaishi*, 990 (1977) [*CA* **87**, 124592 (1977)].
77PHA555	V. E. Limanov, *Pharmazie* **32**, 555 (1977).
77TL3619	H. Hamberger, H. Reinshagen, G. Schulz, and G. Sigmund, *Tetrahedron Lett.*, 3619 (1977).
77ZC333	H. Zaschke, C. Hyna, and H. Schubert, *Z. Chem.* **17**, 333 (1977).
77ZN(C)236	H. K. Lichtenthaler and H. K. Kleudgen, *Z. Naturforsch., C: Biosci.* **32C**, 236 (1977).
78ACS(A)219	T. Ottersen and J. Almløf, *Acta Chem. Scand., Ser. A* **A32**, 219 (1978).
78AF351	E. Winkelmann, W. Raether, and A. Sinharay, *Arzneim.-Forsch.* **28**, 351 (1978).
78AJC389	G. B. Barlin and P. Lakshminarayana, *Aust. J. Chem.* **31**, 389 (1978).
78AJC1889	R. P. Nott and B. K. Selinger, *Aust J. Chem.* **31**, 1889 (1978).
78BBA229	B. J. Graves, D. J. Hodgson, D. J. Katz, D. S. Wise, and L. B. Townsend, *Biochim. Biophys. Acta* **520**, 229 (1978).
78BCJ179	H. Besso, K. Imafuku, and H. Matsumura, *Bull. Chem. Soc. Jpn.* **51**, 179 (1978).

78BCJ3063	Y. Yamamoto and H. Yamada, *Bull. Chem. Soc. Jpn.* **51**, 3063 (1978).
78BCJ3443	M. Ohsaku, A. Imamura, and K. Hirao, *Bull. Chem. Soc. Jpn.* **51**, 3443 (1978).
78CB596	W. Duisman and C. Rüchardt, *Chem. Ber.* **111**, 596 (1978).
78CC304	R. D. Chambers, R. D. Harcliffe, and W. K. R. Musgrave, *J. C. S. Chem. Commun.*, 304 (1978).
78CCC2415	V. Konečny, Š. Varkonda, J. Šustek, and Š. Kovač, *Collect. Czech. Chem. Commun.* **43**, 2415 (1978).
78CPB2428	A. Ohsawa, T. Uezu, and H. Igeta, *Chem. Pharm. Bull.* **26**, 2428 (1978).
78CPB2550	A. Ohsawa, Y. Abe, and H. Igeta, *Chem. Pharm. Bull.* **26**, 2550 (1978).
78CPB3124	T. Hayashi, M. Sato, M. Ohki, and T. Kishikawa, *Chem. Pharm. Bull.* **26**, 3124 (1978).
78CPB3633	A. Ohsawa, T. Uezu, and H. Igeta, *Chem. Pharm. Bull.* **26**, 3633 (1978).
78CPB3884	S. Kamiya, M. Miyahara, S. Sueyoshi, I. Suzuki, and S. Odashima, *Chem. Pharm. Bull.* **26**, 3884 (1978).
78FES99	F. Parravicini, G. Scarpitta, L. Dorigotti, and G. Pifferi, *Farmaco, Ed. Sci.* **33**, 99 (1978).
78FES565	E. Bellasio and A. Campi, *Farmaco, Ed. Sci.* **33**, 565 (1978).
78H(9)243	P. K. Kadaba and J. Triplett, *Heterocycles* **9**, 243 (1978).
78H(9)1367	A. Ohsawa, H. Arai, and H. Igeta, *Heterocycles* **9**, 1367 (1978).
78H(9)1397	Y. Abe, Ohsawa, H. Arai, and H. Igeta, *Heterocycles* **9**, 1397 (1978).
78H(9)1771	M. Quinteiro, C. Seoane, and J. L. Soto, *Heterocycles* **9**, 1771 (1978).
78H(10)269	K. Satoh and T. Miyasaka, *Heterocycles* **10**, 269 (1978).
78H(11)155	A. G. Anderson and T. Tober, *Heterocycles* **11**, 155 (1978).
78H(11)387	K. Hafner, H. J. Lindner, and W. Wassem, *Heterocycles* **11**, 387 (1978).
78HCA589	Y. Nakamura, K. Bachmann, H. Heimgartner, H. Schmid, and J. J. Daly, *Helv. Chim. Acta* **61**, 589 (1978).
78HCA2116	H. Link, K. Bernauer, S. Chaloupka, H. Heimgartner, and H. Schmid, *Helv. Chim. Acta* **61**, 2116 (1978).
78IC1093	K. H. Pannell, M. Guadalupe, De la Paz Saenz Gonzales, H. Leano, and R. Iglesias, *Inorg. Chem.* **17**, 1093 (1978).
78IJC(B)631	N. Latif, A. A. Nada, and N. B. Hanna, *Indian J. Chem., Sect. B* **16B**, 631 (1978).
78IJC(B)936	S. M. Zayed, H. M. El-Namaky, and F. M. A. Moti, *Indian J. Chem., Sect. B* **16B**, 936 (1978).
78IJC(B)980	S. S. Talwar, *Indian J. Chem., Sect. B* **16B**, 980 (1978).
78IJC(B)1000	H. Jahine, H. A. Zaher, Y. Akhnookh, and Z. El-Gendy, *Indian J. Chem., Sect. B* **16B**, 1000 (1978).
78JA4012	S. F. Nelson, E. L. Clennan, and D. H. Evans, *J. Am. Chem. Soc.* **100**, 4012 (1978).
78JCR(S)40	E. V. Dehmlov and Naser-Ud-Din, *J. Chem. Res., Synop.*, 40 (1978).
78JCS(P1)378	M. G. Barlow, R. N. Haszeldine, and J. A. Pickett, *J. C. S. Perkin 1*, 378 (1978).

78JHC637	G. Heinisch, E. Luszczak, and A. Mayrhofer, *J. Heterocycl. Chem.* **15**, 637 (1978).
78JHC749	H. Felbecker, D. Hollenberg, R. Schaaf, T. Bluhm, and H. H. Perkampus, *J. Heterocycl. Chem.* **15**, 749 (1978).
78JHC881	J. D. Albright, F. J. McEvoy, and D. B. Moran, *J. Heterocycl. Chem.* **15**, 881 (1978).
78JHC1057	K. H. Pannell, B. L. Kalsotra, and C. Párkányi, *J. Heterocycl. Chem.* **15**, 1057 (1978).
78JHC1105	M. Drobnič-Košorok, S. Polanc, B. Stanovnik, M. Tišler, and B. Verček, *J. Heterocycl. Chem.* **15**, 1105 (1978).
78JHC1425	M. Iwao and T. Kuraishi, *J. Heterocycl. Chem.* **15**, 1425 (1978).
78JMC1333	J. A. May and A. C. Sartorelli, *J. Med. Chem.* **21**, 1333 (1978).
78JOC1190	R. B. Phillips and C. C. Wamser, *J. Org. Chem.* **43**, 1190 (1978).
78JOC3370	Y. Tamaru, T. Harada, and Z. Yoshida, *J. Org. Chem.* **43**, 3370 (1978).
78JOC3615	J. Dodge, W. Hedges, J. W. Timberlake, L. M. Trefonas, and R. J. Majeste, *J. Org. Chem.* **43**, 3615 (1978).
78JSP394	B. D. Ransom and K. K. Innes, *J. Mol. Spectrosc.* **69**, 394 (1978).
78KGS1120	M. F. Marshalkin, V. A. Azimov, L. F. Lindberg, L. Y. Yakhontov, *Khim. Geterotsikl. Soedin.*, 1120 (1978).
78KGS1684	V. A. Nikitin, V. N. Verbov, and K. N. Zelenin, *Khim. Geterotsikl. Soedin.*, 1684 (1978).
78LA635	G. Steiner, *Liebigs Ann. Chem.*, 635 (1978).
78LA1505	L. Horner and M. Jordan, *Liebigs Ann. Chem.*, 1505 (1978).
78M63	M. Braun, G. Hanel, and G. Heinisch, *Monatsh. Chem.* **109**, 63 (1978).
78MI1	F. S. Babichev, L. M. Fereshivana, and Yu. M. Volovenko, *Dopov. Akad. Nauk Ukr. RSR, Ser. B: Geol., Khim. Biol. Nauki*, 30 (1978).
78MI2	S. Baloniak, R. S. Ludwiczak, and E. Melzer, *Pol. J. Chem.* **52**, 1249 (1978).
78MI3	A. M. Eirin, L. Santana, E. Raviña, F. Fernandez, E. Sanchez-Abarca, and J. M. Calleja, *Eur. J. Med. Chem.—Chim. Ther.* **13**, 533 (1978).
78MI4	D. S. Frear and H. R. Swanson, *J. Agric. Food Chem.* **26**, 660 (1978).
78MI5	S. Groszkowski and J. Wrona, *Pol. J. Chem.* **52**, 1029 (1978).
78MI6	Y. Imai, M. Ueda, and T. Aizawa, *J. Polym. Sci., Polym. Chem. Ed.* **16**, 163 (1978).
78MI7	M. Khan, D. J. Chapman, N. W. Lem, K. R. Chandorkar, and J. P. Williams, *Dev. Plant Biol.* 415 (1978).
78MI8	V. Konečny and Š. Kovač, *Pestic. Sci.* **9**, 571 (1978).
78MI9	R. C. Moreau and P. Loiseau, *Ann. Pharm. Fr.* **36**, 67 (1978).
78MI10	Naser-ud-Din, *Pak. J. Sci. Ind. Res.* **21**, 119 (1978).
78MI11	M. Petrovanu and Mai Van Tri, *Bul. Inst. Politeh. Iasi, Sect. 2: Chim. Ing. Chim.* **24**, 87 (1978) [*CA* **91**, 39412 (1979)].
78MI12	H. Tanigawa, T. Kosazuma, H. Tanaka, and R. Obori, *J. Toxicol. Sci.* **3**, 69 (1978).
78MI13	S. Ueda and T. Amano, *Kyushu Sangyo Daigaku Kogakubu Kenkyu Hokoky* **15**, 127 (1978) [*CA* **91**, 4853 (1979)].

78MI14	M. Sato, H. Tanizawa, T. Fukuda, and T. Yuizono, *Nippon Yakurigaku Zasshi* **74**, 841 (1978) [*CA* **90**, 115089 (1979)].
78MI15	M. Sato, H. Tanizawa, and T. Yuizono, *Oyo Yakuri* **16**, 1011 (1978) [*CA* **90**, 197741 (1979)].
78MI16	T. Seki, *Rinsho Yakuri* **9**, 143 (1978) [*CA* **89**, 15762 (1978)].
78MI17	T. Seki, T. Hayashi, M. Oki, and T. Kishikawa, *Rinsho Yakuri* **9**, 149 (1978) [*CA* **89**, 173334 (1978)].
78MP997	E. J. Pedersen, R. R. Vold, and R. L. Vold, *Mol. Phys.* **35**, 997 (1978).
78RTC116	R. E. van der Stoel and H. C. van der Plas, *Recl. Trav. Chim. Pays-Bas* **97**, 116 (1978).
78T1971	H. Fehn and H. H. Perkampus, *Tetrahedron* **34**, 1971 (1978).
78T2069	A. Hassner, L. R. Krepski, and V. Alexanian, *Tetrahedron* **34**, 2069 (1978).
78T2509	W. Friedrichsen and H. Wallis, *Tetrahedron* **34**, 2509 (1978).
78T3419	W. C. Herndon and C. Parkanyi, *Tetrahedron* **34**, 3419 (1978).
78TL619	G. Heinisch, A. Jentzsch, and I. Kirchner, *Tetrahedron Lett.*, 619 (1978).
78TL1979	A. Ohsawa, T. Akimoto, A. Tsuji, and H. Igeta, *Tetrahedron Lett.*, 1979 (1978).
78TL3059	B. Stanovnik, M. Tišler, M. Kunaver, D. Gabrijelčič, and M. Kočevar, *Tetrahedron Lett.* **33**, 3059 (1978).
78TL4955	C. Degrand and D. Jacquin, *Tetrahedron Lett.*, 4955 (1978).
78YZ67	J. Haginiwa, Y. Higuchi, K. Nishioka, and Y. Yokokawa, *Yakugaku Zasshi* **98**, 67 (1978) [*CA* **88**, 170096 (1978)].
78YZ413	M. Takaya, T. Yamashita, and K. Ozeki, *Yakugaku Zasshi* **98**, 413 (1978) [*CA* **89**, 109321 (1978)].
78YZ1421	M. Takaya, T. Yamada, and H. Shimamura, *Yakugaku Zasshi* **98**, 1421 (1978) [*CA* **90**, 80647 (1979)].
78YZ1472	M. Takaya, K. Terashima, and K. Ozeki, *Yakugaku Zasshi* **98**, 1472 (1978) [*CA* **90**, 152105 (1979)].
78YZ1530	M. Takaya, T. Yamashita, A. Yamaguchi, and H. Kohara, *Yakugaku Zasshi* **98**, 1530 (1978) [*CA* **90**, 87377 (1979)].
79ACS(A)125	J. Sen and H. Taube, *Acta Chem. Scand., Ser. A* **A33**, 125 (1979).
79AHC363	M. Tišler and B. Stanovnik, *Adv. Heterocycl. Chem.* **24**, 363 (1979).
79AJC2297	G. B. Barlin and M. D. Fenn, *Aust. J. Chem.* **32**, 2297 (1979).
79BCJ3420	G. Marcotrigiano, L. Menabue, P. Morini, and G. C. Pellacani, *Bull. Chem. Soc. Jpn.* **52**, 3420 (1979).
79BSB905	G. Beynon, H. P. Figeys, D. Lloyd, and R. K. Mackie, *Bull. Soc. Chim. Belg.* **88**, 905 (1979).
79CC445	R. D. Chambers, C. R. Sargent, and M. Clark, *J. C. S. Chem. Commun.*, 445 (1979).
79CC446	R. D. Chambers and C. R. Sargent, *J. C. S. Chem. Commun.*, 446 (1979).
79CC1019	M. Ahern and G. W. Gokel, *J. C. S. Chem. Commun.*, 1019 (1979).
79CCC1761	V. Konečny and Š. Varkonda, *Collect. Czech. Chem. Commun.* **44**, 1761 (1979).
79CPB916	A. Ohsawa, T. Uezu, and H. Igeta, *Chem. Pharm. Bull.* **27**, 916 (1979).
79CPB1169	T. Tsujimoto, T. Nomura, M. Iifuru, and Y. Sasaki, *Chem. Pharm. Bull.* **27**, 1169 (1979).

79CPB2105	T. Tsujimoto, C. Kobayashi, T. Nomura, M. Iifuru, and Y. Sasaki, *Chem. Pharm. Bull.* **27**, 2105 (1979).
79FES299	F. Parravicini, G. Scarpitta, L. Dorigotti, and G. Pifferi, *Farmaco, Ed. Sci.* **34**, 299 (1979).
79G117	G. Adembri, S. Chimichi, R. Nesi, and M. Scotton, *Gazz. Chim. Ital.* **109**, 117 (1979).
79H(12)457	B. Stanovnik, M. Tišler, J. Bradač, B. Budič, B. Koren, and B. Mozetič-Reščič, *Heterocycles* **12**, 457 (1979).
79H(12)661	K. Matsumoto and T. Uchida, *Heterocycles* **12**, 661 (1979).
79H(12)1157	A. Tomažič, M. Tišler, and B. Stanovnik, *Heterocycles* **12**, 1157 (1979).
79H(13)389	F. G. Contreras and M. Lora-Tamayo, *Heterocycles* **13**, 389 (1979).
79IJC(A)276	J. S. Dwiwedi and U. Agarwala, *Indian J. Chem., Sect. A* **17A**, 276 (1979).
79IJC(B)136	M. A. El-Hashash, M. Y. El-Kady, and M. M. Mohamed, *Indian J. Chem., Sect. B* **18B**, 136 (1979).
79IZV803	D. N. Kursanov, M. I. Kalinkin, S. A. Gridchin, G. V. Shatalov, and Z. N. Parnes, *Izv. Akad. Nauk SSSR, Ser. Khim.*, 803 1979).
79JA766	K. Nicolaou, W. E. Bamette, and R. L. Magolda, *J. Am. Chem. Soc.* **101**, 766 (1979).
79JA2069	P. B. Dervan, T. Uyehara, and D. S. Santilli, *J. Am. Chem. Soc.* **101**, 2069 (1979).
79JA3663	D. S. Santilli and P. B. Dervan, *J. Am. Chem. Soc.* **101**, 3663 (1979).
79JCS(P1)1199	Y. Maki, M. Suzuki, T. Furuta, M. Kawamura, and M. Kuzuya, *J. C. S. Perkin 1*, 1199 (1979).
79JCS(P1)1451	C. H. Hassall, W. H. Johnson, and C. J. Theobald, *J. C. S. Perkin 1*, 1451 (1979).
79JCS(P1)2136	L. Crombie, N. A. Kerton, and G. Pattenden, *J. C. S. Perkin 1*, 2136 (1979).
79JHC245	I. Maeba, M. Ando, S. Yoshina, and R. N. Castle, *J. Heterocycl. Chem.* **16**, 245 (1979).
79JHC249	I. Maeba and R. N. Castle, *J. Heterocycl. Chem.* **16**, 249 (1979).
79JHC855	C. Stam, J. J. Zwinselman, H. C. van der Plas, and S. Baloniak, *J. Heterocycl. Chem.* **16**, 855 (1979).
79JHC861	A. Tomažič, M. Tišler, and B. Stanovnik, *J. Heterocycl. Chem.* **16**, 861 (1979).
79JHC1213	I. Maeba and R. N. Castle, *J. Heterocycl. Chem.* **16**, 1213 (1979).
79JIC749	K. Lal, *Indian J. Chem. Soc.* **56**, 749 (1979).
79JMC53	M. Takaya, M. Sato, K. Terashima, H. Tanizawa, and Y. Maki, *J. Med. Chem.* **22**, 53 (1979).
79JOC218	B. B. Snider, R. S. E. Conn, and S. Sealfon, *J. Org. Chem.* **44**, 218 (1979).
79JOC629	M. J.. Haddadin, S. J. Firsan, and B. S. Nader, *J. Org. Chem.* **44**, 629 (1979).
79JOC3053	S. Gelin, *J. Org. Chem.* **44**, 3053 (1979).
79JOC3524	A. Ohsawa, H. Arai, H. Igeta, T. Akimoto, A. Tsuji, and Y. Iitaka, *J. Org. Chem.* **44**, 3524 (1979).
79JOM97	A. L. Balch and R. D. Cooper, *J. Organomet. Chem.* **169**, 97 (1979).
79JPR71	K. Gewald and J. Oelsner, *J. Prakt. Chem.* **321**, 71 (1979).
79JPR629	H. Zaschke and C. Hyna, *J. Prakt. Chem.* **321**, 629 (1979).
79KFZ34	V. I. Shvedov, N. V. Savitskaya, V. G. Nyrkova, Z. A. Pankina,

	I. N. Fedorova, O. A. Fedotova, L. F. Linberg, and A. I. Polezhaeva, *Khim.-Farm. Zh.* **13**, 34 (1979).
79KGS943	S. G. Agbalyan and R. D. Khachikyan, *Khim. Geterotsikl. Soedin.*, 943 (1979).
79LA675	M. Bachmann and H. Neunhoeffer, *Liebigs Ann. Chem.*, 675 (1979).
79M635	G. Heinisch and I. Kirchner, *Monatsh. Chem.* **110**, 365 (1979).
79MI1	A. Assandri, A. Perazzi, G. Buniva, and V. Pagani, *Eur. J. Drug Metab. Pharmacokinet.* **4**, 75 (1979).
79MI2	A. Aydin and H. Feuer, *Chim. Acta Turc.* **7**, 121 (1979).
79MI3	S. Baloniak and E. Melzer, *Acta Pol. Pharm.* **36**, 147 (1979).
79MI4	S. Baloniak, E. Drodzynska, E. Linkowska, M. Filczewski, A. Mroczkiewicz, K. Oledzka, and I. Zyczynska-Baloniak, *Acta Pol. Pharm.* **36**, 295 (1979).
79MI5	S. Baloniak, E. Linkowska, and I. Zyczynska-Baloniak, *Acta Pol. Pharm.* **36**, 301 (1979).
79MI6	E. Beretta, T. Christina, P. Ferrari, G. Tuan, L. F. Zerilli, and E. Martinelli, *Eur. J. Drug Metab. Pharmacokinet.* **4**, 29 (1979).
79MI7	M. Caprosu, M. Ungureanu, I. Druta, N. Stavri, and M. Petrovanu, *Bul. Inst. Politeh. Iasi, Sect. 2: Chim. Ing. Chim.* **25**, 79 (1979) [*CA* **92**, 198338 (1980)].
79MI8	Y. S. Cho, *Yakhak Hoe Chi* **23**, 125 (1979) [*CA* **92**, 164267 (1980)].
79MI9	M. G. Ckitishvili, P. V. Gororishvili, M. V. Chrelashvili, and A. E. Shvelashvili, *Izv. Akad. Nauk Gruz. SSR, Ser. Khim.* **5**, 13 (1979).
79MI10	H. Fujihara, M. Yoshihara, Y. Matsubara, and T. Maeshima, *J. Macromol. Sci., Chem.* **A13**, 789 (1979).
79MI11	R. K. Goel and S. N. Sharma, *Indian J. Pure Appl. Phys.* **17**, 630 (1979) [*CA* **92**, 13130 (1980)].
79MI12	Y. Gomita, H. Kawasaki, M. Sato, T. Yuizono, and T. Fukuda, *Oyo Yakuri* **18**, 23 (1979) [*CA* **92**, 104391 (1980)].
79MI13	T. Herczeg, E. Lehoczki, and L. Szalay, *FEBS Lett.* **108**, 226 (1979).
79MI14	K. Hokama, S. Yogi, and M. Higa, *Bull. Coll. Sci., Univ. Ryukyus* **28**, 63 (1979) [*CA* **94**, 120420 (1981)].
79MI15	Y. Kanaoka, M. Hasebe, and E. Sato, *Fukusokan Kagaku Toronkai Koen Yoshishu, 12th*, 156 (1979) [*CA* **93**, 71679 (1980)].
79MI16	M. U. Khan, N. W. Lem, K. R. Chandorkar, and J. P. Williams, *Plant Physiol.* **64**, 300 (1979).
79MI17	V. Konečny, Š. Varkonda, and V. Kubala, *Chem. Zvesti* **33**, 669 (1979).
79MI18	V. Konečny, Š. Varkonda, and V. Kubala, *Chem. Zvesti* **33**, 675 (1979).
79MI19	V. Konečny and Š. Varkonda, *Chem. Zvesti* **33**, 683 (1979).
79MI20	V. Konečny and Š. Varkonda, *Chem. Zvesti* **33**, 822 (1979).
79MI21	V. Konečny, Š. Varkonda, J. Kovačičova, V. Batora, and M. Vargova, *Pestic. Sci.* **10**, 139 (1979).
79MI22	G. Mannini, G. Biasoli, E. Perrone, A. Forgione, A. Buttinoni, and M. Ferrari, Eur. J. Med. Chem.—Chim. Ther. **14**, 53 (1979).
79MI23	G. Neikov and N. Tyutyulkov, *Ivz. Khim.* **12**, 27 (1979).
79MI24	C. Onodera, T. Hayashi, I. Makita, T. Hashi, K. Takeda, F. Ozeki, and H. Shimazu, *J. Toxicol. Sci.* **4**, 229 (1979).
79MI25	A. Osawa, I. Wada, H. Igeta, T. Akimoto, and A. Tsuji, *Fukusokan*

Kagaku Toronkai Koen Yoshishu, 12th, 256 (1979) [*CA* **93**, 71680 (1980)].

79MI26   K. H. Pannell and R. Iglesias, *Inorg. Chim. Acta* **33**, L161 (1979).

79MI27   S. M. Ridley and J. Ridley, *Plant Physiol.* **63**, 392 (1979).

79MI28   M. Sato, H. Tanizawa, T. Fukuda, and T. Yuizono, *Jpn. J. Pharmacol.* **29**, 303 (1979) [*CA* **91**, 6848 (1979)].

79MI29   M. Sato, I. Kimura, H. Tanizawa, T. Fukuda, and T. Yuizono, *Oyo Yakuri* **17**, 241 (1979) [*CA* **91**, 83393 (1979)].

79MI30   M. Sato, Y. Ishizuka, H. Tanizawa, T. Fukuda, and T. Yuizono, *Nippon Yakurigaku Zasshi* **75**, 291 (1979) [*CA* **91**, 117292 (1979)].

79MI31   M. Sato, *Nippon Yakurigaku Zasshi* **75**, 695 (1979) [*CA* **92**, 157654 (1980)].

79MI32   H. Satoh, M. Tonegawa, K. Kitahara, and R. Aoyagi, *Tokyo Ika Daigaku Kiyo* **5**, 71 (1979) [*CA* **94**, 84045 (1981)].

79MI33   M. S. De Giambiagi and M. Giambiagi, *Lett. Nuovo Cimento Soc. Ital. Fis.* [2] **25**, 459 (1979).

79MI34   K. Shimpo, N. Mori, M. Takahashi, N. Aoki, H. Togashi, K. Hiroga, T. Kosazuma, H. Shimazu, and T. Tanabe, *J. Toxicol. Sci.* **4**, 255 (1979).

79MI35   E. Stefanescu, M. Caprosu, E. Gafitanu, R. Danet, and M. Petrovanu, *Rev. Med.-Chir.* **83**, 311 (1979) [*CA* **92**, 140510 (1980)].

79MI36   G. Szilagyi, E. Kasztreiner, L. Tardos, L. Jaszlits, E. Kósa, G. Cseh, P. Tolnay, and I. Kovacs-Szabó, *Eur. J. Med. Chem.— Chim. Ther.* **14**, 439 (1979).

79MI37   H. Tanigawa, R. Obori, H. Tanaka, J. Yoshida, and T. Kosazuma, *J. Toxicol. Sci.* **4**, 163 (1979).

79MI38   H. Tanigawa, R. Obori, H. Tanaka, J. Yoshida, and T. Kosazuma, *J. Toxicol. Sci.* **4**, 175 (1979).

79MI39   M. G. Tsikishvili, P. V. Gogovishvili, M. V. Chelashvili, and A. E. Shvelashvili, *Izv. Akad. Nauk. Gruz. SSSR, Ser. Khim.* **5**, 13 (1979) [*CA* **92**, 163917 (1980)].

79MI40   E. Ueda, Y. Udagawa, and M. Ito, *Koen Yoshishu—Bunski Kozo Sogo Toronkai*, 74 (1979) [*CA* **93**, 185210 (1980)].

79MI41   Yu. M. Volovenko, L. M. Pereshivana, and F. S. Babichev, *Dopov. Akad. Nauk Ukr. RSR, Ser. B: Geol., Khim. Biol. Nauk*, 193 (1979) [*CA* **90**, 204004 (1979)].

79MI42   W. Von Niessen, W. P. Kraemer, and G. H. F. Diercksen, *Chem. Phys.* **41**, 113 (1979).

79MI43   C. Willemot, H. J. Hope, and C. J. St.-Pierre, *Can. J. Plant Sci.* **59**, 249 (1979).

79MP1381   E. W. Thulstrup, J. Spanget-Larsen, and R. Gleiter, *Mol. Phys.* **37**, 1381 (1979).

79PIA(A)279   N. K. Sanyal, S. L. Srivastava, R. K. Goel, and S. D. Sharma, *Proc.—Indian Acad. Sci., Sect. A* **88A**, 279 (1979).

79RRC453   N. A. Shams, *Rev. Roum. Chim.* **24**, 453 (1979).

79RRC899   M. F. Ismail, N. A. Shams, S. E. A. Rahman, and A. K. Fateen, *Rev. Roum. Chim.* **24**, 899 (1979).

79RRC1381   M. M. Mohamed, M. A. El-Hashash, and M. El-Kady, *Rev. Roum. Chim.* **24**, 1381 (1979).

79S385   R. Gompper and R. Sobotta, *Synthesis*, 385 (1979).

79S790	C. Venturello and R. D'Aloisio, *Synthesis*, 790 (1979).
79T277	S. Satish, A. Mitra, and M. V. George, *Tetrahedron* **35**, 277 (1979).
79T1267	A. Ohsawa, H. Arai, H. Igeta, T. Akimoto, A. Tsuji, and Y. Iitaka, *Tetrahedron* **35**, 1267 (1979).
79T1595	M. Ohsaku, H. Murata, A. Iwamura, and K. Hirao, *Tetrahedron* **35**, 1595 (1979).
79TL1333	F. Gaviña, P. Gil, and B. Palazon, *Tetrahedron Lett.*, 1333 (1979).
79TL2821	D. M. Holton, P. M. Hoyle, and D. Murphy, *Tetrahedron Lett.*, 2821 (1979).
79TL2921	E. Toja, A. Omodei-Salè, and G. Nathansohn, *Tetrahedron Lett.*, 2921 (1979).
79TL4837	S. Restle and C. G. Wermuth, *Tetrahedron Lett.*, 4837 (1979).
79TL5025	N. Viswanathan and A. R. Sidhaye, *Tetrahedron Lett.*, 5025 (1979).
79YZ211	M. Takaya, T. Yamada, and A. Yamaguchi, *Yakugaku Zasshi* **99**, 211 (1979) [*CA* **91**, 39411 (1979)].
79YZ1091	M. Amino, T. Hayashi, T. Matsuo, and A. Yamaguchi, *Yakugaku Zasshi* **99**, 1091 (1979) [*CA* **82**, 87794 (1980)].
79ZN(C)1052	F. A. Eder, Z. *Naturforsch., C: Biosci.* **34C**, 1052 (1979).
80ACH405	J. Kosáry and P. Sohár, *Acta Chim. Acad. Sci. Hung.* **103**, 405 (1980).
80ACS(B)597	A. A. El Barbary, S. Scheibye, S. O. Lawesson, and H. Fritz, *Acta Chem. Scand., Ser. B* **B32**, 597 (1980).
80AG815	W. Adam and O. De Lucchi, *Angew. Chem.* **92**, 815 (1980).
80AKZ862	G. V. Grigoryan and S. C. Agbalyan, *Arm. Khim. Zh.* **33**, 862 (1980) [*CA* **94**, 139528 (1981)].
80AP53	G. Heinisch and A. Mayrhofer, *Arch. Pharm. (Weinheim, Ger.)* **313**, 53 (1980).
80CCC127	Š. Kovač and V. Konečny, *Collect. Czech. Chem. Commun.* **45**, 127 (1980).
80CCC2247	V. Konečny, Š. Varkonda, and V. Kubala, *Collect. Czech. Chem. Commun.* **45**, 2247 (1980).
80CCC2343	V. Konečny and Š. Varkonda, *Collect. Czech. Chem. Commun.* **45**, 2343 (1980).
80CPB198	T. Jojima, H. Takeshiba, and T. Kinoto, *Chem. Pharm. Bull.* **28**, 198 (1980).
80CPB529	S. Kamiya and M. Tanno, *Chem. Pharm. Bull.* **28**, 529 (1980).
80CPB2676	T. Tsuchiya, J. Kurita, and K. Takayama, *Chem. Pharm. Bull.* **28**, 2676 (1980).
80CPB3488	A. Ohsawa, Y. Abe, and H. Igeta, *Chem. Pharm. Bull.* **28**, 3488 (1980).
80CPB3570	A. Ohsawa, H. Arai, and H. Igeta, *Chem. Pharm. Bull.* **28**, 3570 (1980).
80CRV99	P. S. Engel, *Chem. Rev.* **80**, 99 (1980).
80DOK1392	G. L. Rusinov, I. Ya. Postovskii, and E. G. Kovalev, *Dokl. Akad. Nauk SSSR* **253**, 1392 (1980).
80HCA504	W. Städeli, W. Philipsborn, A. Wick, and I. Kompiš, *Helv. Chim. Acta* **63**, 504 (1980).
80IJC(B)203	M. F. Ismail, A. A. El-Khamry, N. A. Shams, and O. M. El-Sawy, *Indian J. Chem., Sect. B* **19B**, 203 (1980).
80IJC(B)648	D. V. Sule, N. P. Karambelkar, K. D. Deodhar, and R. A. Kulkarni, *Indian J. Chem., Sect. B* **19B**, 648 (1980).

80IJC(B)1038	S. A. Khattab and M. M. Hosny, *Indian J. Chem., Sect. B* **19B**, 1038 (1980).
80JA3863	P. B. Dervan and D. S. Santilli, *J. Am. Chem. Soc.* **102**, 3863 (1980).
80JA7131	W. Adam and F. Mazenod, *J. Am. Chem. Soc.* **102**, 7131 (1980).
80JA7438	A. Schweig, N. Thon, S. F. Nelsen, and L. A. Grezzo, *J. Am. Chem. Soc.* **102**, 7438 (1980).
80JCS(P1)72	M. M. Baradarani and J. A. Joule, *J. C. S. Perkin 1*, 72 (1980).
80JCS(P2)1339	S. Chimichi, R. Nesi, M. Scotton, C. Mannucci, and G. Adembri, *J. C. S. Perkin 2*, 1339 (1980).
80JHC541	I. E. S. El-Kholy, H. M. Fuid-Alla, and M. M. Mishrikey, *J. Heterocycl. Chem.* **17**, 541 (1980).
80JHC781	P. Mátyus, G. Szilágyi, E. Kasztreiner, and P. Sohár, *J. Heterocycl. Chem.* **17**, 781 (1980).
80JHC1465	M. J. Kornet and R. Daniels, *J. Heterocycl. Chem.* **17**, 1465 (1980).
80JHC1501	G. Heinisch and I. Kirchner, *J. Heterocycl. Chem.* **17**, 1501 (1980).
80JHC1527	S. Sunder and N. P. Peet, *J. Heterocycl. Chem.* **17**, 1527 (1980).
80JMC578	B. P. Das, R. A. Wallace, and D. W. Boykin, *J. Med. Chem.* **23**, 578 (1980).
80JMC1398	R. Buchman, J. A. Scozzie, Z. S. Ariyan, R. D. Heilman, D. J. Rippin, W. J. Pyne, L. J. Powers, and J. R. Matthews, *J. Med. Chem.* **23**, 1398 (1980).
80JMC1445	S. W. Fogt, J. A. Scozzie, R. D. Heilman, and L. J. Powers, *J. Med. Chem.* **23**, 1445 (1980).
80JOC1695	A. G. Anderson and T. Y. Tober, *J. Org. Chem.* **45**, 1695 (1980).
80JOC3609	S. F. Nelsen and W. C. Hollinsed, *J. Org. Chem.* **45**, 3609 (1980).
80JOC4584	M. D. Bezoari and W. W. Paudler, *J. Org. Chem.* **45**, 4584 (1980).
80JPR617	A. K. Fateen, M. F. Ismail, A. M. Kaddah, and N. A. Shams, *J. Prakt. Chem.* **322**, 617 (1980).
80KGS228	I. J. Deeva and A. N. Kost, *Khim. Geterotsikl. Soedin.*, 228 (1980).
80KGS394	G. V. Shatalov, S. A. Gridchin, and N. I. Mikhantiev, *Khim. Geterotsikl. Soedin.*, 394 (1980).
80KGS1287	A. N. Kost, L. A. Sviridova, G. A. Golubeva, and A. V. Dovgilevich, *Khim. Geterotsikl. Soedin.*, 1287 (1980).
80LA590	H. Heydt, K. H. Busch, and M. Regitz, *Liebigs Ann. Chem.*, 590 (1980).
80LA1307	R. R. Schmidt and R. Scheibe, *Liebigs Ann. Chem.*, 1307 (1980).
80MI1	A. Aminzadeh, V. Fawcett, and D. A. Long, *J. Raman Spectrosc.* **9**, 214 (1980).
80MI2	A. Aydin and H. Feuer, *Chim. Acta Turc.* **8**, 113 (1980).
80MI3	A. Aydin and H. Feuer, *Chim. Acta Turc.* **8**, 199 (1980).
80MI4	M. Bidlo-Igloy and T. Horvath, *HRC CC, J. High Resolut. Chromatogr. Chromatogr. Commun.* **3**, 421 (1980).
80MI5	G. Boxhoorn, J. M. Ernsting, D. J. Stufkens, and A. Oskam, *Thermochim. Acta* **42**, 315 (1980).
80MI6	T. Eda, Y. Matsubara, M. Yoshihara, and T. Maeshima, *J. Macromol. Sci., Chem.* **A14**, 771 (1980) [*CA* **93**, 26881 (1980)].
80MI7	T. Eda, C. Y. Huang, Y. Matsubara, M. Yoshihara, and T. Maeshima, *J. Macromol. Sci., Chem.* **A14**, 1035 (1980) [*CA* **93**, 132873 (1980)].
80MI8	M. A. El-Hashash, M. Abdalla, A. Essawy, and A. M. El-Gendy, *Pak. J. Sci. Ind. Res.* **23**, 254 (1980) [*CA* **95**, 80870 (1981)].

80MI9	G. Georgiev, G. Vasilev, L. Dryanovska-Noniska, M. Georgieva, O. S. Chizhov, and B. M. Zolotaev, *Dokl. Bolg. Akad. Nauk* **33**, 1385 (1980) [*CA* **95**, 43027 (1981)].
80MI10	T. Herczeg, E. Lehoczki, I. Rojik, I. Vaas, and T. Farkas, *Plant Sci. Lett.* **19**, 285 (1980).
80MI11	B. Hladon, S. Baloniak, P. Szafarek, and I. Zyczynska-Baloniak, *Arch. Immunol. Ther. Exp.* **28**, 427 (1980).
80MI12	E. H. M. Ibrahim, O. Sherif, and M. Saeda, *Egypt. J. Chem.* **23**, 107 (1980) [*CA* **96**, 142785 (1982)].
80MI13	E. Kasztreiner, G. Szilágyi, Z. Husti, P. Mátyus, J. Kosáry, L. Tardos, E. Kósa, L. Jaszlits, and G. Cseh, *Adv. Pharmacol. Res. Pract., Proc. Congr. Hung. Pharmacol. Soc., 3rd, 1979*, 227 (1980) [*CA* **94**, 185394 (1981)].
80MI14	K. Kawashima, T. X. Watanabe, and H. Sokabe, *Jpn. J. Pharmacol.* **30**, 116 (1980) [*CA* **93**, 616 (1980)].
80MI15	E. Kósa, L. Jaszlits, and L. Tardos, *Adv. Pharmacol. Res. Pract., Proc. Congr. Hung. Pharmacol. Soc., 3rd, 1979*, 233 (1980) [*CA* **94**, 185542 (1981)].
80MI16	J. Kosári, *Magy. Kem. Foly.* **86**, 564(1980).
80MI17	G. Laskay, E. Lehoczki, and L. Szalay, *Acta Biol. (Szeged)* **26**, 21 (1980).
80MI18	V. P. Mamaev and O. P. Shkurko, *Izv. Sib. Otd. Akad. Nauk SSSR, Ser. Khim. Nauk*, 22 (1980).
80MI19	W. Molzberger and F. Lingens, *Zentralbl. Bakteriol., Abt. 1, Orig. C* **1**, 306 (1980) [*CA* **95**, 38833 (1981)].
80MI20	A. Mroczkiewicz, *Pol. J. Chem.* **54**, 1095 (1980).
80MI21	A. A. Nada, A. M. A. Emran, and A. M. Mourad, *Egypt. J. Chem.* **23**, 165 (1980) [*CA* **96**, 162625 (1982)].
80MI22	G. Natile, F. Gasparrini, B. Galli, A. M. Manotti-Lanfredi, and A. Tiripicchio, *Inorg. Chim. Acta* **44**, L29 (1980).
80MI23	W. H. Newsome, *J. Agric. Food Chem.* **28**, 270 (1980).
80MI24	S. Okabe, Y. Ishihara, S. Adachi, M. Matsumoto, Y. Kawahara, and K. Arita, *Oyo Yakuri* **19**, 1019 (1980) [*CA* **94**, 76683 (1981)].
80MI25	G. Pifferi and G. Gaviraghi, *Cron. Farm.* **23**, 186 (1986).
80MI26	K. Shimpo, H. Togashi, N. Mori, M. Takahashi, M. Iwata, K. Hiraga, T. Kosazuma, H. Shimazu, and T. Tanabe, *J. Toxicol. Sci.* **4**, 139 (1979)
80MI27	P. Sohár, P. Mátyus, and G. Szilágyi, *Kem. Kozl.* **54**, 313 (1980) [*CA* **95**, 203163 (1981)].
80MI28	H. Tanigawa, R. Obori, H. Tanaka, J. Yoshida, and S. Okuma, *Iyakuhin Kenkyu* **11**, 559 (1980) [*CA* **94**, 132177 (1981)].
80MI29	J. B. Visockii, N. A. Kovač, and O. P. Švaika, *Izv. Sib. Otd. Akad. Nauk SSSR, Ser. Khim. Nauk*, 3 (1980).
80MI30	J. Yoshida, H. Tanaka, R. Obori, and H. Tanigawa, *Iyakuhin Kenkyu* **11**, 566 (1980) [*CA* **94**, 95989 (1981)].
80MP519	S. S. Z. Adnan, S. Bhattacharya, and D. Mukherjee, *Mol. Phys.* **39**, 519 (1980).
80NKK33	T. Eda, Y. Matsubara, M. Yoshihara, and T. Maeshima, *Nippon Kagaku Kaishi*, 33 (1980) [*CA* **93**, 71676 (1980)].
80PHA140	W. Schliemann, A. Büge, and L. Reppel, *Pharmazie* **35**, 140 (1980).

80RRC1375	G. H. Sayed, M. El-Kady, and M. A. El-Hashash, *Rev. Roum. Chim.* **25**, 1375 (1980) [*CA* **94**, 192249 (1981)].
80RTC234	R. E. van der Stoel, H. C. van der Plas, H. Longejan, and L. Hoeve, *Recl. Trav. Chim. Pays-Bas* **99**, 234(1980).
80S129	S. Polanc, B. Stanovnik, and M. Tišler, *Synthesis*, 129 (1980).
80S410	M. F. Ismail, N. A. Shams, and O. M. El-Sawy, *Synthesis*, 410 (1980).
80S457	A. K. Fateen, A. H. Moustafa, A. M. Kaddah, and N. A. Shams, *Synthesis*, 457 (1980).
80S623	C. Deshayes and S. Gelin, *Synthesis*, 623 (1980).
80S746	M. Yoshihara, T. Eda, K. Sakaki, and T. Maeshima, *Synthesis*, 746 (1980).
80S830	S. Polanc, B. Stanovnik, and M. Tišler, *Synthesis*, 830 (1980).
80TL1009	A. Padwa and H. Ku, *Tetrahedron Lett.*, **21**, 1009 (1980).
80TL2939	H. J. Bestmann, G. Schmid, and D. Sandmeier, *Tetrahedron Lett.*, **21**, 2939 (1980).
80TL3723	S. P. J. M. van Nispen, C. Mensink, and A. M. van Leusen, *Tetrahedron Lett.*, **21**, 3723 (1980).
80TL4615	Y. Kobayashi, T. Nakano, K. Shirahashi, A. Takeda, and I. Kumadaki, *Tetrahedron Lett.*, **21**, 4615 (1980).
80YZ774	A. Ohsawa, T. Uezu, and H. Igeta, *Yakugaku Zasshi* **100**, 774 (1980) [*CA* **94**, 47270 (1981)].
80ZN(A)844	I. Novak, L. Klasinc, J. V. Knop, and T. Bluhm, *Z. Naturforsch., A* **35A**, 844 (1980).
81ACH205	P. Mátyus, G. Szilágyi, E. Kasztreiner, and P. Sohár, *Acta Chim. Acad. Sci. Hung.* **106**, 205 (1981).
81ACR131	S. F. Nelsen, *Acc. Chem. Res.* **14**, 131 (1981).
81ACS(B)185	P. Fuchs, U. Hess, H. H. Holst, and H. Lund, *Acta Chem. Scand, Ser B* **B35**, 185 (1981).
81AF1529	F. Reicheneder, T. F. Burger, H. König, R. Kropp, H. Lietz, M. Thyes, and W. W. Wiersdorff, *Arzneim.-Forsch.* **31**, 1529 (1981).
81AF1533	J. Gries, J. Schuster, H. Giertz, H. D. Lehmann, D. Lenke, and W. Wortstmann, *Arzneim.-Forsch.* **31**, 1533 (1981).
81AF1544	H. D. Lehmann, H. Giertz, R. Kretzschmar, D. Lenke, G. Phillipsborn, M. Raschack, and J. Schuster, *Arzneim.-Forsch.* **31**, 1544 (1981).
81AF1558	D. Lenke, J. Gries, and R. Kretzschmar, *Arzneim.-Forsch.* **31**, 1558 (1981).
81AF1566	M. Traut, E. Brode, and H. D. Hoffmann, *Arzneim.-Forsch.* **31**, 1566 (1981).
81AF1574	C. D. Müller, J. Gries, D. Lenke, H. H. Teschendorf, and H. Weifenbach, *Arzneim.-Forsch.* **31**, 1574 (1981).
81AF1580	H. J. Teschendorf, R. Kretschmar, H. Kreiskott, and H. Weifenbach, *Arzneim.-Forsch.* **31**, 1580 (1981).
81AF1589	M. Traut, H. Morgenthaler, and E. Brode, *Arzneim.-Forsch.* **31**, 1589 (1981).
81AF1594	M. Traut, E. Brode, B. Neumann, and H. Kummer, *Arzneim.-Forsch.* **31**, 1594 (1981).
81AF1605	M. Traut and E. Brode, *Arzneim.-Forsch.* **31**, 1605 (1981).
81AF1616	M. Traut, H. Kummer, and E. Brode, *Arzneim.-Forsch.* **31**, 1616 (1981).

81AF1623	T. Moest, H. J. Dechow, and C. H. Pich, *Arzneim.-Forsch.* **31**, 1623 (1981).
81AG1638	K. Wilsmann, G. Neugebauer, R. Kessel, and E. Lang. *Arzneim.-Forsch.* **31**, 1638 (1981).
81AF1647	K. Wilsmann, G. Neugebauer, R. Kessel, and E. Lang, *Arzneim.-Forsch.* **31**, 1647 (1981).
81AF1653	S. Kaumeier, O. H. Kehrhahn, H. Morgenthaler, and G. Neugebauer, *Arzneim.-Forsch.* **31**, 1653 (1981).
81AF1657	S. H. Oloff, *Arzneim.-Forsch.* **31**, 1657 (1981).
81AG567	A. R. Katritzky, R. C. Patel, and F. G. Riddell, *Angew. Chem.* **93**, 567 (1981).
81AJC1223	R. A. Russell, D. E. Marsden, M. Sterns, and R. N. Warrener, *Aust. J. Chem.* **34**, 1223 (1981).
81AJC1729	B. Stanovnik, M. Tišler, A. Hribar, G. D. Barlin, and D. J. Brown, *Aust. J. Chem.* **34**, 1729 (1981).
81AP376	G. Seitz and W. Overheu, *Arch. Pharm. (Weinheim, Ger.)* **314**, 376 (1981).
81AP892	G. Seitz, T. Kämpchen, W. Overheu, and U. Martin, *Arch. Pharm. (Weinheim, Ger.)* **314**, 892 (1981).
81AX(B)1576	B. J. Graves and D. J. Hodgson, *Acta Crystallogr., Sect. B* **B37**, 1576 (1981).
81BSF(1)153	J. Rambaud, R. Roques, J. P. Declercq, G. Germain, and F. Sabon, *Bull. Soc. Chim. Fr.*, Part 1, 153 (1981).
81CB564	H. Stetter and F. Jonas, *Chem. Ber.* **114**, 564 (1981).
81CB3154	L. Birkofer and E. Hänsel, *Chem. Ber.* **114**, 3154 (1981).
81CI(L)365	A. G. Rowley and J. R. F. Steedman, *Chem. Ind. (London)*, 365 (1981).
81CL873	E. Ueda, Y. Udagawa, and M. Ito, *Chem. Lett.*, 873 (1981).
81CL1631	A. Inoue, S. Yoshida, and N. Fbara, *Chem. Lett.*, 1631 (1981).
81CPB2837	H. Yamanaka, S. Konno, T. Sakamoto, S. Niitsuma, and S. Noji, *Chem. Pharm. Bull.* **29**, 2837 (1981).
81CPB3433	T. Yamada, Y. Nobuhara, H. Shimamura, K. Yoshihara, A. Yamaguchi, and M. Ohki, *Chem. Pharm. Bull.* **29**, 3433 (1981).
81H9	B. J. Graves, D. J. Hodgson, S. F. Chen, and R. P. Panzica, *Heterocycles* **16**, 9 (1981).
81H1705	O. Tsuge, M. Tanaka, H. Shimohanada, and M. Noguchi, *Heterocycles* **16**, 1705 (1981).
81H2173	B. Stanovnik, J. Žmitek, and M. Tišler, *Heterocycles* **16**, 2173 (1981).
81HCA1930	M. Roth, *Helv. Chim. Acta* **64**, 1930 (1981).
81IC2084	D. R. Corbin, L. C. Francesconi, D. N. Hendrickson, and G. D. Stucky, *Inorg. Chem.* **20**, 2084 (1981).
81IJC(B)78	M. F. Ismail, N. A. Shams, E. A. Soliman, and O. M. El-Sawy, *Indian J. Chem., Sect. B* **20B**, 78 (1981).
81IJC(B)424	G. H. Sayed and M. S. Abd-Elhalim, *Indian J. Chem., Sect. B* **20B**, 424 (1981).
81IJC(B)502	H. Jahine, H. A. Zaher, O. Sherif, and M. Seada, *Indian J. Chem., Sect. B* **20B**, 502 (1981).
81IJC(B)845	G. H. Sayed, M. Y. El-Kady, and M. S. Abd-Elhalim, *Indian J. Chem., Sect. B* **20B**, 845 (1981).

81IJC(B)1084	A. H. Moustafa, A. M. Kaddah, and N. A. Shams, *Indian J. Chem., Sect. B* **20B**, 1084 (1981).
81IJC(B)1097	M. T. Omar and M. E. Eid, *Indian J. Chem., Sect. B* **20B**, 1097(1981).
81JA3300	S. F. Nelsen and P. M. Gannett, *J. Am. Chem. Soc.* **103**, 3300 (1981).
81JA7011	P. M. Lahti and J. A. Berson, *J. Am. Chem. Soc.* **103**, 7011 (1981).
81JCR(S)103	F. Sauter, P. Stanetty, A. Blaschke, and H. Vyplel, *J. Chem. Res. Synop.*, 103 (1981).
81JCR(S)104	J. Bourguignon, C. Bécue, and G. Quéguiner, *J. Chem. Res., Synop.*, 104 (1981).
81JCS(D)1524	R. W. Balk, G. Boxhoorn, T. L. Snoeck, G. C. Schoemaker, D. J. Stufkens, and A. Oskam, *J. C. S. Dalton*, 1524 (1981).
81JCS(P1)73	K. Matsumoto and T. Uchida, *J. C. S. Perkin 1*, 73 (1981).
81JCS(P1)1071	R. D. Chambers, W. K. R. Musgrave, and C. R. Sargent, *J. C. S. Perkin 1*, 1071 (1981).
81JCS(P1)2059	R. N. Barnes, R. D. Chambers, R. D. Hercliffe, and W. K. R. Musgrave, *J. C. S. Perkin 1*, 2059 (1981).
81JHC255	F. Heresch, G. Allmaier, and G. Heinisch, *J. Heterocycl. Chem.* **18**, 255 (1981).
81JHC293	M. J. Kornet and J. Chu, *J. Heterocycl. Chem.* **18**, 293 (1981).
81JHC303	S. F. Chen and R. P. Panzica, *J. Heterocycl. Chem.* **18**, 303 (1981).
81JHC333	S. Plescia, G. Daidone, J. Fabra, and V. Sprio, *J. Heterocycl. Chem.* **18**, 333 (1981).
81JHC425	J. Bourguignon, C. Bécue, and G. Quéguiner, *J. Heterocycl. Chem.* **18**, 425 (1981).
81JHC443	S. Veeraraghavan, D. Bhattacharjee, and F. D. Popp, *J. Heterocycl. Chem.* **18**, 443 (1981).
81JHC803	A. W. Addison and P. J. Burke, *J. Heterocycl. Chem.* **18**, 803 (1981).
81JHC1109	S. Baloniak and H. C. van der Plas, *J. Heterocycl. Chem.* **18**, 1109 (1981).
81JHC1465	A. Turck, J. F. Brument, and G. Quéguiner, *J. Heterocycl. Chem.* **18**, 1465 (1981).
81JHC1551	R. N. Castle, R. D. Thompson, N. K. Dalley, S. H. Simonsen, and S. B. Larsen, *J. Heterocycl. Chem.* **18**, 1551 (1981).
81JMC59	G. Steiner, J. Gries, and D. Lenke, *J. Med. Chem.* **24**, 59 (1981).
81JMC592	J. D. Albright, D. B. Moran, W. B. Wright, J. B. Collins, C. B. Beer, A. S. Lippa, and E. Greenblatt, *J. Med. Chem.* **24**, 592 (1981).
81JMR387	T. Wamsler, J. T. Nielsen, E. J. Pedersen, and K. Schaumburg, *J. Magn. Reson.* **43**, 387 (1981).
81JOC294	D. R. Carver, A. P. Komin, J. S. Hubbard, and J. F. Wolfe, *J. Org. Chem.* **46**, 294 (1981).
81JOC881	A. J. Ashe, III, W. T. Chan, T. W. Smith, and K. M. Taba, *J. Org. Chem.* **46**, 881 (1981).
81JOC2467	S. F. Chen and R. P. Panzica, *J. Org. Chem.* **46**, 2467 (1981).
81JOC5156	C. Paradisi, M. Prato, U. Quintily, G. Scorrano, and G. Valle, *J. Org. Chem.* **46**, 5156 (1981).
81JPR164	A. I. Hashem and M. E. Shaban, *J. Prakt. Chem.* **323**, 164 (1981).
81JPS334	P. Ventura, F. Parravicini, L. Simonotti, R. Colombo, and G. Pifferi, *J. Pharm. Sci.* **70**, 334 (1981).

81JPS419	L. J. Powers, D. J. Eckert, and L. Gehrlein, *J. Pharm. Sci.* **70**, 419 (1981).
81JST191	M. D. Child, G. A. Foulds, G. C. Percy, and D. A. Thornton, *J. Mol. Struct.* **75**, 191 (1981).
81KGS530	S. A. Stehova, V. V. Lapachev, and V. V. Mamaev, *Khim. Geterotsikl. Soedin.*, 530 (1981).
81MI1	E. N. Ashworth, M. N. Christiansen, J. B. St. John, and G. W. Patterson, *Plant Physiol.* **67**, 711 (1981).
81MI2	G. Boxhoorn, G. C. Schoemaker, D. J. Stufkens, and A. Oskam, *Inorg. Chim. Acta* **53**, L121 (1981).
81MI3	T. Braumann and L. H. Grimme, *J. Chromatogr.* **206**, 7 (1981).
81MI4	R. Buchman, *J. Labelled Compd. Radiopharm.* **18**, 1759 (1981).
81MI5	M. J. Calhorda and A. R. Dias, *Rev. Port. Quim.* **23**, 12 (1981).
81MI6	T. Eda, K. Arai, Y. Matsubara, M. Yoshihara, and I. Maeshima, *J. Macromol. Sci., Chem.* **A15**, 359 (1981) [*CA* **93**, 221153 (1980)].
81MI7	T. Eda, Y. Matsubara, M. Yoshihara, and T. Maeshima, *J. Macromol. Sci., Chem.* **A15**, 69 (1981) [*CA* **93**, 186875 (1980)].
81MI8	F. Fariña, M. V. Martin, and A. Tito, *An. Quim.* **77C**, 188 (1981).
81MI9	F. Fariña, M. V. Martin, and M. C. Paredes, *An. Quim.* **77C**, 213 (1981).
81MI10	J. Gieldanowski, W. Steuden, J. Skowronska, W. Gorczyca, and W. Doscocz, *Arch. Immunol. Ther. Exp.* **29**, 249 (1981).
81MI11	B. Hladon, P. Szafarek, S. Baloniak, and A Mroczkiewicz, *Pol. J. Pharmacol. Pharm.* **33**, 445 (1981).
81MI12	M. F. Ismail and N. A. Shams, *Egypt. J. Chem.* **24**, 365 (1981).
81MI13	M. F. Ismail, N. A. Shams, and O. M. El-Sawy, *Egypt. J. Chem.* **24**, 223 (1981) [*CA* **99**, 88135 (1983)].
81MI14	M. F. Ismail, N. A. Shams, and A. M. Kaddah, *Egypt. J. Chem.* **24**, 375 (1981) [*CA* **99**, 53532 (1983)].
81MI15	N. V. Karapetyan, M. G. Rakhimberdieva, E. Lehoczki, and A. A. Krasnovskii, *Biokhimiya (Moscow)* **46**, 2082 (1981).
81MI16	G. Kempter, D. Henning, G. Zeiger, and H. D. Beerbalk, *Wiss. Z. Paedagog. Hochsch. "Karl Liebknecht" Potsdam* **25**, 23 (1981).
81MI17	Y. J. Lee and Y. J. Park, *Chongi Hakhoe Chi* **25**, 219 (1981) [*CA* **96**, 51523 (1982)].
81MI18	P. Mátyus, G. Szilágyi, and E. Kasztreiner, *Magy. Kem. Foly.* **87**, 1 (1981).
81MI19	M. Morishita and M. Katayanagi, *Bunseki Kagaku* **30**, 460 (1981) [*CA* **95**, 107952 (1981)].
81MI20	A. Mroczkiewicz, *Acta Pol. Pharm.* **38**, 187 (1981).
81MI21	A. Mroczkiewicz, *Pol. J. Chem.* **55**, 2637 (1981).
81MI22	G. Nannini, E. Perrone, D. Severino, F. Casabuona, A. Bedeschi, F. Buzzetti, P. N. Giraldi, G. Meinardi, G. Monti, A. Ceriani, and I. De Carneri, *J. Antibiot.* **34**, 1456 (1981).
81MI23	K. Samula and M. Serwin-Krajewska, *Acta Pol. Pharm.* **38**, 33 (1981).
81MI24	B. Stanovnik, A. Hribar, and M. Tišler, *Vestn. Slov. Kem. Drus.* **28**, 35 (1981).
81MI25	M. Stogniew, L. A. Geelhaar, and P. S. Callery, *J. Labelled Compd. Radiopharm.* **18**, 897 (1981).

81MI26	N. Shimada, E. Morimoto, H. Naganawa, T. Takita, M. Hamada, K. Maeda, T. Tokeuchi, and H. Umezawa, *J. Antibiot.* **34**, 1613 (1981).
81MI27	K. Morimoto, N. Shimada, H. Naganawa, T. Takita, and H. Umezawa, *J. Antibiot.* **34**, 1615 (1981).
81PHA698	R. Bluth, *Pharmazie* **36**, 698 (1981).
81PHA775	R. Bluth, *Pharmazie* **36**, 775 (1981).
81PS323	R. Cremlyn, F. Swinbourne, S. Plant, D. Saunders, and C. Sinderson, *Phosphorus Sulfur* **10**, 323 (1981).
81RRC1285	A. Essawy, M. A. El-Hashash, M. M. Abdalla, and A. M. El-Gendy, *Rev. Roum. Chim.* **26**, 1285 (1981).
81S608	G. Gaviraghi, M. Pinza, and G. Piffer, *Synthesis*, 608 (1981).
81S631	T. Yamada and M. Oki, *Synthesis*, 631 (1981).
81SC655	H. P. Figeys, A. Marthy, and A. Dralants, *Synth. Commun.* **11**, 655 (1981).
81SC835	G. Szilágyi, E. Kasztreiner, P. Mátyus, and K. Csakó, *Synth. Commun.* **11**, 835 (1981).
81T1787	A. Tomažič, M. Tišler, and B. Stanovnik, *Tetrahedron* **37**, 1787(1981).
81TL2277	Y. Ogawa, S. Iwasaki, and S. Okuda, *Tetrahedron Lett.*, **22**, 2277 (1981).
81TL3637	Y. Ogawa, S. Iwasaki, and S. Okuda, *Tetrahedron Lett.*, **22**, 3637 (1981).
81UKZ657	N. V. Kuznetsov and V. E. Makarenko, *Ukr. Khim. Zh. (Russ. Ed.)* **47**, 657 (1981).
81ZOR1909	N. A. Bogoslovskii, G. E. Litvinova, I. A. Titova, G. I. Samokhvalov, V. G. Mairanovskii, V. M. Gurevich, and T. M. Filippova, *Zh. Org. Khim.* **17**, 1909 (1981).
82ABC865	T. Shiroza, N. Ebisawa, K. Furihata, T. Endo, H. Seto, and N. Otake, *Agric. Biol. Chem.* **46**, 865 (1982).
82ABC1885	T. Shiroza, N. Ebisawa, Kojima, K. Furihata, A. Shimazu, T. Endo, H. Seto, and N. Otake, *Agric. Biol. Chem.* **46**, 1885 (1982).
82ABC1891	T. Shiroza, N. Ebisawa, K. Furihata, T. Endo, H. Seto, and N. Otake, *Agric. Biol. Chem.* **46**, 1891 (1982).
82ACH(109)237	P. Mátyus, G. Szilágyi, E. Kasztreiner, M. Soti, and P. Sohár, *Acta Chim. Acad. Sci. Hung.* **109**, 237 (1982).
82ACH(111)387	H. M. Moussa, *Acta Chim. Acad. Sci. Hung.* **111**, 387 (1982).
82ACS(B)1	S. H. Andersen, N. B. Das, R. D. Jørgensen, G. Kjeldsen, J. S. Knudsen, S. C. Sharma, and K. B. G. Torssell, *Acta Chem. Scand., Ser. B* **B36**, 1 (1982).
82AP175	G. Heinisch, A. Mayrhofer, and R. Waglechner, *Arch. Pharm. (Weinheim, Ger.)* **315**, 175 (1982).
82AX(B)135	B. J. Graves and D. J. Hodgson, *Acta Crystallogr., Sect. B* **B38**, 135 (1982).
82BSB49	A. Michel, R. Gustin, G. Evrard, and F. Durant, *Bull. Soc. Chim. Belg.* **91**, 49 (1982).
82BSB123	A. Michel, R. Gustin, G. Evrard, and F. Durant, *Bull. Soc. Chim. Belg.* **91**, 123 (1982).
82BSB153	G. Maury, D. Meziane, D. Srairi, J. P. Paugan, and R. Paugam, *Bull. Soc. Chim. Belg.* **91**, 153 (1982).

82BSF(1)433	G. Rio and F. Pareze, *Bull. Soc. Chim. Fr.*, Part 1, 433 (1982).
82CB2574	L. Birkofer, E. Hänsel, and A. Steigel, *Chem. Ber.* **115**, 2574 (1982).
82CB2965	A. Heydt, H. Heydt, B. Weber, and M. Regitz, *Chem. Ber.* **115**, 2965 (1982).
82CC1003	A. E. Baydar, G. V. Boyd, and P. F. Lindley, *J. C. S. Chem. Commun.*, 1003 (1981)
82CC1412	R. D. Chambers, M. J. Silvester, M. Tamura, and D. E. Wood, *J. C. S. Chem. Commun.*, 1412 (1982).
82CJC2668	J. Bourguignon, S. Chapelle, P. Granger, and G. Quéguiner, *Can. J. Chem.* **60**, 2668 (1982).
82CPB1557	K. Satoh, T. Miyasaka, and K. Arakawa, *Chem. Pharm. Bull.* **30**, 1557 (1982).
82G249	S. Chimichi, R. Nesi, F. De Sio, R. Pepino, and A. Degl'Innocenti, *Gazz. Chim. Ital.* **112**, 249 (1982).
82H(18)175	F. Fariña, M. V. Martin, F. Sanchez, and A. Tito, *Heterocycles* **18**, 175 (1982).
82H(19)93	W. Paudler and M. V. Jovanovic, *Heterocycles* **19**, 93 (1982).
82H(19)1415	N, Suzuki, M. Kato, K. Sano, and Y. Izawa, *Heterocycles* **19**, 1415 (1982).
82H(19)1427	F. De Sio, S. Chimichi, R. Nesi, and L. Cecchi, *Heterocycles* **19**, 1427 (1982).
82H(19)2283	A. Katoh, C. Kashima, and Y. Omote, *Heterocycles* **19**, 2283 (1982).
82IJC(B)273	P. Waykole and R. N. Usgaonkar, *Indian J. Chem., Sect. B*, **21B**, 273 (1982).
82IJC(B)317	N. A. Shams, A. M. Koddah, and A. H. Moustafa, *Indian J. Chem., Sect. B* **21B**, 317 (1982).
82IJC(B)371	M. F. Ismail, N. A. Shams, and O. E. Mostafa, *Indian J. Chem., Sect. B* **21B**, 371 (1982).
82IJC(B)763	A. I. Hashem, M. E. Shaban, and A. F. El-Kafrawy, *Indian J. Chem., Sect. B* **21B**, 763 (1982).
82JA548	M. F. Ahern, A. Leopold, J. R. Beadle, and G. W. Gokel, *J. Am. Chem. Soc.* **104**, 548 (1982).
82JA7658	P. A. Lay, R. H. Magnuson, J. Sen, and H. Taube, *J. Am. Chem. Soc.* **104**, 7658 (1982).
82JCS(P1)989	M. H. Elnagdi, H. A. Elfahham, M. R. H. Elmoghayar, K. U. Sadek, and G. E. H. Elgemeie, *J. C. S. Perkin 1*, 989 (1982).
82JCS(P1)1845	H. C. McNab and I. Stobie, *J. C. S. Perkin 1*, 1845 (1982).
82JHC577	B. Stanovnik, A. Štimac, M. Tišler, and B. Verček, *J. Heterocycl. Chem.* **19**, 577 (1982).
82JHC1285	H. Hara and H. C. van der Plas, *J. Heterocycl. Chem.* **19**, 1285 (1982).
82JMC813	D. J. Katz, D. S. Wise, and L. B. Townsend, *J. Med. Chem.* **25**, 813 (1982).
82JMC975	T. Yamada, Y. Nobuhara, A. Yamaguchi, and M. Ohki, *J. Med. Chem.* **25**, 975 (1982).
82JOC398	M. A. Fox, D. M. Lemal, D. W. Johnson, and J. R. Hohman, *J. Org. Chem.* **47**, 398 (1982).
82JOC2768	E. Schweizer and K. J. Lee, *J. Org. Chem.* **47**, 2768 (1982).
82KGS1536	B. I. Buzykin, A. A. Maksimova, A. P. Stolyarov, S. S. Flegontov, and Yu. P. Kitaev, *Khim. Geterotsikl. Soedin.*, 1536 (1982).

82MI1	M. A. Acuna de Molina, M. N. Loncharich, J. L. Gimenez de Paez, and Y. P. W. Lobo, *An. Asoc. Quim. Argent.* **70**, 1043 (1982).	
82MI2	A. Aydin and H. Feuer, *Chim. Acta Turc.* **10**, 93 (1982).	
82MI3	S. Baloniak and E. Melzer, *Acta Pol. Pharm.* **39**, 71 (1982).	
82MI4	S. Baloniak, E. Melzer, and L. Zaprutko, *Acta Pol. Pharm.* **39**, 297 (1982).	
82MI5	S. Baloniak and I. Zyczynska-Baloniak, *Acta Pol. Pharm.* **39**, 65 (1982).	
82MI6	S. Baloniak, A. Mroczkiewicz, and A. Ostrowicz, *Acta Pol. Pharm.* **39**, 193 (1982).	
82MI7	A. Beno and V. Konecny, *Acta Fac. Rerum Nat. Univ. Comenianae, Chim.* **30**, 145 (1982) [*CA* **98**, 154731 (1983)].	
82MI8	M. Cechura, V. Bekarek, and J. Slouka, *Acta Univ. Palacki Olomuc, Fac. Rerum Nat.* **73**, 51 (1982).	
82MI9	Y. Gondo, Y. Gondo, and Y. Kanda, *Mem. Fac. Sci., Kyushu Univ., Ser. C* **13**, 249 (1982).	
82MI10	S. A. Gridchin, G. V. Shatalov, M. N. Masterova, B. I. Mikhantev, and V. P. Zubov, *Vysokomol. Soedin., Ser. A* **24**, 272 (1982) [*CA* **96**, 181663 (1982)].	
82MI11	G. Heinisch and A. Mayrhofer, *Sci. Pharm.* **50**, 120 (1982).	
82MI12	B. Hladon, W. Gutsche, W. Jungstand, and S. Baloniak, *Pol. J. Pharmacol. Pharm.* **34**, 207 (1982).	
82MI13	E. Lehoczki, M. G. Rakhimberdieva, and N. V. Karapetyan, *Fiziol. Rast. (Moscow)* **29**, 682 (1982) [*CA* **97**, 121976 (1982)].	
82MI14	J. Lehotay, E. Matisova, J. Garaj, and A. Violova, *J. Chromatogr.* **246**, 323 (1982).	
82MI15	M. G. Rakhimberdieva, E. Lehoczki, N. V. Karapetyan, and A. A. Krasnovskii, *Biokhimiya (Moscow)* **47**, 637 (1982).	
82MI16	M. S. De Giambiagi, M. Giambiagi, and A. F. Castro, *Lett. Nuovo Cimento Soc. Ital. Fis.* **33**, 327 (1982).	
82MI17	G. V. Shatalov, S. A. Gridchin, G. V. Kovalev, B. I. Mikhantev, and S. M. Gofman, *Izv. Vyssh. Uchebn. Zaved., Khim. Khim. Tekhnol.* **25**, 1179 (1982) [*CA* **98**, 89286 (1983)].	
82MI18	S. Spyroudis and A. Varvoglis, *Chem. Chron.* **11**, 173 (1982).	
82MI19	C. Willemot, C. R. Slack, J. Browse, and P. G. Roughan, *Plant Physiol.* **70**, 78 (1982).	
82MI20	K. Morimoto, N. Shimada, H. Naganawa, T. Takita, and H. Umezawa, *J. Antibiot.* **35**, 378 (1982).	
82OMR192	W. W. Paudler and M. V. Jovanovic, *Org. Magn. Reson.* **19**, 192 (1982).	
82PHA51	R. Bluth, *Pharmazie,* **37**, 51 (1982).	
82PHA136	R. Bluth, *Pharmazie,* **37**, 136 (1982).	
82PHA208	R. Bluth, *Pharmazie,* **37**, 208 (1982).	
82PHA285	R. Bluth, *Pharmazie,* **37**, 285 (1982).	
82PHA441	R. Bluth, *Pharmazie,* **37**, 441 (1982).	
82PHA502	R. Bluth, *Pharmazie,* **37**, 502(1982).	
82S490	S. M. Fahmy, N. M. Abed, R. M. Mohareb, and M. H. Elnagdi, *Synthesis,* 490 (1982).	
82S958	R. R. Schmidt and A. Wagner, *Synthesis,* 958 (1982).	
82TL3875	D. P. Gapinski and M. F. Ahern, *Tetrahedron Lett.,* 3875 (1982).	

82ZN(C)1092	G. Sandmann and P. Boger, *Z. Naturforsch., C: Biosci.* **37**, 1092 (1982).
82ZOR1557	N. F. Volynets, I. V. Samartseva, and L. A. Pavlova, *Zh. Org. Khim.* **18**, 1557 (1982).
82ZVK591	V. A. Kozlov, T.Yu. Dolnikova, V. I. Ivanchenko, V. V. Negrebetskii, A. F. Gropov, and N. N. Melnikov, *Zh. Vses. Khim. O-va.* **27**, 591 (1982) [*CA* **98**, 89288 (1983)].
83ACH43	J. Fetter, K. Lempert, J. Møller, J. Nyitrai, and K. Zauer, *Acta Chim. Hung.* **112**, 43 (1983).
83AG804	O. Schneider and M. Hanack, *Angew. Chem.* **95**, 804 (1983).
83AP508	G. Heinisch, I. Kirchner, I. Kurzmann, G. Lötsch, and R. Waglechner, *Arch. Pharm. (Weinheim, Ger.)* **316**, 508 (1983).
83AX(C)1415	C. Caristi, M. Gattuso, G. Stagno d'Alcontres, A. Ferlazzo, J. C. J. Bart, and C. Day, *Acta Crystallogr., Sect. C: Cryst. Struct. Commun.* **C39**, 1415 (1983).
83CZ172	G. Seitz, R. Dhar, and R. Mohr, *Chem.-Ztg.* **107**, 172 (1983).
83H551	L. Ceraulo, L. Lamartina, O. Migliara, S. Petruso, and V. Sprio, *Heterocycles* **20**, 551 (1983).
83H765	G. Szilágyi, M. Sóti, P. Mátyus, and E. Kasztreiner, *Heterocycles* **20**, 765 (1983).
83H1591	M. Tišler, *Heterocycles* **20**, 1591 (1983).
83H2385	C. B. Rao, P. V. N. Paju, R. Flammang, A. Maquestiau, and J. Elguero, *Heterocycles* **20**, 2385 (1983).
83IJC(B)512	K. D. Deodhar, D. S. Kanekar, and P. N. Pabrekar, *Indian J. Chem., Sect. B* **22B**, 512 (1983).
83IZV1687	O. P. Petrenko, V. V. Lapachev, and V. P. Mamaev, *Izv. Akad. Nauk SSSR, Ser. Khim.*, 1687 (1983).
83JCP4083	J. I. Selco, P. L. Holt, and R. B. Weisman, *J. Chem. Phys.* **79**, 4083 (1983).
83JCR(S)37	A. H. Moustafa, A. A. Shalaby, and R. A. Jones, *J. Chem. Res., Synop.*, 37 (1983).
83JCS(D)649	M. G. B. Drew, P. J. Varicelli, P. C. H. Mitchell, and A. R. Read, *J. C. S. Dalton*, 649 (1983).
83JCS(P1)1203	H. McNab, *J. C. S. Perkin 1*, 1203 (1983).
83JCS(P1)1601	D. Hunter and D. G. Neilson, *J. C. S. Perkin 1*, 1601 (1983).
83JCS(P1)1803	S. J. Clarke, D. E. Davie, and T. L. Gilchrist, *J. C. S. Perkin 1*, 1803 (1983).
83JCS(P2)1113	A. G. Rowley and J. R. F. Steedman, *J. C. S. Perkin 2*, 1113 (1983).
83JHC101	A. Turck, J. F. Brument, and G. Quéguiner, *J. Heterocycl. Chem.* **20**, 101 (1983).
83JHC193	R. E. Draper and R. N. Castle, *J. Heterocycl. Chem.* **20**, 193 (1983).
83JHC369	D. J. Katz, D. S. Wise, and L. B. Townsend, *J. Heterocycl. Chem.* **20**, 369 (1983).
83JHC855	J. Baker, W. Hedge, J. W. Timberlake, and L. M. Trefonas, *J. Heterocycl. Chem.* **20**, 855 (1983).
83JHC1259	A. Counotte-Potman and H. C. van der Plas, *J. Heterocycl. Chem.* **20**, 1259 (1983).
83JHC1473	I. Sircar, *J. Heterocycl. Chem.* **20**, 1473 (1983).
83JMC373	T. Yamada, Y. Nobuhara, H. Shimamura, Y. Tsukamoto, K. Yoshihara, A. Yamaguchi, and M. Ohki, *J. Med. Chem.* **26**, 373 (1983).

83JMC800	M. Thyes, H. D. Lehmann, J. Gries, H. König, R. Kretzschmar, J. Kunze, R. Lebkücher, and D. Lenke, *J. Med. Chem.* **26**, 800 (1983).
83JMC1144	T. Yamada, H. Shimamura, Y. Tsukamoto, A. Yamaguchi, and M. Ohki, *J. Med. Chem.* **26**, 1144 (1983).
83JOC2998	I. Maeba, K. Iwata, F. Usami, and H. Furukawa, *J. Org. Chem.* **48**, 2998 (1983).
83JOC3765	D. J. Katz, D. S. Wise, and L. B. Townsend, *J. Org. Chem.* **48**, 3765 (1983).
83MI1	N. H. Ayachit, K. S. Rao, and M. A. Shashidhar, *Indian J. Pure Appl. Phys.* **21**, 490 (1983).
83MI2	G. A. Bean, and A. Sonthall, *Appl. Environ. Microbiol.* **46**, 503 (1983).
83MI3	V. P. Beketov, A. K. Trukhmanov, L. S. Subbotina, V. V. Znamenskii, B. A. Titov, P. G. Zherebchenko, and V. P. Evdakov, *Radiobiologiya* **23**, 813 (1983).
83MI4	T. Crolla, L. Citerio, M. Visconti, and G. Pifferi, *HRC CC, J. High Resolut. Chromatogr. Chromatogr. Commun.* **6**, 445 (1983).
83MI5	A. O. Davies and J. L. Harwood, *J. Exp. Bot.* **34**, 1089 (1983) [*CA* **100**, 1795 (1984)].
83MI6	G. De Munno, G. Denti, and P. Dapporto, *Inorg. Chim. Acta* **74**, 199 (1983).
83MI7	A. Essawy, A. A. Hamed, M. El-Garby Younes, and I. A. G. El-Karim, *Egypt. J. Chem.* **46**, 213 (1983) [*CA* **102**, 5665 (1985)].
83MI8	C. W. Howden, H. L. Elliott, C. B. Lawrie, and J. L. Reid, *J. Cardiovasc. Pharmacol.* **5**, 552 (1983).
83MI9	D. C. Humber, P. S. Jones, G. H. Phillips, M. G. Dodds, and P. G. Dolamore, *Steroids* **42**, 171 (1983).
83MI10	D. C. Humber, G. H. Phillips, M. G. Dodds, P. G. Dolamore, and I. Machin, *Steroids* **42**, 189 (1983).
83MI11	Z. Husti, G. Szilágyi, and E. Kasztreiner, *Biochem. Pharmacol.* **32**, 627 (1983).
83MI12	T. Ito, A. Hosono, H. Tachibana, and K. Nitanai, *Nippon Nogei Kagaku Kaishi* **57**, 445 (1983) [*CA* **99**, 158352 (1983)].
83MI13	T. Ito, H. Yamazaki, T. Udagawa, and K. Nitanai, *Nippon Nogei Kagaku Kaishi* **57**, 743 (1983) [*CA* **100**, 6660 (1984)].
83MI14	T. Ito, S. Kajiya, S. Ogawa, and K. Natanai, *Nippon Nogei Kagaku Kaishi* **57**, 749 (1983) [*CA* **100**, 6661 (1984)].
83MI15	V. Konečny and Š. Kovač, *Chem. Zvesti* **37**, 827 (1983).
83MI16	J. Kosáry, G. Szilágyi, P. Mátyus, K. Csakó, and E. Kasztreiner, *Acta Pharm. Hung.* **53**, 106 (1983).
83MI17	H. H. Moussa, *Egypt. J. Chem.* **26**, 417 (1983) [*CA* **101**, 211704 (1984)].
83MI18	A. A. Nada and M. S. A. Abd El-Mottaleb, *Egypt. J. Chem.* **26**, 247 (1983) [*CA* **101**, 170531 (1984)].
83MI19	A. A. Nada, B. Haggag, S. Abdel-Halim, and Z. M. Rifae, *Egypt. J. Chem.* **26**, 371 (1983) [*CA* **101**, 230451 (1984)].
83MI20	O. P. Petrenko, V. V. Lapachev, and V. P. Mamaev, *Izv. Sib. Otd. Akad. Nauk SSSR, Ser. Khim. Nauk*, 87 (1983).
83MI21	A. A. Ramadan, M. H. Seada, and E. N. Rizkalla, *Talanta* **30**, 245 (1983).

83MI22	F. R. Rittig, J. B. St. John, R. Becker, H. Bleiholder, F. Feichtmayr, and E. Haedicke, *Pestic. Sci.* **14**, 299 (1983).
83MI23	M. Schadt, M. Petrzilka, P. R. Gerber, A. Villiger, and G. Trickes, *Mol. Cryst. Liq. Cryst.* **94**, 139 (1983).
83MI24	S. C. Shin and Y. Y. Lee, *Taehan Hwahakhoe Chi* **27**, 382 (1983) [*CA* **100**, 103276 (1984)].
83MI25	E. Sypniewska, A. Tippe, and Z. Eckstein, *Przem. Chem.* **62**, 334 (1983) [*CA* **100**, 6433 (1984)].
83MI26	T. Yamada, Y. Tsukamoto, H. Shimamura, S. Banno, and M. Sato, *Eur. J. Med. Chem.—Chim. Ther.* **18**, 209 (1983).
83NJC667	G. Maghioros, G. Schlewer, C. G. Wermuth, J. Lagrange, and P. Lagrange, *Nouv. J. Chim.* **7**, 667 (1983).
83OMR42	R. Nesi, S. Chimichi, F. De Sio, and M. Scotton, *Org. Magn. Reson.* **21**, 42 (1983).
83S312	T. Sakamoto, M. Shiraiwa, Y. Kondo, and H. Yamanaka, *Synthesis*, 312 (1983).
83SA(A)367	A. Lautié, J. Hervieu, and J. Belloc, *Spectrochim. Acta, Part A* **39A**, 367 (1983).
83T2295	C. Joliveau and C. G. Wermuth, *Tetrahedron* **39**, 2295 (1983).
83T2869	D. L. Boger, *Tetrahedron* **39**, 2869 (1983).
83TL489	E. Giraudi and P. Teisseire, *Tetrahedron Lett.*, **24**, 489 (1983).
83TL1285	C. Caristi, M. Gattuso, A. Ferlazzo, and G. Stagno d'Alcontres, *Tetrahedron Lett.*, **24**, 1285 (1983).
83UKZ1095	F. S. Babichev, Yu. M. Volovenko, and L. M. Pereshivana, *Ukr. Khim. Zh.* **49**, 1095 (1983).
83UKZ1197	F. S. Babichev, Yu. M. Volovenko, and L. M. Pereshivana, *Ukr. Khim. Zh.* **49**, 1197 (1983).
83ZC296	A. Isenberg and H. Zaschke, *Z. Chem.* **23**, 296 (1983).
83ZN(A)415	M. H. Palmer and I. Simpson, *Z. Naturforsch., A* **38**, 415 (1983).
83ZOR465	M. I. Terekhova, E. S. Petrov, O. P. Shkurko, M. A. Mikhaleva, V. P. Mamaev, and A. I. Shatenshtein, *Zh. Org. Khim.* **19**, 465 (1983).
84AKZ572	S. G. Konkova, A. A. Safaryan, and A. N. Akopyan, *Arm. Khim. Zh.* **37**, 572 (1984).
84AX(C)441	U. Engelhardt and B. Stromburg, *Acta Crystallogr., Sect. C: Cryst. Struct. Commun.* **C40**, 441 (1984).
84BSF(2)129	J. Mayrargue, J. L. Avril, and M. Miocque, *Bull. Soc. Chim. Fr.*, Part 2, 129 (1984).
84CB445	P. Eisenbarth and M. Regitz, *Chem. Ber.* **117**, 445 (1984).
84CB1523	H. Vorbrüggen and K. Krolikiewicz, *Chem. Ber.* **117**, 1523 (1984).
84CC295	A. Padwa and M. Tohidi, *J. C. S. Chem. Commun.* 295 (1984).
84CC422	R. E. Bambury, D. T. Feeley, G. C. Lawton, J. M. Weaver, and J. Wemple, *J. C. S. Chem. Commun.*, 422 (1984).
84CC1373	G. R. Brown, A. J. Foubister, and B. Wright, *J. C. S. Chem. Commun.*, 1373 (1984).
84CC1627	M. Satoh, K. Kaneto, and K. Yoshino, *J. C. S. Chem. Commun.*, 1627 (1984).
84CCC2541	H. Pischel, A. Holy, J. Vesely, and G. Wagner, *Collect. Czech. Chem. Commun.* **49**, 2541 (1984).
84CCC2620	A. Kurfürst and J. Kuthan, *Collect. Czech. Chem. Commun.* **49**, 2620 (1984).

84G289	E. M. Becalli, A. Marchesini, and L. Rusconi, *Gazz. Chim. Ital.* **114**, 289 (1984).
84G521	M. T. Cocco, A. Maccioni, and A. Plumitallo, *Gazz. Chim. Ital.* **114**, 521 (1984).
84H513	P. Mátyus, P. Sóhar, and H. Wamhoff, *Heterocycles* **22**, 513 (1984).
84H1395	G. Heinisch and G. Lötsch, *Heterocycles* **22**, 1395 (1984).
84H1545	B. Stanovnik, O. Bajt, B. Belčič, B. Koren, M. Prhavc, A. Štimac, and M. Tišler, *Heterocycles* **22**, 1545 (1984).
84H1801	B. Singh, *Heterocycles* **22**, 1801 (1984).
84H2483	D. Konwar, D. Prajapati, J. S. Sandhu, T. Kametani, and T. Honda, *Heterocycles* **22**, 2483 (1984).
84IJC(B)439	M. Ali and N. H. Khan, *Indian J. Chem., Sect. B* **23B**, 439 (1984).
84JA557	Z. W. Qui, D. M. Grant, and R. J. Pugmire, *J. Am. Chem. Soc.* **106**, 557 (1984).
84JA6552	J. Catalan, J. C. G. De Paz, M. Yañez, and J. Elguero, *J. Am. Chem. Soc.* **106**, 6552 (1984).
84JCS(P1)229	L. J. S. Knutsen, B. D. Judkins, W. L. Mitchell, R. F. Newton, and D. I. C. Scopes, *J. C. S. Perkin 1*, 229 (1984).
84JCS(P1)2491	S. Chimichi, R. Nesi, and M. Neri, *J. C. S. Perkin 1*, 2491 (1984).
84JCS(P2)1465	M. Kuzuya, F. Miyake, and T. Okuda, *J. C. S. Perkin 2*, 1465 (1984).
84JHC305	J. Taoufik, J. D. Couquelet, J. M. Couquelet, and A. Carpy, *J. Heterocycl. Chem.* **21**, 305 (1984).
84JHC435	G. Allmaier and G. Heinisch, *J. Heterocycl. Chem.* **21**, 435 (1984).
84JHC1297	J. Liang, *J. Heterocycl. Chem.* **21**, 1297 (1984).
84JHC1389	N. P. Peet, *J. Heterocycl. Chem.* **21**, 1389 (1984).
84JHC1727	G. Heinisch and R. Waglechner, *J. Heterocycl. Chem.* **21**, 1727 (1984).
84JMC1077	E. Bellasio, A. Campi, N. Di Mola, and E. Baldoli, *J. Med. Chem.* **27**, 1077 (1984).
84JMC1099	J. A. Bristol, I. Sircar, W. H. Moos, D. B. Evans, and R. E. Weishaar, *J. Med. Chem.* **27**, 1099 (1984).
84JMC1613	R. E. Bambury, D. T. Feeley, G. C. Lawton, J. M. Weaver, and J. Wemple, *J. Med. Chem.* **27**, 1613 (1984).
84JMR334	C. D. Buckley and K. A. McLauchlan, *J. Magn. Reson.* **58**, 334 (1984).
84JOC1261	J. A. Moore, O. S. Rothenberger, W. C. Fultz, and A. L. Rheingold, *J. Org. Chem.* **49**, 1261 (1984).
84JOC4405	D. L. Boger, R. S. Coleman, J. S. Panek, and D. Yohannes, *J. Org. Chem.* **49**, 4405 (1984).
84JOC4769	A. G. Anderson and R. P. Ko, *J. Org. Chem.* **49**, 4769 (1984).
84JPC2132	O. Kasende and T. Zeegers-Huyskens, *J. Phys. Chem.* **88**, 2132 (1984).
84JPR599	N. A. Shams, *J. Prakt. Chem.* **326**, 599 (1984).
84KGS832	O. P. Petrenko, S. F. Bychkov, V. V. Lapachev, and V. P. Mamaev, Khim. Geterotsikl. Soedin., 832 (1984).
84M1171	G. Heinisch and R. Waglechner, *Monatsh. Chem.* **115**, 1171 (1984).
84MI1	E. M. Abdel-Moety, F. H. Wahdan, N. A. Sharaby, and S. A. Ismiel, *Acta Pharm. Jugosl.* **34**, 223 (1984).
84MI2	A. J. St. Angelo, R. L. Ory, and F. R. Rittig, *J. Plant Growth Regul.* **3**, 183 (1984).

84MI3	C. D. Buckely, A. I. Grant, K. A. McLauchlan, and A. J. D. Ritchie, *Faraday Discuss. Chem. Soc.* **78,** 257 (1984).
84MI4	H. C. Chiang, K. M. Yao, and K. F. Huang, *Proc. Natl. Sci. Counc., Repub. China* **8,** Part A, 18 (1984) [*CA* **101,** 230371 (1984)].
84MI5	D. T. Clark and M. M. Abu-Shbak, *J. Polym. Sci., Polym. Chem. Ed.* **22,** 1 (1984) [*CA* **100,** 68863 (1984)].
84MI6	D. T. Clark, S. A. Johnson, and W. J. Brennan, *J. Polym. Sci., Polym. Chem. Ed.* **22,** 2145 (1984).
84MI7	P. Dapporto, G. De Munno, A. Sega, and C. Mealli, *Inorg. Chim. Acta* **83,** 171 (1984).
84MI8	M. Dorneanu, E. Carp, E. Stefanescu, and G. Grosu, *Rev. Med.-Chir.* **88,** 712 (1984) [*CA* **103,** 160463 (1985)].
84MI9	M. A. El-Hashash, S. I. El-Nagdy, and R. M. Saleh, *J. Chem. Eng. Data* **29,** 361 (1984).
84MI10	A. Gartner, B. Koren, B. Stanovnik, and M. Tišler, *Vestn. Slov. Kem. Drus.* **31,** 1 (1984).
84MI11	G. M. Gherasim and I. Zugravescu, *Rev. Chim. (Bucharest)* **35,** 24 (1984) [*CA* **101,** 7110 (1984)].
84MI12	G. M. Gherasim, S. Pecincu, and I. Zugravescu, *Rev. Chim. (Bucharest)* **35,** 212 (1984) [*CA* **101,** 722 (1984)].
84MI13	P. Honerjäger, A. Heiss, M. Schäfer-Korting, G. Schönsteiner, and M. Reiter, *Naunyn-Schmiedberg's Arch. Pharmacol.* **325,** 259 (1984).
84MI14	V. Konečny, Š. Kovač, and Š. Varkonda, *Chem. Zvesti* **38,** 239 (1984).
84MI15	G. Laskay, E. Lehoczki, M. Somogyi, and L. Szalay, *Adv. Photosynth. Res., Proc. Int. Congr. Photosynth. 6th, 1983,* Vol. 4, p. 49 (1984) [*CA* **101,** 34448 (1984)].
84MI16	F. J. Lopez Aparicio, M. Plaza Lopez-Espinoza, and R. Robles Diaz, *Carbohydr. Res.* **132,** 233 (1984).
84MI17	F. J. Lopez Aparicio, M. Plaza Lopez-Espinoza, and R. Robles Diaz, *An. Quim., Ser. C* **80,** 156 (1984).
84MI18	H. Mikashima, T. Nakao, K. Goto, H. Ochi, H. Yasuda, and T. Tsumagari, *Thromb. Res.* **35,** 589 (1984).
84MI19	A. Mroczkiewicz, *Acta Pol. Pharm.* **41,** 429 (1984).
84MI20	A. Rodriguez, M. Eschalier, P. Duchene-Marullaz, J. Taoufik, J. Couquelet, and P. Tronche, *IRCS Med. Sci.* **12,** 484 (1984).
84MI21	M. Satoh, K. Kaneto, and Y. Katsumi, *Jpn. J. Appl. Phys.* **23,** Part 2, 875 (1984) [*CA* **102,** 104216 (1985)].
84MI22	G. H. Sayed, A. A. Ismail, and Z. Hashem. *J. Chem. Soc. Pak.* **6,** 95 (1984).
84MI23	R. Schönbeck and E. Kloimstein, *Oesterr. Chem. Z.* **85,** 185 (1984).
84MI24	L. Skibinska, B. Hladon, T. Hermann, and S. Baloniak, *Farm. Pol.* **40,** 535 (1984).
84MI25	N. Suzuki, T. Ueyama, M. Kato, K. Sano, K. Iwasaki, and Y. Izawa, *Res. Rep. Fac. Eng., Mie Univ.* **9,** 37 (1984) [*CA* **103,** 53522 (1985)].
84MI26	G. Szilagyi, E. Kasztreiner, P. Mátyus, J. Kosáry, K. Czakó, G. Cseh, Z. Huszti, L. Tardos, E. Kósa, and L. Jaszlits, *Eur. J. Med. Chem.—Chim. Ther.* **19,** 111 (1984).

84MI27	J. Taoufik, J. D. Couquelet, J. Bastide, P. Bastide, and P. Champiat, *Ann. Pharm. Fr.* **42**, 135 (1984).
84MI28	G. Tian and C. Lin, *Gaodeng Xuexiao Huaxue Xuebao* **5**, 350 (1984) [*CA* **101**, 191820 (1984)].
84MI29	M. Ungureanu, M. Dorneanu, E. Stefanescu, M. Pavelescu, C. Radu, and M. Petrovanu, *Rev. Med.-Chir.* **88**, 709 (1984) [*CA* **103**, 171461 (1985)].
84MI30	R. Uson, L. A. Oro, M. Esteban, D. Carmona, R. M. Claramunt, and J. Elguero, *Polyhedron* **3**, 213 (1984).
84MI31	R. Zabska and T. Jakóbiec, *Arch. Immunol. Ther. Exp.* **32**, 255 (1984).
84MI32	R. Zabska and T. Jakóbiec, *Arch. Immunol. Ther. Exp.* **32**, 263 (1984).
84OMR201	L. Stefaniak, J. D. Roberts, M. Witanowski, and G. A. Webb, *Org. Magn. Reson.* **22**, 201 (1984).
84S62	K. Gewald and U. Hain, *Synthesis*, 62 (1984).
84S786	T. J. Reichelt and H. U. Reissig, *Synthesis*, 786 (1984).
84TL57	F. Bronberger and R. Huisgen, *Tetrahedron Lett.*, **26**, 57 (1984).
84TL1769	A. K. Forrest and R. R. Schmidt, *Tetrahedron Lett.*, **25**, 1769 (1984).
84TL2541	K. Müller and J. Sauer, *Tetrahedron Lett.*, **25**, 2541 (1984).
84TL3101	F. Sannicolo, *Tetrahedron Lett.*, **25**, 3101 (1984).
84ZOR416	R. G. Dubenko, A. I. Dychenko, P. S. Pelkis, and M. O. Lozinskii, *Zh. Org. Khim.* **20**, 416 (1984).
84ZOR1760	N. F. Volynets, I. V. Samartseva, I. V. Khramova, and L. A. Pavlova, *Zh. Org. Khim.* **20**, 1760 (1984).
85ACH177	K. Körmendy, Z. Soltesz, F. Ruff, and I. Kovesdi, *Acta Chim. Hung.* **120**, 177 (1985).
85ACH221	K. Czakó and G. Szilágyi, *Acta Chim. Hung.* **118**, 221 (1985).
85AF508	A. Assandri, E. Bellasio, A. Bernareggi, T. Cristina, A. Perazzi, and G. Odasso, *Arzneim.-Forsch.* **35**, 508 (1985).
85AF784	O. H. Hishmat, S. M. S. Atta, M. M. Atalla, and A. H. Abd El Rahman, *Arzneim.-Forsch.* **35**, 784 (1985).
85AG(E)692	G. Heinisch and G. Lötsch, *Angew. Chem., Int. Ed. Engl.* **24**, 692 (1985).
85AJC1685	C. B. Barlin and S. J. Ireland, *Aust. J. Chem.* **38**, 1685 (1985).
85AKZ720	L. A. Khachatryan, L. A. Saakyan, V. K. Ksipteridis, and M. T. Dangyan, *Arm. Khim. Zh.* **38**, 720 (1985) [*CA* **105**, 60574 (1986)].
85AKZ743	R. S. Vartanyan, L.O. Avetyan, S. A. Karamyan, and S. A. Vartanyan, *Arm. Khim. Zh.* **38**, 743 (1985) [*CA* **105**, 133826 (1986)].
85AX(C)1130	G. Maas, *Acta Crystallogr., Part C* **41**, 1130 (1985).
85AX(C)1277	P. F. Lindley, A. E. Baydar, and G. V. Boyd, *Acta Crystallogr., Sect C: Cryst. Struct. Commun.* **C41**, 1277 (1985).
85AX(C)1532	C. Van der Brempt, F. Durant, B. Norberg, and G. Evrard, *Acta Crystallogr., Sect C: Cryst. Struct. Commun.* **C41**, 1532 (1985).
85AX(C)1807	V. M. Lynch, S. H. Simonsen, M. J. Musmar, and G. E. Martin, *Acta Crystallogr., Sect C: Cryst. Struct. Commun.* **C41**, 1807, (1985).
85BSB261	C. Van der Brempt, G. Evrard, and F. Durant, *Bull. Soc. Chim. Belg.* **94**, 261 (1985).

85BSF865	G. Maghioros, G. Schlewer, and C. G. Wermuth, *Bull. Soc. Chim. Fr.*, 865 (1985).
85CB1709	H. J. Bestmann, G. Schmid, D. Sandmeier, G. Schade, and H. Oechsner, *Chem. Ber.* **118**, 1709 (1985).
85CC1632	N. L. Yang, S. S. Wang, C. J. Hou, L. Rodriguez, J. Jolson, and J. Waggoner, *J. C. S. Chem. Commun.*, 1632 (1985).
85CC1709	S. K. Mandal, L. K. Thompson, and A. W. Hanson, *J. C. S. Chem. Commun.*, 1709 (1985).
85CCC492	V. Konečny, Š. Kovač, and Š. Varkonda, *Collect. Czech. Chem. Commun.* **50**, 492 (1985).
85CI(L)697	B. Robinson and D. G. Hawkins, *Chem. Ind. (London)*, 697 (1985).
85CJC3037	J. T. Gupton, G. Hertel, G. DeCrescenzo, C. Colon, D. Baran, D. Dukesherer, S. Novick, D. Liotta, and J.P. Idoux, *Can. J. Chem.* **63**, 3037 (1985).
85CPB3540	J. Kurita, K. Takayama, and T. Tsuchiya, *Chem. Pharm. Bull.* **33**, 3540 (1985).
85CPL321	M. Terazima, S. Yamauchi, and N. Hirota, *Chem. Phys. Lett.* **120**, 321 (1985).
85FES200	A. Eschalier, M. Rodriguez, P. Duchene-Marullaz, J. Taoufik, J. Couquelet, and P. Tronche, *Farmaco, Ed. Sci.* **40**, 200 (1985).
85FES517	N. Di Mola and E. Bellasio, *Farmaco, Ed. Sci.* **40**, 517 (1985).
85FES921	N. A. Santagati, F. Duro, A. Caruso, S. Trombadore, and M. Amico-Roxas, *Farmaco, Ed. Sci.* **40**, 921 (1985).
85H683	A. T. M. Marcelis and H. C. van der Plas, *Heterocycles* **23**, 683 (1985).
85H1675	K. Kamata and O. Tsuge, *Heterocycles* **23**, 1675 (1985).
85H1999	G. E. H. Elgemeie, H. A. Elfahham, S. Elgamal, and M. H. Elnagdi, *Heterocycles* **23**, 1999 (1985).
85H2603	K. Makino and G. Sakata, *Heterocycles* **23**, 2603 (1985).
85H2651	N. Haider and G. Heinisch, *Heterocycles* **23**, 2651 (1985).
85IJC(B)785	K. R. Raju and P. S. Rao, *Indian J. Chem., Sect. B* **24B**, 785 (1985).
85JA3902	M. J. S. Dewar, E. G. Zoebisch, E. F. Healy, and J. J. P. Stewart, *J. Am. Chem. Soc.* **107**, 3902 (1985).
85JCP5892	Z. Gama, *J. Chem. Phys.* **83**, 5892 (1985).
85JCR(S)310	S. J. Clarke and T. L. Gilchrist, *J. Chem. Res., Synop.*, 310 (1985).
85JCS(P1)1627	S. P. Breukelman, G. D. Meakins, and A. M. Roe, *J. C. S. Perkin 1*, 1627 (1985).
85JCS(P2)417	J. Spanget-Larsen, *J. C. S. Perkin 2*, 417 (1985).
85JCS(P2)1533	Y. Kuroda, Y. Fujiwara, and K. Matsushita, *J. C. S. Perkin 2*, 1533 (1985).
85JHC719	J. P. Monthéard and J. C. Dubois, *J. Heterocycl. Chem.* **22**, 719, (1985).
85JHC927	C. Kashima, A. Katoh, M. Fukusawa, and Y. Omote, *J. Heterocycl. Chem.* **22**, 927 (1985).
85JHC981	D. W. Boykin, P. Balakrishnan, and A. L. Baumstark, *J. Heterocycl. Chem.* **22**, 981 (1985).
85JHC1543	W. Dostal and G. Heinisch, *J. Heterocycl. Chem.* **22**, 1543 (1985).
85JHC1615	J. Taoufik, J. M. Couquelet, J. D. Couquelet, and P. Tronche, *J. Heterocycl. Chem.* **22**, 1615 (1985).
85JMC1405	I. Sircar, B. L. Duell, G. Bobowski, J. A. Bristol, and D. B. Evans, *J. Med. Chem.* **28**, 1405 (1985).

85JOC4749	S. F. Nelsen and N. P. Yumibe, *J. Org. Chem.* **50**, 4749 (1985).	
85JPR536	N. A. Shams, *J. Prakt. Chem.* **327**, 536 (1985).	
85LA167	N. Haider, G. Heinisch, I. Kurzmann-Rauscher, and M. Wolf, *Liebigs Ann. Chem.*, 167 (1985).	
85LA628	H. Gnichtel and C. Gumprecht, *Liebigs Ann. Chem.*, 628 (1985).	
85LA853	H. Neunhoeffer and G. Werner, *Liebigs Ann. Chem.*, 853 (1985).	
85LA1465	M. C. Gomez-Gil, V. Gomez-Parra, F. Sanchez, and T. Torres, *Liebigs Ann. Chem.*, 1465 (1985).	
85LA1492	S. M. Fahmy and R. M. Mohareb, *Liebigs Ann. Chem.*, 1492 (1985).	
85M463	A. A. T. Ramadan, M. H. Seada, and E. N. Rizkalla, *Monatsh. Chem.* **116**, 463 (1985).	
85M1447	M. Merslavič, B. Stanovnik, and M. Tišler, *Monatsh. Chem.* **116**, 1447 (1985).	
85MI1	A. O. Abdelhamid and N. M. Abed, *Rev. Port. Quim.* **27**, 500 (1985).	
85MI2	G. Bertrand and E. Leleu, *Rubber World* **192**, 32 (1985).	
85MI3	J. P. Chambon, J. Brochard, A. Hallot, M. Heaulme, R. Brodin, R. Roncucci, and K. Bizière, *J. Pharmacol. Exp. Ther.* **233**, 836 (1985).	
85MI4	N. Di Mola, E. Bellasio, P. Ferrari, and L. F. Zerilli, *J. Labelled Compd. Radiopharm.* **22**, 1299 (1985).	
85MI5	A. I. Dychenko, P. S. Pelkis, R. G. Dubenko, E. F. Gorbenko, M. O. Lozinskii, E. F. Granin, and L. P. Charuiskaya, *Fiziol. Akt. Veschestva* **17**, 53 (1985) [*CA* **104**, 64061 (1986)].	
85MI6	M. A. El-Hashash, M. M. Mohamed, M. A. Sayed, and O. A. Abo-Baker, *Pak. J. Sci. Ind. Res.* **28**, 229 (1985) [*CA* **106**, 18468 (1987)].	
85MI7	G. Garcia, E. Raviña, L. Santana, F. Orallo, and J. M. Calleja, *An. R. Acad. Farm.* **51**, 23 (1985) [*CA* **103**, 171690 (1985)].	
85MI8	G. Garcia Mera, E. Raviña, L. Santana, F. Orallo, J. A. Fontenla, and J. M. Calleja, *An. Quim. Ser. C* **81**, 110 (1985).	
85MI9	M. Ghedini, F. Neve, F. Morazzoni, and C. Oliva, *Polyhedron* **4**, 497 (1985).	
85MI10	Z. P. Gribova, A. N. Tikhonov, L. A. Postnov, I. K. Irbe, and N. N. Zoz, *Biofizika* **30**, 1035 (1985).	
85MI11	S. P. Gupta, D. G. Shewada, C. Garg, A. Handa, and Y. S. Prabhakar, *Indian J. Biochem. Biophys.* **22**, 122 (1985) [*CA* **103**, 115689 (1985)].	
85MI12	A. M. Huralikoppi, N. H. Ayachit, and M. A. Shashidhar, *Indian J. Pure Appl. Phys.* **23**, 53 (1985).	
85MI13	M. F. Ismail, N. A. Shams, A. M. A. Elkhamry, and O. E. A. Moustafa, *J. Chem. Soc. Pak.* **7**, 41 (1985).	
85MI14	M. F. Ismail, N. A. Shams, A. M. A. Elkhamry, and O. E. A. Moustafa, *Egypt. J. Chem.* **28**, 77 (1985).	
85MI15	V. Konečny, Š. Kovač, and Š. Varkonda, *Chem. Zvesti* **39**, 513 (1985) [*CA* **104**, 207222 (1986)].	
85MI16	R. M. Leech, C. A. Walton, and N. R. Baker, *Planta* **165**, 277 (1985).	
85MI17	D. R. Martin, C. M. Merkel, J. U. Mondal, and C. R. Rushing, *Inorg. Chim. Acta* **99**, 81 (1985).	
85MI18	A. Nada, M. F. Zayed, and W. Gad, *Egypt. J. Chem.* **28**, 505 (1985) [*CA* **108**, 112365 (1988)].	

85MI19	A. Nada, M. F. Zayed, and Z. Refae, *Egypt. J. Chem.* **28**, 439 (1985) [*CA* **108**, 131725 (1988)].
85MI20	F. Orallo, J. M. Calleja, G. Garcia Mera, and E. Raviña, *Arch. Farmacol. Toxicol.* **11**, 171 (1985).
85MI21	T. Otto and K. Galewicz, *Acta Pol. Pharm.* **42**, 331 (1985).
85MI22	N. S. Rao, M. M. Das, G. B. Rao, and D. Ziessow, *Spectrosc. Lett.* **18**, 827 (1985).
85MI23	E. Raviña, G. Garcia Mera, L. Santana, F. Orallo, and J. M. Calleja, *Eur. J. Med. Chem.—Chim. Ther.* **20**, 475 (1985).
85MI24	V. Santucci, M. Fournier, J. P. Chambon, and K. Bizière, *Eur. J. Pharmacol.* **114**, 219 (1985).
85MI25	H. Satoh, M. Tonegawa, and R. Inoue, *Tokyo Ika Daigaku Kiyo* **11**, 1 (1985) [*CA* **103**, 104905 (1985)].
85MI26	D. Saunders and B. H. Warrington, *J. Labelled Compd. Radiopharm.* **22**, 869 (1985).
85MI27	G. H. Sayed, M. Y. El-Kady, I. Abd-Elmawgoud, and M. Hamdy, *J. Chem. Soc. Pak.* **7**, 263 (1985).
85MI28	N. A. Shams, E. A. Soliman, A. A. Hamed, and S. A. Donia, *Egypt. J. Chem.* **28**, 473 (1985) [*CA* **108**, 112364 (1988)].
85MI29	E. A. Soliman, M. A. I. Salem, and F. A. El-Shahed, *Egypt. J. Chem.* **28**, 389 (1985) [*CA* **108**, 130989 (1988)].
85MI30	G. H. Tamam, A. A. Hamed, M. El-Mobyed, and A. Y. Mohamed, *Egypt. J. Chem.* **28**, 331 (1985) [*CA* **108**, 94047 (1988)].
85MI31	G. H. Tamam, E. A. Soliman, A. M. El-Gendy, and A. Y. El-Kady, *Egypt. J. Chem.* **28**, 341 (1985) [*CA* **108**, 131724 (1988)].
85MI32	M. Tišler, *J. Serb. Chem. Soc.* **50**, 161 (1985).
85MI33	C. G. Wermuth, *Actual. Chim. Ther.* **12**, 3 (1985).
85MI34	C. G. Wermuth and D. Rognan, *Actual. Chim. Ther.* **12**, 215 (1985).
85MI35	C. H. Huang and S. T. Crooke, *Cancer Res.* **45**, 3768 (1985).
85MP1	C. D. Buckley and K. A. McLauchlan, *Mol. Phys.* **54**, 1 (1985).
85OMS483	C. Kascheres, A. Kascheres, and R. A. Pilli, *Org. Mass Spectrom.* **20**, 483 (1985).
85OPP107	N. M. Abed, N. S. Ibrahim, S. M. Fahmy, and M. H. Elnagdi, *Org. Prep. Proced. Int.* **17**, 107 (1985).
85P1923	D. J. Murphy, J. L. Harwood, K. A. Lee, F. Roberto, P. K. Stumpf, and J. B. St.John, *Phytochemistry* **24**, 1923 (1985).
85PNA1832	J. P. Chambon, P. Feltz, M. Heaulme, S. Restle, T. Schlichter, K. Bizière, and C. G. Wermuth, *Proc. Natl. Acad. Sci. U.S.A.* **82**, 1832 (1985).
85S1135	S. M. Fahmy and R. M. Mohareb, *Synthesis,* 1135 (1985).
85SA(A)703	M. Asai, K. Noda, and A. Sado, *Spectrochim. Acta, Part A* **41A**, 703 (1985).
85SC939	F. Barba, M. D. Velasco, A. Guirado, and N. Moreno, *Synth. Commun.* **15**, 939 (1985).
85T1199	G. Heinisch and G. Lötsch, *Tetrahedron* **41**, 1199 (1985).
85TL655	T. Harada, E. Akiba, and A. Oku, *Tetrahedron Lett.*, **26**, 655 (1985).
85TL4355	G. Seitz, L. Görge, and S. Dietrich, *Tetrahedron Lett.*, **26**, 4355 (1985).
85ZN(B)664	R. M. Mohareb and S. M. Fahmy, *Z. Naturforsch., B: Anorg. Chem., Org. Chem.* **40**, 664 (1985).

85ZOR1026	M. I. Komendanov, I. B. Logosh, and I. N. Domnin, *Zh. Org. Khim.* **21,** 1026 (1985).
85ZOR2445	V. Ya. Komarov, N. F. Volynets, I. V. Samartseva, L. A. Pavlova, and B. I. Ionin, *Zh. Org. Khim.* **21,** 2445 (1985).
86AF84	G. Di Francesco, E. Baldoli, G. Marchetti, and A. Glasser, *Arzneim.-Forsch.* **36,** 84 (1986).
86AP60	M. C. Gomez-Gil, V. Gomez-Parra, F. Sanchez, and T. Torres, *Arch. Pharm. (Weinheim, Ger.)* **319,** 60 (1986).
86AX(C)214	C. Van der Brempt, F. Durant, and G. Evrard, *Acta Crystallogr., Sect. C: Cryst. Struct. Commun.* **C42,** 214 (1986).
86AX(C)1206	C. Van der Brempt, G. Evrard, and F. Durant, *Acta Crystallogr., Sect. C: Cryst. Struct. Commun.* **C42,** 1206 (1986).
86BCJ2997	Y. Ozono, Y. Nibu, H. Shimada, and R. Shimada, *Bull. Chem. Soc. Jpn.* **59,** 2997 (1986).
86CPB3061	E. Sato, M. Hasebe, and Y. Kanaoka, *Chem. Pharm. Bull.* **34,** 3061 (1986).
86H101	A. O. Abdelhamid and N. M. Bed, *Heterocycles* **24,** 101 (1986).
86H793	W. Dostal and G. Heinisch, *Heterocycles* **24,** 793 (1986).
86H1219	N. S. Ibrahim, F. M. Abdel Galil, R. M. Abdel-Motaleb, and M. H. Elnagdi, *Heterocycles* **24,** 1219 (1986).
86H3059	K. Kamata and O. Tsuge, *Heterocycles* **24,** 3059(1986).
86IJC(A)707	M. Haridas and T. P. Singh, *Indian J. Chem., Sect. A* **25A,** 707 (1986).
86JA3237	R. W. Taft, F. Anvia, M. Taagepere, J. Catalán, and J. Elguero, *J. Am. Chem. Soc.* **108,** 3237 (1986).
86JPC1996	P. L. Holt, J. I. Selco, and R. B. Weisman, *J. Chem. Phys.* **84,** 1996 (1986).
86JCR(S)214	D. Carmona, M. Esteban, L. A. Oro, and M. P. Puebla, *J. Chem. Res., Synop.,* 214 (1986).
86JCS(D)625	L. Rosenberg, L. K. Thompson, E. J. Gabe, and F. L. Lee, *J. C. S. Dalton,* 625 (1986).
86JCS(D)2381	T. C. Woon, R. McDonald, S. K. Mandal, L. K. Thompson, S. P. Connors, and A. W. Addison, *J. C. S. Dalton,* 2381 (1986).
86JCS(P1)169	N. Haider and G. Heinisch, *J. C. S. Perkin 1,* 169 (1986).
86JCS(P2)359	F. Billes and A. Toth, *J. C. S. Perkin 2,* 359 (1986).
86JCS(P2)1255	N. Al-Awadi and R. Taylor, *J. C. S. Perkin 2,* 1255 (1986).
86JHC93	G. Heinisch, W. Holzer, and G. A. M. Nawwar, *J. Heterocycl. Chem.* **23,** 93 (1986).
86JHC193	N. P. Peet, *J. Heterocycl. Chem.* **23,** 193 (1986).
86JHC385	H. Heydt, P. Eisenbarth, K. Feith, and M. Regitz, *J. Heterocycl. Chem.* **23,** 385 (1986).
86JHC621	H. Tondys and H. C. van der Plas, *J. Heterocycl. Chem.* **23,** 621 (1986).
86JHC1515	E. W. Badger and W. H. Moos, *J. Heterocycl. Chem.* **23,** 1515 (1986).
86JMC261	I. Sircar, G. Bobowski, J. A. Bristol, R. E. Weishaar, and D. B. Evans, *J. Med. Chem.* **29,** 261 (1986).
86JMC369	A. Hallot, R. Brodin, J. Merlier, J. Brochard, J. P. Chambon, and K. Bizière, *J. Med. Chem.* **29,** 369 (1986).
86JMC1832	D. W. Robertson, J. H. Krushinski, E. E. Beedle, V. Wyss, G. Don

	Pollock, H. Wilson, R. F. Kauffman, and J. Scott Hayes, *J. Med. Chem.* **29,** 1832 (1986).
86JMC2142	I. Sircar, B. L. Duell, H. H. Cain, S. E. Burke, and J. A. Bristol, *J. Med. Chem.* **29,** 2142 (1986).
86JOC950	L. Maggiora and M. P. Mertes, *J. Org. Chem.* **51,** 950 (1986).
86JOM207	M. J. Bermejo, B. Martinez, and J. Vinaixa, *J. Organomet. Chem.* **304,** 207 (1986).
86JPR522	N. A. Shams, A. F. El-Kafrawy, O. M. El-Sawy, and M. F. Ismail, *J. Prakt. Chem.* **328,** 522 (1986).
86JPR932	M. M. Abbasi, M. Abou-Sekkina, Y. Hafez, and H. H. Zoorob, *J. Prakt. Chem.* **328,** 932 (1986).
86JST33	H. Fernandez, J. J. Lopez, A. Cardenete, and J. F. Arenas, *J. Mol. Struct.* **142,** 33 (1986).
86JST59	F. Billes and M. Gal, *J. Mol. Struct.* **142,** 59 (1986).
86KGS951	O. P. Shkurko, S. A. Kuznecov, A. Yu. Denisov, and V. P. Mamaev, *Khim. Geterotsikl. Soedin.*, 951 (1986).
86M221	M. Merslavič, B. Stanovnik, and M. Tišler, *Monatsh. Chem.* **117,** 221 (1986).
86M231	Y. Akcamur, G. Penn, E. Ziegler, H. Sterk, E. M. Peters, and H. G. Schnering, *Monatsh. Chem.* **117,** 231 (1986).
86MI1	Y. Cui, C. Lin, and G. Tian, *Gaoden Xuexiao Huaxue Xuebao* **7,** 417 (1986) [*CA* **107,** 58959 (1986)].
86MI2	S. A. Gridchin, G. V. Shatalov, M. B. Lachinov, and V. P. Zubov, *Vysokomol. Soedin.*, *Ser. B* **28,** 897 (1986) [*CA* **106,** 120266 (1987)].
86MI3	S. A. Gridchin, M. B. Lachinov, A. V. Kisin, G. V. Shatalova, and V. P. Zubov, *Vysokomol. Soedin.*, *Ser. A* **28,** 2191 (1986) [*CA* **106,** 5490 (1987)].
86MI4	M. Heaulme, J. P. Chambon, R. Leyris, C. G. Wermuth, K. Bizière, *Neuropharmacology* **25,** 1279 (1986).
86MI5	M. Heaulme, J. P. Chambon, R. Leyris, J. C. Molimard, C. G. Wermuth, and K. Bizière, *Brain Res.* **384,** 224 (1986).
86MI6	G. Heinisch, A. Jentzsch, I. Kirchner, and G. Lötsch, *Vestn. Slov. Kem. Drus.* **33,** 197 (1986).
86MI7	H. Heydt, P. Eisenbarth, K. Feith, K. Urgast, G. Maas, and M. Regitz, *Isr. J. Chem.* **27,** 96 (1986).
86MI8	A. V. Kiselev and L. A. Dementeva, *Zh. Fiz. Khim.* **60,** 1951 (1986).
86MI9	J. Kulig, B. Lenarcik, and M. Rzepska, *Pol. J. Chem.* **60,** 715 (1986).
86MI10	G. Laskay and E. Lehoczki, *J. Exp. Bot.* **37,** 1558 (1986).
86MI11	J. C. Liang and J. O. Cross, *Mol. Cryst. Liq. Cryst.* **141,** 25 (1986).
86MI12	J. C. Liang and J. O. Cross, *Mol. Cryst. Liq. Cryst.* **132,** 123 (1986).
86MI13	J. C. Liang and J. O. Cross, *Mol. Cryst. Liq. Cryst.* **133,** 235 (1986).
86MI14	F. J. Lopez Aparicio, R. Robles Diaz, M. T. Plaza Lopez-Espinoza, and A. M. Perez Rogjas, *An. Quim., Ser. C* **82,** 179 (1986).
86MI15	S. K. Mandal, L. K. Thompson, M. J. Newlands, F. L. Lee, Y. Lepage, J. P. Charland, and E. J. Gabe, *Inorg. Chim. Acta* **122,** 199 (1986).
86MI16	R. M. Mannan and S. Bose, *J. Biosci.* **10,** 283 (1986).
86MI17	M. K. May and A. E. Lanzilotti, *J. Labelled Compd. Radiopharm.* **21,** 489 (1986).

86MI18	I. Nabih, A. A. Zayed, J. Metri, M. M. Kamel, and M. S. Motawie, *Egypt. J. Chem.* **29**, 101 (1986) [*CA* **108**, 186703 (1988)].
86MI19	S. Papadopoulos and J. Stephanidou-Stephanatou, *Chem. Chron.* **15**, 49 (1986).
86MI20	A. Perio, J. P. Chambon, R. Calassi, M. Heaulme, and K. Bizière, *J. Pharmacol. Exp. Ther.* **239**, 542 (1986).
86MI21	F. R. Rittig, R. L. Ory, and J. B. St. John, *Dev. Food Sci.* **12**, 609 (1986).
86MI22	D. W. Robertson, J. H. Krushinski, and D. Kau, *J. Labelled Compd. Radiopharm.* **23**, 343 (1986).
86MI23	H. Satoh, R. Inoue, M. Tonegawa, Y. Nishimura, K. Kitahara, and C. Fukusawa, *Tokyo Ika Daigaku Kiyo* **12**, 1 (1986) [*CA* **106**, 102194 (1987)].
86MI24	G. H. Sayed, A. A. Ismail, and Z. Hashem, *Egypt. J. Chem.* **27**, 757 (1984) [*CA* **105**, 226436 (1986)].
86MI25	M. Terazima, S. Yamauchi, N. Hirota, O. Kitao, and H. Nakatsuji, *Chem. Phys.* **107**, 81 (1986).
86MI26	H. C. van der Plas and M. Wozniak, *Croat. Chem. Acta* **59**, 33 (1986).
86MI27	C. G. Wermuth and K. Bizière, *Trends Pharmacol. Sci.* **7**, 421 (1986).
86MI28	Y. J. Yun, *Hwahak Kwa Kongop Ui Chinbo* **26**, 670 (1986) [*CA* **107**, 59332 (1987)].
86PHA460	O. H. Hishmat, S. M. S. Atta, M. M. Atalla, and A. H. Abd El Rahman, *Pharmazie* **40**, 460 (1985).
86RRC387	A. M. A. El-Khamry, M. E. Shaban, and M. M. Habashy, *Rev. Roum. Chim.* **31**, 387 (1986) [*CA* **106**, 119812 (1987)].
86S807	B. Stanovnik, V. Stibilj, and M. Tišler, *Synthesis*, 807 (1986).
86SC543	T. N. Srinivasan, K. R. Rao, and P. B. Sattur, *Synth. Commun.* **16**, 543 (1986).
86TL2747	G. Seitz, S. Dietrich, L. Görge, and J. Richter, *Tetrahedron Lett.*, **27**, 2747 (1986).
86UKZ1087	I. P. Kupchevskaya and A. V. Stetsenko, *Ukr. Khim. Zh. (Russ. Ed.)* **52**, 1087 (1986).
86ZC21	M. P. Burrow, G. W. Gray, D. Lacey, and K. J. Toyne, *Z. Chem.* **26**, 21 (1986).
86ZC438	B. Roloff, W. Jugelt, and W. Duczek, *Z. Chem.* **26**, 438 (1986).
86ZN(B)105	R. M. Mohareb and S. M. Fahmy, *Z. Naturforsch., B* **41**, 105 (1986).
86ZN(C)585	G. Laskay, E. Lehoczki, A. L. Dobi, and L. Szalay, *Z. Naturforsch., C: Biosci.* **41**, 585 (1986).
86ZOR711	V. V. Shevchenko, G. A. Vasilevskaya, N. S. Klimenko, S. N. Lashkareva, and T. S. Khramova, *Zh. Org. Khim.* **22**, 711 (1986).
87AF398	A. Narimatsu, Y. Kitada, N. Satoh, R. Suzuki, and H. Okushima, *Arzneim.-Forsch.* **37**, 398 (1987).
87AP1222	G. Heinisch and D. Lassnig, *Arch. Pharm. (Weinheim, Ger.)* **320**, 1222 (1987).
87BCJ3645	K. Matsumoto, T. Uchida, Y. Ikemi, T. Tanaka, M. Asahi, T. Kato, and H. Konishi, *Bull. Chem. Soc. Jpn.* **60**, 3645 (1987).
87BCJ4486	N. S. Ibrahim. F. M. Abdel-Galil, R. M. Abdel-Motaleb, and M. H. Elnagdi, *Bull. Chem. Soc. Jpn.* **60**, 4486 (1987).

87C125	U. Pinder and L. Pfeuffer, *Chimia* **41**, 125 (1987).
87CB1597	W. Ried and U. Reiher, *Chem. Ber.* **120**, 1597 (1987).
87CCC696	B. Vrábel, F. Pavelčik, E. Hellö, S. Miertuš, V. Konečny, and J. Lokaj, *Collect. Czech. Chem. Commun.* **52**, 696 (1987).
87Cl(L)694	G. L. Goe, C. A. Huss, J. G. Keay, and E. F. V. Scriven, *Chem. Ind. (London),* 694 (1987).
87CPB350	H. Nagashima, K. Ukai, H. Oda, Y. Masaki, and K. Kaji, *Chem. Pharm. Bull.* **35**, 350 (1987).
87CPB421	H. Nagashima, H. Oda, J. I. Hida, and K. Kaji, *Chem. Pharm. Bull.* **35**, 421 (1987).
87CZ16	G. Seitz and L. Görge, *Chem. Ztg.* **111**, 16 (1987).
87CZ81	G. Seitz and R. Mohr, *Chem.-Ztg.* **111**, 81 (1987).
87FES585	D. Barlocco, A. Martini, G. A. Pinna, M. M. Curzu, F. Sala, and M. Germini, *Farmaco, Ed. Sci.* **42**, 585 (1987).
87H1	H. Nagashimna, H. Oda, T. Hayakawa, and K. Kaji, *Heterocycles* **26**, 1 (1987).
87H337	D. R. Borthakur, D. Prajapati, and J. S. Sandhu, *Heterocycles* **26**, 337 (1987).
87H481	G. Heinisch, *Heterocycles* **26**, 481 (1987).
87H689	B. Koren, B. Stanovnik, and M. Tišler, *Heterocycles* **26**, 689 (1987).
87H731	G. Heinisch and G. Lötsch, *Heterocycles* **26**, 731 (1987).
87H899	M. H. Mohamed, N. S. Ibrahim, and M. H. Elnagdi, *Heterocycles* **26**, 899(1987).
87H2101	E. S. H. El-Ashry, Y. El-Kilany, and A. Amer, *Heterocycles* **26**, 2101 (1987).
87H2677	A. Ohsawa, T. Itoh, and H. Igeta, *Heterocycles* **26**, 2677 (1987).
87IC2384	S. K. Mandal, L. K. Thompson, E. J. Gabe, F. C. Lee, and J. P. Charland, *Inorg. Chem.* **26**, 2384 (1987).
87IJC(B)348	S. N. Sawhney, S. Bhutani, and D. Vir. *Indian J. Chem., Sect. B* **26B**, 348 (1987).
87JA2717	D. L. Boger and R. S. Coleman, *J. Am. Chem. Soc.* **109**, 2717 (1987).
87JA5316	S. Alvarez, M. J. Bermejo, and J. Vinaixa, *J. Am. Chem. Soc.* **109**, 5316 (1987).
87JCP6707	J. Wanna and E. R. Bernstein, *J. Chem. Phys.* **86**, 6707 (1987).
87JCR(S)170	S. Papadopoulos and J. Stephanidou-Stephanatou, *J. Chem. Res., Synop.,* 170 (1987).
87JHC23	M. S. Akhtar, V. L. Sharma, and A. P. Bhaduri, *J. Heterocycl. Chem.* **24**, 23 (1987).
87JHC63	E. S. H. El-Ashry, A. Amer, G. H. Labib, M. M. A. Rahman, and A. M. El-Massry, *J. Heterocycl. Chem.* **24**, 63 (1987).
87JHC1285	S. Sueur, M. Lagrenee, F. Abraham, and C. Bremard, *J. Heterocycl. Chem.* **24**, 1285 (1987).
87JHC1745	H. M. F. Allah and R. Soliman, *J. Heterocycl. Chem.* **24**, 1745 (1987).
87JMC239	C. G. Wermuth, J. J. Bourguignon, G. Schlewer, J. P. Giess, A. Schönfelder, A. Melikian, M. J. Bouchet, D. Chanteux, J. C. Molimard, M. Heaulme, J. P. Chambon, and K. Bizière, *J. Med. Chem.* **30**, 239 (1987).
87JMC623	D. W. Robertson, N. D. Jones, J. H. Krushinski, G. Don Pollock,

	J. K. Schwarzendruber, and J. Scott Hayes, *J. Med. Chem.* **30**, 623 (1987).
87JMC824	D. W. Robertson, J. H. Krushinski, G. Don Pollock, H. Wilson, R. F. Kauffman, and J. Scott Hayes, *J. Med. Chem.* **30**, 824 (1987).
87JMC1157	H. Okushima, A. Narimatsu, M. Kobayashi, R. Furuya, K. Tsuda, and Y. Kitada, *J. Med. Chem.* **30**, 1157 (1987).
87JMC1955	I. Sircar, R. E. Weishaar, D. Kobylarz, W. H. Moos, and J. A. Bristol, *J. Med. Chem.* **30**, 1955 (1987).
87JMC1963	W. H. Moos, C. C. Humblet, I. Sircar, C. Rithner, R. E. Weishaar, J. A. Bristol, and A. T. McPhail, *J. Med. Chem.* **30**, 1963 (1987).
87JOC1043	G. Lunn, *J. Org. Chem.* **52**, 1043 (1987).
87JOC2026	M. Balogh, P. Laszlo, and K. Simon, *J. Org. Chem.* **52**, 2026 (1987).
87JOC2277	T. Shimizu, Y. Hayashi, M. Miki, and K. Teramura, *J. Org. Chem.* **52**, 2277 (1987).
87JOC4521	I. Maeba, M. Suzuki, O. Hara, T. Takeuchi, T. Iijimo, and H. Furukawa, *J. Org. Chem.* **52**, 4521 (1987).
87JOC5498	W. Adam and M. A. Miranda, *J. Org. Chem.* **52**, 5498 (1987).
87JPC11	M. Takahashi, M. Niwa, and M. Ito, *J. Phys. Chem.* **91**, 11 (1987).
87JPR525	M. M. Abbasi, A. Akelah, Y. Hafez, and E. S. Ismail, *J. Prakt. Chem.* **329**, 525 (1987).
87KGS672	O. P. Shkurko and V. P. Mamaev, *Khim. Geterotsikl. Soedin.* 672 (1987).
87LA889	M. Mittelbach, U. Wagner, and C. Kratky, *Liebigs Ann. Chem.*, 889 (1987).
87MI1	F. Barba, *Stud. Org. Chem. (Amsterdam)* **30**, 195 (1987).
87MI2	M. J. Bermejo, J. I. Ruiz, and J. Vinaixa, *Transition Met. Chem.* **12**, 245 (1987).
87MI3	E. Birkás-Faigl, G. Zólyomi, N. Makk, and E. Kasztreiner, *J. Labelled Compd. Radiopharm.* **24**, 1317 (1987).
87MI4	T. Boulanger, D. P. Vercauteren, G. Evrard, and F. Durant, *J. Crystallogr. Spectrosc. Res.* **17**, 561 (1987).
87MI5	I. Ciocoiu, S. Cilianu, A. Cirstea, M. Caprosu, E. Stefanescu, and M. Petrovanu, *Rev. Med.-Chir.* **91**, 137 (1987).
87MI6	Z. P. Gribova, T. M. Kerimov, A. N. Tikhonov, and I. K. Irbe, *Izv. Akad. Nauk SSSR, Ser. Biol.*, 628 (1987) [*CA* **107**, 149216 (1987)].
87MI7	S. Gu, R. Deng, and J. Wu, *Gaodeng Huexino Huaxue Xuebao* **8**, 575 (1987) [*CA* **108**, 67789 (1988)].
87MI8	A. L. Guy and J. E. Pemberton, *Langmuir* **3**, 125 (1987).
87MI9	S. J. Hays, J. L. Hicks, and C. C. Huang, *J. Labelled Compd. Radiopharm.* **24**, 1185 (1987).
87MI10	M. Hosono and N. Taira, *J. Cardiovasc. Pharmacol.* **10**, 692 (1987).
87MI11	E. T. Kang, K. G. Neoh, T. C. Tan, and H. C. Ti, *Mol. Cryst. Liq. Cryst.* **147**, 199 (1987).
87MI12	Y. Kitada, A. Narimatsu, N. Matsuura, and M. Endo, *Eur. J. Pharmacol.* **134**, 229 (1987).
87MI13	A. Melikian, G. Schlewer, S. Hurt, D. Chantreux, and C. G. Wermuth, *J. Labelled Compd. Radiopharm.* **24**, 267 (1987).
87MI14	J. C. Michaud, J. M. Mienville, J. P. Chambon, and K. Bizière, *Neuropharmacology* **25**, 1197 (1986).

87MI15	A. I. Moneer, A. Salem, and A. M. El-Kasaby, *J. Chem. Soc. Pak* **9**, 245 (1987).
87MI16	H. Oda, Y. Masaki, and H. Nagashima, *Gifu Yakka Daigaku Kiyo*, 42 (1987).
87MI17	R. Rajasekharan and P. S. Sastry, *Pestic. Biochem. Physiol.* **29**, 163 (1987) [*CA* **108**, 33545 (1988)].
87MI18	K. Samuel and S. Bose, *J. Biosci.* **12**, 211 (1987).
87MI19	S. Shashi Lata, P. P. Sharma, and G. C. Shivahare, *J. Inst. Chem. (India)* **59**, 107 (1987).
87MI20	L. Skibinska and B. Stefanska, *Farm. Pol.* **43**, 203 (1987).
87MI21	G. H. Tamam, E. A. Soliman, A. M. El-Gendy, and A. Y. El-Kady, *J. Chem. Soc. Pak.* **9**, 1 (1987).
87MI22	L. K. Thomson, S. K. Mandal, L. Rosenberg, F. L. Lee, and E. J. Gabe, *Inorg. Chim. Acta* **133**, 81 (1987).
87MI23	M. Tonegawa, Y. Nishimura, C. Fukasawa, K. Kitahara, and H. Satoh, *Tokyo Ika Daigaku Kiyo* **13**, 11 (1987) [*CA* **108**, 150399 (1988)].
87MI24	X. M. Wang and D. F. Hildebrand, *Plant Sci.* **51**, 29 (1987).
87MI25	C. G. Wermuth, J. P. Chambon, M. Heaulme, A. Melikian, G. Schlewer, R. Leyris, and K. Bizière, *Eur. J. Pharmacol.* **144**, 375 (1987).
87MI26	P. Worms, J. P. Chambon, and K. Bizière, *Eur. J. Pharmacol.* **138**, 343 (1987).
87MI27	Y. J. Yoon, D. S. Wise, and L. B. Townsend, *Bull. Korean Chem. Soc.* **8**, 171 (1987) [*CA* **108**, 204962 (1988)].
87MI28	D. L. Boger and S. M. Weinreb, "Hetero Diels-Alder Methodology in Organic Synthesis," p. 313. Academic Press, San Diego, California, 1987.
87OMS477	S. N. Sawhney, S. Bhutani, and D. Vir, *Org. Mass Spectrom.* **22**, 477 (1987).
87OPP80	T. N. Srinivasan, K. R. Rao, and P. B. Sattur, *Org. Prep. Proced. Int.* **19**, 80 (1987).
87PHA695	A. A. A. El Bannany, S. A. S. Ghozlan, and L. I. Ibraheim, *Pharmazie* **42**, 695 (1987).
87PIA415	K. Samuel and S. Bose, *Proc.—Indian Acad. Sci., Plant Sci.* **97**, 415 (1987).
87SC1253	H. A. Dowlatshahi, *Synth. Commun.* **17**, 1253 (1987).
87SC1907	S. Jones, N. Lewis, C. O'Farrell, and P. Wallbank, *Synth. Commun.* **17**, 1907 (1987).
87T2443	T. H. Fischer, J. C. Crook, and S. Chang, *Tetrahedron* **43**, 2443 (1987).
87TL6027	U. Grusek and M. Heuschmann, *Tetrahedron Lett.*, 6027 (1987).
87UKZ1099	A. V. Stetsenko, I. P. Kupchevskaya, N. D. Mikhnovskaya, V. Ya. Podlipskii, and N. V. Klyauz, *Ukr. Khim. Zh. (Russ. Ed.)* **53**, 1099 (1987).
87YZ819	M. Takaya, *Yakugaku Zasshi* **107**, 819 (1987) [*CA* **109**, 54725 (1988)].
87ZN(A)341	K. Hensen and J. Gaede, *Z. Naturforsch., A: Phys., Phys. Chem. Kosmophys.* **42A**, 341 (1987).
87ZN(B)210	Heydt, H. W. Breiner, V. Hell, and M. Regitz, *Z. Naturforsch., B: Anog. Chem., Org. Chem.*, **42**, 210 (1987).

87ZN(C)808	J. A. Graf, R. J. Strasser, and U. Kull, *Z. Naturforsch., C: Biosci.* **42C**, 808 (1987).
87ZOR1063	Yu. V. Shurukin, N. A. Klynev, and I. I. Grandberg, *Zh. Org. Khim.* **23**, 1063 (1987).
88AF237	Y. Terauchi, S. Watari, S. Ishikawa, K. Takayama, Y. Sekine, M. Hashimoto, and T. Hayashi, *Arzneim.-Forsch.* **38**, 237 (1988).
88AHC127	M. R. Grimmett and B. R. T. Keene, *Adv. Heterocycl. Chem.* **43**, 127 (1988).
88AHC173	R. Gallo, C. Roussel, and U. Berg, *Adv. Heterocycl. Chem.* **43**, 173 (1988).
88AX(C)1267	F. Abraham, B. Mernari, M. Lagrenee, and S. Sueur, *Acta Crystallogr., Sect. C: Cryst. Struct. Commun.* **44**, 1267 (1988).
88CB573	K. Gewald, U. Hain, and M. Gruner, *Chem. Ber.* **121**, 573 (1988).
88CJC348	C. K. Thompson, S. K. Mandal, J. P. Charland, and E. J. Gabe, *Can. J. Chem.* **66**, 348 (1988).
88CPL286	M. Terazima and T. Azumi, *Chem. Phys. Lett.* **145**, 286 (1988).
88H1431	F. Fariña, M. V. Martin, M. Romañach, and F. Sánchez, *Heterocycles* **27**, 1431 (1988).
88H1579	S. M. R. Sherif, R. M. Mohareb, G. E. H. Elgemeie, and R. P. Singh, *Heterocycles* **27**, 1579 (1988).
88JA6297	M. Meot-Ner (Mautner) and S. S. Kafafi, *J. Am. Chem. Soc.* **110**, 6297 (1988).
88JA6303	S. F. Nelsen, D. T. Rumack, L. W. Sieck, and M. Meot-Ner (Mautner), *J. Am. Chem. Soc.* **110**, 6303 (1988).
88JCS(P1)401	N. Haider and G. Heinisch, *J. C. S. Perkin 1*, 401 (1988).
88JCS(P1)1179	W. Dmowski and A. Haas, *J. C. S. Perkin 1*, 1179 (1988).
88JCS(P2)1259	M. C. B. L. Shohoji, H. M. Novais, and B. J. Herold, *J. C. S. Perkin 2*, 1259 (1988).
88JHC119	N. Haider, G. Heinisch, and D. Lassnigg, *J. Heterocycl. Chem.* **25**, 119 (1988).
88JHC799	P. Coudert, J. Couquelet, and P. Tronche, *J. Heterocycl. Chem.* **25**, 799 (1988).
88JHC831	A. T. M. Marcelis, H. Tondijs, and H. C. van der Plas, *J. Heterocycl. Chem.* **25**, 831 (1988).
88JMC345	R. A. Slater, W. Howson, G. T. G. Swayne, E. M. Taylor, and D. R. Reavill, *J. Med. Chem.* **31**, 345 (1988).
88JMC352	W. Howson, J. Kitteringham, J. Mistry, M. B. Mitchell, R. Novelli, R. A. Slater, and G. T. G. Swayne, *J. Med. Chem.* **31**, 352 (1988).
88JMC461	D. W. Robertson, J. H. Krushinski, G. Don Pollock, and J. Scott Hayes, *J. Med. Chem.* **31**, 461 (1988).
88JOC1415	D. L. Boger and S. M. Sakya, *J. Org. Chem.* **53**, 1415 (1988).
88JPC954	M. Muniz-Miranda, N. Neto, and G. Sbrana, *J. Phys. Chem.* **92**, 954 (1988).
88M83	B. Koren, B. Stanovnik, and M. Tišler, *Monatsh. Chem.* **119**, 83 (1988).
88MI1	M. Ghedini, F. Neve, and M. C. Bruno, *Inorg. Chim. Acta* **143**, 89 (1988).
88MI2	R. Grote, Y. Chen, A. Zeeck, Z. Chen, H. Zähner, P. Mischnick-Lübbecke, and W. A. König, *J. Antibiot.* **41**, 595 (1988).
88MI3	J. P. Kan, R. Steinberg, J. Leclercq, P. Worms, and K. Bizière, *J. Neurochem.* **50**, 1137 (1988).

88MI4	S. Shibata, N. Satake, R. K. Hester, and K. Kurahashi, *Eur. J. Pharmacol.* **145**, 113 (1988).
88MI5	T. J. Sutton, J. S. H. Luke, and H. B. Jones, *Hum. Toxicol.* **7**, 243 (1988).
88OPP117	J. G. Samaritoni, *Org. Prep. Proced. Int.* **20**, 117 (1988).
88T2449	M. Gebauer, G. Heinisch, and G. Lötsch, *Tetrahedron* **44**, 2449 (1988).
88T3149	D. E. Pereira, A. Petrič, and N. J. Leonard, *Tetrahedron* **44**, 3149 (1988).
88T3281	F. Effenberger and J. König, *Tetrahedron* **44**, 3281 (1988).
88ZN(C)418	A. Hloušek-Radojčič and N. Ljubešič, *Z. Naturforsch., C: Biosci.* **43C**, 418 (1988).

OCT 12 1990